LAMINAR FLOW
FORCED CONVECTION
IN DUCTS

Advances in
HEAT TRANSFER

Edited by

Thomas F. Irvine, Jr.
Department of Mechanics
State University of New York
at Stony Brook
Stony Brook, New York

James P. Hartnett
Energy Resources Center
University of Illinois
at Chicago Circle
Chicago, Illinois

Supplement 1

LAMINAR FLOW FORCED CONVECTION IN DUCTS
A Source Book for Compact Heat Exchanger Analytical Data
R. K. Shah and A. L. London

LAMINAR FLOW FORCED CONVECTION IN DUCTS

A Source Book for Compact Heat Exchanger Analytical Data

R. K. Shah
Harrison Radiator Division
General Motors Corporation
Lockport, New York 14094

A. L. London
Mechanical Engineering Department
Stanford University
Stanford, California 94305

 1978

ACADEMIC PRESS · New York · San Francisco · London
A Subsidiary of Harcourt Brace Jovanovich, Publishers

ACADEMIC PRESS, INC.
111 Fifth Avenue, New York, New York 10003

United Kingdom Edition published by
ACADEMIC PRESS, INC. (LONDON) LTD.
24/28 Oval Road, London NW1 7DX

LIBRARY OF CONGRESS CATALOG CARD NUMBER: 78–62274

ISBN 0–12–020051–1

PRINTED IN THE UNITED STATES OF AMERICA

To Rekha Shah
Charlotte London

CONTENTS

PREFACE

The purpose of writing this monograph is to present the available analytical solutions for laminar fluid flow and forced convection heat transfer in circular and noncircular pipes. The subject has importance in a large variety of traditional engineering disciplines, such as heating and cooling devices used in electronics, biomechanics, aerospace, instrumentation, and pipelines for oil, water and other fluids, and, in particular, for compact heat exchanger and solar collector designs. Generally, one way to reduce heat exchanger costs is to use more compact surfaces, as both the cost per unit area and heat transfer flux per unit temperature difference are simultaneously improved.

With the advancement of science and technology, a variety of passage geometries are utilized for internal flow forced convection heat transfer applications. Analytical laminar flow solutions for many passage geometries are available in the heat transfer literature. However, these solutions are scattered throughout the world-wide technical journals, reports, and student theses. These solutions, often generated by applied mathematicians, are not always accessible to the practicing engineers and researchers who are generally faced with a limited availability of time. Also, these solutions may be difficult to interpret by the engineer, as they may be presented with all the details of analyses but without the final results which an engineer would like to use. The aim of this monograph is to present all the analytical solutions known to the authors for laminar internal flow, in a readily usable, coordinated, and unified format. In the interest of brevity, derivations of formulas and detailed explanations, such as are appropriate in a textbook, are omitted; only the final results are presented in graphical and tabular forms. The extensive bibliography will provide the detailed information needed by scholars in this field of analysis.

This monograph is the result of the authors' need for a compendium of solutions for investigating improved surface geometries. It is an outgrowth of an initial compilation of laminar flow solutions prepared five years ago as the dissertation of the senior author. Hopefully, it will also serve as a useful reference to the engineers and researchers in the field.

Since it would take many volumes to do justice to all aspects of internal laminar flows, the scope of this monograph is restricted to laminar flow and forced convection for a Newtonian fluid with constant properties, passing through stationary, straight, nonporous ducts of constant cross section.

Except for the twisted-tape flows, all forms of body forces are omitted. Also, magnetohydrodynamic flows, electrical conducting flows, heat radiating flows, and the effects of natural convection, change of phase, mass transfer and chemical reactions are all excluded. Although some of the geometries described may not be visualized for the compact heat exchanger application, they are included for the completeness of available solutions for the laminar flow problem.

Emphasis is given to the summary of analytical solutions. Except for the circular tube, whenever experimental results are available to support or contradict the theory, they are also described in the text. For the circular tube, a vast amount of analytical and experimental results are available for the laminar flow forced convection. These experimental results generally support the analytical results. However, no compilation is provided for these experimental results because of space limitations.

We have not attempted a scientific or technological history. Although an effort was made to compile laminar flow analytical solutions from all available sources, it is quite probable that several important sources may not have come to the authors' attention.

The first four chapters describe the basic problems, solution techniques, dimensionless groups and generalized solutions. Chapters V through XVI describe the solutions for 39 duct geometries. Chapter XVII provides an overview of these solutions. Table 136 provides a ready reference for locating a particular solution. We recommend reading Chapters I and XVII first.

We are grateful to many researchers who have furnished the information requested. Without their whole-hearted assistance, this monograph would not be of great value. Their assistance is acknowledged as a personal communication in the references. The authors are thankful to Prof. T. F. Irvine, Jr., the editor, Prof. H. C. Perkins and Dr. M. R. Doshi who read the manuscript and made many helpful suggestions. Prof. K. P. Johannsen reviewed Chapter XV. Dr. D. B. Taulbee reviewed sections on the hydrodynamically developing flow. The authors are grateful to these researchers for their constructive suggestions. The institutional support of Stanford University, the Office of Naval Research, and the Harrison Radiator Division of General Motors Corporation is gratefully acknowledged. Lastly, the first author would like to express sincere appreciation to his wife Rekha for her patience, understanding and encouragement during the preparation of this monograph.

R. K. SHAH
A. L. LONDON

Chapter I

Introduction

Interest in heat exchanger surfaces with a high ratio of heat transfer area to core volume is increasing at an accelerated pace. The primary reasons for the use of these more compact surfaces is that smaller, lighter-weight, and lower-cost heat exchangers are the result. These gains are brought about both by the direct geometric advantage of higher "area density" and also a higher heat transfer coefficient for the smaller flow passages.

Because of the smaller flow passage hydraulic radius, with gas flows particularly, the heat exchanger design range for the Reynolds number usually falls well within the laminar flow regime. It follows then that theoretically derived laminar flow solutions for fluid friction and heat transfer in ducts of various flow cross section geometries become important. These solutions are the subject matter of this monograph. A direct application of these results may be in the development of new heat exchanger surfaces with improved characteristics. A critical examination of the theoretical solutions may prove to be fruitful because there is a wide range for the heat transfer coefficient at a given friction power for geometries of different cross sections. In addition to compact heat exchangers, applications of laminar flow theory are also of interest in the aerospace, nuclear, biomedical, electronics, and instrumentation fields.

It has long been realized that laminar flow heat transfer is dependent on the duct geometry, flow inlet velocity and temperature profiles, and wall temperature and/or heat flux boundary conditions. These conditions are difficult to produce even in the laboratory; nevertheless, there is substantial ongoing experimental research effort being devoted to this task. A theoretical base is needed in order to interpret the experimental results and extrapolate them for the design of practical heat exchanger systems. However, it is recognized that this theory is founded on idealizations of geometry and boundary conditions that are not necessarily well duplicated either in the application or even in the laboratory. The development of this theoretical base has been a fertile field for applied mathematicians since the early days of the science of heat transfer. Today, by the application of modern computer technology, analysis has exceeded, to some degree, the experimental verification. Since there are a large number of theoretical solutions available in the literature, a compilation and comparison of these solutions, employing a uniform format, should be of value to the designer as well as the researcher.

The study of heat transfer in laminar flow through a closed conduit was first made by Graetz [1] in 1883, and later independently by Nusselt [2] in 1910. Drew [3] in 1931 prepared a compilation of existing theoretical results for heat transfer. Dryden *et al.* [4] in 1932 compiled fully developed laminar flow solutions for ducts of various geometries. Later, several literature surveys were made for particular duct geometries. Rohsenow and Choi [5] in 1961 presented a limited compilation of solutions for simple cylindrical ducts. Kays and London [6] published a compilation in 1964 pertinent to compact heat exchangers. The theoretical development and the details of analysis for the laminar as well as turbulent flow problems were described in depth by Kays [7] in 1966. Such analyses are also available in other heat transfer text books. Petukhov [8] in 1967 compiled laminar duct flow solutions primarily for liquids. He presented mathematical details for the solutions including the consideration of temperature-dependent viscosity. Only a few solutions are covered by Petukhov compared to the present scope of investigation. Martynenko and Eichhorn [9] in 1968 compiled laminar hydrodynamic and thermal entry length solutions primarily for circular tube and parallel plates. With the cooperation of 30 British industries, Porter [10] in 1971 compiled the laminar flow solutions for Newtonian and non-Newtonian liquids with constant and variable fluid properties. The purpose of Porter's survey was to identify those areas which presented difficulties in thermal designs of chemical, plastic, and food-processing problems. He suggested the best design equations available and made specific recommendations for future investigation. Kooijman [11] in 1973 compiled laminar thermal entrance solutions for fully developed, constant-property Newtonian fluid flow through circular and noncircular ducts. Kays and Perkins [12] in 1973 presented a compilation of analytical forced-convection solutions for laminar and turbulent flow through circular and noncircular ducts. The heat transfer literature up to 1967 was reviewed by Kays and Perkins, who provided available results in terms of equations, tables, and graphs for design purposes.

Shah and London [13] in 1971 made a literature survey up to December 1970 for laminar forced-convection heat transfer and flow friction through straight and curved ducts. Available analytical solutions for constant-property Newtonian fluids, both liquids and gases, were compiled in contrast to liquids only, as considered by Porter [10]. The area investigated by Shah and London was much more exhaustive for constant-property Newtonian fluids and thus complemented the work of Porter. The present work augments and extends this first effort [13].

The specific objectives of the present work are as follows:

(1) To provide an up-to-date compilation of analytical laminar internal flow solutions with results in numerical and graphical dimensionless forms.

(2) To clarify and systemize various thermal boundary conditions and dimensionless groups.

(3) To indicate those areas where further contributions may be made.

Primarily, the English language literature up to December 1975 is reviewed. Emphasis is given to the analytical solutions for developed and developing velocity and temperature profiles through axisymmetric and two-dimensional constant cross section straight ducts. Only the forced convection *steady-state* laminar flow of a constant-property Newtonian fluid through a stationary, nonporous duct is considered. The effects of thermal energy sources, viscous dissipation, flow work, and axial heat conduction within the fluid have been considered in the literature for laminar flow through some duct geometries. This limited available information is included at appropriate places. All forms of body forces are neglected, except for the centrifugal force for the twisted-tape flow. Magnetohydrodynamic flows, electrically conducting flows, high-temperature (heat-radiating) flows, etc., are not considered. Also omitted are the effects of natural convection, change of phase, mass transfer, and chemical reaction. The solutions are applicable for ducts with smooth or rough walls, as long as the surface roughness does not significantly affect the cross section of the duct geometry.

The analytical solutions for the laminar flow problem were obtained from worldwide technical journals (as outlined in the Appendix), reports, and student theses. In reviewing the literature for each geometry, a chronological history is not preserved. Instead, material is classified according to boundary conditions and geometry. No attempt has been made to present the solutions in detail. Only those final results which are useful to a heat transfer designer are presented in tabular and graphical forms. Presented for each geometry is the information available for u, u_m, f Re, $K_d(\infty)$, $K_e(\infty)$, $K(\infty)$, L_{hy}^+, L_{th}^*, and Nusselt numbers corresponding to the boundary conditions of Table 2. The originally available tabular and graphical information has been augmented by (1) more information obtained from correspondence with authors, (2) the knowledge of limiting cases of boundary conditions or duct geometries, and (3) results computed by the present authors for a limited number of solutions in more detailed than was heretofore available.

Differential equations and boundary conditions for hydrodynamically and thermally developing and developed laminar flows are described in Chapter II. Definitions of important variables and dimensionless groups associated with the laminar flow problem are presented in Chapter III. General methods used in the heat transfer and fluid mechanics literature to solve the problems formulated in Chapter II are outlined in Chapter IV. The analytical solutions to the aforementioned laminar flow problem are summarized for 40 duct geometries in Chapters V–XVI. For the designer, an overview of the subject is provided in Chapter XVII. This overview includes comparisons of the solutions, approximate solutions for heat exchangers with multigeometry passages in parallel, influence of superimposed free convection, influence of temperature-dependent fluid properties, comments on the format of published papers, the complete solution for the laminar flow problem for constant cross-sectional ducts, and suggested areas of future research.

Chapter II

Differential Equations
and Boundary Conditions

The applicable differential equations and boundary conditions for the velocity and temperature problems are described in this chapter. These are for steady-state laminar flow through constant cross section ducts. Before proceeding to the details, the following terms are defined: laminar developed and developing flows, forced convection heat transfer, and thermally developed and developing flows.

Flow is *laminar* when the velocities are free of macroscopic fluctuations at any point in the flow field. For steady-state laminar flow, all velocities at a stationary point in the flow field remain constant with respect to "time," but velocities may be different at different points. Laminar flow, also referred to as viscous or streamline flow, is characteristic of a viscous fluid flow at low Reynolds number.

Laminar flow in a two-dimensional stationary straight duct is designated as *hydrodynamically fully developed* (or established) when the fluid velocity distribution at a cross section is of an invariant form, i.e., independent of the axial distance x, as shown in Fig. 1:

$$u = u(y, z) \quad \text{or} \quad u(r, \theta) \qquad \text{only} \tag{1a}$$

$$v, w = 0 \tag{1b}$$

The fluid particles move in definite paths called streamlines, and there are no components of fluid velocity normal to the duct axis. In a fully developed laminar flow, the fluid appears to move by sliding laminae of infinitesimal thickness relative to adjacent layers. Depending upon the smoothness of the

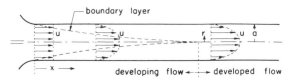

FIG. 1. Developed and developing laminar flow.

tube inlet and tube inside wall, fully developed laminar flow persists up to Re $\lesssim$ 2300 for a duct length L greater than the hydrodynamic entry length L_{hy}; otherwise for Re $\gtrsim$ 2300, the developing flow, as described below, could exist.

The hydrodynamic entrance region of the duct is that region where the velocity boundary layer is developing, for example, from zero thickness at the entrance to a thickness equal to the pipe radius far downstream. In this region, the fluid velocity profile changes from the initial profile at the entrance to an invariant form downstream. The flow in this region, as a result of the viscous fluid behavior, is designated as *hydrodynamically developing* (or establishing) flow, and is also shown in Fig. 1.

The hydrodynamic entrance length L_{hy} is defined as the duct length required to achieve a maximum duct section velocity of 99% of that for fully developed flow when the entering fluid velocity profile is uniform. The maximum velocity occurs at the centroid for the ducts symmetrical about two axes (e.g., circular tube and rectangular ducts). The maximum velocity occurs away from the centroid on the axis of symmetry for isosceles triangular, trapezoidal, and sine ducts. For nonsymmetrical ducts, no general statement can be made for the location of u_{max}. There are a number of other definitions used in the literature for L_{hy}. However, unless specified, the foregoing definition is used throughout this monograph.

In a short duct, such that the dimensionless length L_{hy}^{+} ($= L_{hy}/D_h \, \mathrm{Re}$) is less than 10^{-3}, developing laminar flow could exist even up to Re $\simeq 10^{5}$ [14,15]. In this short duct, this laminar boundary layer soon becomes turbulent, and fully developed turbulent flow exists at this high Reynolds number downstream from the developing region.

In *convection heat transfer*, a combination of mechanisms is active. Pure conduction exists at the wall, but away from the wall internal thermal energy transport takes place by fluid mass motion as well as conduction. Thus the convection heat transfer process requires a knowledge of both heat conduction and fluid flow. If the motion of the fluid arises solely due to external *force fields* such as gravity, centrifugal, or Coriolis body forces, the process is referred to as *natural* or *free convection*. If the fluid motion is induced by some external *means* such as pump, blower, fan, wind, or vehicle motion, the process is referred to as *forced convection*. Throughout this monograph, only pure forced convection heat transfer is considered.

Laminar flow in a two-dimensional stationary duct is designated as *thermally fully developed* (or established) when, according to Seban and Shimazaki [16], the dimensionless fluid temperature distribution, as expressed below in brackets, at a cross section is invariant, i.e., independent of x:

$$\frac{\partial}{\partial x}\left[\frac{t_{w,m} - t}{t_{w,m} - t_m}\right] = 0 \tag{2}$$

However, note that t is a function of (y, z) as well as x, unlike u, which is a function of (y, z) only and is independent of x for fully developed flow. Hydrodynamically and thermally developed flow is designated throughout simply as *fully developed* flow.

The thermal entrance region of the duct is that region where the temperature boundary layer is developing. For this region, the dimensionless temperature profile $(t_{w,m} - t)/(t_{w,m} - t_m)$ of the fluid changes from the initial profile at a point where the heating is started to an invariant form downstream. The flow in this region is designated as *thermally developing* flow. The velocity profile in this region could be either developed or developing. Thermally developing flow with a developing velocity profile is referred to as *simultaneously developing* flow.

The thermal entrance length L_{th} is defined, somewhat arbitrarily, as the duct length required to achieve a value of local $Nu_{x,bc}$ equal to $1.05 \, Nu_{bc}$ for fully developed flow, when the entering fluid temperature profile is uniform.

A. Velocity Problem

The differential equations and boundary conditions for the velocity problem are presented separately below for developed and developing flows.

1. HYDRODYNAMICALLY DEVELOPED FLOW

Consider a fully developed, steady state laminar flow in a two-dimensional singly (as in Fig. 2) or multiply connected stationary duct with the boundary Γ. The fluid is idealized as liquid or low-speed gas with the fluid properties ρ, μ, c_p, and k constant (independent of fluid temperature). Moreover, body forces such as gravity, centrifugal, Coriolis, and electromagnetic do not exist. The applicable momentum equation is [7]

$$\nabla^2 u = \frac{g_c}{\mu}\frac{dp}{dx} = c_1 \tag{3}$$

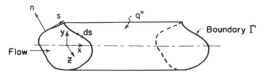

Fig. 2. A singly connected duct of constant cross-sectional area.

where x is the axial coordinate along the flow length of the duct and c_1 is defined as a pressure drop parameter. ∇^2 is the two-dimensional Laplacian operator. Note that the right-hand side of Eq. (3) is independent of (y, z) or (r, θ), and so it is designated as a constant c_1. Equation (3), in Cartesian coordinates, is

$$\frac{\partial^2 u}{\partial y^2} + \frac{\partial^2 u}{\partial z^2} = c_1 \tag{4}$$

In cylindrical coordinates, it is

$$\frac{1}{r}\frac{\partial}{\partial r}\left(r\frac{\partial u}{\partial r}\right) + \frac{1}{r^2}\frac{\partial^2 u}{\partial \theta^2} = c_1 \tag{5}$$

The boundary condition for the velocity problem is the no-slip condition, namely,

$$u = 0 \quad \text{on} \quad \Gamma \tag{6}$$

By the description of fully developed laminar flow of an incompressible fluid, the solution of the continuity equation (conservation of mass) is implicitly given by Eq. (1). Moreover, the continuity equation is utilized in deriving Eq. (3). Consequently, only the solution to the momentum equation[†] is required for the fully developed laminar fluid flow problem described below.

2. Hydrodynamically Developing Flow

As described on p. 6, the hydrodynamic entrance region is where the velocity boundary layer is developing. However, the hydrodynamic entrance flow problem is not strictly a boundary layer problem. This is because very near the entrance the axial molecular momentum transport $\mu(\partial^2 u/\partial x^2)$ is not a negligible quantity, and far from the entrance, the boundary layer thickness is not negligible compared to the characteristic dimension of the duct. It is also possible that very close to the entry the transverse pressure

---

[†] The continuity equation, in addition to the momentum equation, is required separately for the exact solution of developing velocity profiles in the duct entrance region.

gradient across the section may not be negligible. To take these effects into account, a complete set of Navier–Stokes equations needs to be solved. However, except for very close to the entry, $\mu(\partial^2 u/\partial x^2)$, $(\partial p/\partial y)$, and/or $(\partial p/\partial z)$ terms are negligible. Even though the physical concept[†] of a boundary layer introduced by Prandtl is not strictly applicable to the entrance flow problem, the Prandtl boundary layer idealizations

$$u \gg v, w \tag{7a}$$

$$\frac{\partial u}{\partial y}, \frac{\partial u}{\partial z} \gg \frac{\partial u}{\partial x}, \frac{\partial v}{\partial x}, \frac{\partial v}{\partial y}, \frac{\partial v}{\partial z}, \frac{\partial w}{\partial x}, \frac{\partial w}{\partial y}, \frac{\partial w}{\partial z} \tag{7b}$$

are good approximations for laminar flow in ducts. As a result, it is found that the terms of the y and z momentum equations are one order of magnitude smaller than the corresponding terms of the x momentum equation and hence may be neglected. In this case, the fluid pressure is a function of x only. Additionally, if all the idealizations made for the fully developed flow are invoked, the governing boundary layer type x momentum equation, for axially symmetric flow, in cylindrical coordinates from [7] is

$$u \frac{\partial u}{\partial x} + v \frac{\partial u}{\partial r} = -\frac{g_c}{\rho} \frac{dp}{dx} + v \left(\frac{\partial^2 u}{\partial r^2} + \frac{1}{r} \frac{\partial u}{\partial r} \right) \tag{8}$$

and in Cartesian coordinates, it is

$$u \frac{\partial u}{\partial x} + v \frac{\partial u}{\partial y} + w \frac{\partial u}{\partial z} = -\frac{g_c}{\rho} \frac{dp}{dx} + v \left(\frac{\partial^2 u}{\partial y^2} + \frac{\partial^2 u}{\partial z^2} \right) \tag{9}$$

The no-slip boundary condition for this case is

$$u, v, w - 0 \quad \text{on} \quad \Gamma \tag{10}$$

An initial condition is also required, and usually uniform velocity profile is assumed at the entrance:

$$u = u_e = u_m \quad \text{at} \quad x = 0 \tag{11}$$

In addition, the continuity equation needs to be solved simultaneously. In cylindrical coordinates [7], it is

$$\frac{\partial u}{\partial x} + \frac{\partial v}{\partial r} + \frac{v}{r} = 0 \tag{12}$$

[†] A momentum or velocity boundary layer of the Prandtl type is a thin region very close to the body surface or wall where the influence of fluid viscosity is predominant. The remainder of the flow field can to a good approximation be treated as inviscid and can be analyzed by the potential flow theory.

and in Cartesian coordinates [7], it is

$$\frac{\partial u}{\partial x} + \frac{\partial v}{\partial y} + \frac{\partial w}{\partial z} = 0 \tag{13}$$

The solution to the hydrodynamic entry length problem is obtained by solving Eqs. (8) and (12) or (9) and (13) simultaneously with the boundary and initial conditions of Eqs. (10) and (11).

For axisymmetric ducts, two unknowns u and v are obtained by solving Eqs. (8) and (12). Since Eq. (8) is nonlinear, various approximate methods have been used to obtain the solution. These methods are described briefly in Chapter IV. The axial pressure distribution is then obtained from another physical constraint such as from the solution of the mechanical energy integral equation[†] or integral continuity equation. For a two-dimensional duct, in which case both v and w components exist in the entrance region, a third equation [in addition to Eqs. (9) and (13)] is essential for a rigorous solution for u, v, and w. Such an equation for the square and eccentric annular ducts is briefly described on pp. 211 and 333.

B. Temperature Problem

The solution to the temperature problem involves the determination of the fluid and wall temperature distributions and/or the heat transfer rate between the wall and the fluid. Such a problem has mathematically complex features, such as heat conduction in normal, peripheral, and axial directions; variable heat transfer coefficients along the periphery and in the axial direction; and invariant or changing dimensionless velocity and temperature profiles along the flow length. The theoretical solutions of such problems provide the quantitative design information on the controlling dimensionless groups and indicate when to neglect certain effects. To illustrate this interplay, some idealized problems are formulated in this section for laminar internal flow forced convection heat transfer.

Attention is focused primarily on the forced convection heat transfer rate from the wall to the fluid (or vice versa). The determination of this heat transfer involves the solution of either (a) the conventional convection problem or (b) the conjugated problem.

In the *conventional convection problem*, heat transfer through the wall is characterized by a thermal boundary condition directly or indirectly specified

---

[†] The mechanical energy integral equation is obtained by multiplying Eq. (8) or (9) by u and integrating over the flow cross-sectional area. The momentum integral equation is obtained by integrating Eq. (8) or (9) over the flow cross-sectional area.

at the wall–fluid interface. The velocity and forced convection temperature problems are solved only for the fluid region. The solution to the temperature problem for the solid wall is not needed; but in its application it is implicitly assumed that the duct has a uniform wall thickness and there does not exist simultaneous heat conduction in the wall in axial, peripheral, and normal directions. The heat transfer rate through the wall, normal to the flow, is subsequently determined from the solution of the fluid temperature problem for the specified wall–fluid interface boundary conditions. The dimensionless temperature profile of the fluid may be either fully developed or developing along the flow length for this problem class.

However, in a broader view, the heat transfer through the duct wall by conduction may have significant normal and/or peripheral as well as axial components; or the fluid heating (or cooling) flux may be nonuniform around the duct periphery; or the duct wall may be of nonuniform thickness. As a result, the temperature problem for the solid wall needs to be analyzed simultaneously with that for the fluid in order to establish the actual wall–fluid interface heat transfer flux distribution. This combination is referred to as the *conjugated problem*. The simultaneous solutions of the energy equations for the solid and fluid media are obtained by considering temperature and heat flux as continuous functions at the solid wall–fluid interface. The velocity distribution for the fluid medium must first be found by solving the applicable continuity and momentum equations. The dimensionless temperature profile is always variant (never fully developed) for this class of problems.

Some idealized conjugated and conventional convection problems are now formulated.

1. Conjugated Problem

As described above, the formulation of the conjugated problem involves the application of the energy equations for both the fluid and solid regions. The temperature and heat fluxes at the solid–fluid interface are considered continuous.

As an illustration, the conjugated problem is formulated for a thick-walled circular tube and the fluid on the inside. The heat transfer from the other fluid side is represented by a thermal boundary condition. Consider a steady-state, laminar, constant-properties flow of a Newtonian fluid in the duct of constant cross-sectional area. Thermal energy sources, viscous dissipation effects, and flow work within the fluid are neglected. In the absence of free convection, mass diffusion, chemical reaction, change of phase, and electromagnetic effects, the governing differential equations and boundary conditions are as follows [17] (refer to Fig. 3 for the system coordinates).

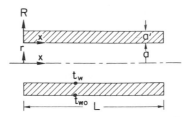

FIG. 3. Coordinate system for the circular tube conjugated problem.

Fluid

D.E. $\dfrac{\partial^2 t_f}{\partial r^2} + \dfrac{1}{r}\dfrac{\partial t_f}{\partial r} = \dfrac{u}{\alpha}\dfrac{\partial t_f}{\partial x}$ (14)

B.C. $t_f = t_w$ at $r = a$ (15)

$k_f \dfrac{\partial t_f}{\partial r} = k_w \dfrac{\partial t_w}{\partial r}$ at $r = a$ (16)

I.C. $t_f = t_e$ at $x = 0$ (17)

Solid

D.E. $\dfrac{\partial^2 t_w}{\partial r^2} + \dfrac{1}{r}\dfrac{\partial t_w}{\partial r} + \dfrac{\partial^2 t_w}{\partial x^2} = 0$ (18)

B.C. $\dfrac{\partial t_w}{\partial x} = 0$ at $x = 0, L$ (19)

q' or t_{wo} is specified at $r = a + a'$ (20)

Introduce $x^* = x/D_h$ Pe, $r^* = r/a$, $X = x/L$, $u^* = u/u_m$, and $R^* = R/a' = (r - a)/a'$. Equations (14), (16), and (18) reduce to the forms

$$\frac{\partial^2 t_f}{\partial r^{*2}} + \frac{1}{r^*}\frac{\partial t_f}{\partial r^*} = \frac{1}{4} u^* \frac{\partial t_f}{\partial x^*} \tag{21}$$

$$\left.\frac{\partial t_w}{\partial R^*}\right|_{R^* = 0} = 2R_w \left.\frac{\partial t_f}{\partial r^*}\right|_{r^* = 1} \tag{22}$$

$$\frac{\partial^2 t_w}{\partial R^{*2}} + \frac{1}{R^* + (a/a')}\frac{\partial t_w}{\partial R^*} + \left(\frac{a'}{L}\right)^2 \frac{\partial^2 t_w}{\partial X^2} = 0 \tag{23}$$

where $R_w = k_f/U_w D_h$ and $U_w = k_w/a'$.

As can be seen, the energy equations for the solid and fluid media are coupled by the boundary conditions, Eqs. (15) and (16).

The conjugated problem for the circular tube [18–20] and parallel plates [18,19,21–23] has been analyzed by employing highly sophisticated mathematical techniques. The solution to the conjugated problem for a circular tube by Mori *et al.* [20] is the most comprehensive.

2. CONVENTIONAL CONVECTION PROBLEM

As described earlier, the formulation of this class of temperature problem involves the specification of the applicable energy equation and the thermal boundary condition at the duct wall. The differential energy equations are presented in the following text for two kinds of flows: (a) fully developed flow [both hydrodynamically and thermally, Eqs. (1) and (2)], and (b) thermally developing flows (with developed or developing velocity profiles). In contrast, note that the flow is always thermally developing for a conjugated problem.

a. *Thermally Developed Flow*

Consider a steady-state, laminar, constant-properties (μ, c_p, k) flow in a duct of constant cross-sectional area. In the absence of free convection, mass diffusion, chemical reaction, change of phase, and electromagnetic effects, the governing differential energy equation for a *perfect gas* is [7]

$$k \nabla^2 t = k \frac{\partial^2 t}{\partial x^2} + k\left(\frac{\partial^2 t}{\partial y^2} + \frac{\partial^2 t}{\partial z^2}\right)$$

$$= \rho c_p u \frac{\partial t}{\partial x} - S - \frac{\mu}{g_c J}\left[\left(\frac{\partial u}{\partial y}\right)^2 + \left(\frac{\partial u}{\partial z}\right)^2\right] - \frac{u}{J}\frac{dp}{dx} \tag{24}$$

The third term on the right-hand side of Eq. (24) represents part of the work done by the fluid on adjacent layers due to action of shear forces and is usually referred to as "viscous dissipation" in the English language literature and as "internal friction" in the Russian literature. The fourth term on the right is referred to as the "flow work" and "work done by pressure forces" by English and Russian authors, respectively. Sometimes flow work is also referred to as "gas compression work." This appears in the energy equation when the equations for energy conservation and momentum are manipulated so as to eliminate the kinetic energy term [7].

The corresponding differential energy equation for an *incompressible liquid* is the same as Eq. (24) with the omission of the $u(dp/dx)/J$ term, because in the derivation this term cancels with the pressure component of enthalpy for an incompressible liquid [7]. For some liquids, under special circumstances, the flow work term may not be negligible, as discussed on p. 81.

Note that if Eq. (24) is operated on by ∇^2, the right-hand side of Eq. (24) will contain $\nabla^2 u$, which equals c_1 from Eq. (3). The resulting equation will be a fourth-order differential equation for the dependent variable t.

Equation (24) in the following development is made dimensionless, except for t, in a particular manner so that when the effects of axial heat conduction, thermal energy sources, and viscous dissipation are negligible, the resulting energy equation is parameter-free. For thermally developed or developing flow, introduce the dimensionless x^* ($=x/D_h$ Pe), y^* ($=y/D_h$), z^* ($=z/D_h$), and u^* ($=u/u_m$) in Eq. (24). After some rearrangements, it reduces to the form

$$\frac{1}{\text{Pe}^2}\frac{\partial^2 t}{\partial x^{*2}} + \frac{\partial^2 t}{\partial y^{*2}} + \frac{\partial^2 t}{\partial z^{*2}}$$

$$= u^* \frac{\partial t}{\partial x^*} - \frac{SD_h^2}{k} - \frac{\mu u_m^2}{g_c Jk}\left[\left(\frac{\partial u^*}{\partial y^*}\right)^2 + \left(\frac{\partial u^*}{\partial z^*}\right)^2 - 2u^*(f\,\text{Re})\right] \quad (25)$$

where for hydrodynamically developed flow

$$f\,\text{Re} = -\frac{g_c D_h^2}{2\mu u_m}\frac{dp}{dx} \quad (26)$$

is a constant, dependent on the duct geometry. For hydrodynamically developing flow, the right-hand side of Eq. (26) is first obtained from the solution of the velocity problem and then is used in Eq. (25) in place of $f\,\text{Re}$.

For *constant axial wall temperature boundary conditions*, the dimensionless temperature θ is defined as

$$\theta = \frac{t_w - t}{t_w - t_e} \quad (27)$$

Then Eq. (25) is reduced to the dimensionless form

$$\frac{1}{\text{Pe}^2}\frac{\partial^2 \theta}{\partial x^{*2}} + \frac{\partial^2 \theta}{\partial y^{*2}} + \frac{\partial^2 \theta}{\partial z^{*2}} = u^* \frac{\partial \theta}{\partial x^*} + S^* + \text{Br}\,\psi \quad (28)$$

where the dissipation number Br is defined by Eq. (139), and ψ designates the bracketed term of Eq. (25) which includes viscous dissipation and flow work terms.

For *constant axial wall heat flux boundary conditions*, the dimensionless temperature Θ may be defined as

$$\Theta = \frac{t - t_e}{q'' D_h / k} \quad (29)$$

Substitution of this into Eq. (25) results in a dimensionless equation with the dissipation number Br' defined by Eq. (140) as a parameter for the viscous dissipation (and for gases also flow work) effects.

The thermal boundary conditions associated with Eq. (24) will be discussed separately later. It should be emphasized that only the *fluid* medium needs to be analyzed for the solution of the posed (fully developed) temperature problem and the associated thermal boundary conditions. The solution to the temperature problem for the solid medium (duct wall) is not required for the analysis of *fully developed flows* regardless of thick or thin duct walls.

b. *Thermally Developing Flow*

All the idealizations made in the fully developed case are still applicable except that thermal energy sources, viscous dissipation, and flow work within the fluid are neglected. Also, the boundary layer type idealizations, Eq. (7), and

$$\frac{\partial t}{\partial y}, \frac{\partial t}{\partial z} \gg \frac{\partial t}{\partial x} \tag{30}$$

are invoked. (Refer to the footnote on p. 9 and associated discussion.) The governing boundary layer type energy equation for the developing laminar temperature profile of a perfect gas or an incompressible liquid is [7]

$$u\frac{\partial t}{\partial x} + v\frac{\partial t}{\partial r} = \alpha\left(\frac{\partial^2 t}{\partial r^2} + \frac{1}{r}\frac{\partial t}{\partial r} + \frac{\partial^2 t}{\partial x^2}\right) \tag{31}$$

in cylindrical coordinates, and

$$u\frac{\partial t}{\partial x} + v\frac{\partial t}{\partial y} + w\frac{\partial t}{\partial z} = \alpha\left(\frac{\partial^2 t}{\partial y^2} + \frac{\partial^2 t}{\partial z^2} + \frac{\partial^2 t}{\partial x^2}\right) \tag{32}$$

in Cartesian coordinates. Equation (31) includes the idealization that the heating is axially symmetrical.

The effects of thermal energy sources, viscous dissipation and flow work can be included in the energy equation by adding corresponding terms on the right-hand side of Eq. (31) or (32). These terms for the boundary layer type idealizations, when multiplied by ρc_p, are presented in Eq. (24).

In addition to the thermal boundary conditions to be described in the following section, an initial condition is also required and is normally employed as the uniform temperature at the point where heating (or cooling) is started:

$$t = t_e \qquad \text{at} \quad x = 0 \tag{33}$$

For the exact solution to the thermal entry length problem, the continuity and momentum equations need to be solved first to determine u, v, and w.

Thermal entry length problems are of three categories: (1) the velocity profile is fully developed and remains fixed while the temperature profile develops, (2) the velocity and temperature profiles develop simultaneously, and (3) the temperature profile starts developing at some point in the hydrodynamic entry region.

The thermal entry length problem of the second category, simultaneously developing velocity and temperature profiles, is also referred to as the *combined entry length problem*. The rate of developments of velocity and temperature profiles in the entrance region depends upon the fluid Prandtl number. For $Pr = 1$, the velocity and temperature profiles develop at the same rate, if both are uniform at the entrance. For $Pr > 1$, the velocity profile develops more rapidly than the temperature profile. For $Pr < 1$, the temperature profile develops more rapidly than the velocity profile. For the limiting case of $Pr = \infty$, the velocity profile is developed before the temperature profile starts developing. For the other limiting case of $Pr = 0$, the velocity profile never develops (remains uniform) while the temperature profile is developing. The idealized $Pr = \infty$ and 0 cases are good approximations for highly viscous fluids and liquid metals, respectively.

It may be noted that the thermal entry length problem of the first category, which is valid for any fluid (any Pr), is identical to the combined entry length problem for $Pr = \infty$. Thus, the combined entry length solution for $Pr = \infty$ is also a solution for any fluid with a developed velocity profile and a developing temperature profile. Similarly, the combined entry length solution for $Pr = 0$ also refers to the solution for any fluid with uniform velocity profile (*slug flow*) and developing temperature profile. One way of presenting the combined entry length solution is to present the Nusselt number versus $x^* = x/(D_h \, Re \, Pr)$ with Pr as a parameter. In such a plot, however, the parameters $Pr = \infty$ and 0 designate the nature of the velocity profile in the thermal entrance region as mentioned above, and should not be confused with Pr in x^*.

C. Thermal Boundary Conditions

The thermal boundary condition is the set of specifications describing temperature and/or heat flux conditions at the *inside* wall of the duct. The peripheral average heat transfer flux is strongly dependent on the thermal boundary condition in the laminar flow regime, while very much less dependent in the turbulent flow regime for fluids with $Pr \gtrsim 1$. Hence, the following classification of thermal boundary conditions, though applicable also to turbulent flows, is useful mainly for laminar flows.

A large variety of thermal boundary conditions can be specified for the temperature problem. Generally, these boundary conditions are not clearly and consistently defined in the literature, and therefore these highly sophisticated results are difficult to interpret by a designer. Shah and London [24] attempted to systemize the thermal boundary conditions. These results are now summarized for singly, doubly, and multiply connected ducts.

1. THERMAL BOUNDARY CONDITIONS FOR SINGLY CONNECTED DUCTS

A region Ω is called singly connected if any simple closed curve that lies in Ω can be shrunk to a point without leaving Ω. A region Ω that is not singly connected is called *multiply connected*. The duct cross sections considered in Chapters V–XI and again in Chapter XVI are singly connected.

Numerous thermal boundary conditions can be specified for a singly connected duct of constant cross-sectional area (Fig. 2). These are categorized in three groups in the first column of Table 1: (1) a specified axial wall temperature distribution t_w, (2) a specified axial wall heat flux distribution q', and (3) a specified combination of axial wall temperature and heat flux distributions. Around the periphery of the duct Γ, any combination of t_w, q'', or ($q'' \propto t_w^n$) may be specified. These boundary conditions may be applicable to thermally developed and/or thermally developing flows. A general classification of thermal boundary conditions and their application to thermally developed or developing flows is presented in Table 1. Those thermal boundary and initial conditions with finite axial heat conduction within the fluid are described in Figs. 17 and 23 and the associated discussion. Axial heat conduction in a "thin" wall could be important for axially constant heat flux boundary conditions. The resulting thermal boundary condition is described on p. 30.

For the case of specified arbitrary variations in wall temperature or wall heat flux, the solution to the energy equation may be obtained by superposition methods for constant property flow. Literature sources describing the superposition methods for arbitrary variations in t_w or q'' axially or peripherally are indicated in Table 1. Since so many boundary conditions can be generated by the superposition techniques, specific names are not recommended for axially variable thermal boundary conditions.

In the Soviet literature [18,33], conjugated problems and boundary conditions of four kinds are widely used for unsteady heat conduction problems. These are also applied to the duct flow forced convection problems. In a most general form, these boundary conditions are

$$\text{first kind:} \qquad t_w = t_w(x, y, z, \tau) \tag{34}$$

$$\text{second kind:} \qquad q_w'' = q_w''(x, y, z, \tau) \tag{35}$$

$$\text{third kind:} \qquad q_w'' = U_w(t_{wo} - t_w) \tag{36}$$

$$\text{fourth kind:} \qquad t_w(\tau) = [t_f(\tau)]_w \tag{37}$$

$$q_w''(\tau) = [q_f''(\tau)]_w \tag{38}$$

TABLE 1

GENERAL CLASSIFICATION OF THERMAL BOUNDARY CONDITIONS

General classification	Specified t_w, q'', or $q'' = f(t_w)$ axially and/or peripherally, as noted in columns (a) through (d)			
	(a) Constant axially, constant peripherally	(b) Constant axially, variable peripherally	(c) Variable axially, constant peripherally	(d) Variable axially, variable peripherally
(1) Specified t_w axially	(T); applicable to both thermally developed and developing flows	Only (T3), (T4) analyzed; applicable to both thermally developed and developing flows	Analyzed by superposition methods [7,25,26] from the solution of case 1a; also (ΔT); applicable to thermally developing flows only	A few solutions;[†] applicable to thermally developing flows only
(2) Specified q' axially	(H1), (H2); applicable to both thermally developed and developing flows	Analyzed by superposition methods [27,28] from the solution of case 2a; also (H3), (H4)	Analyzed by superposition methods [7,29–32] from the solution of case 2a; applicable to thermally developing flows only; also (H5) analyzed for thermally developed flows only	A few solutions;[†] applicable to thermally developing flows only
(3) Specified $q' = f(t_w)$ axially	No known solution; applicable to both thermally developed and developing flows	No known solution; applicable to both thermally developed and developing flows	No known solution; applicable to developing flows only	No known solution; applicable to thermally developing flows only

† Some specific solutions, using superposition techniques, are available for thermally developing and hydrodynamically developing flows. These are described in the text at appropriate places.

Boundary conditions of the first, second, and third kinds are, respectively, the cases 1d, 2d, and the restricted 3d of Table 1 [corresponding to the linear functional relationship of Eq. (41)], except that in steady laminar flow no time dependence is involved.

Since the above boundary conditions do not make any distinction between axial and peripheral conditions, and since the solutions appear in the literature with this distinction, some specific thermal boundary conditions are proposed and systemized for *thermally developed and developing flows.* They are summarized in Table 2 and described in the following section in further detail for the singly connected duct of Fig. 2. From the solutions for these boundary conditions, a solution may be obtained for arbitrary variations in corresponding parameters by superposition methods. These basic boundary conditions are relatively simple to analyze mathematically, are realized approximately in practical systems, and have been analyzed to a varying degree for ducts having different cross sections, as described in Chapters V–XVI. Thermal boundary conditions that yield thermally fully developed flow for a long duct are limited in number. According to the authors' knowledge, Table 2 appears to be a complete list of such thermal boundary conditions of practical interest.

The following nomenclature scheme is used for these boundary conditions. Generally, two characters are enclosed in a circle; the first character (T or H) represents axially constant wall temperature or heat transfer rate, respectively; the second character (1, 2, or 3) indicates, respectively, the foregoing boundary conditions of the first, second, or third kind in the peripheral direction.[†] For example, (H1) represents a constant axial heat transfer rate q' with a constant peripheral surface temperature. The numerals 4 and 5 designate other specialized axial or peripheral boundary conditions.

In reporting the forced convection heat transfer results, it is recommended that the dimensionless variables be presented with either one or two sets of subscripts, as shown here in a specific example for the Nusselt number:

$$Nu_{x,bc}, \quad Nu_{m,bc}, \quad Nu_{p,bc}, \quad Nu_{bc}$$

The subscripts x and m represent the local (peripheral average) and mean (flow length average) values, respectively, in the thermal entrance region; p represents the peripheral local value; Nu_{bc} denotes the peripheral average Nusselt number; $Nu_{p,bc}$ and Nu_{bc} are for the fully developed region; the subscript bc designates the thermal boundary condition for which the solution has been obtained. Since the peripheral local Nusselt number in

---

[†] For boundary conditions of the first, second, and third kind (1, 2, and 3), the constant temperature, constant heat flux, and a linear combination of heat flux and wall temperature are specified respectively at the duct cross section boundary.

TABLE 2

Thermal Boundary Conditions for Developed and Developing Flows through Singly Connected Ducts[†]

Designation	Description	Equations	Applications
Ⓣ	Constant wall temperature peripherally as well as axially	$t|_\Gamma = t_w = $ constant, independent of (x, y, z)	Condensers, evaporators, automotive radiators (at high flows), with negligible wall thermal resistance
Ⓣ3	Constant axial wall temperature with finite normal wall thermal resistance	$t_{wo}(x, y, z) = t_{wo}(y, z)$, independent of x $$\left.\frac{\partial t}{\partial n^*}\right\|_\Gamma = \frac{1}{R_w}(t_{wo} - t\|_\Gamma)$$	Same as those for Ⓣ with finite wall thermal resistance
Ⓣ4	Nonlinear radiant-flux boundary condition	$T_a(x, y, z) = T_a(y, z)$, independent of x $$\left.\frac{\partial T^*}{\partial n^*}\right\|_\Gamma = -\gamma[T^*\|_\Gamma^4 - T_a^{*4}]$$	Radiators in space power systems, high-temperature liquid-metal facilities, high-temperature gas flow systems
Ⓗ1	Constant axial wall heat flux with constant peripheral wall temperature	$q'(x) = $ constant, independent of x $t|_\Gamma = t_w = $ constant, independent of (y, z)	Same as those for Ⓗ4 for highly conductive materials
Ⓗ2	Constant axial wall heat flux with uniform peripheral wall heat flux	$q'(x) = $ constant, independent of x $\left. k \frac{\partial t}{\partial n^*}\right|_\Gamma = $ constant, independent of (y, z)	Same as those for Ⓗ4 for very low conductive materials with the duct having uniform wall thickness

H3	Constant axial wall heat flux with finite normal wall thermal resistance	$q'(x) =$ constant, independent of x $$\left.\frac{\partial t}{\partial n^*}\right|_\Gamma = \frac{1}{R_w}(t_{wo} - t|_\Gamma)$$	Same as those for H4 with finite normal wall thermal resistance and negligible peripheral wall heat conduction
H4	Constant axial wall heat flux with finite peripheral wall heat conduction	$q'(x) =$ constant, independent of x $$\frac{q_p'' D_h}{k} - \left.\frac{\partial t}{\partial n^*}\right|_\Gamma + \left.K_p \frac{\partial^2 t}{\partial s^{*2}}\right|_\Gamma = 0$$	Electric resistance heating, nuclear heating, gas turbine regenerator, counterflow heat exchanger with $C_{min}/C_{max} \simeq 1$, all with negligible normal wall thermal resistance
H5	Exponential axial wall heat flux	$q_x' = q_e' e^{mx^*}$ $t|_\Gamma = t_w =$ constant, independent of (y, z)	Parallel and counterflow heat exchangers with appropriate values of m
Δt	Constant axial wall to fluid bulk temperature difference	$\Delta t(x) = t_{w,m} - t_m$ $=$ constant, independent of x $t|_\Gamma = t_w =$ constant, independent of (y, z)	Gas turbine regenerator

---

[†] The Δt boundary condition is primarily applied to thermally developing flow. For fully developed flow, it is the same as the H1 boundary condition (refer to text).

A boundary condition in which the wall temperature varies linearly along the flow direction and remains uniform peripherally is not included in the above list. The local Nusselt numbers in the thermal entrance region for this boundary condition are higher than those for the T and H1 boundary conditions. As $x^* \to \infty$, this boundary condition approaches the H1 boundary condition. It has been analyzed for a circular tube [25,34] and rectangular and elliptical ducts [35].

the thermal entrance region is not usually required in the design of a heat exchanger (except for the hot or cold spots), and since this information is currently not available in the literature, no specific nomenclature is recommended for it.

The thermal boundary conditions, which are summarized in Table 2, are described below. In all these boundary conditions, axial heat conduction in the duct wall is considered as zero.

a. Constant Axial Wall Temperature, Ⓣ, Ⓣ3, and Ⓣ4

The thermal boundary condition of approximately constant axial wall temperature is realized in many practical applications, for example, condensers, evaporators, and automotive radiators having high liquid flow rates. If the wall temperature is also uniform peripherally, the boundary conditions will be referred to as Ⓣ. According to the foregoing proposed scheme, it should have been designated as Ⓣ1; but since the wall temperature is the same everywhere, it is designated for convenience as Ⓣ. The peripheral uniform wall heat flux case Ⓣ2 is not realized in practical applications and hence is of less importance. Moreover, solutions are also not available in the literature. The Ⓣ3 and Ⓣ4 have linear and nonlinear combinations of wall heat flux and temperature on the duct periphery as described below.

(i) Ⓣ *Boundary Condition.* The wall temperature of the duct is uniform and constant both axially and peripherally for this boundary condition:

$$t|_\Gamma = t_w = \text{constant}, \qquad \text{independent of } (x, y, z) \tag{39}$$

The constant wall temperature condition can be pictured in two ways: (1) With reference to Fig. 4, the thermal resistance of both the wall and t_a fluid is zero and t_a is constant. In this case, the axial wall thermal conductivity can be arbitrary, but the normal (radial) thermal conductivity is implicitly assumed to be infinite. (2) The wall thermal conductivity is idealized to be infinite in the axial and peripheral directions; therefore, the normal thermal conductivity or normal wall temperature profile is not involved in this analysis.

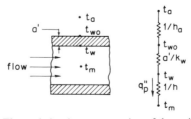

FIG. 4. Thermal circuit representation of the resistances.

Some of the technical applications of the T boundary condition are outlined in Table 2. The T boundary condition has been analyzed for a circular tube, parallel plates, and rectangular, isosceles triangular, sine, elliptical, and concentric annular ducts.

(ii) T3 *Boundary Condition.* The outside wall temperature is constant axially, while the heat flux is linearly proportional to wall temperature peripherally at every cross section x for the T3 boundary condition:

$$t_{wo}(x, y, z) = t_{wo}(y, z), \qquad \text{independent of } x \tag{40}$$

$$k \left. \frac{\partial t}{\partial n} \right|_{\Gamma} = U_w(t_{wo} - t|_{\Gamma}) \tag{41}$$

Note that the inside wall temperature $t_w - t|_{\Gamma}$ for this boundary condition is a function of (y, z) as well as x, unlike the outside wall temperature t_{wo}, which can vary around the periphery only and does not depend on x.

The reciprocal of the proportionality constant in Eq. (41), $1/U_w$, will be referred to as the *wall thermal resistance.* It is based on the inside duct wall area. For a straight thin wall of uniform thickness a', it may be expressed as

$$\frac{1}{U_w} = \frac{a'}{k_w} \tag{42}$$

For a circular tube of radius a and wall thickness a', it would be

$$\frac{1}{U_w} = \frac{a \ln[1 + (a'/a)]}{k_w} \tag{43}$$

If the outside wall to fluid thermal resistance is included in $1/U_w$, as is done sometimes in a condenser design,

$$\frac{1}{U_w} = \frac{a'}{k_w} + \frac{1}{h_a} \tag{44}$$

The corresponding wall heat flux of Eq. (41) would be (from Fig. 4)

$$q'' = U_w(t_a - t|_{\Gamma}) \tag{45}$$

In the literature, t_a is generally considered as a constant. After dividing n by the characteristic dimension D_h, Eq. (41) reduces to

$$\left. \frac{\partial t}{\partial n^*} \right|_{\Gamma} = \frac{1}{R_w}(t_{wo} - t|_{\Gamma}) \tag{46}$$

where $R_w = k/U_w D_h$ is the dimensionless wall thermal resistance.

Implicit idealizations made for wall thermal conductivity for the T3 boundary condition are described in Table 3. There are two limiting cases

TABLE 3

IDEALIZATIONS OF WALL THERMAL
CONDUCTIVITY IMPLIED FOR THE THERMAL
BOUNDARY CONDITIONS OF TABLE 2

Thermal boundary condition	Axial k_w	Peripheral k_w	Normal k_w
(T)	arbitrary infinite	arbitrary infinite	infinite arbitrary
(T3)	zero	zero	finite
(T4)	zero	zero	infinite
(H1)	zero	infinite	infinite
(H2)	zero	zero	infinite
(H3)	zero	zero	finite
(H4)	zero	finite	infinite
(H5)	zero	infinite	infinite
(Δt)	zero	infinite	arbitrary

of the (T3) boundary condition: (1) When the wall thermal resistance R_w approaches zero, (T3) reduces to the (T) boundary condition; (2) when the wall thermal resistance R_w approaches infinity, $q'' = 0$ axially as well as peripherally, which is a special case of the (H2) boundary condition described below (q'' constant everywhere). Thus when $R_w \to \infty$, the (T3) solution goes to (H2).[†] When the peripheral wall temperature is constant,[‡] the (T3) boundary condition is a special case of the (H5) boundary condition for *fully developed flow* with the exponent m of Eq. (62) as $-4 \, \mathrm{Nu_o}$. Here $\mathrm{Nu_o}$ is the overall Nusselt number defined by Eq. (120).

Some of the technical applications of the (T3) boundary condition are outlined in Table 2. The (T3) boundary condition has been analyzed for the circular tube, parallel plates, an elliptical duct, a sine duct, and concentric annular ducts.

(iii) (T4) *Boundary Condition.* For this boundary condition, wall thermal resistance R_w is considered as zero, so that $t_{wo} = t_w$ in Fig. 4. The outside

[†] $\mathrm{Nu_{H2}}$ is higher than $\mathrm{Nu_T}$ for the circular tube, parallel plates, and elliptical ducts, while $\mathrm{Nu_{H2}}$ is lower than $\mathrm{Nu_T}$ for sine ducts. Hence, increasing the value of R_w, $\mathrm{Nu_{T3}}$ will increase for the circular tube, parallel plates, and elliptical ducts, while $\mathrm{Nu_{T3}}$ will decrease for sine ducts. This is the reason for the anomalous behavior of $\mathrm{Nu_{T3}}$ for sine ducts not explained by Sherony and Solbrig [36] in reporting their results.

[‡] As in the case of the circular tube and parallel plates.

fluid or the environment temperature T_a is idealized as axially constant. Peripherally, wall heat flux is nonlinearly proportional to the wall temperature—a situation existing when the duct wall is radiating thermal energy to the environment:

$$T_a(x, y, z) = T_a(y, z), \qquad \text{independent of } x \tag{47}$$

$$-k \left. \frac{\partial T}{\partial n} \right|_\Gamma = \varepsilon_w \sigma [T|_\Gamma{}^4 - T_a{}^4] \tag{48}$$

Here the equivalent temperature T_a of the environment is defined by

$$\alpha_w H = \varepsilon_w \sigma T_a{}^4 \tag{49}$$

where α_w is the absorptivity of wall material and H the incident radiant energy. $T_w = T|_\Gamma$ is a function of x for the (T4) boundary condition, as it is for the (T3) boundary condition.

Nondimensionalizing temperature and normal direction with T_e and D_h, respectively, and after some rearrangement, Eq. (48) reduces to

$$\left. \frac{\partial T^*}{\partial n^*} \right|_\Gamma = -\gamma [T^*|_\Gamma{}^4 - T_a^{*4}] \tag{50}$$

where $\gamma = \varepsilon_w \sigma T_e{}^3 D_h / k$ is the radiation parameter.

Implicit idealizations made for the wall thermal conductivity for the (T4) boundary condition are described in Table 3. There are two limiting cases of the (T4) boundary condition: (1) When γ approaches ∞, (T4) reduces to the (T) boundary condition; (2) when γ approaches 0, (T4) reduces to the (H2) boundary condition.

Some of the technical applications of the (T4) boundary condition are outlined in Table 2. The (T4) boundary condition has been analyzed only for the circular tube with T_a as zero.

b. Constant Axial Wall Heat Flux, (H1), (H2), (H3), and (H4)

The thermal boundary condition of approximately constant axial heat rate per unit duct length ($q' \simeq \text{constant}$) is realized in many practical applications: electric resistance heating, nuclear heating, and in a counterflow heat exchanger with equal thermal capacity rates (Wc_p), etc. This boundary condition will be referred to as (H) when there is no peripheral variation of wall temperature (or wall heat flux). This is the case for symmetrically heated straight ducts having constant peripheral curvature and no corners, e.g., the circular duct, parallel plates, and concentric annular ducts.

For symmetrically heated noncircular ducts (with corners or variable peripheral curvature), the peripheral wall temperature and/or heat flux are variable. There are four simplified cases of these variations: (1) the peripheral wall temperature constant with variable peripheral heat flux; (2) the peripheral wall heat flux constant with variable peripheral temperature; (3) the peripheral local heat flux linearly proportional to the wall temperature at that point, and (4) finite peripheral heat conduction where both the peripheral wall temperature and heat flux vary. These four boundary conditions for noncircular ducts (e.g., rectangular, triangular, elliptical) will be referred to as (H1), (H2), (H3), and (H4), respectively. They are described below.

(i) (H1) *Boundary Condition.* The wall heat transfer rate is constant in the axial direction while the wall temperature at any cross section x is constant in the peripheral direction for the (H1) boundary condition:

$$q' = Wc_p \frac{dt_m}{dx} = \text{constant}, \qquad \text{independent of } x \tag{51}^\dagger$$

$$t|_\Gamma = t_w = \text{constant}, \qquad \text{independent of } (y, z) \tag{52}$$

Implicit idealizations made for the wall thermal conductivity for the (H1) boundary condition are described in Table 3.

The (H1) boundary condition is a limiting case of a generalized boundary condition (H4) to be described later. For a heat exchanger with highly conductive materials (e.g., copper, aluminum), the (H1) boundary conditions may apply. In practice, it may be difficult to achieve this boundary condition for noncircular ducts, as discussed by Irvine [37]. However, since it is mathematically amenable, it is the most frequently investigated boundary condition in the literature, and solutions are available for over 30 duct geometries (Table 136) considered in Chapters V–XVI. Nu_{H1} is always higher than Nu_T for a given duct geometry, as explained in Chapter XVII, p. 391.

Note that q' is related to average and peripheral local wall heat fluxes q'' and q_p'' as

$$\frac{q'}{P} = q'' = \frac{1}{P} \int_\Gamma q_p'' \, ds \tag{53}$$

(ii) (H2) *Boundary Condition.* The wall heat flux q'' is constant in the axial as well as the peripheral direction for this boundary condition:

$$q''P = q' = Wc_p \frac{dt_m}{dx} = \text{constant}, \qquad \text{independent of } x \tag{54}$$

$$k \frac{\partial t}{\partial n}\bigg|_\Gamma = q_p'' = \frac{q'}{P} = \text{constant}, \qquad \text{independent of } (y, z) \text{ as well as } x \tag{55}$$

---

† The equality on the left represents an energy balance in the absence of thermal energy sources, viscous dissipation, and axial heat conduction, if any, within the fluid. Also, free convection, mass diffusion, chemical reactions, change of phase, and electromagnetic effects are neglected.

Implicit idealizations made for the wall thermal conductivity are described in Table 3.

For a heat exchanger with constant q' and low thermal conductivity materials (e.g., glass–ceramic, Teflon), the (H2) boundary condition may be realized if the wall thickness all around the duct periphery is uniform. This boundary condition is also amenable to mathematical analysis. It has been analyzed for rectangular, isosceles triangular, elliptical, sine, circular sector, circular segment, regular polygonal, corrugated, and cardioid ducts, and is described later.

Nu_{H2} is lower than Nu_{H1} for noncircular ducts, due to corner or curvature effects, as explained in Chapter XVII, p. 390. The (H2) boundary condition is a limiting case of the (H4) boundary condition described below.

(iii) (H3) *Boundary Condition.* The wall heat transfer rate is constant in the axial direction, the same as Eq. (51), while the heat flux is linearly proportional to the wall temperature in the peripheral direction, the same as Eq. (41) or (46). For this boundary condition.

$$q'(x) = Wc_p \frac{dt_m}{dx} = \text{constant}, \qquad \text{independent of } x \tag{56}$$

$$\left.\frac{\partial t}{\partial n^*}\right|_\Gamma = \frac{1}{R_w}(t_{wo} - t|_\Gamma) \tag{57}$$

Implicit idealizations made for the wall thermal conductivity for the (H3) boundary condition are described in Table 3. There are two limiting cases of the (H3) boundary condition: (1) when the wall thermal resistance R_w approaches zero, (H3) reduces to the (H1) boundary condition; (2) when R_w approaches infinity, (H3) reduces to the (H2) boundary condition.

This boundary condition appears to be of less practical importance, and has not been analyzed for noncircular ducts.

(iv) (H4) *Boundary Condition.* The wall heat transfer rate q' is constant in the axial direction while finite wall heat conduction is specified in the peripheral direction.

Consider a steady-state energy balance on the wall element ds of unit depth in Fig. 5. The temperature distribution in the wall is related to the wall heat flux along the periphery as

$$q_p'' - k\left.\frac{\partial t}{\partial n}\right|_\Gamma + k_w \frac{\partial^2}{\partial s^2}\int_n^{n+a'} t_w\, dn = 0 \tag{58}$$

The temperature across any cross section for a *thin wall* may be taken as uniform. This is equivalent to idealizing zero wall thermal resistance in the normal direction. If the wall thickness a' is uniform, then

$$\int_n^{n+a'} t_w\, dn \simeq a't|_\Gamma \tag{59}$$

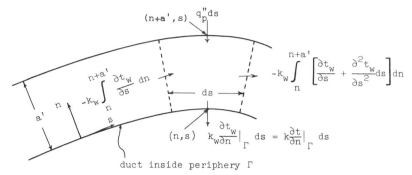

FIG. 5. Energy transfer terms in the duct wall cross section for finite peripheral heat conduction.

After dividing n and s by the characteristic dimension D_h and some rearrangement, Eq. (58) reduces to

$$\frac{q_p'' D_h}{k} - \left.\frac{\partial t}{\partial n^*}\right|_\Gamma + K_p \left.\frac{\partial^2 t}{\partial s^{*2}}\right|_\Gamma = 0 \qquad (60)$$

where $K_p = k_w a'/k D_h$ is the peripheral heat conduction parameter. Additionally, along the axial direction,

$$q'(x) = W c_p \frac{dt_m}{dx} = \text{constant}, \qquad \text{independent of } x \qquad (61)$$

The equations set (60, 61) constitutes the (H4) boundary condition.

Implicit idealizations made for the wall thermal conductivity for the (H4) boundary condition are described in Table 3. The peripheral k_w is used in Eq. (58) and in the parameter K_p. The special cases $K_p = \infty$ and 0 yield the (H1) and (H2) boundary conditions, respectively (see Table 3). It should be emphasized that K_p does not involve h, but consists of duct geometrical properties and thermal conductivities. Hence, it is a constant when thermal conductivities are treated as constants.

Some of the more important technical applications of the (H4) boundary conditions are described in Table 2; but the solutions are available only for a square duct [38] and rectangular ducts [39] as presented in Table 46.

A more generalized boundary condition, which takes into account the wall heat conduction in the peripheral direction and a finite wall thermal resistance in the normal direction, has been discussed by Lyczkowski et al. [38].

c. Exponential Axial Wall Heat Flux, (H5)

The axial wall heat flux varies exponentially along the flow length, while the peripheral wall temperature is constant at a particular section x for this boundary condition:

$$q_x' = q_e' e^{mx^*} \tag{62}$$

$$t|_\Gamma = t_w = \text{constant}, \qquad \text{independent of } (y, z) \text{ but varies with } x \tag{63}$$

Implicit idealizations made for the wall thermal conductivity for the (H5) boundary condition are described in Table 3.

The wall and fluid bulk mean temperatures are shown in Fig. 6 for varying values of the exponent m for the circular tube. It can be shown [13, 24] for

$$m = -\frac{4r_j^*}{1 + r^*} \text{Nu}_T, \qquad \text{and } m = 0 \tag{64}$$

the (H5) boundary condition reduces to the (T) and (H) boundary conditions, respectively, for the symmetrically heated concentric annular ducts. It also reduces to the (T3) boundary condition for *fully developed flow* for

$$m = -4 \text{Nu}_o \tag{65}$$

for the circular tube and parallel plates.

The (H5) boundary condition with an appropriate value of m (Fig. 6) can be used to approximate either a parallel or counterflow heat exchanger in which the fluid bulk mean temperature varies exponentially along the duct. Solutions are available for the circular tube, parallel plates, concentric annular ducts, and longitudinal flow over circular cylinders.

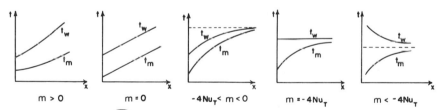

FIG. 6. (H5) temperature variations along the circular tube length.

d. *Constant Axial Wall-to-Fluid Temperature Difference,* (Δt)

The wall-to-fluid bulk mean temperature difference is constant in the axial direction, while the wall temperature at any cross section is uniform for the (Δt) boundary condition:

$$\Delta t(x) = t_w - t_m = \text{constant}, \qquad \text{independent of } x \tag{66}$$

$$t|_\Gamma = t_w = \text{constant}, \qquad \text{independent of } (y, z) \tag{67}$$

It is implied that the wall thermal conductivity is zero in the axial direction, infinite in the peripheral direction, and arbitrary in the normal direction, as summarized in Table 3.

The two limiting cases of the (Δt) boundary condition are: (1) when x approaches zero (i.e., in the thermal entrance region very close to the point of step change in wall temperature), then (Δt) approaches the (T) boundary condition; and (2) when x approaches infinity (i.e., near the fully developed region), (Δt) approaches the $(H1)$ boundary condition.[†]

The (Δt) boundary condition has been analyzed only for a circular tube [40].

The counterflow heat exchanger with $C_{min}/C_{max} = 1$ (e.g., the gas turbine regenerator) has a boundary condition between $(H4)$ and (Δt). In the gas turbine regenerator literature, the $(H1)$ boundary condition of Eqs. (51) and (52) is exclusively employed or implied, except for the case of the circular tube considered by Kays [40].

As shown in this section, (T), $(H1)$, and $(H2)$ are limiting thermal boundary conditions of the more general boundary conditions $(T3)$, $(T4)$, $(H3)$, and $(H4)$. The latter boundary conditions take into account normal wall thermal resistance or peripheral wall heat conduction.

It is important to know the magnitudes of Nu_T, Nu_{H1}, and Nu_{H2} for different duct geometries. Information from the available literature is summarized in later sections. These magnitudes for some of the duct geometries with more practical importance are condensed in Table 138, p. 394.

e. Finite Wall Axial Heat Conduction Boundary Condition

In all of the foregoing thermal boundary conditions, axial heat conduction in the duct wall was neglected. Axial heat conduction in the duct wall can be important for the axially constant heat flux boundary conditions. An example of a finite axial wall heat conduction boundary condition is derived below for a "thin" circular tube. For a "thick" wall, a conjugated problem, as described on p. 11, needs to be solved.

An energy balance on a differential element of a uniform thickness wall of a circular tube, heated from the outside with a constant heat flux q'' and convectively cooled from the inside wall, yields

$$-k_w a' \frac{\partial^2 t_w}{\partial x^2} + k_f \frac{\partial t_f}{\partial r}\bigg|_{r=a} - q'' = 0, \qquad t_w = t_f|_{r=a} \qquad (68)$$

where a' is the uniform thickness of a "thin" wall and a the inside radius. Introducing $x^* = (x/L)(L/D_h\ Pe)$, $r^* = r/a$, $R_w = (k_f a')/(k_w D_h)$, and $\Theta = (t_f - t_e)/$

[†] This is the case only when axial heat conduction, viscous dissipation, and thermal energy sources are neglected within the fluid.

$(q''D_h/k_f)$, the foregoing equation reduces to a dimensionless form:

$$\frac{1}{R_w}\left(\frac{a'}{L}\right)^2\left(\frac{L}{D_h\,Pe}\right)^2\frac{\partial^2\Theta}{\partial x^{*2}}\bigg|_{r^*=1} - \frac{\partial\Theta}{\partial r^*}\bigg|_{r^*=1} + 1 = 0 \qquad (69)$$

Thus the important parameter for wall axial heat conduction arising from the boundary condition is

$$\frac{R_w}{(a'/D_h\,Pe)^2} = \frac{K_p}{Pe^2} = \frac{16\lambda N_{tu}}{Nu} \qquad (70)$$

where K_p is the wall axial heat conduction parameter, the inverse of Pe represents the effect of fluid axial heat conduction relative to the thermal energy convected in the fluid, λ is the longitudinal wall heat conduction parameter conventionally used in the heat exchanger design theory, and N_{tu} is the number of heat transfer units [6].

2. THERMAL BOUNDARY CONDITIONS FOR DOUBLY CONNECTED DUCTS

All boundary conditions in Tables 1 and 2 are also applicable to a doubly connected duct (see Fig. 7),[†] but the problem is further complicated by the fact that there are two walls, and on each wall the boundary conditions of singly connected ducts can be separately applied. Thus there are many more combinations of boundary conditions that can be analyzed for doubly connected ducts. Some specific boundary conditions are now presented.

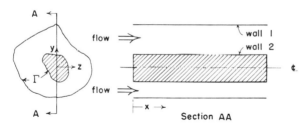

FIG. 7. A doubly connected duct of constant cross-sectional area.

a. Fundamental Boundary Conditions

Kays [7] and Lundberg et al. [41, 42] have described four kinds of fundamental boundary conditions for thermally developing flow through concentric annular ducts *heated symmetrically* (see Fig. 8 and Table 4). Once the solution to the energy equation (linear and homogeneous) is obtained for each of the four kinds of fundamental boundary conditions, the solution

---

[†] Such as an annular duct formed by concentric or eccentric cylinders.

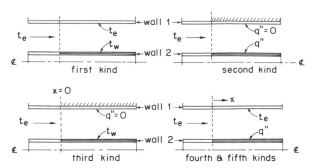

Fig. 8. Five kinds of fundamental boundary conditions for doubly connected ducts.

for any combination of these boundary conditions can be obtained by super-position techniques [7,41,42]. The fundamental boundary conditions of the fifth kind are added here to Kays' classification because they are mathematically most amenable for fully developed flow through doubly connected ducts. For the concentric annular duct family, boundary conditions of the fourth and fifth kinds are the same. They are different for other doubly connected ducts. Each kind of fundamental boundary conditions can be applied to inner–outer walls or outer–inner walls combinations; thus there are two solutions to the energy equation for each kind of fundamental boundary conditions. These boundary conditions for doubly connected ducts are described below. They are depicted in Fig. 8 and are summarized in Table 4. This table is by no means an exhaustive list of all possible fundamental boundary conditions. It may be extended as needed.

(1) *Fundamental boundary conditions of the first kind.* On one wall constant temperature, different from the entering fluid temperature, is specified. The other wall is at the constant temperature of the entering fluid.

(2) *Fundamental boundary conditions of the second kind.* On one wall constant axial and peripheral wall heat flux is specified. The other wall is insulated (adiabatic, zero heat flux).

(3) *Fundamental boundary conditions of the third kind.* On one wall constant temperature, different from the entering fluid, is specified. The other wall is insulated.

(4) *Fundamental boundary conditions of the fourth kind.* On one wall constant axial and peripheral wall heat flux is specified. The other wall is at the constant temperature of the entering fluid.

(5) *Fundamental boundary conditions of the fifth kind.* On one wall constant axial wall heat flux with constant peripheral surface temperature is specified. The other wall is insulated, but the peripheral temperature is constant.

TABLE 4

FUNDAMENTAL THERMAL BOUNDARY CONDITIONS FOR DOUBLY CONNECTED DUCTS[†]

Fundamental boundary condition		Axial boundary condition	Peripheral boundary condition
First kind	(i)	$t_o = t_e = $ constant, $t_i = $ constant	$t\vert_{r_o} = t_e = $ constant, $t\vert_{r_i} = t_i = $ constant
	(ii)	$t_i = t_e = $ constant, $t_o = $ constant	$t\vert_{r_i} = t_e = $ constant, $t\vert_{r_o} = t_o = $ constant
Second kind	(iii)	$q_o' = 0$, $q_i' = $ constant	$k(\partial t/\partial n)\vert_{r_o} = 0$, $k(\partial t/\partial n)\vert_{r_i} = q_i'' = $ constant
	(iv)	$q_i' = 0$, $q_o' = $ constant	$k(\partial t/\partial n)\vert_{r_i} = 0$, $k(\partial t/\partial n)\vert_{r_o} = q_o'' = $ constant
Third kind[‡]	(v)	$q_o' = 0$, $t_i = $ constant	$k(\partial t/\partial n)\vert_{r_o} = 0$, $t\vert_{r_i} = t_i = $ constant
	(vi)	$q_i' = 0$, $t_o = $ constant	$k(\partial t/\partial n)\vert_{r_i} = 0$, $t\vert_{r_o} = t_o = $ constant
Fourth kind	(vii)	$t_o = t_e = $ constant, $q_i' = $ constant	$t\vert_{r_o} = t_e = $ constant, $k(\partial t/\partial n)\vert_{r_i} = q_i'' = $ constant
	(viii)	$t_i = t_e = $ constant, $q_o' = $ constant	$t\vert_{r_i} = t_e = $ constant, $k(\partial t/\partial n)\vert_{r_o} = q_o'' = $ constant
Fifth kind	(xi)	$q_o' = 0$, $q_i' = $ constant	$t\vert_{r_o} = $ constant, $t\vert_{r_i} = $ constant
	(x)	$q_i' = 0$, $q_o' = $ constant	$t\vert_{r_i} = $ constant, $t\vert_{r_o} = $ constant

[†] In case of thermally fully developed flow, t_e is considered as a constant temperature (irrespective of the fluid inlet temperature) of the designated (inner or outer) wall. Thus for the fully developed case, there is essentially only one fundamental problem of the first kind.
[‡] The solution for the fully developed case may only be determined as an asymptotic solution from the thermally developing case.

b. Nomenclature Scheme

The nomenclature as adopted by Kays et al. [7,41,42] is recommended for the solution of doubly connected duct heat transfer problems. The superscript k ($k = 1, \ldots, 5$) on the dimensionless fluid temperature θ or heat flux Φ represents the kind of fundamental solution corresponding to the fundamental boundary conditions. The subscript j ($= i$ or o) stands for the heated surface j (having a nonzero boundary condition). θ and Φ are now defined for each of the five kinds of fundamental boundary conditions just mentioned:

$$\theta_j^{(1)}, \theta_j^{(3)} = \frac{t - t_e}{t_j - t_e} \tag{71}$$

$$\theta_j^{(2)}, \theta_j^{(4)}, \theta_j^{(5)} = \frac{t - t_e}{(q'/P)_j D_h/k} \tag{72}$$

$$\Phi_j^{(k)} = -D_h \frac{\partial \theta_j^{(k)}}{\partial n} \tag{73}$$

The dimensionless wall temperature, wall heat flux, and Nusselt numbers at either wall for any fundamental case k are designated in a manner similar to the example of the Nusselt number described as follows:

$$\text{Nu}_{x,lj}^{(k)}, \quad \text{Nu}_{m,lj}^{(k)}, \quad \text{Nu}_{p,lj}^{(k)}, \quad \text{Nu}_{lj}^{(k)}, \tag{74}$$

where $l = \text{i}$ or o, referring to the wall at which the Nusselt number (or the dimensionless θ or Φ) is evaluated, $j = \text{i}$ or o, referring to the wall at which the nonzero boundary condition is applied. In the double subscript notation, the subscripts x and m represent the local (peripheral average) and mean (flow length average) values in the thermal entrance region; p represents the peripheral local value; $\text{Nu}_{lj}^{(k)}$ denotes the peripheral average Nusselt number; $\text{Nu}_{p,lj}^{(k)}$ and $\text{Nu}_{lj}^{(k)}$ are for the fully developed region.

c. *Applicability of Superposition Methods and*
   Specific Boundary Conditions

The energy equation is linear and homogeneous when the thermal energy source functions, viscous dissipation, and flow work within the fluid are neglected [7]. In that case, a sum of solutions is again a solution to that energy equation. It is then possible to construct a solution for an arbitrarily specified axial variation of wall temperature (or heat flux) by merely breaking up the surface temperature (or heat flux) into a number of constant temperature (or heat flux) steps and summing or superposing the corresponding thermal entry length solutions for each step.

If the velocity profile is considered fully developed in the thermal entrance region, the solution for each step (temperature or heat flux) is the same as the conventional thermal entrance solution for the hydrodynamically developed flow. Thus for each duct geometry, only one thermal entrance solution for hydrodynamically developed flow, used in conjunction with the superposition method, is required for a complete solution to arbitrary variations in the wall temperature or heat flux.

However, for the simultaneously developing flow,[†] the situation is different. At each step of the axial wall temperature and/or heat flux, the velocity profile, developing in the entrance region, is different. Hence a large number of thermal entrance solutions, with heating started at different locations in the hydrodynamic entry length, would be required for the superposition for a complete solution to arbitrary variations in the wall temperature or heat flux. The above discussion applies either to singly or multiply connected ducts. Since it is impractical to carry out such a large number of

---

[†] The hydrodynamically and thermally developing flow is referred to as the simultaneously developing flow.

solutions for simultaneously developing flow, this problem is solved only for the circular tube [43] and to a limited extent for the concentric annular ducts [44–46]. For a doubly connected duct, the total number of thermal entrance solutions required is enormous because of at least five possible fundamental boundary conditions.

A number of thermal boundary conditions with practical importance are outlined in Table 5 for doubly connected ducts. Solutions for these boundary conditions can be obtained by the principle of superposition once the solutions for each of the five kinds of fundamental boundary conditions (Table 4) are derived. A different set of nomenclature for the Nusselt number is also outlined in Table 5. Similar nomenclature may be used for θ and Φ.

TABLE 5

Some Specific Thermal Boundary Conditions for Doubly Connected Ducts

		Special boundary condition	Superimpose b.c. as follows from Table 4	Sample nomenclature
(1)	(a)	constant and uniform but different wall temperatures on each wall	(i) + (ii)	$Nu_{x,i}^{(1a)}$, $Nu_{x,o}^{(1a)}$
	(b)	constant, uniform, and equal wall temperature on both walls, $\textcircled{T}$[†]	(i) + (ii)	$Nu_{x,i}^{(1b)}$, $Nu_{x,o}^{(1b)}$
(2)	(a)	constant, uniform, and equal or different axial q' such that uniform but different t_w along both peripheries	(ix) + (x)	$Nu_{x,i}^{(2a)}$, $Nu_{x,o}^{(2a)}$
	(b)	constant, uniform but different axial q' such that uniform and equal t_w along both peripheries, $\textcircled{H1}$	(ix) + (x)	$Nu_{x,i}^{(2b)}$, $Nu_{x,o}^{(2b)}$
(3)	(a)	constant, uniform, and equal or different axial q' such that uniform but different q'' along both peripheries	(iii) + (iv)	$Nu_{x,i}^{(3a)}$, $Nu_{x,o}^{(3a)}$
	(b)	constant, uniform but different axial q' such that uniform and different q'' but equal $t_{w,m}$ along both peripheries, $\textcircled{H2}$	(iii) + (iv)	$Nu_{x,i}^{(3b)}$, $Nu_{x,o}^{(3b)}$
(4)	(a)	constant and uniform temperature on outer wall, constant and uniform axial as well as peripheral heat flux on inner wall	(vi) + (vii)	$Nu_{x,i}^{(4a)}$, $Nu_{x,o}^{(4a)}$
	(b)	constant and uniform temperature on inner wall, constant and uniform axial as well as peripheral heat flux on outer wall	(v) + (viii)	$Nu_{x,i}^{(4b)}$, $Nu_{x,o}^{(4b)}$

[†] The solution for the fully developed case may only be determined as an asymptotic solution from the thermally developing case.

For cases 1b, 2b, and 3b of Table 5, t_w (or $t_{w,m}$) is the same for both inner and outer walls at a given cross section. The total heat transfer through both walls or t_m can readily be calculated for these cases, if the perimeter average heat transfer coefficient h is defined as follows:

$$q_i' + q_o' = (h_i P_i + h_o P_o)(t_w - t_m) = h(P_i + P_o)(t_w - t_m) \tag{75}$$

so that

$$h = \frac{h_i P_i + h_o P_o}{P_i + P_o} \tag{76}$$

Nusselt numbers based on the above definition of h for cases 1b, 2b, and 3b will be referred to as Nu_T, Nu_{H1}, and Nu_{H2} for doubly connected ducts. For each of the cases 2b and 3b, the ratio q_i''/q_o' is a constant uniquely related to the doubly connected duct geometry.

3. THERMAL BOUNDARY CONDITIONS FOR MULTIPLY CONNECTED DUCTS

The concept of thermal boundary conditions for singly and doubly connected ducts can be extended to multiply connected ducts, but then the problem is even more complicated because of the choice of boundary conditions on each wall of multiply connected ducts. No classification of boundary conditions for multiply connected ducts is presented here.

Practically, the most important multiply connected duct is a bundle of parallel rods where the fluid is flowing longitudinally over the rods. The (H1) and (H2) boundary conditions, applied to each rod of circular cross section, have been analyzed in the literature and are described in Chapter XV.

Chapter III

Dimensionless Groups
and Generalized Solutions

In Chapter II, the appropriate differential equations and boundary conditions were outlined for laminar fluid flow and heat transfer problems associated with singly and multiply connected ducts. Solutions to these equations are generally presented in terms of dimensionless groups so as to be more amenable to practical use and tabular/graphical presentations.

In this section, some physical quantities involved in the dimensionless groups are described first. Then dimensionless groups are defined. Finally, solutions of the momentum and energy equations, in terms of general functional relationships (implicit form), are presented. The physical quantities, dimensionless groups, and solutions are outlined separately for fluid flow and heat transfer.

A. Fluid Flow

From an engineering point of view, it is important to know how much power will be needed to pump the fluid through the heat exchanger. The fluid pumping power is proportional to the pressure drop in the fluid across the heat exchanger. The pressure drop in fully developed flow is caused by the wall shear. In the developing flow, however, it results from the wall shear and also the change in momentum flow rate as the velocity profile develops. Throughout this monograph, considerations of abrupt contraction and expansion losses at the duct entrance and exit are omitted, as are form drag and flow acceleration pressure effects due to density changes. Addi-

37

tionally, the pressure drop or rise due to area change at the duct entrance and exit is also omitted. In design applications, these factors need to be included [6]. Conveniently, these factors are additive for the evaluation of total pressure drop.

1. Physical Quantities

Fluid mean axial velocity and wall shear stress are two important physical quantities in the laminar fluid flow problem. They are defined next.

a. Mean Velocity u_m

The fluid mean axial velocity is defined as the integrated average axial velocity with respect to the flow area A_c:

$$u_m = \frac{1}{A_c} \int_{A_c} u \, dA_c \tag{77}$$

where the velocity distribution u for a given duct geometry is determined from the applicable Eq. (3), (8), or (9). Note that $u_m = u_e$ for a uniform velocity profile at the entrance.

b. Wall Shear Stress τ

The local shear stress at the wall for a Newtonian fluid flowing through the duct is expressed as [7][†]

$$\tau_p = -\frac{\mu}{g_c} \left(\frac{\partial u}{\partial n} \right)_p \tag{78}$$

Except for those cases where more detailed information is needed, the axially local wall shear stress is consistently defined as the average wall shear stress with respect to the perimeter of the duct:

$$\tau_x = -\frac{\mu}{g_c} \left(\frac{\partial u}{\partial n} \right)_{w,m} \tag{79}$$

For example, the wall shear stress for the axisymmetric flow in a rectangular duct, Fig. 9, at any cross section x, is expressed as

$$\tau_x = -\frac{\mu}{2g_c(a + b)} \left[\int_{-a}^{a} \left(\frac{\partial u}{\partial y} \right)_{y=b} dz + \int_{-b}^{b} \left(\frac{\partial u}{\partial z} \right)_{z=a} dy \right] \tag{80}$$

[†] The dynamic viscosity coefficient μ defined here is g_c times the usual fluid mechanics dynamic viscosity coefficient. Thus Newton's second law of motion is not invoked in Eq. (78), even though g_c appears in the equation.

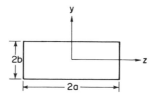

FIG. 9. A rectangular duct.

The flow length average wall shear stress is then defined as

$$\tau_m = \frac{1}{x} \int_0^x \tau_x \, dx \tag{81}$$

2. DIMENSIONLESS GROUPS

All dimensionless groups that appear in solutions to the laminar duct flow problem are described next, except for the Reynolds number Re.

a. Fanning Friction Factor f

The ratio of wall shear stress τ to the flow kinetic energy per unit volume $\rho u_m^2 / 2g_c$[†] is defined as the Fanning friction factor. The peripheral average axially local Fanning friction factor is then expressed as

$$f_x = \frac{\tau_x}{\rho u_m^2 / 2g_c} \tag{82}$$

The mean (flow length average) Fanning friction factor in the hydrodynamic entrance region is defined as

$$f_m = \frac{1}{x} \int_0^x f_x \, dx = \frac{\tau_m}{\rho u_m^2 / 2g_c} \tag{83}$$

In the hydrodynamic entrance region, pressure drops result from the wall shear and the change in momentum flow rate across the two relevant duct sections. The pressure drop for constant density flow for the duct of Fig. 2, from section $x = 0$ to x, can be found by applying Newton's second law of motion and conservation of matter principles as

$$\frac{\Delta p}{\rho u_m^2 / 2g_c} = f_m \frac{x}{r_h} + \frac{2}{A_c} \int_{A_c} \left(\frac{u}{u_m}\right)^2 dA_c - 2 \tag{84}$$

where f_m is given by Eq. (83) and u is the velocity profile at the cross section x. The third term on the right-hand side of Eq. (84) represents a ratio of the

---

[†] Also referred to as the velocity "pressure."

momentum flow rate at the duct inlet $\rho u_m^2/g_c$ (where the velocity profile is idealized as uniform) to $\rho u_m^2/2g_c$. However, this representation of pressure drop is not useful for engineering calculations, because f_m and the velocity profile u are needed as a function of x and the integral must also be evaluated. Consequently, the solution to the velocity problem is also presented for operational convenience in terms of an *apparent* Fanning friction factor:

$$\Delta p^* = \frac{\Delta p}{\rho u_m^2/2g_c} = f_{app}\frac{x}{r_h} \qquad (85)^\dagger$$

f_{app} is thus based on the total pressure drop from $x = 0$ to x. It takes into account both the skin friction and the change in momentum rate (due to change in the shape of the velocity profile) in the hydrodynamic entrance region.

For the case of fully developed flow through a duct, the velocity profile is invariant across any flow cross section. Consequently, the wall shear stress does not change axially, and the average friction factor is the same as the local friction factor for that part of the duct beyond the hydrodynamic entry length. In this case, the constant-density pressure drop across two flow cross sections, separated by a distance L, takes the following form instead of Eq. (84):[‡]

$$\frac{\Delta p}{\rho u_m^2/2g_c} = f\frac{L}{r_h} \qquad (86)$$

Fanning friction factor for fully developed flow is presented as either f without a subscript or f with fd as a subscript at some places.

In the fully developed region, Eq. (86) may be rearranged, using the definitions of Re and c_1, so that

$$f\,\text{Re} = -\frac{c_1 D_h^2}{2u_m} = -2\left[\frac{\partial(u/u_m)}{\partial(n/D_h)}\right]_{w,m} \qquad (87)$$

Also, based on the solution of differential equation (3), it can be shown that

$$f\,\text{Re} = K_f \qquad (88)$$

[†] In the literature, the dimensionless pressure drop is also presented by an *Euler Number* $Eu = \Delta p/(\rho u_m^2/g_c) = \Delta p^*/2$.

[‡] In the literature, the "large" or Darcy friction factor is also used. It is defined such that $f_D = 4f$, and the right-hand side of Eq. (86) becomes $f_D L/D_h$. The f_D is also referred to as the Darcy–Weisbach friction factor.

where K_f is a constant dependent on the geometry of the duct cross section, and Re is the Reynolds number based on the hydraulic diameter D_h.

b. *Dimensionless Axial Distance x^+*

The dimensionless axial distance in the flow direction for the hydrodynamic entrance region is defined as

$$x^+ = x/D_h \, \mathrm{Re} \tag{89}$$

For the hydrodynamic entrance region problem, many variations of x^+ are used as the dimensionless axial distance in the literature. To avoid confusing definitions, the above definition of x^+ is recommended for the dimensionless axial distance in the solution of the hydrodynamic entry length problem.

c. *Hydrodynamic Entrance Length L_{hy}^+*

The hydrodynamic entrance length L_{hy} is defined, somewhat arbitrarily, as the duct length required to achieve a duct section maximum velocity of 99% of the corresponding fully developed magnitude when the entering flow is uniform. The hydrodynamic entrance length is also referred to as the *settling length*. There are a number of other definitions used in the literature for L_{hy}; however, the foregoing definition is used throughout this monograph unless clearly indicated. The dimensionless hydrodynamic entrance length is expressed as $L_{hy}^+ = L_{hy}/(D_h \, \mathrm{Re})$.

Lakshmana Rao [47] outlined a method to determine the approximate hydrodynamic entrance length when the velocity profiles at $x = 0$ and L_{hy}, having nonzero first derivatives at the duct wall, are known. Lakshmana Rao's method incorporates many simplifications, and produces an L_{hy} that is too low compared to the one based on the rigorous entry length solution.

McComas [48] carried out an analysis to determine the approximate hydrodynamic entry length. He considered the Schiller velocity profile [49] (see p. 69) but only dealt with the inviscid core, i.e., the flow along the centerline of the tube was assumed to be inviscid up to the axial location where the boundary layer had completely filled the duct. He also imposed the condition that the centerline was the line of maximum velocity for the fully developed flow. The hydrodynamic entrance length, from his analysis, is given as

$$L_{hy}^+ = \frac{L_{hy}}{D_h \, \mathrm{Re}} = \frac{(u_{max}/u_m)^2 - 1 - K(\infty)}{4f \, \mathrm{Re}} \tag{90}$$

McComas' approximate method to determine L_{hy}^+ was first proposed by Olson in the discussion of a paper by Sparrow et al. [50]. Except very near the entrance, the assumption of an inviscid core by McComas is not realistic. This may be the reason for McComas' L_{hy}^+ being substantially shorter than those from other analyses and experiments. It may be noted that the definition of McComas' L_{hy} is different and is not based on the duct section maximum velocity as a certain percentage of the fully developed value.

McComas presented L_{hy}^+, u_{max}/u_m, $K(\infty)$, and f Re in tabular form for fully developed laminar flow through circular, elliptical, annular, rectangular, and isosceles triangular ducts.

d. Incremental Pressure Drop Number $K(x)$

As noted before, the pressure drop evaluation from Eq. (84) for the hydrodynamic entrance region is operationally inconvenient. As a result, the apparent Fanning friction factor was introduced and defined. An alternative way of evaluating pressure drop in the hydrodynamic entrance region treats the pressure drop as consisting of two components: (1) the pressure drop based on the fully developed flow, and (2) additional pressure drop due to momentum change and accumulated increment in wall shear between developing flow and developed flow. The second contribution to the pressure drop, in a dimensionless form, is designated as the incremental pressure drop number $K(x)$, defined by

$$\Delta p^* = \frac{\Delta p}{\rho u_m{}^2/2g_c} = f_{fd}\frac{x}{r_h} + K(x) = \left|\frac{dp^*}{dx^+}\right|_{fd} x^+ + K(x) \tag{91}$$

$K(x)$ is also sometimes referred to as the *pressure defect*. $K(x)$ increases monotonically from 0 at $x = 0$ to a constant value in the fully developed region. Note that $K(x)$ and f_{app} Re are related as

$$\Delta p^* = (f_{app}\,Re)(4x^+) = (f\,Re)(4x^+) + K(x) \tag{92}$$

Note that sometimes in the literature Δp^* is considered to have one more component on the right-hand side of Eq. (91); that component is $+1$ and it corresponds to one velocity head pressure drop at $x = 0$ for fluid acceleration from zero velocity upstream ($x \ll 0$) to uniform velocity u_m at $x = 0$. This changes the magnitude of Δp^* without affecting $K(x)$.

$K(x)$ is designated as $K(\infty)$ in the fully developed region. In a "long duct" heat exchanger analysis, the knowledge of f_{fd} and $K(\infty)$ is sufficient to establish the total pressure drop; a detailed investigation of velocity and pressure distribution in the flow field is not needed. Lundgren et al. [51] devised an approximate analytical method to determine $K(\infty)$ for the ducts of arbitrary cross section. They obtained

$$K(\infty) = \frac{2}{A_c} \int_{A_c} \left[\left(\frac{u_{fd}}{u_m} \right)^3 - \left(\frac{u_{fd}}{u_m} \right)^2 \right] dA_c \qquad (93)$$

Thus the fully developed velocity profile u_{fd}/u_m is needed from the solution of Eq. (3) to evaluate $K(\infty)$. The $f_D \, Re$ and $K(\infty)$ have been determined for the circular tube, elliptical ducts, rectangular ducts, isosceles triangular ducts, and concentric annular ducts by Lundgren et al. [51]. $K(\infty)$ determined by Eq. (93) is generally higher than that measured experimentally [50]. As noted above, McComas [48] extended the analysis of Lundgren et al. to determine approximate L_{hy}^+.

Sastry applied the conformal mapping method to determine $K(\infty)$ as expressed by Eq. (93) and obtained $K(\infty)$ for elliptical ducts [52] and the duct of Booth's lemniscate section [53].

e. *Momentum Flux Correction Factor $K_d(x)$*

This is defined as

$$K_d(x) = \frac{1}{A_c} \int_{A_c} \left(\frac{u}{u_m} \right)^2 dA_c \qquad (94)$$

The momentum rate at any cross section of the duct is then given by

$$\text{momentum rate} = K_d(x) \frac{\rho u_m^2}{g_c} A_c \qquad (95)$$

K_d is greater than unity for any nonuniform velocity distribution across the duct cross section. In the hydrodynamic entrance region, $K_d(x)$ increases monotonically from unity for a uniform entrance velocity profile to an asymptotic constant value in the fully developed region.

f. *Kinetic Energy Correction Factor $K_e(x)$*

This is defined as

$$K_e(x) = \frac{1}{A_c} \int_{A_c} \left(\frac{u}{u_m} \right)^3 dA_c \qquad (96)$$

The kinetic energy of the fluid at any cross section of the duct is given by

$$\text{kinetic energy} = K_e(x) \frac{\rho u_m^3}{2g_c} A_c \qquad (97)$$

K_e is greater than unity for any nonuniform velocity distribution across the section. In the hydrodynamic entrance region, $K_e(x)$ increases monotonically from unity for a uniform entrance velocity profile to an asymptotic constant value in the fully developed region.[†]

3. SOLUTIONS IN TERMS OF GENERAL FUNCTIONAL RELATIONSHIPS

The solutions to the velocity problem involve the determination of velocity components u, v and w in the whole flow field, wall shear stress distribution, and pressure drop. However, such detailed information is usually not required by a designer, and the solutions are usually preferred in dimensionless form. These solutions to the velocity problem are presented for the following set of dimensionless groups: $f_{app} \, Re$, $K(x)$, L_{hy}^+, K_d and K_e. $f_{app} \, Re$, based on the solution of Eq. (8) or (9), is a function of x^+,

$$f_{app} \, Re = f_{app} \, Re(x^+) \tag{98}$$

If the boundary layer type idealizations are not invoked, but instead the complete Navier–Stokes equations are solved, the Reynolds number Re would be an additional parameter on which $f_{app} \, Re$ would depend.

$$f_{app} \, Re = f_{app} \, Re(x^+, Re) \tag{99}$$

The incremental pressure drop number $K(x)$ is also dependent on x^+ as well as on Re if the complete set of Navier–Stokes equations is solved. However, $K(\infty)$ and L_{hy}^+ are independent of x^+.

B. Heat Transfer

The solution to the duct fluid flow problem involving heat transfer requires the determination of q'' distribution, and the fluid and wall temperatures. If the fluid inlet conditions are given, the outlet conditions can be determined, provided the wall heat flux distribution is known along with the flow path geometry. Or, if the inlet and outlet conditions are known, the length of the duct (or a heat exchanger) of a given cross-sectional geometry can be determined from the wall heat flux distribution. This heat flux is related to wall and fluid bulk mean temperatures $t_{w,m}$ and t_m through the concept of a heat transfer coefficient h. These physical quantities are defined first; then the dimensionless groups associated with the laminar duct flow forced convection

---

[†] Note that K_e used for the abrupt expansion loss in a heat exchanger [6] is not the same as $K_e(x)$ of Eq. (97). It is defined by

$$\Delta p = K_e \, \frac{\rho u_m{}^2}{2g_c} \tag{100}$$

where Δp is the pressure drop associated with the abrupt expansion.

problem are presented; and finally, the generalized functional relationships between the dimensionless groups are formulated.

1. PHYSICAL QUANTITIES

a. *Mean Temperatures* $t_{w,m}$ *and* t_m

The peripheral mean wall temperature $t_{w,m}$ and the fluid bulk mean temperature t_m at an arbitrary duct cross section x are defined as

$$t_{w,m} = \frac{1}{P} \int_\Gamma t_w \, ds \qquad (101)$$

$$t_m = \frac{1}{A_c u_m} \int_{A_c} ut \, dA_c \qquad (102)$$

The fluid bulk mean temperature t_m is also sometimes referred to as the "mixing cup" or the "flow average" temperature. Conceptually, t_m is the temperature one would measure if the duct were cut off at a section, and escaping fluid were collected and thoroughly mixed in an adiabatic container.

b. *Heat Transfer Coefficient* h

The heat transfer coefficient h is operationally convenient for the description of heat flux at a local point on the wall. h relates to the temperature difference $(t_w - t_m)$ in Fig. 4 as

$$q_p'' = h_p(t_w - t_m) \qquad (103)$$

where the suffix p denotes the peripheral local quantities. Since h signifies a conductance in the thermal circuit representation of Fig. 4, it is also referred to as the *convection conductance*. The convection heat transfer mechanism is in some measure dependent on the fluid motion; thus h may also depend upon the fluid motion. In constant-property flow (linear problem), q'' is directly proportional to the temperature difference $(t_w - t_m)$; thereby h is found to be independent of this temperature difference. For nonlinear problems, h may also be dependent on $(t_w - t_m)$. In such situations, the operational convenience of the conductance concept is diminished, although the definition of Eq. (103) is perfectly valid as such.

The peripherally average, but axially local, heat transfer coefficient h_x is defined by

$$q_x'' = h_x(t_{w,m} - t_m) \qquad (104)$$

where h_x may or may not be the peripherally integrated average of h_p because of $t_{w,m}$. The flow length average heat transfer coefficient h_m is the integrated average of h_x from $x = 0$ to x:

$$h_{\mathrm{m}} = \frac{1}{x} \int_0^x h_x \, dx \tag{105}$$

In Eq. (103) or (104), the heat flux and temperature difference are vector quantities. The direction of heat transfer was considered from the wall to the fluid, and as a result, the temperature difference employed was also in the same direction. If the heat transfer rate q_x'' had been from the fluid to the wall, the definition of h_x would have been

$$q_x'' = h_x(t_{\mathrm{m}} - t_{\mathrm{w,m}}) \tag{106}$$

In the definition of h_x, only the wall temperature and fluid bulk mean temperature are involved. If the fluid temperature profile at a flow cross section does not have both a "dip" and a "bulge," the preceding definition provides a value of h_x as positive for both the fluid heating and cooling situations. However, under rapidly changing boundary conditions, nonuniform internal thermal energy generation, or nonuniform heating or cooling for doubly connected ducts, the fluid temperature profile may well have both a dip and a bulge. This could result in zero or negative value of $(t_{\mathrm{w,m}} - t_{\mathrm{m}})$ when q_x'' is from the wall to the fluid. Mathematically for this case, h_x will be infinite or negative respectively. In such situations, the conductance concept of the heat transfer coefficient losses its significance. Examples of anomalous behavior of h_x are provided in the discussion of Figs. 16 and 22.

In the case of finite wall thermal resistance boundary conditions, e.g., (T3) and (H3), the overall heat transfer coefficient U_{o} has considerable practical importance. $U_{\mathrm{o},x}$ is defined by

$$q_x'' = U_{\mathrm{o},x}(t_{\mathrm{a}} - t_{\mathrm{m}}) \tag{107}$$

where from Fig. 4,

$$\frac{1}{U_{\mathrm{o},x}} = \frac{1}{h_{x,\mathrm{T3}}} + \frac{1}{U_{\mathrm{w}}} \tag{108}$$

2. Dimensionless Groups

Dimensionless groups are generally deduced from normalizing the differential equations and boundary conditions in a special way. Consequently many different sets of dimensionless groups can be formulated for a particular duct. Such results may not be useful to a designer because he cannot readily interpret them in terms of the familiar conventional set. A standardization of the dimensionless groups arising from the differential equations is recommended by Shah and London [54]. These recommendations are more in line with the conventional definitions of Boucher and Alves [55] and McAdams [56], and are summarized below. The dimensionless groups arising from the boundary conditions are defined in Chapter II, with the appropriate thermal boundary conditions. These groups are R_w, K_p, γ, m, and $R_w/(a'/D_h \, Pe)^2$. Whenever applicable, the hydraulic diameter D_h should be used as a characteristic dimension representative of the flow cross section.

a. *Nusselt Number* Nu

The ratio of the convective conductance h to the pure molecular thermal conductance k/D_h is defined as a Nusselt number. At a local peripheral point on the duct boundary, the Nusselt number is defined as

$$\text{Nu}_p = \frac{h_p}{k/D_h} = \frac{h_p D_h}{k} = \frac{q_p'' D_h}{k(t_w - t_m)} = \frac{D_h(\partial t/\partial n)_p}{(t_w - t_m)} \tag{109}$$

The Nusselt number has a physical significance in the sense that the heat transfer coefficient h in Nu represents the convective conductance in a thermal circuit representation (Fig. 4), with the heat flux q_p'' as the current and $(t_w - t_m)$ as the potential.

The peripheral average local Nusselt number for a noncircular duct is based on h_x and is defined as

$$\text{Nu}_x = \frac{h_x D_h}{k} = \frac{q_x'' D_h}{k(t_{w,m} - t_m)} = \frac{D_h[(\partial t/\partial n)_{w,m}]_x}{(t_{w,m} - t_m)} \tag{110}$$

where $(\partial t/\partial n)_{w,m}$ represents a peripheral average temperature gradient at wall. This definition of a perimeter average local Nusselt number *may* or *may not* represent the integrated (around the periphery) average of Nu_p because of $t_{w,m}$. Nu_x could be positive, infinite, or negative depending upon the magnitude of h_x as discussed on the preceding page.

The mean (flow length average) Nusselt number in the thermal entrance region is based on h_m and is defined as

$$\text{Nu}_m = \frac{h_m D_h}{k} = \frac{1}{x} \int_0^x \text{Nu}_x \, dx \tag{111}$$

When the effects of axial heat conduction, viscous dissipation, and flow work are neglected within the fluid, it can be shown that

$$\text{Nu}_m = q_m'' D_h/k(\Delta t)_{ave} \tag{112}$$

where

$$(\Delta t)_{ave} = (\Delta t)_{lm} = \frac{[t_w - t_e] - [t_w - t_m(x)]}{\ln[(t_w - t_e)/(t_w - t_m)]} \quad \text{for} \quad ⓣ \tag{113}$$

and

$$(\Delta t)_{ave} = \left[\frac{1}{x} \int_0^x \frac{1}{t_w(x) - t_m(x)} \, dx \right]^{-1} \qquad \text{for} \quad \text{(H1)} \qquad (114)$$

The nomenclature scheme for the Nusselt number for a specific thermal boundary condition is described on pp. 19 and 34.

Colburn [57] and McAdams [56] correlated the (T) experimental data, using a mean Nusselt number based on an arithmetic instead of a logarithmic mean temperature difference, defined as

$$\text{Nu}_{a,T} = \frac{h_{a,T} D_h}{k} = \frac{q_m'' D_h}{k[t_w - \frac{1}{2}(t_m + t_e)]} \qquad (115)$$

$\text{Nu}_{a,T}$ and $\text{Nu}_{m,T}$ are related by

$$\text{Nu}_{a,T} = \frac{2(1 - \theta_m)}{(1 + \theta_m) \ln(1/\theta_m)} \text{Nu}_{m,T} \qquad (116)$$

where

$$\theta_m = \frac{t_w - t_m}{t_w - t_e} \qquad (117)$$

For the (T3) boundary condition, an overall local Nusselt number $\text{Nu}_{o,x}$ is defined as

$$\text{Nu}_{o,x} = \frac{U_{o,x} D_h}{k} \qquad (118)$$

where $U_{o,x}$ is given by Eq. (108). From Eqs. (107) and (108), it is seen that the temperature difference involved for $\text{Nu}_{o,x}$ is $(t_a - t_m)$. The flow length average $\text{Nu}_{o,m}$ may be defined in two ways:

$$\text{Nu}_{o,m} = \frac{1}{x} \int_0^x \text{Nu}_{o,x} \, dx \qquad (119)$$

or

$$\frac{1}{\text{Nu}_{o,m}} = \frac{1}{\text{Nu}_{m,T3}} + R_w \qquad (120)$$

Although the determination of $\text{Nu}_{o,m}$ by Eq. (119) is straightforward from the mathematical point of view, the definition of Eq. (120) is more useful from the heat exchanger design point of view. In heat exchanger design, one usually treats the integrated average Nusselt number on each fluid side as constant and then includes a wall thermal resistance in order to arrive at the overall conductance U.

The thermal entrance local Nusselt number, as defined by Eq. (110), approaches the fully developed flow value for large x^*, while the local Nu (in terms of h) based on other temperature differences does not have this behavior and therefore *should not* be designated as a Nusselt number. The other most common temperature difference used for the dimensionless heat transfer modulus in the thermal entrance regime is $(t_w - t_e)$. This modulus will be referred to as the dimensionless heat flux and is defined next.

b. *Dimensionless Heat Flux* Φ

In the thermal entrance region, the local and mean wall heat fluxes for the Ⓣ boundary condition cannot be evaluated from the corresponding local and mean Nusselt numbers unless $t_m(x)$ and $(\Delta t)_{lm}$ are known. Additionally, if axial heat conduction within the fluid is considered, the simple relationships of Eqs. (112) and (113) are not valid. To determine the wall heat flux directly, the analytical solutions are sometimes presented in terms of a dimensionless heat transfer modulus based on the $(t_w - t_e)$ temperature difference. This dimensionless modulus is also referred to as a Nusselt number! To avoid confusion, these local and mean dimensionless numbers will be referred to as the dimensionless heat fluxes $\Phi_{x,T}$ and $\Phi_{m,T}$:

$$\Phi_{x,T} = \frac{q_x'' D_h}{k(t_w - t_e)} \tag{121}$$

$$\Phi_{m,T} = \frac{q_m'' D_h}{k(t_w - t_e)} \tag{122}$$

For the Ⓣ3 and Ⓣ4 boundary conditions, t_w is replaced by $t_{w,m}$. The "heat transfer coefficient" in Φ_x does not represent a thermal conductance in a thermal circuit, nor does it approach the fully developed thermal conductance at large x^* (when $x^* \to \infty$).

For hydrodynamically developing or developed and thermally developing flows, the local and average Nusselt numbers and dimensionless heat fluxes are related as follows:

$$\Phi_{x,T} = \theta_m \, \mathrm{Nu}_{x,T} \tag{123}$$

$$\Phi_{m,T} = \frac{1 - \theta_m}{\ln(1/\theta_m)} \, \mathrm{Nu}_{m,T} = \frac{1 - \theta_m}{4x^*} \tag{124}$$

for a duct of arbitrary constant cross section, when axial heat conduction, viscous dissipation, and thermal energy sources within the fluid are neglected.

c. *Dimensionless Axial Distance x* and Graetz Number* Gz

The dimensionless distance in the flow direction for the thermal entrance region heat transfer is specified as

$$x^* = \frac{x}{D_h \, \mathrm{Pe}} = \frac{x}{D_h \, \mathrm{Re} \, \mathrm{Pr}}$$ (125)

Throughout the thermal entrance heat transfer literature, the dimensionless axial distance is defined variously as x^*, $1/x^*$, cx^* (where c is a constant), etc. These dimensionless axial variables are all designated as a Graetz number! McAdams [56] defines the Graetz number Gz as

$$\mathrm{Gz} = \frac{Wc_p}{kL} = \frac{\mathrm{Pe} \, P}{4L} = \frac{\mathrm{Re} \, \mathrm{Pr} \, P}{4L}$$ (126)

This definition of the Graetz number is conventionally used in the chemical engineering literature. To avoid confusion of various definitions used for the dimensionless axial distance, it is recommended that (1) McAdams' definition of Graetz number be retained as it is, (2) the dimensionless axial distance $x^* = x/D_h \, \mathrm{Pe}$ be used for the thermal entry length problem, and (3) x^* should not be designated as the Graetz number. Then if the flow length L in Gz is treated as a length variable,

$$x^* = \frac{P}{4D_h} \frac{1}{\mathrm{Gz}}$$ (127)

which reduces to $x^* = \pi/(4 \, \mathrm{Gz})$ for the circular tube.

d. *Thermal Entrance Length* L_{th}^*

The thermal entrance length L_{th} is defined, somewhat arbitrarily, as the duct length required to achieve a value of local Nu_x equal to 1.05 Nu for fully developed flow. The dimensionless thermal entrance length is expressed as

$$L_{th}^* = L_{th}/D_h \, \mathrm{Pe}$$ (128)

There are a number of other definitions used in the literature for L_{th}; however, the foregoing definition is used here throughout unless explicitly mentioned. Note that L_{th} is defined differently than L_{hy}.

Sandall and Hanna [58] are the first investigators who devised an approximate analytical method to determine $L_{th, H1}^*$ for a laminar or turbulent flow of a power law non-Newtonian fluid flowing through a circular or noncircular duct. They considered constant fluid properties and hydrodynamically developed flow at the entrance of a duct. They made an overall energy balance between two duct cross sections, $x = -L'$ and $x = L_{th}$, where $-L'$ is the upstream duct length at which axial heat conduction effects are no longer felt and L_{th} is duct length from the point of duct heating ($x = 0$) at which the temperature profile first becomes fully developed. Based on this energy balance, they derived the following formula for $L^*_{th,H1}$ for laminar flow of a Newtonian fluid through a circular tube:

$$L^*_{th,H1} \geq \frac{1}{2u_m{}^2 a^4} \int_{r=0}^a ur \int_{\xi=0}^r \frac{1}{\xi} \int_{\zeta=0}^\xi u\zeta \, d\zeta \, d\xi \, dr \qquad (129)$$

where ξ and ζ are dummy variables of integration and u is the fully developed laminar velocity profile. A comparison with the more rigorous values of $L^*_{th,H}$ for the circular tube and parallel plates reveals that the above formula provides a lower bound for $L^*_{th,H}$.

Hanna et al. [59] extended the approximate method of Sandall and Hanna [58] to determine (1) $L^*_{th,H1}$ for rectangular and concentric annular ducts for developed velocity profiles, and (2) $L^*_{th,H}$ and $L^*_{th,T}$ for a circular tube for simultaneously developing flow. They included the effect of fluid axial heat conduction in the analysis for all geometries, and also included the effect of viscous dissipation for the foregoing (2) case.

To derive L_{th}, the above investigators made the principal idealization as follows: The temperature at the duct centerline where the fully developed temperature profile starts (the thermal boundary layer merges) is the same as the inlet fluid temperature for the fluid heating situation. However, in reality, the radial heat conduction becomes important near the region where the boundary layers merge, and hence the centerline temperature may be significantly higher than the inlet fluid temperature. Thus L_{th} calculated by Sandall and Hanna, and Hanna et al. is significantly lower than that based on the rigorous solution.

When the effect of fluid axial heat conduction is considered, the thermal entrance length increases for increasing fluid axial heat conduction (i.e., decreasing the Péclet number Pe). For example, in the case of the fluid heating situation, finite axial heat conduction would reduce the temperature of the fluid at any cross section, and hence a longer duct length would be required to achieve the fully developed temperature profile. The analysis of Hanna et al. [59] does not take a proper account of fluid axial heat conduction; they predict L_{th} decreasing with decreasing Péclet numbers.

The thermal entrance length required for the simultaneously developing flow should be greater than the developed velocity profile case. The analysis of Hanna et al., in the absence of viscous dissipation, shows that the calculated thermal entrance lengths are the same for the circular tube.

e. Incremental Heat Transfer Number $N_{bc}(x)$

The effect of a thermal entrance region is to increase the heat transfer rate and/or Nusselt number compared to that of a thermally fully developed region. It is proposed that the incremental increase in heat transfer rate be designated as the incremental heat transfer number $N_{bc}(x)$, defined by

$$\text{Nu}_{m,bc}x^* = \text{Nu}_{bc}x^* + N_{bc}(x) \tag{130}$$

The suffix bc represents the associated thermal boundary condition (e.g., of Table 2). If the thermal energy sources, viscous dissipation, and axial heat conduction within the fluid are neglected, $N_{bc}(x)$ is a function of x^* for hydrodynamically developed flow, as shown in Eq. (130); however, it is a function of x^* and Pr for hydrodynamically developing flow. Thus $N_{bc}(x)$ is defined in a manner similar to the incremental pressure drop number $K(x)$ defined by Eq. (92).

f. Reynolds Number Re

This is defined as

$$\text{Re} = \rho u_m D_h / \mu = G D_h / \mu \tag{131}$$

For internal flow, Re is proportional to the ratio of flow momentum rate ("inertia force") to viscous force for a specified duct geometry. For example, it can be shown that the ratio of flow momentum rate to viscous force for fully developed laminar flow through a circular tube and parallel plates of length L is $(D_h/24L)\,\text{Re}$ and $(D_h/40L)\,\text{Re}$, respectively. Thus, Reynolds number is a flow modulus. If it is the same for two systems that are geometrically and kinematically similar, then dynamic similarity is also realized irrespective of the fluid.

g. Prandtl Number Pr

This is defined as the ratio of momentum diffusivity to thermal diffusivity of the fluid:

$$\text{Pr} = v/\alpha = \mu c_p / k \tag{132}$$

The Prandtl number is solely a fluid property modulus. Its range for several fluids is as follows: 0.001–0.03 for liquid metals, 0.2–1 for gases, 1–10 for water, 5–50 for light organic liquids, and 50–2000 for oils.

h. Péclet Number Pe

This is defined as

$$\text{Pe} = \rho c_p u_m D_h / k = u_m D_h / \alpha \tag{133}$$

On multiplying the numerator and denominator of Eq. (133) by the axial fluid bulk mean temperature gradient (dt_m/dx), it can be shown that

$$\text{Pe} = D_h \frac{W c_p (dt_m/dx)}{k A_c (dt_m/dx)} \tag{134}$$

Thus the Péclet number represents the relative magnitude of the thermal energy convected to the fluid (fluid enthalpy rise) to the thermal energy axially conducted within the fluid. The inverse of the Péclet number is representative of the relative importance of fluid axial heat conduction. It can be shown that

$$Pe = Re\,Pr \qquad (135)$$

i. *Stanton Number* St

This is defined as

$$St = h/Gc_p \qquad (136)$$

By multiplying both the numerator and denominator by $(t_w - t_m)$, it is apparent that St is the ratio of convected heat transfer (per unit duct surface area) to the amount virtually transferable, as if temperature equalization were attained (per unit of flow cross-sectional area). Clearly,

$$Nu = St\,Pe - St\,Re\,Pr \qquad (137)$$

When fluid axial heat conduction is negligible (large Pe), St is frequently preferred, instead of Nu, as a dimensionless modulus for the convective heat transfer correlation. This is because it relates more directly to the designer's task of establishing the number of exchanger transfer units N_{tu}. Moreover, it has a behavior with Re that parallels that of the flow friction factor f.

j. *Colburn Factor j*

A dimensionless group correlating experimental forced convective heat transfer with Reynolds number, proposed by Colburn [57], is referred to as a Colburn modulus:

$$j = St\,Pr^{2/3} = Nu\,Pr^{-1/3}/Re \qquad (138)$$

The j vs. Re characteristic correlates fairly well the experimental data of turbulent flow forced convection heat transfer through circular and non-circular ducts for $0.5 \lesssim Pr \lesssim 10$. Further, for $Pr \simeq 1$, as for gases, the j–Re provides a correlation for the designer over the range from fully developed laminar flow to fully developed turbulent flow [6].

k. *Brinkman Numbers* Br *and* Br' *and Eckert Numbers* Ec *and* Ec'

The effect of viscous dissipation on heat transfer is significant for (1) high-velocity flow, (2) highly viscous fluids at moderate velocities, or (3) fluids with a moderate Prandtl number and moderate velocities, but having small wall-to-fluid temperature difference or having low wall heat fluxes and flowing through very small ducts (capillary flow).

Based on dimensional analysis, Dryden [60] showed that the dimensionless group accounting for the effect of viscous dissipation is

$$\text{Br} = \frac{\mu u_m{}^2}{g_c J k (t_{w,m} - t_e)} \tag{139}$$

for the constant axial wall temperature boundary condition. Brinkman [61], while solving the problem of flow in a circular tube with viscous dissipation effects, arrived implicitly at the above dimensionless group[†] with the temperature difference $(t_m - t_{w,m})$ instead of $(t_{w,m} - t_e)$. Bird *et al.* [55,62] designated this dimensionless group as the Brinkman number.

For constant axial wall heat flux boundary conditions, a new Brinkman number Br′ is proposed as follows:

$$\text{Br}' = \mu u_m{}^2 / g_c J q'' D_h \tag{140}$$

Now the corresponding dimensionless energy equation will have only one additional parameter due to viscous dissipation, as shown on p. 14.

Also shown on p. 14 is that the Brinkman numbers Br and Br′ take into account the viscous dissipation effects for incompressible liquids, and viscous dissipation plus flow work effects for a perfect gas.

Br′ is related to η' of Tyagi [63,64] for the fully developed (H1) heat transfer duct flow problem as

$$\eta' = -\frac{2f \, \text{Re}}{S^* + 4} \, \text{Br}' \tag{141}$$

The effect of viscous dissipation on heat transfer in external high velocity flows is taken into consideration by introducing the dimensionless group referred to as an Eckert number, Ec [55,65].

$$\text{Ec} = \frac{u_\infty{}^2}{g_c J c_p (t_w - t_\infty)} \tag{142}$$

An equivalent definition for internal flow (T), (T3), and (T4) boundary conditions with thermally developed or developing flow would be

$$\text{Ec} = \frac{u_m{}^2}{g_c J c_p (t_{w,m} - t_e)} \tag{143}$$

For the (Δt) boundary condition with thermally developed or developing flow, or for the (H1)–(H4) boundary conditions with thermally developed

---

[†] In the present terminology, Brinkman's dimensionless temperature was defined as $[(t - t_w)/(t_m - t_w)]/(4 \, \text{Br})$. Note that $t_w = t_{w,m}$ for the circular tube problem.

flow, a new Eckert Number Ec' is proposed:

$$\text{Ec}' = \frac{u_m^{2}}{g_c J c_p (t_{w,m} - t_m)} \tag{144}$$

Thus it is seen that the Eckert numbers for internal flow are related to Brinkman numbers Br and Br' as

$$\text{Ec} = \text{Br/Pr} \tag{145}$$

$$\text{Ec}' = (\text{Nu/Pr}) \, \text{Br}' \tag{146}$$

Since Pr appears as an additional parameter in the energy equation when the Eckert number is used, the Brinkman number Br or Br' is recommended when considering the effects of viscous dissipation in internal flows.

1. Brun Number Bn and Biot Number Bi

This number is defined as the ratio of temperature difference in a duct wall $(\Delta t)_w$ to that in a boundary layer $(\Delta t)_f$ (a temperature potential ratio). It is also expressed as [18,19]

$$\text{Bn} = \frac{(\Delta t)_w}{(\Delta t)_f} = \frac{k}{k_w} \frac{a'}{\delta_T} = \frac{h}{k_w/a'} \tag{147}$$

where a' is the duct wall thickness and δ_T the thickness of the thermal boundary layer.

At this point, it is appropriate to compare the Brun number with its equivalent *Biot Number* Bi and the *wall thermal resistance parameter* R_w. For internal flow, they are defined as

$$\text{Bi} = h/(k_w/a') = \text{Bn} \tag{148}$$

$$R_w = (k/D_h)/U_w \tag{149}$$

where U_w is defined by Eqs. (42)–(44). The Biot number is a measure of the relative importance of the thermal resistance within the solid body to convective thermal resistance. Usually the Biot number is used in transient conduction heat transfer problems involving convective boundary conditions. If Bi is small, the temperature differences within the solid are small relative to the convective film temperature difference, and the wall temperature at a given instant can be treated as uniform throughout the wall thickness dimension.

The thermal resistance parameter R_w is the ratio of wall thermal resistance to the pure conductive fluid thermal resistance in a conduction path length D_h. This number is used in the boundary condition for either the forced convection conventional problem [Eq. (46)] or the conjugated problem

[Eq. (22)]. For the conventional problem, if $R_w < 0.05$, the effect of wall resistance on laminar fully developed forced convection Nu_T is less than 5% for a circular tube or parallel plates. In this case, the conventional laminar flow convection problem can be treated as independent of the small wall resistance.

Luikov [18] recommended the use of the Brun number in the conjugated problem for the external flow. If the outside convective resistance is neglected in R_w, then Bn, Bi, and R_w are related as

$$\text{Bn, Bi} = R_w D_h/\delta_T = R_w \, Nu \tag{150}$$

However, for the internal flow, as evident from Eq. (22), the use of R_w is recommended over the Brun number or Biot number for the conjugated problem in order to preserve Nu as a separate entity. Based on the results of the external flow conjugated problem [18], it is recommended that the convection problem be solved as the cojugated problem for $R_w > 0.05$; otherwise it should be solved as the conventional problem with the boundary condition of the third kind.

m. Thermal Energy Source Number S*

This is defined as

$$S^* = SD_h/q'' \tag{151}$$

for axially constant wall heat flux boundary conditions, and

$$S^* = SD_h{}^2/k(t_{w,m} - t_e) \tag{152}$$

for constant wall temperature boundary conditions, where S represents the uniform intensity thermal energy source functions per unit volume.

n. Energy Content δ

The total wall heat transfer rate is specified in the case of axially constant wall heat flux boundary conditions, or it may be determined from Nu_m or Φ_m for constant wall temperature boundary conditions. When fluid axial heat conduction is considered, part of this wall thermal energy transfer appears as the fluid enthalpy rise, and the rest is conducted by the fluid to the end header of the duct. The energy transfer terms and the fluid bulk mean temperature distribution for finite fluid axial heat conduction are shown schematically in Fig. 10. The dashed line represents t_m distribution when the fluid axial heat conduction is zero. The dimensionless energy content of the fluid is defined as the fraction of heat transfer rate that is not conducted away

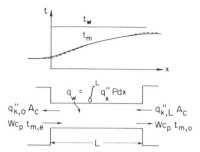

FIG. 10. Energy transfer terms and temperature distribution with fluid axial heat conduction.

axially by the fluid; it appears as an enthalpy rise of the fluid:

$$\delta = \frac{Wc_p(t_{m,o} - t_{m,e})}{q_w} \tag{153}$$

where δ depends on Pe and x^*. Schmidt and Zeldin [66] determined δ for the circular tube and parallel plates Graetz problems.

3. SOLUTIONS IN TERMS OF GENERAL FUNCTIONAL RELATIONSHIPS

The solutions to the internal flow forced convection problem may be presented in a number of different ways. One way is to present them in terms of the set of dimensionless groups Nu, $N_{bc}(x)$, and L_{th}^*. The general functional relationships for the Nusselt number are presented in detail for thermally developing and developed flows through a straight duct [54]. $N_{bc}(x)$ is also dependent on the same set of independent parameters. $N_{bc}(\infty)$ and L_{th}^* are independent of x^*, but depend on the rest of the parameters.

The generalized solutions are presented for the conventional convection problem and separately for the conjugated problem.

a. Conventional Convection Problem

The Nusselt number is usually employed as the dependent parameter in the presentation of the analytical solutions for the conventional convection problem. Baehr and Hicken [67] proposed the use of a dimensionless temperature for thermally developing flows. It will be shown that this parameter is related to the conventional Nu and thus does not appear to offer any additional advantages to the designer.

(i) *Nusselt Number Correlation.* The analytically derived Nusselt numbers are functions of dimensionless groups characterizing duct geometry, boundary conditions, and various parameters affecting heat transfer. This functional dependence of the laminar flow Nu may be illustrated as follows.

For thermally and hydrodynamically developed flow,

$$\text{Nu}_{bc} = \text{Nu (duct geometry, bc)} \qquad (154)$$

Here bc designates the associated thermal boundary condition, as outlined in Table 2. The Nusselt number for fully developed flow, Nu_{bc}, is also referred to as an *asymptotic* or *limiting* Nusselt number.

For thermally developing and hydrodynamically developed flow,

$$\text{Nu}_{x,bc} = \text{Nu (duct geometry, bc, } x^*) \qquad (155)$$

For simultaneously developing flow,

$$\text{Nu}_{x,bc} = \text{Nu (duct geometry, bc, } x^*, \text{Pr)} \qquad (156)$$

The boundary conditions usually have associated dimensionless groups. For example, these groups are R_w, γ, R_w, K_p, and m, respectively, for the (T3), (T4), (H3), (H4), and (H5) boundary conditions. $R_w/(a'/D_h \text{ Pe})^2$, as shown in Eq. (69), appears for the axially constant wall heat flux boundary conditions with finite axial heat conduction in the wall. Moreover, the Nusselt number will depend upon additional dimensionless groups, if other effects are included. For example, Nu depends respectively upon Pe, Br, and S^* if axial heat conduction, viscous dissipation, and uniform thermal energy sources within the fluid are considered.

The reciprocal of the Péclet number represents the magnitude of axial heat conduction within the fluid. Hennecke [68] showed that for Pe > 50 or Pe > 10 (with $x^* \geq 0.01$), fluid axial heat conduction can be neglected for the (T) and (H1) boundary conditions, respectively. For the axially constant wall heat flux boundary conditions, fluid axial heat conduction is constant for fully developed flow, and hence Nu_{bc} does not depend upon Pe for these cases.

Br or Br' represents the effect of viscous dissipation in the fluid. Tyagi [64] showed that for $|\text{Br}'| < 0.01$, the effect of viscous dissipation may be neglected. Cheng [69] showed that the effect of viscous dissipation is greatest for the circular duct, and decreases moderately for the regular polygonal duct with a decreasing number of sides.

The above nonspecific functional relationships cover a broad class of laminar duct flow forced convection heat transfer problems. Unfortunately, a single explicit functional relationship for different geometries is not available for Nu when the conventional D_h is used as a characteristic dimension. This challenges theoreticians and experimentalists to discover a new characteristic dimension that will unify Nusselt numbers for different geometries—at least for the simpler problems such as the (H1) and (T) boundary conditions.

(ii) *Dimensionless Temperature Correlation.* Baehr and Hicken [67] proposed $\vartheta_m(x^*)$ for (T) and $\Theta_{w-m}(x^*)$ for the (H1) boundary condition as new correlating parameters, instead of the conventional Nusselt number. In their development, they neglected the effect of axial heat conduction, viscous dissipation and thermal energy sources within the fluid. $\vartheta_m(x^*)$ and $\Theta_{w-m}(x^*)$ are defined and related to the Nusselt number as follows:

$$\vartheta_m(x^*) = \frac{t_m(x^*) - t_e}{t_w - t_e} = 1 - \exp(-N_{tu}) = 1 - \exp(-4x^* \, Nu_{m,T}) \quad (157)$$

$$\Theta_{w-m}(x^*) = \frac{t_w(x^*) - t_m(x^*)}{q_x'' D_h / k} = \frac{1}{Nu_{x,H1}} \quad (158)$$

They present explicit functional relationships for ϑ_m and Θ_{w-m} for circular and rectangular ducts:

$$\vartheta_m(x^*) = 1 - \exp[-B_1(x^*)^{1/2} - B_2 x^*] \quad (159)$$

$$\Theta_{w-m}(x^*) = \Theta_{w-m}(\infty)\{1 - \exp[-B_1(x^*)^{1/2} - B_2 x^*]\} \quad (160)$$

Values of B_1, B_2, and $\Theta_{w-m}(\infty)$, based on the best curve fit to analytical solutions, are tabulated in [67].

ϑ_m is the same as the heat exchanger effectiveness defined by Kays and London [6] and is generally familar to a designer. Baehr and Hicken compared their scheme with the $(\Delta t)_{lm}$ approach and showed that their proposed procedure is operationally more convenient for the calculations of heat transfer surface area. This view is supported by Kays and London [6].

Θ_{w-m} is related to $Nu_{x,H1}$, as expressed in Eq. (158). Clearly, the newly proposed correlation method by Baehr and Hicken is simply related to the conventional Nusselt number correlation method, so that no additional advantages are to be gained by their use.

b. *Conjugated Problem*

Conjugated problems are sufficiently more complex than the conventional problem and it is not feasible to present the solutions obtained in terms of generalized functional relationships. In order to assess the importance of the related dimensionless groups, the generalized solution is presented in the following text for a simple case, a circular tube having a fully developed fluid velocity profile and either a temperature or heat flux constant at the outside wall. The associated conjugated problem was formulated in Eqs. (14)–(20). The axially local but peripherally average Nusselt number is dependent on five dimensionless groups, as evident from Eqs. (21)–(23):[†]

$$Nu_x = Nu\left(x^*, X, R_w, \frac{a'}{L}, \frac{a'}{D_h}\right) \quad (161)$$

Note that only x^* appears in the conventional convection problem and it is this increased number of parameters that makes the conjugated problem so complex. To eliminate two axial dimensionless groups, Eq. (161) may be

---

[†] a/a' of Eq. (23) is replaced by a'/D_h here.

presented alternatively as

$$Nu_x = Nu\left(\frac{x}{L}, \frac{Pe\,D_h}{L}, R_w, \frac{a'}{L}, \frac{a'}{D_h}\right) \tag{162}$$

It is interesting to note that even though axial heat conduction in the fluid is neglected, the solution depends upon the $Pe\,D_h/L$ group. A comparison of Eq. (161) or (162) with Eq. (155) reveals that even for one of the simplest conjugated problems, four additional parameters are introduced into the solution. It may be noticed that the parameter a'/D_h originates from the second term of Eq. (23) and represents the tube curvature effect. Since there is no such curvature effect for parallel plates, Nu_x for the conjugated problem of parallel plates does not depend upon a'/D_h; it depends only on the remaining four groups of Eq. (162), with $Pe\,D_h/L$ replaced by Pe. Hence, for the parallel plates conjugated problem,

$$Nu_x = Nu\left(\frac{x}{L}, Pe, R_w, \frac{a'}{L}\right) \tag{163}$$

For a noncircular duct, additional geometrical parameters may appear in the solution. If the effects of viscous dissipation, axial heat conduction, and thermal energy sources within the fluid are considered, Nu_x would be dependent on the additional related groups. For simultaneously developing flow, Pr would be an additional parameter.

Chapter IV

General Methods for Solutions

The governing differential equations and associated boundary conditions for laminar duct flow velocity and temperature problems are outlined in Chapter II. A convenient set of dimensionless groups and the resulting generalized solutions are described in Chapter III. In this chapter, the general methods of solutions, as available in the heat transfer literature, are briefly described for the problems formulated in Chapter II. For quick reference, a summary is provided of the investigators who employed particular methods, along with the duct geometry that they analyzed.

A. Fully Developed Flow

A variety of duct configurations is encountered in the existing wide spectrum of thermal systems. No single method of analysis is uniformly applicable. Nine methods have been the primary ones employed for the analysis of hydrodynamically and thermally fully developed laminar fluid flow and heat transfer through ducts of different geometries. There are several other methods that are also summarized under the miscellaneous category. For the fully developed laminar velocity problems, the methods of analysis are not categorized separately, because the solution to the velocity problem is generally found by the same method as that employed for the solution to the temperature problem.

1. EXACT SOLUTIONS

The exact closed-form solutions of the momentum and energy equations are obtained for some simple geometries, such as the circular tube, parallel plates, elliptical ducts, concentric and eccentric annular ducts, and confocal elliptical ducts. The results for various boundary conditions are summarized in their respective sections.

2. Exact or Approximate Solutions by Analogy Method

Marco and Han [70] and Cheng [71] pointed out that the following four Dirichlet problems are related to each other:

    (a) torsion of prismatical bars,
    (b) fully developed laminar flow velocity,
    (c) uniformly loaded, simply supported polygonal plates, and
    (d) fully developed (H1) laminar heat transfer.

Consider (a) and (b). The stress function in torsion theory [72] has the identical differential equation and boundary condition as that of the laminar velocity field for a singly connected duct. Hence, based on the known solutions to torsion of prismatical bars, the velocity distribution and f Re can be determined, as is done for the case of rectangular and moon-shaped ducts in this presentation. For the doubly and multiply connected ducts, however, the torsion and laminar flow problems differ because of the requirement that the displacement in the elastic material be single valued. This introduces boundary conditions for the torsion problem that are not identical to those for the flow problem. Essentially, the flow problem is the easier one. As an example, the solutions of the torsion and flow problems in eccentric annuli differ by a supplementary term Z of Eq. (510), as noted by Caldwell [73] and Piercy et al. [74].

Consider (c) and (d). The governing differential equation is a fourth-order one for small deflections of thin polygonal plates under uniform lateral load and simply supported along all edges [75]. This differential equation and the associated boundary condition for the thin plate problem are identical to the (H1) laminar temperature field for the duct flow, provided that axial heat conduction, thermal energy sources, and viscous dissipation within the fluid are neglected. For plates with curvilinear boundaries, the adjustment of Poisson's ratio is required so that the boundary condition of no-slip flow may be satisfied [70,71]. As noted on p. 13, the solution to a fourth-order differential equation for the (H1) temperature problem will simultaneously provide the solution to the velocity problem. Consequently, based on the known solutions of the thin plate problem, the (H1) temperature and velocity distributions, Nu_{H1}, and f Re can be determined. Nu_{H1} for rectangular, equilateral triangular, right-angled isosceles triangular, semicircular [70], circular sector [27], and rhombic [76] ducts have been evaluated by this method.

As mentioned above, once the solution to the thin plate problem is known, f Re and Nu_{H1}, as well as velocity, stress, temperature distribution, etc., can be calculated. Cheng [71,77] has shown how to apply the Moiré and point-matching methods (used to obtain approximate solutions to the thin plate problem) to the laminar ⟨H1⟩ heat transfer problem.

Aggarwala and Iqbal [78] applied an analogy of the membrane vibration problem to the laminar duct flow problem and determined Nu_{H1} and f Re for the equilateral triangular, right-angled isosceles triangular, and 30–60–90° triangular ducts.

3. EXACT SOLUTIONS BY THE METHOD OF COMPLEX VARIABLES

This method is limited to cases where the velocity and temperature fields are deducible directly from the equations of the boundary curves such as those of equilateral triangular and elliptical ducts. Tao [79] first considered this method for the ⟨H1⟩ boundary condition with thermal energy sources included within the fluid. Sastry [80] applied Tao's method to confocal ellipses. Tyagi [81] generalized Tao's method by including viscous dissipation and flow work within the fluid, and analyzed the equilateral triangular and elliptical ducts.

4. EXACT OR APPROXIMATE SOLUTIONS BY CONFORMAL MAPPING

Any duct cross section that can be mapped exactly or approximately, by a conformal transformation onto a unit circle or concentric circles, can be analyzed by this method. Tao has considered the cardioid duct [82,83], hexagonal duct [82], and Pascal's limaçon [83] for the ⟨H1⟩ boundary condition, and the cardioid duct [84] for the ⟨H2⟩ boundary condition. Tao, in his analysis, included the effect of thermal energy sources, but neglected the effects of viscous dissipation and flow work within the fluid.

In a series of papers [85–89], Sastry employed the exact and approximate methods of conformal mapping to solve the laminar velocity and ⟨H1⟩ temperature problems. In [85], he applied Tao's method [83] for curvilinear polygonal ducts. In [86], he considered a general power series mapping function with specfic examples of cardioid and ovaloid ducts. In [87,88], he employed the Schwarz–Neumann alternating method (an approximate method of conformal mapping). Examples were worked out for two doubly connected duct geometries: (1) outer boundary a circle, inner boundary an ellipse [87]; and (2) outer boundary a circle, inner boundary a square with rounded corners [88]. In the foregoing work, both velocity and ⟨H1⟩ temperature problems were solved. In [89], Sastry presented the solution to the velocity problem for concentric and eccentric annular ducts, confocal elliptical ducts, and an elliptical duct with a central circular core by employing a mapping function of a double infinite series form.

Tyagi extended Tao's work by including the effect of viscous dissipation for the (H1) boundary condition for equilateral triangular and elliptical ducts [63], and the (H2) boundary condition for the cardioid duct [90]. Tyagi also extended Tao's work by including viscous dissipation and flow work for the (H1) [64] and (H2) [91] boundary conditions for the cardioid duct. Iqbal et al. [92] solved the (H2) temperature problem for elliptical ducts by employing the conformal mapping technique.

Casarella et al. [93] proposed an approximate conformal mapping technique for the thermal entrance $Nu_{x,T}$ for a duct of arbitrary cross section with the (T) boundary condition. As an asymptote, Nu_T can be calculated from the results for $Nu_{x,T}$. Only the slug flow problems and the Graetz problem (as defined on p. 99) were solved by Casarella et al.

5. Approximate Solutions by Finite Difference Methods

The differential momentum and energy equations are put into a finite difference form, and the solutions to the resulting system of linear algebraic equations are carried out by a hand calculation or by a digital computer using standard techniques such as relaxation or Gaussian elimination. Clark and Kays [94] employed the relaxation technique and obtained Nu_T and Nu_{H1} for rectangular ducts and the equilateral triangular duct. Schmidt and Newell [95] employed the Gaussian elimination technique with an iterative refinement and obtained Nu_T and Nu_{H1} for rectangular and isosceles triangular ducts. Lyczkowski et al. [38] utilized the extended Dufort–Frankel method and determined Nu_T for rectangular ducts and Nu_{H4} for the square duct. Sherony and Solbrig [36] also used the Dufort–Frankel method and evaluated $f\,Re$ and Nu_T for sine ducts. Dwyer and Berry [96] employed the Gaussian elimination method and determined Nu_{H2} for longitudinal laminar flow over a rod bundle arranged in an equilateral triangular array. Iqbal et al. [76,92] also used the Gaussian elimination method and determined Nu_{H1} and Nu_{H2} for rhombic ducts.

In the foregoing solutions, either a rectangular or a square grid was employed for the finite difference calculations. Nakamura et al. [97,98] employed a uniform triangular grid of arbitrary shape and obtained Nu_{H1} and Nu_T for arbitrary triangular ducts and quadrilateral ducts. They employed the extrapolated Liebmann method to solve the system of linear algebraic equations resulting from a finite difference formulation of the momentum and energy equations.

Date and Singham [99] and Date [100,101] employed an orthogonal grid made up of radial and concentric lines for a finite difference solution of twisted-tape flow. The finite difference equations were formulated by the upwind difference scheme; the solution was obtained by the Gauss–Seidel method.

Meyder [102] employed a general curvilinear orthogonal mesh grid to represent the flow area for a finite difference method. The mesh lines for rectangular or triangular grids are given by simple explicit functions. However, the mesh lines of Meyder are given by solving the Laplace equation twice in the flow area of interest, and by obtaining the orthogonal streamlines and equipotential lines. The intersection of these lines represents the mesh points. An awkward shape duct geometry can be accurately represented by this method with relatively fewer grid points. Meyder then analyzed longitudinal flow over a rod bundle by this method.

Methods 6 and 7, which follow, are referred to as *weighted residual methods*. They are powerful, computationally fast, and accurate to any desired degree for the (H1) and (H2) boundary conditions. Hence they are now preferred over most of the finite difference methods.

6. Approximate Solutions by Point-Matching Methods

After an appropriate change of dependent variables, the momentum and energy equations can be represented as two-dimensional Laplace equations. The general solution to Laplace equations is obtained by a linear combination of harmonic functions in the form of an infinite series. This series is truncated at a finite number of terms n. The n points are selected on the periphery Γ either equidistant or equiangular. The boundary condition is satisfied exactly at these n prechosen points to determine the n unknown coefficients of the truncated series. The velocity and temperature distributions are then obtained in a closed-form series with these coefficients obtained from the solutions of n linear algebraic equations. (H1) and (H2) are the only thermal boundary conditions treated to date by the point-matching method, which is also referred to as the *boundary collocation technique*. The limitations of the point-matching method are discussed by Sparrow in Cheng's paper [71]. The well-known algebraic-trigonometric polynomials, which satisfy the Laplace equation, were employed in the following cases: f Re and Nu_{H1} for longitudinal flow over cylinders arranged in regular arrays by Sparrow *et al.* [103,104]; f Re for isosceles triangular ducts by Sparrow [105]; f Re, Nu_{H1}, and Nu_{H2} for regular polygonal ducts by Cheng [106,107]; f Re for cusped, regular polygonal, equilateral triangular, elliptical, and rectangular ducts by Shih [108]; f Re and Nu_{H1} for regular polygonal ducts with a central circular core by Cheng and Jamil [109]; Nu_{H1} for eccentric annuli by Cheng and Hwang [110]; and f Re and Nu_{H1} for cylindrical ducts with diametrically opposite flat sides by Cheng and Jamil [111].

7. Approximate Solutions by Least Squares Methods

Two methods, the discrete least squares and continuous least squares, have been used to analyze the fully developed laminar forced convection problem. The discrete least squares method is also referred to as the *least squares point-matching method*.

The discrete least squares method differs from the point-matching method in that more than n points (usually $2n$ to $3n$ points) along the boundary are employed to determine n unknown coefficients in the truncated series harmonic function solution for the Laplace equation. The coefficients of the series are evaluated by solving m linear algebraic equation by minimizing the mean squared error of the boundary conditions at m chosen points ($m > n$). Thus the exact fit to these prechosen points is sacrificed in favor of a better fit to the boundary as a whole. The boundary points can be chosen at a regular or irregular interval to analyze the sharp variations around corner regions. The f Re factors were obtained by this method for regular polygonal ducts with a central circular core by Ratkowsky and Epstein [112] and for circular ducts with central regular polygonal cores by Hagen and Ratkowsky [113]. In both cases, these workers [112,113] used the so-called *normal equation* approach for the least squares solution. Sparrow and Haji-Sheikh [114] determined f Re, Nu_{H1}, and Nu_{H2} for circular segment ducts by employing the Gram–Schmidt orthonormalization method. Shah [115,116] obtained f Re, $K(\infty)$, Nu_{H1}, and Nu_{H2} for rectangular, isosceles triangular, sine, trapezoidal, rhombic, and equilateral triangular ducts with rounded corners. Shah employed Golub's method (using Householder reflections) of solving the linear least squares problem. Trombetta [117] obtained solutions for the fundamental boundary conditions of the first, second, and fourth kinds (see p. 32) for eccentric annular ducts. The least squares approximation was provided in [117] by the Gram–Schmidt orthonormalization procedure as employed by Sparrow and Haji-Sheikh [114]. Algebraic–trigonometric polynomials were used for the solution of Laplace equations in all of the above solutions, except for the solution by Sparrow and Haji-Sheikh [114], who employed real and imaginary parts of a complex variable in Cartesian coordinates.

In a continuous least squares method, the boundary condition is continuously satisfied. The n unknown coefficients of the series are evaluated, by minimizing the total integrated mean square error, from the simultaneous solutions of n integral equations. This method has been employed by Ullrich [118] to analyze the longitudinal laminar flow over an n-rod bundle.

8. Approximate Solutions by Variational Methods

A variational method for fully developed laminar flow in noncircular ducts was proposed by Sparrow and Siegel [119] for the (H1) and (H2) boundary conditions. This method requires developing *a priori* a temperature function to satisfy the boundary condition exactly. Examples were worked out for a square duct, a rectangular duct with $\alpha^* = 0.1$, and a circular sector duct. Gupta [120] formulated a variational approach for the (H1) laminar heat transfer problem with the effect of fluid axial heat conduction included. Stewart [121] determined the accuracy of variational flow calculations for a square duct by applying a reciprocal variational principle. Iqbal *et al.* [76,92] presented a variational method for the (H1) and (H2) boundary conditions and determined Nusselt numbers for rectangular, isosceles triangular, right triangular, elliptical, and rhombic ducts. A variational approach for the (T) boundary condition was presented by Pnueli [122], where the upper and lower bounds for Nusselt numbers were obtained for the circular and square ducts.

Finlayson and Scriven [123] discussed and critically reviewed the different variational methods used for the transport and transformation processes. They concluded that the different methods used in the literature did not possess the advantages associated with the genuine variational principles, and that no general variational principle could be devised for the transport and transformation processes. They showed that the variational methods of approximation, used in the literature, were equivalent to the more straightforward Galerkin method, another closely related version of weighted residuals.

9. Approximate Methods for Small Aspect Ratio Ducts

Purday [124] presented an approximate method for calculating fully developed laminar flow u_m and $Q\,(= A_c u_m)$ for ducts with low height-to-width ratio and without abrupt variations in height across their width. The method is similar to the one used for the torsion of a narrow rectangular bar by Timoshenko and Goodier [72]. Maclaine-Cross [125] independently outlined this method and applied it to small aspect ratio ducts. As an example, he calculated $f\,\mathrm{Re}$, $K(\infty)$, and $\mathrm{Nu_{H1}}$ for a hexagonal duct with a small height-to-width ratio. James [126] presented a method to determine the fully developed $\mathrm{Nu_T}$ for narrow ducts, and computed $\mathrm{Nu_T}$ for an elliptical duct with $\alpha^* = 0$. All of these methods are summarized in Chapter XI.

10. Miscellaneous Methods

Masliyah and Nandakumar [127,128] employed a finite element method to analyze the velocity and (H1) temperature problems for a circular duct with longitudinal triangular internal fins.

Hu [129] employed a perturbation method, an interior Galerkin variational method, and a method of boundary integral equations to analyze the velocity and (H1) and (H2) temperature problems for a corrugated duct.

He then compared the results with those of a more accurate method of conformal mapping with Green's function. A brief discussion on the comparisons is presented on p. 276.

While most of the foregoing methods involve a solution to the differential equations, the methods of boundary integral equations and Green's function involve a solution to the integral equations. Aggarwala and Gangal [130,131] analyzed the (H1) temperature problem for rectangular ducts having longitudinal thin fins by reducing it to Fredholm integral equations of the second kind. Kun [132] employed Green's function and analyzed the velocity and (H1) temperature problems for a circular duct having longitudinal thin V-shaped internal fins.

A combination of two or more foregoing methods can be applied to the duct geometries of complicated shapes. A method of conformal mapping and Green's function was employed by Hu [129] for a circular duct with longitudinal internal fins, and by Chen [133] for square and hexagonal ducts with longitudinal internal fins for the (H2) boundary condition. A method combining the Schwarz–Neumann successive approximations technique with a discrete least squares method was employed by Zarling [134] to analyze the (H1) problem for a rectangular duct with semicircular ends.

B. Hydrodynamically Developing Flow

The hydrodynamic entry length problem is solved either by employing the boundary layer type idealizations of Eq. (7) or by considering the complete set of Navier–Stokes equations. Four methods, as described below, have been used to solve the momentum equation (8) or (9), which incorporates the boundary layer type idealizations. Classical boundary layer methods, which have been developed for external flows and which are applicable only near the duct inlet, are not considered here. Three methods are used for the solution of the complete set of Navier–Stokes equations. An excellent review of these methods up to 1970 has been provided by Lakshmana Rao and Sridharan [135].

1. METHODS INCORPORATING BOUNDARY LAYER TYPE IDEALIZATIONS

a. *Matching Method*

The entrance region is divided into two zones. Near the entrance, a boundary layer near the wall is growing with an accelerated external stream. In this section, an approximate solution is obtained in terms of a perturbation of the Blasius external flow boundary layer solution. Far downstream, where the flow is nearly fully developed, the solution is obtained in terms of small perturbations of the fully developed solution, as first suggested by Boussinesq [136]. These solutions are "matched" (joined smoothly) at some appropriate axial location to obtain a complete entrance region solution. This method was first used by Schlichting [65,137] for the parallel plate channel, and later by Atkinson and Goldstein [138] for the circular tube. Gillis and Shimshoni [139] computed Atkinson and Goldstein's solution in greater detail. Collins and Schowalter [140] and Roidt and Cess [141] refined Schlichting's method by retaining more terms in the series of upstream and downstream solutions. Van Dyke [142] showed that Schlichting's upstream and downstream series solutions apply only to conditions far downstream and match another expansion (carried out by Van Dyke) valid near the inlet. Wilson [143] reported complementary aspects of the problem studied by Van Dyke [142]. He presented the criticism of earlier boundary layer type investigations from a more mathematical point of view; in contrast, Van Dyke's criticism was largely from a physical viewpoint. Kapila *et al.* [144] further extended the work of Van Dyke and Wilson by employing the second-order boundary-layer theory to analyze the entrance region of parallel plates.

b. *Integral Method*

The flow cross section is treated as having two regions, a boundary layer developing near the wall and a central inviscid fluid core. In this approach, integral forms of the continuity and momentum equations are applied to the above flow model. The velocity profile in the developing boundary layer is expressed as a polynomial function of the boundary layer thickness δ, as in the standard Kármán–Pohlhausen method. In addition to the integral continuity and momentum equations, a third equation is needed to determine the three unknowns δ, u, and p. The Bernoulli equation (applied to the tube centerline), the mechanical energy integral equation (applied to whole flow cross section), or the differential form of momentum equation (evaluated at the duct wall) may be taken as the third equation. The latter two equations take into account the effect of viscous dissipation in the fluid. The former (Bernoulli) equation neglects this effect. This method was first employed by Schiller [49] for flow in a circular tube and a parallel plate channel. He used a parabolic velocity distribution in the boundary layer and Bernoulli's equation (potential flow dynamical equation) in the inviscid core to determine the pressure distribution in the axial direction. Schiller thus neglected the effect of viscous dissipation in the flow cross section. Shapiro *et al.* [14] modified this method by employing cubic and quartic velocity profiles in the boundary layer. Naito and Hishida [145] and Naito [146] refined Schiller's solution by employing quartic velocity profiles in the boundary layer. Gubin and Levin [147] assumed a logarithmic velocity profile for the boundary layer and deduced the velocity distribution for the circular tube.

Campbell and Slattery [148] further refined the solution by taking into account viscous dissipation in the entire flow cross section. They used Schiller's parabolic profile, but applied the mechanical energy equation to all of the fluid in the pipe. Govinda Rao et al. [149] also used the momentum integral equation for a circular tube with the inclusion of viscous dissipation in the flow cross section. Gupta [150] applied the method of Campbell and Slattery [148] to parallel plates. Williamson [151] also used the same approach of [148] for parallel plates, but with a power law velocity profile, which resulted in a more rapid flow development. Fargie and Martin [152] employed a parabolic velocity distribution in the boundary layer, but applied the differential momentum equation at the wall to determine the pressure gradient. Their use of both differential and integral momentum equations led to simplified closed form expressions for f_{app} Re, $K(\infty)$, and Δp^*.

c. *Linearization Methods*

The above two methods yield discontinuous solutions for the gradients of velocity and pressure distributions in the hydrodynamic entrance region. Langhaar [153] proposed a method that yields a continuous solution in the entrance region of a circular tube. He lincarized the nonlinear inertia term, the left-hand side of Eq. (8), as follows:

$$u\frac{\partial u}{\partial x} + v\frac{\partial u}{\partial r} = \nu\beta^2(x)u \qquad (164)$$

where β is a function of x only. Equation (164) is subsequently solved with the appropriate boundary and initial conditions, and u/u_m is obtained as a function of r and β. Then β is determined as follows: (a) the momentum and continuity equations (8) and (12) are integrated over the duct cross section and combined; (b) the momentum equation (8) is evaluated on the duct centerline and the result is substracted from that of (a) to eliminate the pressure gradient term; (c) finally, when the velocity profile from the solution of Eq. (164) is substituted in the resultant equation of (b), a relationship between β and x is obtained. Subsequently, the pressure drop is determined either from the momentum equation evaluated at the centerline or from the mechanical energy integral equation.

Langhaar's linearization is a good approximation in the central potential core. His solution appears to be satisfactory at the tube centerline and either very near the entrance or far from it. Langhaar's approach was used by Han for rectangular ducts [154] and for parallel plates [155], by Han and Cooper

[156] for the equilateral triangular duct, and by Sugino [157] and Heaton et al. [44] for annular ducts.[†] Miller and Han [158] refined the work of Han [154,156] to determine $K(x)$ in closer agreement with the experimental results. Shumway and McEligot [45] pointed out an error in the analysis of Heaton et al. [44] for the pressure drop and correctly solved the problem using Langhaar's approach for an annular duct with $r^* = 0.25$.

Targ [159,159a] linearized the inertia term of the momentum equation (8) for the circular tube:

$$u \frac{\partial u}{\partial x} + v \frac{\partial u}{\partial r} = u_m \frac{\partial u}{\partial x} \tag{165}$$

Combining the subsequent momentum equation with the continuity equation and integrating it across the flow cross-sectional area, Targ obtained the pressure gradient for the circular tube in terms of a velocity gradient at the wall:

$$\frac{g_c}{\rho} \frac{dp}{dx} = \frac{2v}{a} \left[\frac{\partial u}{\partial r} \right]_{r=a} \tag{166}$$

Targ then solved momentum equation (8) after employing Eqs. (165) and (166). Chang and Atabek [160] and Roy [161] used Targ's linearization approach to annular ducts. In the case of the circular tube, Targ's solution, which neglects a partial contribution of momentum change to pressure gradient, provides a velocity profile that develops too slowly near the entrance.

It may be noted that Langhaar's and Targ's linearizations are exact (a) at the entrance if a uniform velocity is assumed, (b) at all points on the wall of the duct, since $\partial u/\partial x$ and v both vanish there, and (c) in the fully developed region, where $\partial u/\partial x$ and v are again zero.

Sparrow et al. [162] employed the following linearization to the momentum equation by "stretching" the x coordinate.

$$\varepsilon(x) u_m \frac{\partial u}{\partial x} = \Lambda(x) + v \left[\frac{1}{r} \frac{\partial}{\partial r} \left(r \frac{\partial u}{\partial r} \right) \right] \tag{167}$$

where $\varepsilon(x)$ is the mean velocity weighting factor and $\Lambda(x)$ stands for the $g_c(dp/dx)/\rho$ term plus the residual inertia terms. This linearization is similar to Targ's but embodies a more general linearization of the inertia term. Sparrow et al. [162] matched the pressure gradients from the momentum and mechanical energy equations to evaluate $\varepsilon(x)$. This linearization is

---

[†] Unless otherwise specified, an annular duct is a concentric circular annular duct.

restricted to the case where the velocity profile depends only on one cross-sectional coordinate, e.g., a circular tube, parallel plates [162], or an annular duct [163,164].

Wiginton and Wendt [165], Fleming and Sparrow [166], and Aggarwala and Gangal [167] independently generalized this linearization to the case where the velocity profiles depend upon two cross-sectional coordinates, such as for rectangular and isosceles triangular ducts. The solution to the x momentum equation, in terms of the stretched coordinate, was obtained by superimposing two solutions: (1) corresponding to the fully developed velocity profile, and (2) corresponding to the difference (entrance length correction) velocity profile. The solution to the difference velocity was obtained by expanding it in an infinite series containing eigenfunctions, constants, and exponential functions having eigenvalues in the exponent. The resulting eigenvalue problem for the difference velocity was analyzed by employing different methods by the aforementioned investigators. Wiginton and Wendt proposed a method for arbitrary cross sections that can be mapped onto a rectangle. In this case, a closed-form solution is obtained, and no approximation is involved to satisfy the boundary condition. However, approximation may be involved in the mapping. Fleming and Sparrow considered the product-form solution involving trigonometric functions for an arbitrary duct. The unknown coefficients of the series were obtained by matching a finite number of boundary points in a least squares sense. Aggarwala and Gangal obtained the exact closed-form solution for the eigenvalue problem in terms of the solution to another eigenvalue problem, namely, $\nabla^2 \phi + \lambda \phi = 0$ with $\phi = 0$ on the duct boundary. Wiginton and Dalton [168] studied the entrance flow region of rectangular ducts with aspect ratios $\alpha^* = 0.2$, 0.5, and 1. Fleming and Sparrow analyzed the rectangular ducts with $\alpha^* = 0.2$ and 0.5, and isosceles triangular ducts with the apex angle $2\phi = 30$ and $60°$. Aggarwala and Gangal investigated isosceles triangular ducts with $2\phi = 60$ and $90°$.

Wiginton [169] extended the method of Wiginton and Wendt [165] wherein he expanded each eigenfunction (of the difference velocity problem) in terms of exactly those eigenfunctions of the Helmholtz equation that are *not* in the orthogonal complement of the constant functions. The problem reduces to the determination of zeros of a subsequent eigenvalue series. Wendt and Wiginton [170] employed Wiginton's method to solve the hydrodynamic entrance problem for a circular sector duct with $2\phi = 45°$.

The linearization scheme of Eq. (167) is very difficult to employ for anything other than a uniform entrance velocity profile, because the mean velocity weighting factor $\varepsilon(x)$ depends on the solution. Savkar [171] applied the linearization of Eq. (167) with $\varepsilon(x) = 1$ and analyzed the hydrodynamic entrance region for parallel plates with two types of entrance velocity profiles: cosine variations and a step change.

It may be pointed out that the foregoing linearization schemes do not take into account the effect of the transverse flow rigorously, although they predict fairly well the axial velocity distribution and axial pressure drop. Consequently, such linearization schemes are restricted to simple flow geometries with simplified idealizations so that there are no complications of flow asymmetry, strong property variations, and other effects.

d. Finite Difference Methods

The continuity and momentum equations are reduced to a finite difference form and the numerical solution is carried out by a "marching" procedure for the initial value problem. This method is used by Bodoia and Osterle [172] and Naito and Hishida [145] for parallel plates, by Hornbeck [173], Christiansen and Lemmon [174], and Manohar [175] for a circular tube, by Patankar and Spalding [176] and Curr et al. [177] for rectangular ducts, by Carlson and Hornbeck [178,179] for a square duct, by Manohar [180], Shah and Farnia [181], Liu [182], and Coney and El-Shaarawi [46,183] for concentric annular ducts, and by Feldman [184] for eccentric annular ducts.

If the fully developed velocity profile, the f Re factor, and the hydrodynamic entrance length L_{hy}^{+} are known for a given duct geometry, the initial value problem for the hydrodynamic entrance region can be transformed into a boundary value problem. The detailed velocity distribution, friction factor, and pressure distribution can then be obtained for a given Reynolds number by numerically solving the boundary value problem. Miller [185] employed this technique and obtained finite difference hydrodynamic entry length solutions for the square and equilateral triangular ducts at Re – 10^3 using hydrodynamic entrance lengths of McComas [48].

2. METHODS OF SOLVING NAVIER–STOKES EQUATIONS

a. Finite Difference Methods

The Navier–Stokes equations are written in terms of either u, v, and p or the stream function and vorticity. Subsequent partial differential equations are put into a finite difference form, and the solution is obtained by iterative methods. Vrentas et al. [186], Friedmann et al. [187], and Schmidt and Zeldin [188,189] solved the hydrodynamic entry length problem for the circular tube; Wendel and Whitaker [190], Wang and Longwell [191], Gillis and Brandt [192], McDonald et al. [193], and Morihara and Cheng [194] solved it for parallel plates; Fuller and Samuels [195] analyzed it for a concentric annular duct with $r^* = 0.5$. In the above analyses, it was found that if the velocity profile at the inlet is uniform, it subsequently develops overshoots. To investigate whether or not these peculiar profiles are due to numerical errors, Abarbanel et al. [196] obtained an exact solution for Re = 0 and found similar overshoots. The possible reasons and situations for such overshoots are discussed on p. 91.

b. *Integral Method*

This is an approximate momentum integral method. It is similar to the one outlined in the preceding section but includes the effect of axial molecular transport of momentum and the variation of pressure across the flow cross section. Chen [197] showed how this rapid approximate method can be applied to solve the Navier–Stokes equations. He solved the hydrodynamic entry length problem for the circular tube and parallel plates.

c. *Linearization Method*

Narang and Krishnamoorthy [198] proposed a method to solve the complete set of Navier–Stokes equation by linearizing the inertia force term $\mathbf{V} \cdot \nabla \mathbf{V}$ by $\langle \mathbf{V} \rangle \cdot \nabla \mathbf{V}$ for parallel plates; here $\mathbf{V}$ is the velocity vector, which has u and v orthogonal velocity components, and $\langle \mathbf{V} \rangle$ is the axial average velocity over the cross section. Narang and Krishnamoorthy obtained solutions for parallel plates for Re of 2 to 4000.

C. Thermally Developing Flow

Thermally developing flows are categorized according to developed or developing velocity profiles (see p. 15). The methods used in the literature to solve both types of problems are described separately below.

1. HYDRODYNAMICALLY DEVELOPED FLOW

The thermal entrance solutions for hydrodynamically developed flow are obtained primarily by the following seven methods.

a. *Separation of Variables (Eigenvalue) and Similarity Transformation Methods*

The separation of variables method was used in the first study conducted by Graetz [1] of the thermal entrance region of a circular tube. Many of the subsequent solutions are obtained by a similar method employing the separation of variables in the energy equation and determining the eigenvalues and constants of the resulting ordinary differential equations by various approaches. The solution is presented in terms of an infinite series involving eigenvalues, eigenfunctions, and constants. The thermal entrance solutions are obtained by this method for the circular tube, parallel plates, rectangular ducts, elliptical ducts, and concentric annular ducts. These are described later in the appropriate sections.

The infinite series solution, obtained by employing the above method, to the thermal entrance problem converges very slowly near $x^* = 0$. Lévêque [199] provided a solution for the thermal entrance region near $x^* = 0$ by employing the method of similarity transformation. Worsøe-Schmidt [200] refined Lévêque's solution by retaining the first seven terms of the Lévêque series; only the first term was provided by Lévêque. Nunge et al. [201] also obtained the Lévêque solution by employing idealizations different from those of Worsøe-Schmidt. Newman [202] independently refined Lévêque's solution by retaining the first three terms of the Lévêque series. Burghardt and Dubis [203] also formulated a simple method for the thermal entrance region near $x^* = 0$.

b. Variational Methods

Sparrow and Siegel [204] developed a variational method for the analysis of the thermal entrance region for the (H1) boundary condition. They neglected the effect of axial heat conduction, thermal energy sources, and viscous dissipation in the fluid. Examples were worked out for a circular tube, parallel plates, and a square duct. Tao [205] proposed a variational approach to the thermal entrance region for the prescribed wall temperature and wall temperature gradient boundary conditions. He included the effect of viscous dissipation and internal thermal energy generation. He obtained the thermal entrance temperature distribution for the circular and elliptical ducts. Savkar [206] formulated a variational method for solving the thermal entrance problem for fully developed, unsteady laminar flow through a circular tube with a specified surface temperature distribution. He worked out the classical (steady-state) Graetz–Nusselt problem and the problem of parallel plates having an axially sinusoidal temperature distribution. Javeri employed the Kantorowich variational method to analyze the (T3) thermal entrance for a circular tube [207], and linearly varying wall temperature problem for rectangular and elliptical ducts [35].

c. Conformal Mapping Method

Casarella et al. [93] proposed an approximate conformal mapping method for the thermal entrance $Nu_{x,T}$ for ducts of arbitrary cross section. Slug flow solutions were carried out for circular, cardioid, corrugated, square, and hexagonal ducts. A laminar flow solution was carried out for the circular tube only.

d. Simplified Energy Equation

In this approximate method, the variable coefficients energy equation is reduced to a constant coefficient linear equation prior to being solved. No momentum or continuity equation is needed. Sadikov [15] simplified energy equation (32) as follows, after neglecting $\alpha(\partial^2 t/\partial x^2)$:

$$\varepsilon u_m \frac{\partial t}{\partial x} = \alpha\left(\frac{\partial^2 t}{\partial y^2} + \frac{\partial^2 t}{\partial z^2}\right) \tag{168}$$

where the correction factor $\varepsilon = 0.346\,\mathrm{Pr}^{-1/3}$ is based on his analysis of heat transfer for laminar flow over a flat plate. Based on the simplification procedure, this equation is valid in the entrance region where the boundary layer on a wall is not significantly affected by the opposing wall. Using this simplified equation, he solved the thermal entry length problem for parallel plates with the entering fluid at uniform [208] and nonuniform [209] temperatures. In both cases, the wall temperature was assumed to be linearly varying with regard to the axial distance. Sadikov [15] also studied the thermal entry length problem, based on the simplified energy equation, for rectangular ducts with a specified axial wall heat flux boundary condition.

e. *Finite Difference Methods*

The energy equation is expressed in a finite difference form, and the solution is carried out on a computer. Grigull and Tratz [34] employed this technique for the circular tube, and Montgomery and Wibulswas [210] employed it for rectangular ducts, to obtain the Ⓣ and Ⓗ1 temperature distribution and Nusselt numbers. Hong and Bergles [211] employed this technique for the Ⓗ1 problem of a semicircular duct.

f. *Monte Carlo Method*

The Monte Carlo method has been characterized as "the technique of solving a problem by putting in random numbers and getting out random numbers." The energy equation is first put into a finite difference form and is then given a probabilistic interpretation. With this interpretation, the standard random walk calculations or the Exodus method are used to determine the temperature distribution in the thermal entrance region. Chandler *et al.* [212] showed how the Monte Carlo method can be applied to the laminar forced convection heat transfer problem. They considered the thermal entry problem for parallel plates with the Ⓣ boundary condition.

g. *Finite Element Method*

In this method, first the temperature problem governed by the energy equation is expressed as an extremum problem by the methods of the calculus of variations. The extremum problem is the formulation and subsequent minimization of a functional (or integral) that consists of the determination of surface and volume integrals over the continuum space (region of interest). In order to perform the minimization numerically, the problem region is divided into "finite elements" by imaginary surfaces. The shape of these finite elements may be chosen arbitrarily to fit the geometry under consideration. The desired temperature distribution is that for which the minimization of the functional is achieved. Tay and De Vahl Davis [213] showed how this method can be applied to the thermal entrance region by solving the four fundamental problems of the parallel plates geometry. They have also discussed qualitatively the applicability of the finite element method to other duct geometries and its comparison with the finite difference method.

2. SIMULTANEOUSLY DEVELOPING FLOW

The thermal entrance solutions are obtained primarily by the following two methods.

a. *Seminumerical Methods*

The velocity profile is obtained from the linearization of the momentum equation. Subsequently, by employing this velocity profile, the temperature distribution is obtained by numerical methods. Employing a Langhaar type velocity profile [153], numerical solutions were carried out for the circular tube by Kays [40], McMordie and Emery [214], and Butterworth and Hazell [43], for parallel plates by Han [155], and for annular ducts by Heaton et al. [44]. Employing Sparrow's velocity profile [162], the numerical solution was carried out for the circular tube by Kakaç and Özgü [215].

b. *Complete Numerical Methods*

Both the velocity and temperature distributions are obtained by solving the corresponding momentum and energy equations numerically. The thermal entry solutions were carried out for the circular tube by Hornbeck [216], Bender [217] and Manohar [175], for parallel plates by Hwang and Fan [218], for rectangular ducts by Montgomery and Wibulswas [219], for equilateral triangular and right-angled isosceles triangular ducts by Wibulswas [220], for concentric annular ducts by Shumway and McEligot [45], Fuller and Samuels [195], and Coney and El-Shaarawi [46], and for eccentric annular ducts by Feldman [184].

Chapter V

Circular Duct

After parallel plates, the circular tube is the simplest cylindrical geometry mathematically. It is the geometry most commonly used in fluid flow and heat transfer devices. Circular tube laminar fully developed and developing flows have been analyzed in great detail for various boundary conditions, including the effect of thermal energy sources, viscous dissipation, fluid axial heat conduction, and momentum diffusion. The results are outlined below.

A. Fully Developed Flow

Steady state fully developed laminar flow of an incompressible fluid through a stationary circular or parallel plates duct is referred to as *Poiseuille or Hagen–Poiseuille flow* after Hagen [221] and Poiseuille [222]. The viscosity of the fluid is specified as constant and body forces are specified as absent. The invariant velocity profile obtained for the Poiseuille flow is parabolic at any cross section of the duct. For a circular tube of radius a and coordinate axes located at the center of the tube, it is given as [7]

$$u = \tfrac{1}{4}c_1(r^2 - a^2) \tag{169}$$

$$u_{\mathrm{m}} = -\tfrac{1}{8}c_1 a^2 \tag{170}$$

The friction factor–Reynolds number product for this flow is

$$f\,\mathrm{Re} = 16 \tag{171}$$

Heat transfer results are now described separately for each boundary condition.

1. CONSTANT SURFACE TEMPERATURE, (T)

The circular tube thermal entrance problem, known as the Graetz problem, was first investigated by Graetz in 1883 [1], and later quite independently by Nusselt in 1910 [2]. These heat transfer theory pioneers evaluated the first three terms of an infinite series solution for hydrodynamically developed and thermally developing flow for the (T) boundary condition. Their asymptotic Nusselt number for fully developed flow was presented as $Nu_T = 3.66$, very close to the more precise magnitude

$$Nu_T = 3.6567935 \tag{172}$$

for the case of negligible axial heat conduction, viscous dissipation, flow work, and thermal energy sources within the fluid. The temperature distribution may be inferred from Sellers et al. [25].

Pahor and Strand [223] included the effect of axial heat conduction in the fluid and presented graphically the fully developed Nusselt number as a function of the Péclet number. They also formulated asymptotic formulas, similar to those of Eqs. (173) and (174). Labuntsov [224] considered the same problem and presented the asymptotic Nu_T as a function of the Péclet number. Ash and Heinbockel [225] refined the work of Pahor and Strand [223] to obtain the temperature distribution in terms of the confluent hypergeometric function. They presented graphically Nu_T as a function of Pe. Their numerical results are listed in Table 6 [226].

Michelsen and Villadsen [227] employed the method of orthogonal collocation and accurately determined Nu_T for various Péclet numbers. These Nu_T are in excellent agreement with those of Table 6. They also provided the asymptotic formulas

$$Nu_T = 3.6568\left(1 + \frac{1.227}{Pe^2} + \cdots\right), \qquad \text{for} \quad Pe > 5 \tag{173}$$

$$Nu_T = 4.1807(1 - 0.0439\,Pe + \cdots), \qquad \text{for} \quad Pe < 1.5 \tag{174}$$

TABLE 6

CIRCULAR DUCT: Nu_T AS A FUNCTION OF Pe FOR FULLY
DEVELOPED LAMINAR FLOW
(FROM ASH [226])

Pe	Nu_T	Pe	Nu_T	Pe	Nu_T
∞	3.6568	6	3.744	0.5	4.098
60	3.660	5	3.769	0.4	4.118
50	3.660	4	3.805	0.3	4.134
40	3.661	3	3.852	0.2	4.150
30	3.663	2	3.925	0.1	4.167
20	3.670	1	4.030	0.04	4.170
10	3.697	0.9	4.043	0.03	4.175
9	3.705	0.8	4.059	0.02	4.175
8	3.714	0.7	4.071	0.01	4.175
7	3.728	0.6	4.086	0.001	4.182
				0	4.1807

Equations (172)–(174) are valid for the case when viscous dissipation and flow work within the fluid are negligible. For liquid flow in very long pipes, viscous dissipation may not be neglected, regardless of how small it is. Ou and Cheng [228] analyzed the (T) problem with finite viscous dissipation and showed that the asymptotic Nusselt number approaches 48/5, independent of the Brinkman number, and θ_m approaches $-(5/6)$Br. For gases, the flow work term is of the same order of magnitude as the viscous dissipation term. Based on the work of Ou and Cheng for parallel plates [229], it is inferred that the asymptotic Nusselt number is zero when both viscous dissipation and flow work terms are included in the analysis.

2. FINITE WALL THERMAL RESISTANCE, (T3)

Based on the results of Sideman et al. [230] for the thermal entry length solution for the (T3) boundary condition, the fully developed Nu_{T3} and Nu_o are calculated and presented in Table 7 for the circular duct. McKillop et al. [231] and Javeri [207] analyzed the (T3) thermal entrance problem for the circular tube. Their tabulated asymptotic Nu_{T3} agree well with the values in Table 7. As noted on p. 29, the (T3) boundary condition is a special case of the (H5) boundary condition with the exponent m in Eq. (62) as $-4Nu_o$.

TABLE 7

CIRCULAR DUCT: Nu_{T3}
AND Nu_o FOR FULLY
DEVELOPED LAMINAR FLOW
(FROM SIDEMAN et al. [230])

R_w	Nu_{T3}	Nu_o
0	3.657	3.657
0.005	3.666	3.600
0.025	3.713	3.398
0.05	3.763	3.167
0.10	3.844	2.777
0.15	3.908	2.464
0.25	4.000	2.000
0.50	4.124	1.347
1.00	4.223	0.8085
∞	4.364	0

Hickman [232] analyzed the (T3) thermal entry length problem by a Laplace transform technique and presented the following formula for the asymptotic Nusselt number:

$$Nu_{T3} = \frac{1 + (48/11)R_w}{(59/220) + R_w} \tag{175}$$

This formula provides Nu_{T3} higher than those of Table 7, ranging from 0% for $R_w = \infty$ to 1.9% for $R_w = 0$.

Hsu [233] included fluid axial heat conduction for the (T3) boundary condition. Based on his results, Nu_{T3} were calculated by the present authors and are presented as a function of the Péclet number in Table 8. Nu_{T3} in Table 8 may be in slight error at low Péclet numbers, as discussed on p. 121.

TABLE 8

CIRCULAR DUCT: Nu_{T3} AS A FUNCTION OF R_w AND Pe FOR FULLY DEVELOPED LAMINAR FLOW (FROM HSU [233])

	Nu_{T3}							
R_w	Pe = 1	Pe = 5	Pe = 10	Pe = 20	Pe = 30	Pe = 50	Pe = 100	Pe = ∞
0	4.030	3.769	3.697	3.670	3.663	3.660	3.658	3.657
0.005	4.058	3.784	3.709	3.680	3.674	3.671	3.669	3.669
0.05	4.296	3.911	3.813	3.777	3.769	3.765	3.763	3.763
0.25	—	4.179	4.056	4.015	4.007	4.002	4.001	4.000
∞	4.364	4.364	4.364	4.364	4.364	4.364	4.364	4.364

It is noted on p. 13 that the flow work term $u(dp/dx)/J$ in Eq. (24) exists because of the specification of perfect gas behavior. For an incompressible liquid, however, this term does not remain. Real liquids cannot always be idealized as incompressible, particularly if the thermal energy generated due to viscous dissipation is significant. Toor [234] showed that the flow work term is generally $T\beta u(dp/dx)/J$, where β is the coefficient of thermal expansion. For a perfect gas $\beta = 1/T$. For an incompressible liquid $\beta = 0$. There are liquids for which the cross-sectional average of $T\beta$ is too large to be treated as zero.

To investigate the combined effect of viscous dissipation and flow work for liquids in long pipes, Toor [234] solved Eq. (24), after setting the following terms to zero: $k(\partial^2 t/\partial x^2)$, $\rho c_p u(\partial t/\partial x)$, and S. He employed the (T3) boundary condition. The net effect of viscous dissipation and flow work is to reduce the temperature of the liquid at the tube center and to increase it away from the centerline relative to the $\beta = 0$ situation. The net effect of flow work for real liquids is to reduce the bulk average fluid temperature below that for the incompressible liquid situation.

3. RADIANT FLUX BOUNDARY CONDITION, (T4)

The fully developed Nu_{T4} may be inferred from the thermal entry length solution of Chen [235] as a function of γ. Kadaner et al. [236] also analyzed the radiant wall heat flux boundary condition. They presented the following approximate formula for Nu_{T4}, which agrees within 0.5% of their solution:

$$Nu_{T4} = \frac{8.728 + 3.66\gamma T_a^3}{2 + \gamma T_a^3} \tag{176}$$

where T_a is the absolute outside fluid temperature normalized by the absolute fluid inlet temperature T_e.

4. Specified Wall Heat Flux Distribution and (H)

As mentioned earlier, the circular tube geometry is unique[†] because the (H1)–(H4) boundary conditions yield the same heat transfer results for both developed and developing velocity and temperature profiles for the symmetrically heated duct. Hence the constant heat flux boundary condition is simply designated as the (H) boundary condition. Glaser [237] provided a solution for fully developed laminar flow with the (H) boundary condition. Tao [79] outlined the fully developed solution for the (H) boundary condition using a complex variables technique; he included thermal energy sources in the fluid. Madejski [238] considered the effect of flow work on the temperature distribution when $q'' = 0$, and observed an effect of temperature drop, similar to the Ranque effect. Tyagi [63] extended Tao's work by including the influence of viscous dissipation on heat transfer. His results for the temperature profile and Nusselt number are given by

$$(t - t_w)_H = \frac{c_4}{64}(r^2 - a^2)\{(r^2 - 3a^2 - 16a^2 c_5)$$

$$+ c_6[r^2 - 3a^2 - 2(r^2 - a^2)]\} \tag{177}$$

$$(t_m - t_w)_H = \frac{11c_4 a^4}{384}\left(1 + \frac{64}{11}c_5 + \frac{5}{11}c_6\right) \tag{178}$$

$$Nu_H = \frac{48}{11}\left(\frac{1}{1 + (3/44)S^* + (48/11)Br'}\right) \tag{179}$$

where the constants c_1, c_2, c_3, and c_4 are defined in the nomenclature, and

$$c_5 = \frac{c_3}{c_4 a^2} = -\frac{S^*}{8[S^* + 4(1 + 8\ Br')]} \tag{180}$$

$$c_6 = \frac{\mu c_1}{k c_2} = -\frac{32\ Br'}{S^* + 4(1 + 8\ Br')} \tag{181}$$

Note that the expressions for c_5 and c_6 are incorrectly formulated by Perkins in [63]. In the presence of thermal energy sources and viscous dissipation, the enthalpy rise (or drop) of the fluid, based on Tyagi's results, is given by

$$Wc_p \frac{dt_m}{dx} = q' + S\pi a^2 + \frac{8\pi\mu u_m^2}{g_c J} \tag{182}$$

[†] The parallel plate and concentric annular duct geometries are also unique for the identical reason.

Ou and Cheng [239] also derived Eq. (179) with $S^* = 0$ for finite viscous dissipation within the fluid.

In the absence of thermal energy sources, viscous dissipation, and flow work, Eq. (179) reduces to

$$Nu_H = 48/11 = 4.36364 \qquad (183)$$

Axial heat conduction within the fluid is constant for the (H) boundary condition. Consequently, it does not affect the Nusselt number. Thus, Nu_H is independent of the Péclet number.

The heat transfer coefficient is treated as constant for the (H) temperature problem [7] and the Nusselt number of Eq. (183) is the result. Gołos [240] rigorously analyzed the (H) temperature problem for the parallel plates with slug flow, and showed that the constant h is a good approximation only when $q'' D_h/kt_e \leq 20$ and $L/(D_h Pe) \geq 0.25$. Based on the magnitudes of eigenvalues, Gołos recommended that the constant h is a good approximation for fully developed laminar and turbulent flows for the above range of parameters.

Reynolds [241] considered the effect of arbitrary peripheral heat flux on fully developed Nusselt numbers with constant axial heat flux per unit length. Of particular interest is the cosine heat flux variation as shown in Fig. 11. The peripheral fully developed Nusselt number for this heat flux variation is

$$Nu_H(\theta) = \frac{1 + b\cos\theta}{(11/48) + \frac{1}{2}b\cos\theta} \qquad (184)$$

Reynolds [28] extended his work by including the effect of heat conduction in the peripheral direction with arbitrary peripheral heat flux and constant axial wall heat flux.

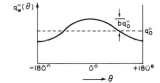

FIG. 11. Cosine heat flux variation along the circular tube periphery.

5. EXPONENTIAL WALL HEAT FLUX, (H5)

Hall et al. [242] first showed that the axially invariant fluid temperature profile, Eq. (2), occurs for an exponential wall heat flux distribution in the axial direction, Eq. (62). Hasegawa and Fujita [243] also found independently that the fully developed situation arises when the axial wall heat flux is exponential. They solved the energy equation numerically, and determined Nu_{H5} as a function of the exponent m. Their Nu_{H5} are in good agreement with those of Table 9. They demonstrated that the Ⓗ and Ⓣ boundary conditions are special cases of the Ⓗ5 boundary condition, with $m = 0$ and -14.627, respectively. The wall and fluid bulk mean temperatures for various values of m are presented in Fig. 6.

Gräber [244] also studied the axial exponential wall heat flux distribution in the circular tube, parallel plates, and annular ducts. He introduced a parameter $F_0 = dt_w/dt_0$. The F_0 is the ratio of the temperature gradients dt_w/dx along the wall transferring heat and dt_0/dx in the fluid at the point where the temperature gradient normal to the wall reduces to zero. F_0 and m

TABLE 9

CIRCULAR DUCT: Nu_{H5} FOR
FULLY DEVELOPED LAMINAR FLOW
[FROM EQ. (186)]

m	Nu_{H5}	m	Nu_{H5}
		10	4.77
		20	5.11
−70	−8.84	30	5.42
−60	−2.29	40	5.71
−50	0.256	50	5.97
−40	1.68		
		60	6.21
−30	2.64	70	6.43
−20	3.34	80	6.64
−14.63	3.657	90	6.84
−10	3.90	100	7.02
0	4.364		

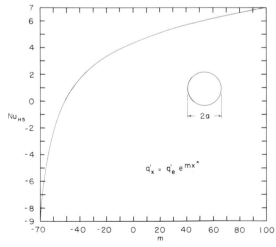

FIG. 12. Circular duct: Nu_{H5} for fully developed laminar flow (from Table 9).

are related by

$$\frac{1}{F_o} = 1 + \frac{1}{t_m^*}\left\{\frac{4Nu_{H5}}{4Nu_{H5} + [(1 + r^*)/r_j^*]m} - 1\right\} \tag{185}$$

Here r^* and r_j^* equal zero and one, respectively, for the circular tube, and $t_m^* = (t_m - t_w)/(t_o - t_w)$, where t_o is the fluid temperature at the point where the temperature gradient normal to the wall $(\partial t/\partial r)$ is zero. The 0 and 1 values of F_o correspond to the boundary conditions ⓉT and Ⓗ respectively. The Nu_{H5}/Nu_{H1} ratio was presented graphically by Gräber as a function of F_o (range from -2 to 8) with r^* (range 0 to 1) as a parameter.

Some experimental verification of the Ⓗ5 boundary condition for the turbulent flow has been presented in [242,245,246].

The fully developed Nu_{H5} can also be obtained in a closed form from the Ⓗ5 thermal entrance solution of Eq. (250) by letting $x^* \to \infty$.

$$Nu_{H5} = \left[\sum_{n=1}^{\infty} \frac{-C_n R_n(1)\beta_n^2}{2\beta_n^2 + m}\right]^{-1} \tag{186}$$

where β_n^2, $C_n R_n(1)$ for n up to 20 are obtained from Table 17 and higher values are determined from Eqs. (236) and (237). The above series is extremely slowly convergent. The present authors computed Nu_{H5} from this equation by taking the first 10^5 terms of the series to get two decimal point accuracy! These Nu_{H5} are listed in Table 9 and graphed in Fig. 12. The Nu_{H5} magnitude for $-14.63 \le m \le 0$, calculated from Eq. (186), are up to 0.2% higher than Nu_{T3} of Table 7 for $m = -4Nu_o$.

B. Hydrodynamically Developing Flow

1. Axially Matched Solutions

The hydrodynamic entry length problem was analyzed first by Boussinesq [136,138] by considering a perturbation about the fully developed Poiseuille profile. His infinite series solution was fairly adequate downstream, but poor near the entrance. Atkinson and Goldstein [138] presented the stream function by a power series to find a solution to Eq. (8) for axial positions close to the inlet. They then joined this solution with the one obtained by Boussinesq's method to obtain a velocity distribution in the entire entrance region. Gillis and Shimshoni [139] computed the Atkinson and Goldstein solution in greater detail. They calculated and tabulated the first seven functions and constants of the generalized Blasius equation.

2. Integral Solutions

Schiller [49] solved the problem by an integral method assuming a developing boundary layer (of parabolic arc) in the entrance region with an impressed pressure gradient and a straight potential core in the remaining central cross section. As noted on p. 69, Schiller thus neglected viscous dissipation within the fluid. This method provides good results at the entrance, but poor downstream as confirmed by experimental results.

Shapiro *et al.* [14] modified Schiller's solution by using cubic and quartic velocity profiles in the boundary layer. They also carried out a careful experimental investigation of the hydrodynamic entrance region. Their experimental data in the range $10^{-5} \le x^{+} \le 10^{-3}$ are best correlated as follows, in terms of Re $(=GD_{h}/\mu)$ or $\text{Re}_{x} (=Gx/\mu)$:

$$\Delta p^{*} = \frac{\Delta p}{(\rho u_{m}^{2}/2g_{c})} = 13.74(x^{+})^{1/2} \tag{187}$$

or

$$f_{\text{app}} = \frac{3.44}{(\text{Re}_{x})^{1/2}} = \frac{3.44}{\text{Re}(x^{+})^{1/2}} \tag{188}$$

where the range of Re was from 51,200 to 113,400. Even at this high Reynolds number, for which fully developed flow is turbulent, a laminar boundary layer is formed at the entry of a tube. This layer increases in thickness along the length of the tube and eventually at $\text{Re}_{x} \simeq 5 \times 10^{5}$ (depending on the entrance condition of the fluid and the tube) undergoes transition and becomes turbulent.

The mean friction factor for laminar flow over a flat plate with zero pressure gradient is given by Kays [7] as

$$f_{m} = 1.328/(\text{Re}_{x})^{1/2} \tag{189}$$

A comparison of Eqs. (188) and (189) reveals that the apparent friction factor for a pipe entry is 2.59 times as great as the mean friction factor for a flat plate. Shapiro *et al.* [14] showed that when the boundary layer is thin compared to the pipe radius, the boundary layer behavior in a tube entry is substantially identical to that on a flat plate. Consequently, the skin friction coefficients due to wall shear stress are the same for the tube entry and the flat plate. The only difference then is that a negative pressure gradient exists in the tube entry, whereas a zero pressure gradient is specified for the flat plate. In the tube entrance with $x^{+} < 10^{-3}$, 39% (1.328/3.44) of the pressure drop is caused by wall friction and the remaining 61% by an increase in momentum flux. The fractional contribution of momentum flux change to the total pressure drop decreases with increasing x^{+} and approaches zero in fully developed flow. The pressure drop determined by the most accurate numerical method is in excellent agreement with Δp^{*} of Eq. (187), as will be discussed later.

Gubin and Levin [147] employed a logarithmic velocity profile in the boundary layer and solved the hydrodynamic entry length problem by an integral method. They presented the velocity distribution in tabular and graphical forms.

Campbell and Slattery [148] refined Schiller's solution by considering energy loss due to viscous dissipation within the fluid. They stated that since an average pressure is to be considered at any cross section, it would be more correct to find the pressure by the application of the macroscopic mechanical energy balance to all of the fluid in the pipe, taking into account the energy loss due to viscous dissipation. Because the boundary layer grows thicker, this correction becomes important as the fluid moves downstream. This improvement due to Campbell and Slattery shows considerably better agreement with the experimental data than the earlier theoretical results of Shapiro et al. [14]. Govinda Rao et al. [149] also used a momentum integral equation including the energy loss due to viscous dissipation within the fluid.

Fargie and Martin [152] considered a parabolic velocity distribution in the boundary layer and solved the integral continuity and momentum equations. They employed the differential momentum equation evaluated at the wall to eliminate the pressure gradient term. This led to a relatively straightforward solution capable of closed-form expressions for f_{app} Re, $K(x)$, and Δp^*. Fargie and Martin also presented tabulated values of these variables as a function of x^+. Their f_{app} Re are generally higher than those by Hornbeck [173], as described later, ranging from about 6% at $x^+ = 0.0002$ to 2% at $x^+ = 0.0488$.

3. SOLUTIONS BY LINEARIZED MOMENTUM EQUATION

All of the above solutions exhibit a discontinuity in the gradients of velocity and pressure distributions in the entrance region. Langhaar [153] proposed a linearization of the momentum equation [see Eq. (164)] and subsequently solved it by an integral method. His linearization is a good approximation in the central potential core. His solution appears to be satisfactory at the tube centerline and either very near the entrance or far from it.

Targ [159, 159a] introduced another approach of linearization, Eq. (165), which is a special case of the linearization [Eq. (167)] performed by Sparrow et al. [162]. In Targ's solution, which neglects the partial contribution of momentum change to the pressure gradient, the velocity profile develops more slowly near the entrance than is predicted by the more accurate numerical solution.

Sparrow et al. [162] employed the linearization approach described by Eq. (167). Their calculated velocity profiles are nearly identical with the numerical solutions described next.

4. SOLUTIONS BY FINITE DIFFERENCE METHODS

Hornbeck [173] introduced a finite difference scheme for the hydro-dynamic entry length problem of the circular tube. He linearized the momentum equation at any cross section x_1 by means of the velocity at $x = x_1 - \Delta x$. Hornbeck's results for u_{max}/u_m, f_{app} Re, and $K(x)$ are presented in Table 10.

TABLE 10

CIRCULAR DUCT: u_{max}/u_m, f_{app} Re, AND $K(x)$ FOR DEVELOPING LAMINAR FLOW

	Hornbeck [173]				Liu [182]		
x^+	$\dfrac{u_{max}}{u_m}$	f_{app}Re	$K(x)$	x^+	$\dfrac{u_{max}}{u_m}$	f_{app}Re	$K(x)$
0.00050	1.150	161.0	0.2900	0.0002116	1.100	230.7	0.1817
0.00125	1.227	100.68	0.4234	0.0004232	1.138	164.3	0.2510
0.00250	1.313	72.04	0.5604	0.0006349	1.166	134.9	0.3020
0.00375	1.378	59.73	0.6559	0.0008467	1.190	117.3	0.3431
0.0050	1.433	52.53	0.7306	0.001058	1.210	105.4	0.3784
0.0075	1.524	44.04	0.8412	0.001900	1.273	79.66	0.4839
0.0100	1.598	39.02	0.9210	0.002747	1.324	67.16	0.5622
0.0125	1.660	35.64	0.9822	0.003593	1.367	59.44	0.6244
0.0175	1.756	31.29	1.070	0.004440	1.405	54.08	0.6764
0.0225	1.824	28.55	1.129	0.005288	1.439	50.08	0.7209
0.0300	1.892	25.89	1.186	0.008658	1.553	40.58	0.8513
0.0400	1.943	23.68	1.229	0.01204	1.644	35.52	0.9404
0.0500	1.970	22.26	1.252	0.01543	1.718	32.28	1.005
0.0625	1.986	21.07	1.269	0.01882	1.776	30.00	1.054
∞	2.000	16.00	–	0.02221	1.822	28.29	1.092
				0.03569	1.928	24.26	1.179
				0.04924	1.971	22.16	1.213
				0.06281	1.989	20.88	1.226
				0.07634	1.996	20.03	1.231
				0.08993	1.999	19.43	1.234

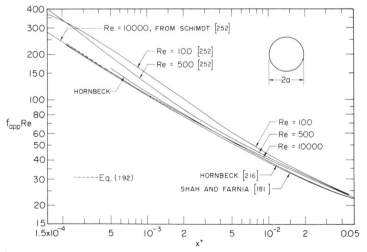

FIG. 13. Circular duct: f_{app} Re for developing laminar flow (from Tables 10 and 11).

Hornbeck's f_{app} Re are about 4.6% higher than those of Eq. (188) for $x^+ < 0.001$. Christiansen and Lemmon [174] and Manohar [175] solved the nonlinear momentum equation iteratively. The results of Christiansen and Lemmon and of Hornbeck are in excellent agreement. The hydrodynamic entrance region results of Manohar [247] are tabulated by Shah and London [13]. Manohar's f_{app} Re are lower than those of Hornbeck, ranging from 9% at $x^+ = 0.0005$ to 1% at $x^+ = 0.05$.

Shah and Farnia [181] employed the finite difference method of Patankar and Spalding [248][†] and obtained the hydrodynamic entry length solution for concentric annuli. Their u_{max}/u_m, f_{app} Re, and $K(x)$ for the circular tube, as tabulated by Liu [182], are presented in Table 10. f_{app} Re of Liu are lower than those of Eq. (188), ranging from 2.4% at $x^+ = 0.0002$ to 0.3% at $x^+ = 0.001$. f_{app} Re of Shah and Farnia are lower than those of Hornbeck by 5.6, 1.2, 1.9, and 0.86% at $x^+ = 0.001$, 0.005, 0.01, and 0.05, respectively. f_{app} Re of Hornbeck and of Shah and Farnia are compared in Fig. 13.

5. SOLUTIONS OF NAVIER–STOKES EQUATIONS

All of the foregoing hydrodynamic entry length solutions involve the idealizations of Eq. (7) (boundary layer type assumptions), i.e., axial diffusion of momentum and radial pressure gradient are neglected. The flow development, as a result of such solutions, is independent of both the inlet profiles of Fig. 14 and the Reynolds number. The boundary layer type hydrodynamic entry length solution for the uniform entrance profile (Fig. 14a) appears to be valid away from the immediate entrance ($x^+ > 0.005$), regardless of the true entrance, providing Re $\gtrsim 400$. The solutions of the complete Navier–Stokes equations take into account the effects of axial diffusion of momentum and radial pressure gradient. The resulting flow development is then dependent upon both the entrance profile (Fig. 14) and the Reynolds number for Re $\lesssim 400$ when $x^+ \lesssim 0.005$.

Axial diffusion of momentum and radial pressure gradient produce second or higher order effects and hence are neglected in the boundary layer treatment. They influence the entry length solution only in a limited region, approximately one pipe diameter both upstream and downstream of the $x^+ = 0$ section; however, the affected dimensionless distance increases as the Reynolds number decreases.

[†] Patankar and Spalding employed the flow direction x and dimensionless stream function as independent coordinates, and transformed the momentum and energy equations into a finite difference form using a fully implicit scheme. Calculations were proceeded by a marching process. The set of linear algebraic equations were solved by straightforward successive-substitution formulas.

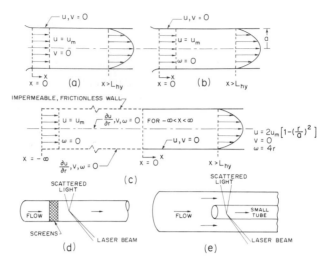

FIG. 14. Initial and boundary conditions for developing flow in a circular tube. (a) Uniform flow entry, (b) irrotational flow entry, (c) stream tube uniform flow entry, (d) uniform entry generation, (e) stream tube approximation. These figures also apply to parallel plates, if the tube radius a is replaced by the half distance between parallel plates, b, and if the fully developed velocity profile in (c) is replaced by $u = 1.5u_m[1 - (y/b)^2]$.

Figure 14a represents the velocity profile uniform at the entrance such that $u = u_m$, $v = 0$. The vorticity[†] ω at the entrance is not zero for this case. This is the initial and boundary condition that is most frequently analyzed for the hydrodynamic entry length problem. Figure 14b represents the velocity profile uniform and parallel at the entrance such that $u = u_m$ and $\partial v/\partial x = 0$ or $\omega = 0$. Since the vorticity is zero at the inlet for this case, it is also referred to as the irrotational flow entry. An initial condition of a uniform velocity profile cannot exist when axial diffusion of momentum or vorticity is included in the analysis. A mathematically more appropriate initial condition is shown in Fig. 14c, in which the velocity profile is idealized as uniform and parallel at $x = -\infty$. Figure 14d,e shows the experimental setup to approximate the idealized initial and boundary conditions of Fig. 14a,c, respectively.

In the following discussion, the peculiarities in the entrance velocity profile, the velocity overshoots, are described first. Then the hydrodynamic

---

[†] In a given plane, the vorticity is the sum of the angular velocities of any pair of mutually perpendicular infinitesimal fluid lines passing through the point in question. Consider a point $P(x, y, z)$ in the flow field having velocity components u, v, and w. The three components of the local vorticity at this point, in Cartesian coordinates, are

$$\omega_1 = \frac{\partial w}{\partial y} - \frac{\partial v}{\partial z}, \qquad \omega_2 = \frac{\partial u}{\partial z} - \frac{\partial w}{\partial x}, \qquad \omega_3 = \frac{\partial v}{\partial x} - \frac{\partial u}{\partial y}$$

entry length solutions of the complete set of Navier–Stokes equations are summarized for the three sets of idealized initial conditions of Fig. 14a–c. After the solutions are discussed for (1) the general case of Re > 0 and (2) the special case of *creeping flow* (Re = 0),† some experimental results are presented to corroborate the analytical results.

a. *Velocity Overshoots*

When the effects of axial diffusion of momentum and radial pressure gradient are included in the analysis, the velocity profiles have peculiar behavior in the entrance region ($x^+ \lesssim 0.005$) for Re $\lesssim 400$. The velocity profiles have a local minimum at the tube centerline and symmetrically located maxima on either side of the centerline near the walls. The normal expectation is either a uniform or a convex profile at the centerline, not a concave profile. This phenomenon has been referred to variously as velocity overshoots, axial velocity inflections, and bulges, kinks, or concavity in the velocity profile. The velocity overshoots were first reported by Wang and Longwell [191] in their numerical solution for parallel plates. The velocity overshoots are found at all Reynolds numbers, but they decrease in magnitude at increasing Re, and disappear at lower x^+ for increasing Re. They are more pronounced for the initial condition of a uniform velocity profile, Fig. 14a, b, in comparison to the initial condition of Fig. 14c. The phenomenon leading to the velocity overshoots may be explained as follows.

Near the entrance where the fluid particles first meet the wall, viscous friction rapidly decelerates the flow to zero velocity at the wall. The high velocity gradient at the wall results in a high shear stress and the high pressure gradient needed to produce the high acceleration in the near wall region. Theoretically, the magnitudes of the pressure gradients $\partial p/\partial x$ and $\partial p/\partial y$ at the wall at $x = 0$ (the point of singularity) approach infinity positively. For low values of Re, the effect of this high positive gradient at the singular point reaches the centerline region immediately. A negative pressure gradient $\partial p/\partial x$ establishes near the wall region a very short distance downstream of the $x = 0$ section; however, a positive (adverse) pressure zone exists in a region near the centerline. Thus, the flow near the centerline is not accelerated immediately, whereas the flow near the wall is forced to be stationary as soon as it enters the inlet region. To satisfy the continuity equation, the velocity overshoots are thus formed. At higher Reynolds

---

† Creeping flow is the generic name given to flows with a very low Reynolds number approaching zero. It represents a limiting case of laminar flow and occurs when viscous forces completely dominate inertia forces. The creeping flow problem is also referred to as the Stokes flow problem. Lubrication, viscometry, and polymer processing are some of the areas of applications of creeping flow.

number, however, the presence of the wall is not felt immediately by fluid particles near the centerline and no positive pressure gradient develops in the centerline region. Instead, this flow is accelerated gradually by the viscous displacement effect due to the presence of the wall. Thus, the development of the velocity overshoots and the existence of an adverse pressure gradient zone are much less pronounced at higher Reynolds numbers.

Friedmann *et al.* [187] showed for the circular tube that the velocity overshoots are significant only for the case of exactly uniform entrance velocity profile, which involves an abrupt change to a no-slip condition at the wall. The velocity overshoots are much weaker for the entrance velocity profile $[(n + 2)/n][1 - (r - a)^n]$ with $n = 50$. The $n = \infty$ represents the case of exactly unifrom entrance velocity profile. In order to ensure further that the predicted velocity overshoots are not due to numerical errors, Abarbanel *et al.* [196] analyzed theoretically the Stokes flow problem ($Re = 0$) in a quarter-plane and did find velocity overshoots. Furthermore, they showed that the velocity overshoots are not merely a mathematical consequence of nonphysical singularity in the velocity profile at the inlet ($x = 0$). The velocity overshoots, although weaker, exist even when the initial conditions are modified so as to remove the singularity at the inlet.

The velocity overshoots for the irrotational entry (Fig. 14b) are not as strong as those for the uniform entry (Fig. 14a) at a specified values of x^+ and Re, as shown by McDonald *et al.* [193]. As a result, the centerline or the "uniform core" velocity is higher for the irrotational entry at any x^+ value.

When the velocity profile is considered uniform at $x = -\infty$, as in Fig. 14c, the velocity profile at $x = 0$ is already partially developed. Hence, for this case, one would expect weaker velocity overshoots near the $x^+ = 0$ region for low Re. This has been confirmed by the results of Vrentas *et al.* [186].

To demonstrate experimentally that the velocity overshoots are just not a mathematical oddity, Berman, Santos, and Burke [249–251] conducted tests using the apparatus described by Fig. 14d, e. They measured velocity profiles in the entrance region by a laser-Doppler velocimeter, and found the velocity overshoots as discussed above.

b. *Solutions for* Re > 0

Friedmann *et al.* [187] and McDonald *et al.* [193] analyzed the initial and boundary conditions of Fig. 14a. Schmidt and Zeldin [188,189] and McDonald *et al.* [193] investigated the conditions of Fig. 14b. Vrentas *et al.* [186] studied the case of Fig. 14c.

Vrentas *et al.* [186] were the first investigators to solve the complete set of Navier–Stokes equations for a circular tube. They obtained a numerical solution of the two coupled elliptical partial differential equations for the initial condition of Fig. 14c, and presented velocity and pressure distributions, $K(x)$ and L_{hy} for $Re = 0, 1, 50, 150, 250$, and ∞. The momentum equation for $Re = \infty$ corresponds to the boundary layer type equation, Eq. (8). They found that for $Re < \infty$, the velocity profiles were partially developed at $x^+ = 0$. One interesting result of their analysis was that the velocity overshoots, as described in the preceding section, were found in the entrance near the $x^+ = 0$ region. The velocity overshoots (concavity) were small, with the maximum velocity never being more than 0.05% higher than the centerline velocity. Vrentas et al. concluded that the concavity increased as the Reynolds number was increased, and occurred only for $Re > 50$. The reason for not finding any concavity at low Reynolds numbers may be due to a relatively coarse grid employed in the entrance region for the numerical analysis, as is discussed by Burke and Berman [249].

At higher Reynolds number, the axial diffusion of vorticity causes some velocity profile development upstream ($x < 0$). This results in a tube centerline velocity development for the Fig. 14c case faster than that of the boundary layer type solutions of [173–175]. The resultant entrance length L_{hy} is smaller than that predicted by the boundary layer analysis, with a maximum difference of 14% at $Re = 150$. At very low Reynolds numbers ($Re < 20$), the axial diffusion of vorticity causes the velocity development to spread downstream and the resulting entrance length is larger than that predicted by the boundary layer analysis. As Re approaches zero, L_{hy}^+ approaches an asymptotic value of $0.31/Re$.

Friedmann et al. [187] numerically solved the complete set of Navier–Stokes equations for a uniform velocity profile at the tube entrance (Fig. 14a). They reported the velocity distribution in tabular and graphical forms as functions of Re and x/D_h, and the hydrodynamic entry length as a function of Re for $Re = 0$ to 500. Because of the different idealizations for the entrance velocity profile, they found more pronounced velocity overshoots than that of Vrentas et al. [186]. Friedmann et al. obtained the maximum velocity as 7.5–18% higher than the centerline velocity for $Re \le 1$ to $Re > 100$. They explained that in the boundary layer approximations, the equations of motion have a similarity solution in terms of x^+ and r/a. In these coordinates, the axial range of the kinked velocity profiles becomes vanishingly small and hence such concavity in the velocity profile is not seen in boundary layer type solutions. As reported on p. 92, they found that the velocity overshoots were more pronounced only for the case of exactly uniform entrance velocity profile and were much weaker for the entrance velocity deviating from an exactly uniform one.

The centerline velocity development of the solution by Friedmann et al. (Fig. 14a) is slower than that of the boundary layer type solution. Their hydrodynamic entrance length L_{hy} (or L_{hy}^+) consequently is longer than that of the boundary layer type solution. As Re approaches zero, L_{hy}^+ approaches 0.59/Re. At Re = 150, the L_{hy}^+ of Friedmann *et al.* [Eq. (195)] is 1% higher than that predicted by the boundary layer type solution.

Notice that the initial conditions of Fig. 14a and 14c provide opposite results for the u_c/u_m development and the dependence of L_{hy} on Re, for Re $\gtrsim$ 20, when compared to the results of the boundary layer type solution.

Chen [197] employed the approximate integral method to solve the Navier–Stokes equations for the initial condition of Fig. 14a. He presented a closed-form formula for the pressure drop. His centerline velocity and L_{hy}^+ are higher than those of Friedmann *et al.* [187], and $K(x)$ is generally lower than those by numerical solutions. The limiting cases, Re = 0 and ∞, are in excellent agreement with the values from exact solutions.

McDonald *et al.* [193] also solved the complete set of Navier–Stokes equations for the circular tube and parallel plates for two sets of initial conditions at the entrance ($x = 0$): (1) uniform flow, Fig. 14a, and (2) irrotational flow, Fig. 14b. They compared the numerical methods of Wang and Longwell [191], Friedmann *et al.* [187], Gillis and Brandt [192], and others. They showed that each of these methods has some limitations and their proposed numerical method is better for solving the complete set of Navier–Stokes equations. They obtained a detailed solution for Re = 400 and found overshoots in the velocity profiles. Their centerline velocity distribution is in excellent agreement with that of Friedmann *et al.* for the initial condition of uniform flow, Fig. 14a. However, for the irrotational flow initial condition, the centerline velocity is higher than that for the uniform flow, ranging from 8% at $x^+ = 0.002$ to 0% for $x^+ > 0.05$.

Schmidt and Zeldin [188, 189] numerically analyzed the complete set of Navier–Stokes equations for the circular tube and parallel plates, for the irrotational flow condition of Fig. 14b, at Re = 100, 500, and 10,000. They also observed overshoots in the entrance region velocity profiles. They presented dimensionless velocity and pressure distributions as well as the cross-sectional area average $\bar{K}(x)$ for Re = 100, 500, and 10,000. Their results for $\bar{K}(x)$ and the calculated f_{app} Re are reported in Table 11 and f_{app} Re are presented in Fig. 13 [252]. The f_{app} Re versus x^+ curve in Fig. 13 for Re = 100 has a different shape for $x^+ < 0.0002$. It is not clear whether this shape is due to the nature of the finite difference equations and numerical method at startup or to the overshoots in the velocity profile. Their centerline velocity for the irrotational flow initial condition is consistently about 4% higher (for $0.0005 \leq x^+ \leq 0.011$ at Re = 500) than that by Friedmann *et al.* [187] for the uniform flow initial condition. The centerline velocity developments of Schmidt and Zeldin (for Re = 500) and McDonald *et al.* [193] (for Re = 400) for the irrotational flow initial condition agree with each other within 4%.

TABLE 11

CIRCULAR DUCT: f_{app} Re AND $\bar{K}(x)$ AS FUNCTIONS OF x^+ AND Re FOR
DEVELOPING LAMINAR FLOW, BASED ON THE SOLUTION TO THE COMPLETE
NAVIER–STOKES EQUATIONS (FROM SCHMIDT [252])

x^+	Re= 10000		Re = 500		Re = 100	
	f_{app}Re	$\bar{K}(x)$	f_{app}Re	$\bar{K}(x)$	f_{app}Re	$\bar{K}(x)$
0.0000614	424.8	0.1004	653.6	0.1566	382.6	0.09003
0.0000946	348.2	0.1257	519.7	0.1906	403.4	0.1466
0.0001296	297.7	0.1461	434.8	0.2172	386.0	0.1919
0.0002059	236.5	0.1816	331.8	0.2601	331.9	0.2602
0.0003387	184.7	0.2286	246.1	0.3117	263.6	0.3355
0.0005000	152.5	0.2730	194.0	0.3560	215.9	0.3997
0.0007000	129.4	0.3176	157.9	0.3974	180.6	0.4608
0.0008625	117.0	0.3485	139.2	0.4250	161.3	0.5013
0.001056	106.4	0.3811	123.5	0.4537	144.5	0.5426
0.001579	88.00	0.4546	98.10	0.5184	115.6	0.6291
0.002167	76.14	0.5212	82.77	0.5786	96.75	0.6998
0.003076	65.20	0.6053	69.38	0.6568	79.35	0.7794
0.004019	58.16	0.6777	61.14	0.7255	68.38	0.8420
0.005500	51.14	0.7731	53.15	0.8172	57.91	0.9220
0.006610	47.56	0.8345	49.16	0.8768	52.83	0.9739
0.008165	43.85	0.9095	45.09	0.9500	47.78	1.038
0.01050	39.88	1.003	40.79	1.041	42.67	1.120
0.01439	35.37	1.115	36.00	1.151	37.18	1.219
0.02217	29.92	1.234	30.30	1.268	30.94	1.325
0.04550	23.22	1.314	23.40	1.346	23.68	1.397

c. *Solutions for* Re = 0

Friedmann *et al.* [187] employed the initial condition of Fig. 14a to solve the Navier–Stokes equations. They obtained a numerical and an analytical solution in terms of Bessel functions for Re = 0 and observed velocity overshoots. The fit between their analytical and numerical results was not good near the tube inlet and near the tube axis because their analytical solutions for the stream function converged very slowly.

Lew and Fung [253] obtained an analytical solution for Re = 0 by a series approach for the initial condition of Fig. 14a modified so that $u(0, r) = u_m$ for $0 \leq r \leq 0.9a$ and it decreases from u_m to 0 for $0.9a \leq r \leq a$. They extensively presented u, v, and Δp in tabular and graphical forms for x/a varying from 0 to 2.9. They found that the maximum value of the radial velocity is about $0.30u_m$ at $x/a = 0.2$. They determined L_{hy}/D_h as 0.65.

While Friedmann *et al.* [187] obtained an accurate numerical solution for Re = 0 for the initial condition of Fig. 14a, they did not present the results for $\bar{p}$ (average pressure), ω, v, and their numerical convergence criteria. Lew and Fung [253] obtained an approximate solution for the same problem. Barbee and Mikkelsen [254] obtained an accurate numerical solution for Re = 0 and graphically presented u, v, $\bar{p}$, and ω, and discussed the numerical convergence criteria. Their centerline velocity development is in excellent agreement with that of Friedmann *et al.* They showed the maximum radial velocity as $0.39u_m$ at $x/a = 0.1905$. Lew and Fung [253] showed it to be $0.30u_m$.

A numerical and an analytical solution for Re = 0 for the initial condition of Fig. 14c was obtained by Vrentas et al. [186] and Vrentas and Duda [255], respectively. They showed an excellent agreement between the two solutions for Re = 0, thus providing at least some confidence in the numerical results for higher Re of [186]. They found L_{hy}/D_h as 0.31 for Re = 0.

Atkinson et al. [256] employed a finite element method and analyzed the same problem as that of Vrentas et al. [186] for Re = 0. They found their velocity profiles in the entrance region in excellent agreement with those of Vrentas et al. obtained by a finite difference method. Atkinson et al. [256] also analyzed the hydrodynamic entry length problem for the initial condition of Fig. 14a for Re = 0 by the finite element method. They presented graphically u/u_m as functions of r/a and x/a. The hydrodynamic entry length from their solution agreed well with that of Eq. (194).

d. Experimental Results

The solution of Navier–Stokes equations provides some unexpected and interesting results for the hydrodynamic entry length problem. To prove that these results are just not a mathematical oddity, Berman, Santos, and Burke [249–251] conducted a series of experiments using a laser-Doppler velocimeter to measure velocity profiles in the entrance region.

Any of the three initial conditions of Fig. 14a–c is difficult to achieve in reality. Berman et al. [249,250] employed a porous plug of stacked screen as shown in Fig. 14d to generate a uniform velocity profile downstream of the plug so as to approximate the idealized initial condition of Fig. 14a. The screens did not allow measurements to be made at very small downstream axial distances since strong vortices existed at Reynolds number over 300. Clearly a true uniform velocity profile without fluctuations was not obtained and the flow development downstream depended on the nature of the starting flow situation. They found that the entrance lengths were longer than the boundary layer theory results and increased with decreasing Reynolds numbers. Their measurements at Re of 108 showed velocity dips of 21, 13, and 7% at $4x^+ = 0.011$, 0.015, and 0.020, respectively. The velocity dip is defined as the ratio of the difference between the maximum and centerline velocity to the maximum velocity at the same axial location.

To approximate the initial condition of Fig. 14c, Berman et al. employed the concentric tube arrangement with a uniform entry face of Fig. 14e. In most of their tests, the flow through the annulus (large tube) was blocked off so that the fluid was rapidly accelerating and a solid boundary existed in the radial direction at the entrance. Hence, their experimental conditions did not adequately represent the idealized condition of Fig. 14c for small

x^+. Nevertheless the experimental results were found in good agreement with those of the stream tube (Fig. 14c) analytical results for larger x^+.

At Re = 304, 206, and 108, Burke and Berman [249] did find overshoots in the velocity profile for $x^+ \lesssim 0.001$. The peak magnitude of the velocity dip was 5.2% at $x^+ = 0.00045$ and did not change with the Reynolds number. They also found that the centerline velocity development was faster for the test section compared to Hornbeck's boundary layer solution [173]. This results in a shorter entrance length for the initial condition of Fig. 14e.

Both the experimental and analytical results support the existence of overshoots in the velocity profile at the entrance. The abrupt change to zero velocity at the wall leads to an overshoot of maximum velocity near the wall. The radial location of this maximum moves toward the center as the velocity profile matures with increasing x^+. A well-rounded inlet will show no overshoots, since the development of a velocity gradient at the wall is not sufficiently abrupt.

6. PRESSURE DROP AND APPARENT FRICTION FACTORS

The apparent Fanning friction factor–Reynolds number product and incremental pressure drop number are presented in Table 10 for the hydrodynamic entrance region of the circular tube. It was shown that the f_{app} Re factors of this table are in excellent agreement with those by the empirical equation (188) for a "short duct" with $x^+ \leq 10^{-3}$. Consequently, Δp^* values calculated from f_{app} Re of Table 10 and Eq. (92) for $x^+ \leq 10^{-3}$ are in excellent agreement with those of the empirical equation (187).

For a "long duct" with $x^+ \gtrsim 0.06$, Δp^* or f_{app} Re can be determined by Eq. (92) for the known values of f Re ($= 16$) and the best established value for $K(\infty)$. Many investigators, while theoretically investigating the entry flow problem, have determined $K(\infty)$, the value of which is reported as varying from 1.08 to 1.41; the experimental values vary from 1.20 to 1.32. Comparisons of $K(\infty)$ are provided by Lakshmana Rao and Sridharan [135], Christiansen and Lemmon [174], Schmidt and Zeldin [189], and Atkinson et al. [257]. Bender [258] obtained theoretically $K(\infty) = 1.25$, which approximates a mean of the experimental values and a mean of the numerical values derived from f_{app} Re of Table 10. Thus, for $x^+ \gtrsim 0.06$, Eq. (92) yields

$$\Delta p^* = 1.25 + 64x^+ \qquad (190)$$

Bender then combined Eqs. (187) and (190) to cover the complete range of x^+ as follows:

$$\Delta p^* = 13.74(x^+)^{1/2} + \frac{1.25 + 64x^+ - 13.74(x^+)^{1/2}}{1 + 0.00018(x^+)^{-2}} \qquad (191)$$

Shah [259] modified the constant 0.00018 to 0.00021 so that the calculated Δp^* provided the least rms error when compared with Δp^* of Liu [182] derived from the results of Table 10. Equation (191) with the modified constant yields f_{app} Re factors as

$$f_{app}\,Re = \frac{3.44}{(x^+)^{1/2}} + \frac{1.25/(4x^+) + 16 - 3.44/(x^+)^{1/2}}{1 + 0.00021(x^+)^{-2}} \tag{192}$$

This relation is recommended for the entire range of x^+ for the circular tube. f_{app} Re of Eq. (192) agree with those of Liu within $\pm 1.9\%$ for $x^+ > 0.00022$, while Hornbeck's f_{app} Re of Table 10 are higher by up to 2.8% for $x^+ > 0.0005$. Equation (192) is shown in Fig. 13 as a dotted line.

$K(\infty)$ and hence f_{app} Re is also dependent upon Re for low Re, as shown by solutions of the Navier–Stokes equations. Chen [197] presented the following equation for $K(\infty)$, derived from his integral method results:

$$K(\infty) = 1.20 + (38/Re) \tag{193}$$

7. HYDRODYNAMIC ENTRANCE LENGTH L_{hy}^+

The hydrodynamic entrance length L_{hy}^+, defined on p. 41 (u_{max} approaching $0.99u_{max,fd}$), is constant when the boundary layer type momentum equation (8) is solved. Hornbeck [173] and Liu [182] obtained L_{hy}^+ as 0.056 and 0.054, respectively. Kays [7] reports L_{hy}^+ as 0.050, where hc defined it as the duct length required to achieve the tube centerline velocity to approach 98% of the fully developed value. The values of L_{hy}^+ obtained by different investigators are compared by Christiansen and Lemmon [174] and Friedmann et al. [187].

L_{hy}^+ is a function of Re for low Reynolds number flows as found by solving the complete set of Navier–Stokes equations. Atkinson et al. [257] presented equations that related L_{hy}^+ to Re for the circular tube and parallel plates for the case of uniform entrance velocity profile (Fig. 14a). These equations are a linear combination of creeping flow[†] and boundary layer type solutions. The creeping flow solution was obtained by minimization of viscous dissipation using a finite element method. For the circular tube, they presented

$$L_{hy}/D_h = 0.59 + 0.056\,Re \tag{194}$$

The L_{hy}^+ predicted by the above equation is somewhat higher than L_{hy}^+ observed experimentally, where the definitions of both L_{hy}^+ are the same (a dimensionless duct length required to achieve u_{max} as $0.99u_{max,fd}$). The tabulated values of L_{hy}^+ by Friedmann et al. [187], which are based on the solution of the Navier–Stokes equations for a uniform entrance velocity profile (Fig. 14a), are in better agreement with the experimental values. The

---

[†] See footnote on p. 91 for the definition of creeping flow.

results of Friedmann *et al.* are best approximated by Chen [197] as

$$\frac{L_{\text{hy}}}{D_{\text{h}}} = \frac{0.60}{0.035\,\text{Re} + 1} + 0.056\,\text{Re} \tag{195}$$

As mentioned in the preceding section, L_{hy} for the initial condition of Fig. 14c is shorter than that for the initial condition of Fig. 14a. $L_{\text{hy}}/D_{\text{h}}$ approaches 0.31 and 0.59 as Re tends to 0 for the initial conditions of Fig. 14c and 14a, respectively. A plot of $2L_{\text{hy}}/D_{\text{h}}$ vs. Re for the Fig. 14c conditions is presented by Vrentes *et al.* [186].

For the case of a uniform entrance velocity profile (Fig. 14a), it is observed that the entrance length is about 0.6 tube diameters for Re $\simeq$ 0 (as in blood vessels). The dimensional L_{hy} increases linearly with Re for laminar flow in engineering applications of Re > 0, because it requires a longer duct length to accelerate the fluid particles to achieve fully developed velocities at increasing Reynolds number.

C. Thermally Developing and Hydrodynamically Developed Flow

Thermal entrance problems with a developed velocity profile are divided into five categories depending upon the thermal boundary conditions: (1) specified axial wall temperature distribution with zero wall thermal resistance, which includes the (T) problem; (2) the (T3) problem; (3) the (T4) problem; (4) specified axial wall heat flux distribution, which includes the (H) and (H5) problems; and (5) conjugated problems.

1. Specified Axial Wall Temperature Distribution

For this class of problems, the temperature distribution is specified along the tube wall. The wall thermal resistance normal to the fluid flow direction is assumed to be zero. This class is further subdivided as the Graetz problem, the Lévêque problem, extensions to the Graetz problem, and problems with arbitrary axial wall temperature distribution. A separate discussion follows on each of these problems.

a. *Graetz Problem*

As mentioned in Chapter I, the study of heat transfer in laminar flow through a closed conduit was first made by Graetz [1] in 1883 and independently by Nusselt [2] in 1910. They considered an incompressible fluid flowing through a circular tube, with constant physical properties, having a fully developed laminar velocity profile and a developing laminar temperature profile. The tube was maintained at a constant and uniform temperature (i.e., (T)) different from the uniform temperature of the fluid at the entrance. Axial heat conduction, viscous dissipation, flow work, and thermal energy sources within the fluid were neglected. The resulting energy equation is

$$\frac{\partial^2 t}{\partial r^2} + \frac{1}{r}\frac{\partial t}{\partial r} = \frac{u}{\alpha}\frac{\partial t}{\partial x} \tag{196}$$

with initial and boundary conditions

$$t = t_e = \text{constant} \quad \text{for} \quad x \leq 0 \tag{197}$$

$$t = t_w = \text{constant} \quad \text{at} \quad r = a \tag{198}$$

$$\frac{\partial t}{\partial r} = 0 \quad \text{at} \quad r = 0 \tag{199}$$

and the velocity profile u is given by Eq. (169). The above problem is now well-known as the *Graetz problem*, but is sometimes also referred to as the *Graetz–Nusselt problem*. If the duct in question is not a circular tube, the problem is usually characterized as the Graetz-type problem.

The closed-form solution to the Graetz problem has been obtained primarily by two methods: (a) Graetz method and (b) Lévêque method. The Graetz method uses the separation of variables technique and as a result the governing differential equation is reduced to the Sturm–Liouville type. The solution is then obtained in the form of an infinite series expansion in terms of eigenvalues and eigenfunctions. The unattractive feature of this approach is that the number of terms required for a desired accuracy increases sharply as x^* approaches zero. The Lévêque method employs the similarity transformation technique. The solution to the resulting equation is valid only near the $x^* = 0$ region, leading to large errors far from the $x^* = 0$. The Lévêque method has been extended by a perturbation analysis to obtain an accurate solution for intermediate values of x^*. The Graetz and Lévêque solutions are discussed next.

The solution to the Graetz problem, the fluid temperature distribution, is presented in an infinite series (known as a *Graetz series*) form:

$$\theta = \frac{t - t_w}{t_e - t_w} = \sum_{n=0}^{\infty} C_n R_n \exp(-2\lambda_n^2 x^*) \tag{200}$$

The dimensionless fluid bulk mean temperature θ_m and local and mean Nusselt numbers $\text{Nu}_{x,\text{T}}$ and $\text{Nu}_{m,\text{T}}$ are expressed as

$$\theta_m = \frac{t_m - t_w}{t_e - t_w} = 8 \sum_{n=0}^{\infty} \frac{G_n}{\lambda_n^2} \exp(-2\lambda_n^2 x^*) \tag{201}$$

$$\text{Nu}_{x,\text{T}} = \frac{\sum_{n=0}^{\infty} G_n \exp(-2\lambda_n^2 x^*)}{2\sum_{n=0}^{\infty} (G_n/\lambda_n^2) \exp(-2\lambda_n^2 x^*)} \tag{202}$$

$$\text{Nu}_{m,\text{T}} = \frac{1}{4x^*} \ln\left(\frac{1}{\theta_m}\right) \tag{203}$$

Here λ_n, R_n, and C_n are eigenvalues, eigenfunctions, and constants, respectively, and $G_n = -(C_n/2)R_n{}'(1)$, as listed in Table 12.

TABLE 12

CIRCULAR DUCT: INFINITE SERIES SOLUTION
FUNCTIONS FOR THE GRAETZ PROBLEM
(BASED ON THE RESULTS OF BROWN [260])

n	λ_n	G_n
0	2.70436 44199	0.74877 4555
1	6.67903 14493	0.54382 7956
2	10.67337 95381	0.46286 1060
3	14.67107 84627	0.41541 8455
4	18.66987 18645	0.38291 9188
5	22.66914 33588	0.35868 5566
6	26.66866 19960	0.33962 2164
7	30.66832 33409	0.32406 2211
8	34.66807 38224	0.31101 4074
9	38.66788 33469	0.29984 4038
10	42.66773 38055	0.29012 4676

Graetz and Nusselt obtained only the first two and three terms, respectively, of the infinite series solution for the fluid temperature and the local Nusselt number as a function of dimensionless axial distance x^*, a grouping sometimes known as the *Graetz number*. The dimensionless independent variable for this problem is usually designated as the Graetz number Gz. However, to avoid confusion with various definitions used in the literature, the dimensionless axial distance x^*, instead of Gz, is used in the above expressions. Refer to p. 50 for further clarification of the Graetz number.

An excellent review of the earlier work on the Graetz problem, with all the mathematical details, has been made by Drew [3] and Jakob [261].

Abramowitz [262] and Lipkis [263] employed a fairly rapidly converging series solution for the Graetz problem, and obtained the first five eigenvalues and constants. Asymptotic expressions were presented for the higher eigenvalues and constants. Sellers *et al.* [25] independently extended the original work of Graetz by determining the first ten eigenvalues and constants, and presented asymptotic formulas for the higher eigenvalues and constants. They considered the (T) boundary condition, arbitrary axial wall temperature (specialized to linear variations in the wall temperature), and prescribed wall heat flux boundary conditions. The (H) heat transfer problem was worked out by an inversion method. Brown [260] provided a comprehensive literature survey for the Graetz problem. He evaluated and tabulated more accurately (10 decimal point accuracy) the first 11 eigenvalues and constants for the Graetz problem. He also presented the first six eigenfunctions. Larkin [264] extended the tabulation of Brown by presenting the sixth through fifteenth eigenfunctions. Table 12 is based on Brown's results.

Munakata [265] substituted the asymptotic formulas of Sellers *et al.* [25] in the expression of θ_m, Eq. (201), and converted the subsequent infinite series into an integral formula. The integration was carried out approximately by a Taylor series expansion about $x^* = 0$. The subsequent formula for $\Phi_{x,T}$, which is valid for very small values of x^*, has the first term identical to that of the Lévêque solution, Eq. (214), and the second term as -1.21 instead of the exact value of -1.2. Thus Munakata bridged the gap between Graetz and Lévêque solutions.

Newman [266] analyzed the Graetz problem and provided the more precise implicit equations for the higher eigenvalues and constants. He also presented explicit formulas, derived from his implicit formulas, for the higher eigenvalues and constants:

$$\lambda_n = \lambda + S_1 \lambda^{-4/3} + S_2 \lambda^{-8/3} + S_3 \lambda^{-10/3} + S_4 \lambda^{-11/3} + O(\lambda^{-14/3}) \quad (204)$$

$$G_n = \frac{C}{\lambda_n^{1/3}} \left[1 + \frac{L_1}{\lambda^{4/3}} + \frac{L_2}{\lambda^{6/3}} + \frac{L_3}{\lambda^{7/3}} + \frac{L_4}{\lambda^{10/3}} + \frac{L_5}{\lambda^{11/3}} + O(\lambda^{-4}) \right] \quad (205)$$

where

$$\lambda = 4n + \tfrac{8}{3}, \qquad n = 0, 1, 2, \ldots \quad (206)$$

$$S_1 = 0.159152288, \qquad S_2 = 0.0114856354, \qquad S_3 = -0.224731440$$

$$S_4 = -0.033772601, \qquad C = 1.012787288$$

$$L_1 = 0.144335160, \qquad L_2 = 0.115555556, \qquad L_3 = -0.21220305$$

$$L_4 = -0.187130142, \qquad L_5 = 0.0918850832 \quad (207)$$

Lauwerier [267] provided the first three terms of λ_n, although with a wrong sign for S_2. Sellers et al. [25] provided only the first term of Eqs. (204) and (205).

Kuga [268] solved the Graetz problem by reducing the Sturm–Liouville type governing differential equation to a Fredholm integral equation, and obtained numerically the first ten eigenvalues and constants, and six eigenfunctions.

Tao [205] solved the Graetz problem by a variational method, including the effects of viscous dissipation and thermal energy sources. Savkar [206] solved the Graetz problem by formulating a variational problem in the Laplace transformed domain, used in conjunction with the Ritz–Galerkin technique. $\Phi_{x,T}$ obtained by the second approximation are in fair agreement with the Graetz solution for $x^* > 0.005$. Note that the abscissas of Figs. 5 and 6 of Savkar [206] are mislabeled as x^* instead of $1/x^*$.

Grigull and Tratz [34] solved the Graetz problem numerically by a finite difference method. The dimensionless temperature distribution and local and mean Nusselt numbers were presented graphically as a function of x^*. For $x^* \geq 0.001$, they approximated $Nu_{x,T}$ by Eq. (217). They also presented an approximate equation for $Nu_{m,T}$. However, the calculated values of $Nu_{m,T}$ from their equation are as much as 16% low compared to those in Table 13.

Koyama *et al.* [269] experimentally verified the temperature distribution for the Graetz problem for Newtonian and non-Newtonian fluids.

Javeri [207] analyzed the Graetz problem by the Galerkin-Kantorowich method of variational calculus. His $Nu_{x,T}$ for $x^* > 0.0005$ are within 1% of those in Table 13. The accuracy degenerates for lower values of x^*.

TABLE 13

CIRCULAR DUCT: θ_m, $Nu_{x,T}$, $Nu_{m,T}$, AND $N_T(x)$ (THE GRAETZ
SOLUTION) FOR A FULLY DEVELOPED VELOCITY PROFILE
(FROM SHAH [270])

x^*	θ_m	$Nu_{x,T}$	$Nu_{m,T}$	$N_T(x)$
0.000001	0.99936	106.538	160.358	0.5382(−4)
0.0000015	0.99916	92.935	139.947	0.7052(−4)
0.000002	0.99898	84.341	127.051	0.8542(−4)
0.000003	0.99867	73.549	110.854	0.1119(−3)
0.000004	0.99839	66.731	100.621	0.1356(−3)
0.000005	0.99814	61.877	93.334	0.1573(−3)
0.000006	0.99790	58.170	87.769	0.1776(−3)
0.000007	0.99767	55.208	83.322	0.1968(−3)
0.000008	0.99745	52.763	79.651	0.2151(−3)
0.000009	0.99725	50.695	76.545	0.2326(−3)
0.00001	0.99705	48.914	73.869	0.2496(−3)
0.000015	0.99614	42.614	64.406	0.3269(−3)
0.00002	0.99534	38.637	58.429	0.3958(−3)
0.00003	0.99391	33.645	50.925	0.5184(−3)
0.00004	0.99264	30.495	46.186	0.6276(−3)
0.00005	0.99147	28.254	42.813	0.7280(−3)
0.00006	0.99039	26.544	40.238	0.8217(−3)
0.00007	0.98937	25.178	38.181	0.9102(−3)
0.00008	0.98839	24.051	36.483	0.9945(−3)
0.00009	0.98746	23.099	35.047	0.1075(−2)
0.0001	0.98657	22.275	33.815	0.1154(−2)
0.00015	0.98249	19.381	29.442	0.1509(−2)
0.0002	0.97888	17.558	26.685	0.1825(−2)
0.0003	0.97251	15.277	23.228	0.2305(−2)
0.0004	0.96688	13.842	21.049	0.2883(−2)
0.0005	0.96175	12.824	19.501	0.3338(−2)
0.0006	0.95698	12.050	18.321	0.3763(−2)
0.0007	0.95250	11.433	17.379	0.4162(−2)
0.0008	0.94826	10.926	16.603	0.4542(−2)
0.0009	0.94420	10.498	15.948	0.4905(−2)
0.001	0.94032	10.130	15.384	0.5254(−2)
0.0015	0.92276	8.8404	13.398	0.6837(−2)
0.002	0.90736	8.0362	12.152	0.8231(−2)
0.003	0.88057	7.0432	10.599	0.1067(−1)
0.004	0.85723	6.4296	9.6280	0.1279(−1)
0.005	0.83622	6.0015	8.9432	0.1471(−1)
0.006	0.81693	5.6812	8.4251	0.1646(−1)
0.007	0.79899	5.4301	8.0145	0.1809(−1)
0.008	0.78215	5.2269	7.6783	0.1961(−1)
0.009	0.76623	5.0584	7.3963	0.2104(−1)
0.01	0.75111	4.9161	7.1552	0.2239(−1)
0.015	0.68436	4.4406	6.3211	0.2821(−1)
0.02	0.62803	4.1724	5.8146	0.3284(−1)
0.03	0.53487	3.8942	5.2145	0.3961(−1)
0.04	0.45901	3.7689	4.8668	0.4391(−1)
0.05	0.39530	3.7100	4.6406	0.4653(−1)
0.06	0.34100	3.6820	4.4829	0.4805(−1)
0.07	0.29438	3.6688	4.3674	0.4890(−1)
0.08	0.25424	3.6624	4.2796	0.4937(−1)
0.09	0.21961	3.6595	4.2109	0.4962(−1)
0.1	0.18971	3.6580	4.1556	0.4976(−1)
0.15	0.09129	3.6568	3.9895	0.4990(−1)
0.2	0.04393	3.6568	3.9063	0.4990(−1)

The Graetz problem is one of the fundamental problems for internal flow convection heat transfer. In addition to its great practical importance, it has induced many applied mathematicians to apply and test different mathematical methods and approaches to solve the same problem. Consequently, using the best information available, Table 12 and Eqs. (204)–(207), the numerical values for the Graetz solution, in terms of θ_m, $Nu_{x,T}$, $Nu_{m,T}$ and $N_T(x)$ as a function of x^*, were accurately established by Shah [270], and are presented in Table 13. Equations (201) through (203) were solved using the first 121 terms of the series; λ_n and G_n up to $n = 10$ were taken from Table 12, and from $n = 11$ to 120 were determined from Eqs. (204) and (205). $N_T(x)$ was calculated using its definition, Eq. (130). The results are listed in Table 13 for $x^* > 10^{-4}$. For $x^* \leq 10^{-4}$, the extended Lévêque solution was used as described in the following subsection. This is because even the first 121 terms of the series are not sufficient in this range of x^*. These $Nu_{x,T}$ and $Nu_{m,T}$ are plotted in Fig. 15. The related variables $\Phi_{x,T}$, $\Phi_{m,T}$, and $Nu_{a,T}$ for the Graetz problem can be evaluated, using the results of Table 13, from Eqs. (123), (124), and (116), respectively.

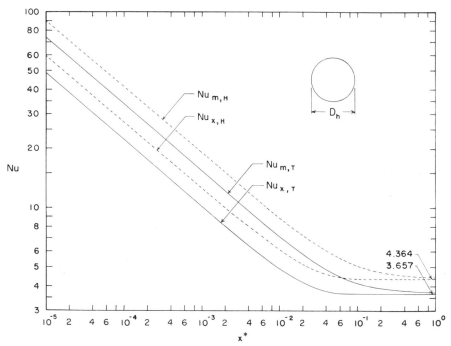

FIG. 15. Circular duct: $Nu_{x,T}$, $Nu_{m,T}$, $Nu_{x,H}$, and $Nu_{m,H}$ for a fully developed velocity profile (from Tables 13 and 18).

The thermal entrance length L_{th}^* defined by Eq. (128) and the incremental heat transfer number $N_T(x)$ defined by Eq. (130) at $x^* = \infty$ are as follows [270]:

$$L_{th,T}^* = 0.0334654 \qquad (208)$$

$$N_T(\infty) = 0.0499024 \qquad (209)$$

b. *Lévêque Solution*

The series of Eq. (202) converges uniformly for all nonzero x^*, but the convergence is exceedingly slow as x^* approaches zero. As mentioned in the preceding subsection, the first 121 terms of the series were insufficient to accurately determine the Graetz solution of Table 13 for $x^* < 10^{-4}$. Lévêque [3,199] alleviated this difficulty in 1928, when only the first three terms of the Graetz series had been evaluated. He employed the "flat plate" solution as an asymptotic approximation near the point where the step change in temperature occurred.

The principal idealization made in the Lévêque analysis is as follows: In the downstream region very close to the point of step change, the temperature changes are confined to a boundary layer that is thin compared to the momentum boundary layer. The following three idealizations result from the foregoing principal idealization. (1) In the thermal boundary layer region, it is assumed that the velocity distribution is linear and has the same slope as that at the wall with $u = 0$ at wall. Thus, the axial velocity u in Eq. (196) is replaced by a linear velocity distribution in the radial direction having the same slope as the actual velocity variation at the wall. (2) For the region close to the wall, the effect of curvature is small; hence the term $(\partial t/\partial r)/r$ is omitted from Eq. (196). The resulting energy equation for the thin boundary layer in a concentric annular duct is

$$\alpha \frac{\partial^2 t}{\partial r^2} = \left[\left(\frac{\partial u}{\partial r} \right)_w (r - r_w) \right] \frac{\partial t}{\partial x} \qquad (210)$$

For the circular tube, $r_w = a$ and $(\partial u/\partial r)_w = 4u_m/a$.

The solution to this equation is found by the method of similarity transformation, where it is further assumed that (3) far from the wall, the fluid temperature approaches the inlet temperature of the fluid. Hence, the symmetry boundary condition at the centerline, Eq. (199), is replaced by $t = t_e$ far from the wall. This latter condition is referred to as the *infinity condition*. This idealization also implies that the temperature gradient approaches zero far from the wall. Consequently, the Lévêque-type solution does not allow a distinction between different kinds of boundary conditions at the opposite wall of a multiply connected duct. With these idealizations, Lévêque obtained the following solution for the circular tube:

$$\theta = \frac{t_w - t(x^*, r/a)}{t_w - t_e} = \frac{1}{\Gamma(\frac{4}{3})} \int_0^\eta \exp(-\sigma^3)\, d\sigma \qquad (211)$$

with the similarity variable $\eta = (1 - r/a)/(9x^*)^{1/3}$ and σ a dummy variable of integration. Lundberg et al. [42] obtained the Lévêque-type solution for concentric annular ducts.

The Lévêque solution is valid only in a very restricted thermal entrance region where the depth of temperature penetration is of the same order of magnitude as the hydrodynamic boundary layer over which the velocity distribution may be considered linear.

Mercer [271], Worsøe-Schmidt [200], and Newman [202] extended the Lévêque solution by a perturbation method solving Eq. (196) directly, thus relaxing the first two idealizations of the Lévêque method, the linear velocity profile and neglect of the curvature effect.

The corresponding solution is valid for the intermediate values of x^* where both the Graetz and Lévêque solutions are not accurate. The extended Lévêque solution is presented in series form as

$$\theta(\xi, \eta) = \frac{t - t_w}{t_e - t_w} = \sum_{n=0}^{N} \xi^n \theta_n(\eta) \qquad (212)$$

$$\Phi_{x,T} = 2 \sum_{n=0}^{N} \xi^{(n-1)/3} \theta_n'(0) \qquad (213)$$

where $\xi = (9x^*)^{1/3}$ and $\eta = (1 - r/a)/(9x^*)^{1/3}$ for the circular tube. This series, Eq. (212), is generally referred to as the *Lévêque series*. Equation (211), obtained by Lévêque, represents the first term $\theta_0(\eta)$ of Eq. (212).

Mercer [271] obtained θ_0 and θ_1 of Eq. (212) in a closed form, θ_2 and θ_3 numerically, and tabulated these variables as a function of η. However, he did not compute the derivatives at the wall, $\theta_n'(0)$, and did not present $\Phi_{x,T}$. Worsøe-Schmidt [200] also presented $(1 - \theta_0)$ and $-\theta_1$ in a closed form and obtained a finite difference solution for higher n. He tabulated $-\theta_n'(0)$ for n up to 6. Newman [202] also solved the same problem independently and obtained θ_n and $\theta_n'(0)$ in a closed form for n up to 2.

Nunge et al. [201] investigated the limitations imposed by the idealizations of the Lévêque-type solutions. Lévêque [199] and Lundberg et al. [42] considered the linear velocity profile and neglected the effect of curvature in the energy equation for the region close to a step change in the boundary condition; Worsøe-Schmidt [200] considered curvature and a fully developed velocity profile, while Nunge et al. [201] assumed curvature and a linear velocity profile. They concluded that the improvement brought about in the original Lévêque solution by including the effect of curvature is important; however the linear approximation to the velocity profile near the wall is excellent for concentric annular ducts.

For the concentric annular duct family, a very accurate thermal entrance solution can be obtained for x^* up to 0.01 with the first seven terms in the above series obtained by Worsøe-Schmidt.

Doshi [272] further refined the extended Lévêque solution by relaxing the infinity condition. He thus solved Eqs. (196)–(199) by the perturbation method, with the introduction of a new transverse coordinate that varied from 0 to ∞ when the physical radial coordinate varied from 0 to a. He obtained the first three terms of Eqs. (212) and (213). His temperature profiles θ_1 and θ_2 were different and the wall gradients $\theta_0'(0)$, $\theta_1'(0)$, and $\theta_2'(0)$ were identical to those of Worsøe-Schmidt and Newman.

The dimensionless wall heat fluxes $\Phi_{x,T}$ and $\Phi_{m,T}$ defined by Eqs. (121) and (122), based on the extended Lévêque solution by Worsøe-Schmidt, are expressed as

$$\Phi_{x,T} = 1.076732109(x^*)^{-1/3} - 1.2 - 0.374094620(x^*)^{1/3}$$
$$- 0.37323(x^*)^{2/3} - 0.48546x^* - 0.72749(x^*)^{4/3} - 1.1986(x^*)^{5/3}$$

$$(214)$$

$$\Phi_{m,T} = 1.615098163(x^*)^{-1/3} - 1.2 - 0.280570965(x^*)^{1/3}$$
$$- 0.22394(x^*)^{2/3} - 0.24273x^* - 0.31178(x^*)^{4/3} - 0.44948(x^*)^{5/3}$$

$$(215)$$

θ_m, $Nu_{x,T}$, and $Nu_{m,T}$ are then determined by the relationships of Eqs. (123) and (124).

Based on Eqs. (214) and (215), θ_m, $Nu_{x,T}$, and $Nu_{m,T}$ were calculated by Shah [270] for the Graetz problem for $10^{-6} \leq x^* \leq 2 \times 10^{-2}$. $Nu_{x,T}$ of Graetz and Lévêque solutions are identical up to five or more digits for $0.00002 < x^* < 0.005$. The results for the Ⓣ problem in Table 13 for $x^* \leq 10^{-4}$ are based on the extended Lévêque solution.

For design purposes, the foregoing thermal entry length solutions are needed in a simplified equation form. The following equations are formulated by Shah [270][†] to predict the results of Table 13 within $\pm 3\%$ accuracy. The temperature difference used in the definition of $Nu_{m,T}$ is $(\Delta t)_{lm}$ as shown by Eq. (113). However, Colburn [57] suggested the use of the mean Nusselt

---

[†] Equation (217) is from Grigull and Tratz [34], with Nu_T modified as 3.657 from their value of 3.655.

number $Nu_{a,T}$, based on the arithmetic average temperature difference, to correlate the experimental data for heat exchanger surfaces in laminar flow. Hence, equations for $Nu_{a,T}$ are also presented:

$$Nu_{x,T} = \begin{cases} 1.077(x^*)^{-1/3} - 0.7 & \text{for } x^* \leq 0.01 \quad (216) \\ 3.657 + 6.874(10^3 x^*)^{-0.488} e^{-57.2x^*} & \text{for } x^* > 0.01 \quad (217) \end{cases}$$

$$Nu_{m,T} = \begin{cases} 1.615(x^*)^{-1/3} - 0.7 & \text{for } x^* \leq 0.005 \quad (218) \\ 1.615(x^*)^{-1/3} - 0.2 & \text{for } 0.005 < x^* < 0.03 \quad (219) \\ 3.657 + \dfrac{0.0499}{x^*} & \text{for } x^* \geq 0.03 \quad (220) \end{cases}$$

$$Nu_{a,T} = \begin{cases} 1.615(x^*)^{-1/3} - 0.7 & \text{for } x^* \leq 0.005 \quad (221) \\ 1.615(x^*)^{-1/3} - 0.2 & \text{for } 0.005 \leq x^* < 0.02 \quad (222) \\ \dfrac{0.5}{x^*} & \text{for } x^* \geq 0.02 \quad (223) \end{cases}$$

Additionally, for large x^*, the θ_m, $\Phi_{x,T}$, and $\Phi_{m,T}$ approach asymptotically the values 0, 0, and $0.25/x^*$, respectively.

The values calculated from Eqs. (216)–(223) are higher ($+$) or lower ($-$) than those of Table 13 by ± 0.5, ± 0.2, ± 2.2. ± 3.0, $+2.0$, ± 2.0, ± 3.7, and $+2.1\%$, respectively. Except for the endpoints of the range, the error is much less than specified.

McAdams [56] suggested the following equation for $Nu_{a,T}$.

$$Nu_{a,T} = 1.62(x^*)^{-1/3} \quad (224)$$

Colburn [57] modified the constant 1.62 to 1.50. Colburn's equation is quite adequate (with at most 3.0% error when compared with the exact solution) for $5 \times 10^{-4} < x^* < 2 \times 10^{-2}$.

Equations (216) and (217) together cover a complete range of x^*. Instead of these two equations, Churchill and Ozoe [273] proposed a single relation

$$\frac{Nu_{x,T} + 1.7}{5.357} = \left[1 + \left(\frac{388}{\pi} x^*\right)^{-8/9}\right]^{3/8} \quad (225)$$

They formulated this expression based on asymptotic formulas for small and large values of x^*. This equation predicts $Nu_{x,T}$ lower than those of Table 13 for $x^* \leq 0.007$ with a maximum error of 4.7% at $x^* = 0.001$. It predicts $Nu_{x,T}$ higher than those of Table 13 for $x^* > 0.007$ with a maximum error of 9.1% at $x^* = 0.04$.

c. *Extension of the Graetz Problem with Nonuniform*
   Inlet Temperature Profiles, Nonparabolic Velocity Profiles,
   and Internal Thermal Energy Generation

Hicken [274] investigated the Graetz problem with nonuniform inlet temperature. He considered five different one-period sinusoidal fluid temperature profiles at $x^* = 0$. He tabulated the eigenvalues and constants, and presented the dimensionless fluid bulk mean temperature $(1 - \theta_m)$ as a function of x^* for these five cases. For the case of heat transfer from the fluid to the wall (i.e., $t_{m,e} > t_w$), Hicken found the following results. For the case of sinusoidal variations in the inlet fluid temperature with the minimum temperature at the center of the pipe, the Nusselt numbers were higher than those for the uniform inlet temperature case. Decreasing this temperature at the pipe center (for the same $t_{m,e}$) increased the magnitude of the Nusselt number in the thermal entrance region. For the fluid temperature having maximum value at the pipe center, the reverse effect on Nusselt number was found.

Instead of using the parabolic velocity distribution, Lyche and Bird [275] and Whiteman and Drake [276] solved the Graetz problem for non-Newtonian fluids having a general power law velocity distribution.

For a fluid having a vanishing viscosity with finite thermal conductivity and Prandtl number, if the flow is radically different from the parabolic profile at the inlet, the velocity profile will be maintained over a considerable distance from the entrance. To investigate the effect of the nonuniform inlet velocity profiles, Barrow and Humphreys [277] analyzed the Graetz problem with slug, inverted conical, and inverted parabolic velocity profiles, instead of the laminar parabolic profile, for the thermal entrance region. They graphically presented $\mathrm{Nu}_{x,T}$ as a function of x^* for different velocity profiles. Their results show that, for a given flow rate, the increase in velocities near the wall results in a shortening of the thermal entry length and an increase in the heat transfer coefficient, as expected.

The Graetz problem with internal thermal energy sources has been solved by Topper [278] and Toor [279].

d. *Extended Graetz Problem with Finite Viscous Dissipation*

While the effect of viscous dissipation is negligible for low-velocity gas flows, it is important for high-velocity gas flows and liquids. The effect of viscous dissipation is of the same order of magnitude as the effect of flow work for gases. The effect of viscous dissipation on the Ⓣ thermal entrance problem has been investigated by Brinkman [61] and Ou and Cheng [228], considering small fluid temperature variations so that the viscosity is treated as constant.

The Graetz problem, as outlined on p. 99, is valid, when the fluid flowing through the tube is being either heated or cooled. However, when viscous dissipation is included (Brinkman number, Br $\neq$ 0), viscous heating effects are produced and the solutions are different depending upon gross fluid heating $(t_w > t_e)$ or cooling $(t_w < t_e)$ situations. Fluid heating or cooling results in positive or negative values of Br, respectively. As will be discussed in detail on p. 129, the thermal energy generated due to viscous dissipation is constant for fully developed flow with constant properties, and does not depend upon the wall heat transfer rate. Locally, the thermal energy generated due to viscous dissipation, Eq. (247), is more pronounced near the wall region compared to the central region of the tube. Hence, the fluid temperature near the wall region is affected more than the fluid bulk mean temperature t_m (as will be discussed in detail on p. 130).

Brinkman [61] solved the ⓉT thermal entrance problem considering the wall temperature maintained at the fluid inlet temperature. Brinkman's main interest was to determine the fluid temperature distribution for capillary flow in the presence of finite viscous dissipation. He presented graphically the dimensionless fluid temperature $(t - t_e)/(4\mu u_m^2/g_c Jk)$ across the tube cross section for $x^* = 0.0015$, 0.005, and 0.15. He found, as expected, the fluid temperature highest near the tube wall where the rate of shear is highest.

Ou and Cheng [228][†] studied the Graetz problem with finite viscous dissipation, Br $\gtrless$ 0. They considered the boundary condition of constant wall temperature different from the fluid entrance temperature and obtained a solution by the eigenvalue method. They tabulated the first ten constants associated with the problem and presented graphically θ_m and $\mathrm{Nu}_{x,T}$ as functions of x^* and Br.

For the fluid heating situation (Br > 0, $t_w > t_e$), the fluid bulk mean temperature rises axially near the entrance due to two effects: (1) heat transfer from the wall to the fluid, and (2) viscous heating effect as shown in Fig. 16. This figure shows the fluid temperature profile development in the thermal entrance region relative to t_m and t_w. For the constant wall temperature and Br > 0 situation, heat is transfered from the wall to the fluid near the tube entrance (Fig. 16a,b). Viscous dissipation affects more strongly the fluid temperature near the wall region. Hence, at some $x^* > 0$, while $t_w > t_m$, the temperature gradient at the wall becomes zero, as in Fig. 16c, and the heat transfer flux q'' as a consequence also goes to zero. Beyond this location, the temperature gradient at the wall is negative, producing q'' from the fluid to the wall. However, $t_m < t_w$, as shown in Fig. 16d, and this results in negative h_x and $\mathrm{Nu}_{x,T}$ (see p. 46 for the definition of h_x). Since t_m is continuously increasing due to viscous dissipation, eventually $t_m(x) = t_w(x)$, as in Fig. 16e. This condition results in an infinite $\mathrm{Nu}_{x,T}$. Beyond this location (referred to as a point of singularity), $t_m > t_w$ and heat transfers from the fluid to the

[†] The Brinkman number defined by Ou and Cheng, in the present terminology, is -Br.

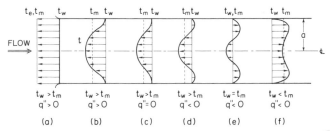

FIG. 16. Fluid temperature development in the thermal entrance region for the (T) boundary condition for Br > 0. The fluid is being heated at the inlet. The arrows show relative magnitudes of the fluid temperature in comparison to the wall temperature. $t < t_w$ for the arrow pointing left; $t > t_w$ for the arrow pointing right. The dotted line represents the fluid bulk mean temperature. Temperature profiles are drawn for increasing x^* from left to right.

wall (Fig. 16f). Ou and Cheng [228] showed that the locations for $q'' = 0$ and $t_m = t_w$ occur at decreasing x^* for increasing magnitudes of Br. For Br > 0, except for the point of singularity, $Nu_{x,T}$ decreases for increasing x^* at a given Br; also $Nu_{x,T}$ decreases for increasing Br at a given x^*.

For the fluid cooling situation (Br < 0, $t_w < t_e$), the heat transfer rate and Nusselt number in the entrance region depend upon the magnitude of Br. For Br < −6/5, the viscous heating effect is always predominant, and $t_m(x) > t_e > t_w$ throughout the entrance and fully developed regions. This results in $Nu_{x,T}$ monotonically decreasing with increasing x^*. At a given x^*, Ou and Cheng [228] showed that $Nu_{x,T}$ is higher with decreasing value of Br for Br < −6/5. However, for −6/5 < Br < 0, the cooling effect dominates over viscous heating in the "lower" x^* region; while the viscous heating effect dominates over cooling in the "higher" x^* region. This results in a minimum in the $Nu_{x,T}$ vs. x^* curve. The minimum occurs at increasing x^* for decreasing absolute values of Br. At a given x^*, $Nu_{x,T}$ decreases with decreasing absolute value of Br. Note that $t_e > t_m(x) > t_w$ for this case.

However, regardless of the heating and cooling situations, $Nu_{x,T}$ asymptotically approaches 9.6 for all values of Br for Br ≠ 0. This result needs an experimental verification.

e. Extended Graetz Problem with Fluid Axial Heat Conduction

This problem has been solved in considerable depth. The dimensionless group to account for axial heat conduction in the fluid is the Péclet number Pe. It will be shown that, except for the immediate neighborhood of $x^* = 0$, axial heat conduction within the fluid is negligible for Pe > 50.

The applicable energy equation for the extended Graetz problem with finite fluid axial heat conduction is Eq. (31) with $v = 0$ (as the velocity profile is considered fully developed). Three sets of initial and boundary conditions are considered for this problem; they are shown in Fig. 17. A

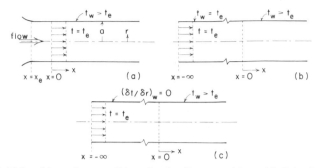

FIG. 17. Initial and boundary conditions for the Graetz problem with finite fluid axial heat conduction.

uniform temperature profile for the entering fluid is considered at $x = 0$ in Fig. 17a and at $x = -\infty$ in Fig. 17b,c. The wall for $-\infty \leq x \leq 0$ is isothermal in Fig. 17b and adiabatic in Fig. 17c.

In practice, heat exchangers are of finite lengths with headers at the inlet and outlet. If good mixing of fluid is obtained in the headers, the initial and boundary conditions of Fig. 17a would be a valid idealization. In the above formulation, $x = 0$ is the section along the tube length where the heat transfer at wall starts and the velocity profile is fully developed.

In laminar flow, the hydrodynamic entrance length L_{hy} is relatively long compared to that for the turbulent flow. If axial heat conduction is important, a considerable amount of thermal energy may be conducted into the hydrodynamically developing region $x_e < x < 0$, affecting the temperature distribution at $x = 0$. Thus, the initial condition of uniform temperature profile is not consistent with the inclusion of fluid axial heat conduction. For this situation, a mathematically more appropriate initial condition, having uniform temperature at $x = -\infty$, is shown in Fig. 17b. Another possible boundary condition, consistent with the inclusion of fluid axial heat conduction is shown in Fig. 17c. Here, the wall from $x = -\infty$ to $x = 0$ is considered adiabatic, in constrast to isothermal, as in Fig. 17b.

Millsaps and Pohlhausen [280], Singh [281], Shapovalov [282], Schmidt and Zeldin [66], Taitel and Tamir [283], and Kader [284] extended the Graetz problem by considering axial heat conduction within the fluid. They employed the initial and boundary conditions of Fig. 17a for the analysis.

Millsaps and Pohlhausen [280] and Singh [281] independently extended the Graetz problem by including axial heat conduction and viscous dissipation. Singh also included a prescribed internal thermal energy generation. These authors solved the associated Sturm–Liouville problem for the fluid temperature by expanding it in an infinite series of Bessel functions of zero order. They tabulated the first four eigenvalues and eigenfunctions for

Pe = 200 and 2000. Singh presented the first six eigenvalues by an approximate method (within 2%) for Pe = 2, 10, 20, 100, 200, 2000, and ∞. He tabulated θ_m and $Nu_{x,T}$ (as a function of $2x^*$) for Pe = 200 and ∞ for the case of zero viscous dissipation and no internal thermal energy generation. Singh also analyzed the same problem for slug flow. However, neither Millsaps and Pohlhausen nor Singh calculated the fluid bulk mean temperature or the Nusselt numbers for low Péclet numbers. Munakata [265] employed the results of these investigators and determined θ_m, $Nu_{x,T}$, $Nu_{m,T}$, and $Nu_{a,T}$ for Pe = 1 and 10. He showed that fluid axial heat conduction is not negligible for Pe < 10.

Shapovalov [282] included the effect of fluid axial heat conduction in the Graetz problem, and obtained the fluid temperature distribution in terms of degenerate hypergeometric functions.

Schmidt and Zeldin [66] solved the same problem by using a finite difference numerical method. They graphically presented $Nu_{x,T}$, $\Phi_{m,T}$, and θ_m as a function of $2x^*$ for Pe = 1, 10, 25, 100, 200, 10^4, and 10^6. Good agreement was found (1) between their results for Pe = 10^6 and the Graetz solution (Table 13) for $x^* \geq 0.0025$, and (2) between their results for Pe = 200 and $Nu_{x,T}$ of [280,281]. However, for $x^* < 0.0025$, their $Nu_{x,T}$ for Pe = 10^6 appears to be diverging from the Graetz solution (Table 13). For example, at $x^* = 0.0005$, their $Nu_{x,T}$ is 25% higher than $Nu_{x,T}$ in Table 13. Thus, Schmidt and Zeldin's results for $x^* < 0.0025$ are questionable. Their results for other Pe may also be questionable for low values of x^*. This may be due to the nature of the finite difference equations and numerical method at startup. In most practical applications of the thermal entrance solution, however, $x^* > 0.0025$, and Schmidt and Zeldin's results are valid.

For the fluid heating situation ($t_f < t_w$), when fluid axial heat conduction is present, part of the energy transfered from the wall shows up as a local enthalpy rise of the fluid and the rest is conducted upstream through the fluid to the inlet header. Even though this thermal energy would preheat the incoming flow, it is excluded in the analysis because the initial condition of Fig. 17a does not allow preheating. Consequently, the fluid bulk mean temperature t_m at any cross section is lower and $(t_w - t_m)$, and the wall gradients are higher than for the local situation with no axial conduction (Pe = ∞), but with the same $t = t_e$ at $x = 0$. Thus for finite fluid axial heat conduction, $Nu_{x,T}$ (and hence $Nu_{m,T}$) increases for a specified x^* as Pe decreases. This means that the thermal entrance length $L_{th,T}^*$ also increases as Pe decreases. $Nu_{x,T}$ is infinite at $x^* = 0$ for all Péclet numbers because the fluid temperature gradient at the wall is infinite. $Nu_{x,T}$ approach the values of Table 6 for $x^* \to \infty$.

Another way to demonstrate the effect of axial heat conduction is to present the ratio of enthalpy rise of the fluid within the duct to total wall heat transfer as a function of x^* for different Pe. This ratio, designated as the dimensionless energy content of the fluid,[†] δ on p. 56, decreases with decreasing Pe as shown in Fig. 18. The corresponding tabular information is presented in Table 14 [252]. Figure 18 may be useful for the design of a high-effectiveness heat exchanger, as one can now establish the criteria as to when fluid axial heat conduction effects are negligible, e.g., at $x^* > 0.4$ and Pe > 200 its neglect results in less than 2% error. Much of the foregoing discussion for fluid heating ($t_f < t_w$) applies also to fluid cooling ($t_w < t_f$).

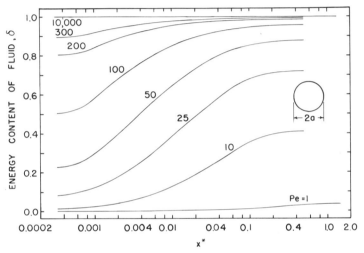

FIG. 18. Circular duct: Energy content of the fluid for the Graetz problem with finite fluid axial heat conduction (from Table 14).

Taitel and Tamir [283] analyzed the fluid axial heat conduction problem for the initial and boundary conditions of Fig. 17a by the integral method. They employed the second-, third-, and fourth-order polynomial approximations for the temperature profile and determined the necessary coefficients to present the solution in a closed form. They defined the heat transfer coefficient and subsequently the Nusselt number based on the *enthalpy rise (or drop)* of the fluid from the entrance. Consequently, their fully developed Nusselt number monotonically increases with Pe, ranging from 0 at Pe $= 0$ to 3.52 at Pe $= \infty$. The last value should have been 3.66, reflecting the order of accuracy obtained by the integral method. Note that if the Nusselt number were based on the *wall heat transfer*, Nu_T would have monotonically decreased from 4.18 at Pe $= 0$ to 3.66 at Pe $= \infty$ as shown in Table 6.

[†] Small δ and small Pe means large fluid axial heat conduction.

TABLE 14

CIRCULAR DUCT: ENERGY CONTENT OF THE FLUID FOR THE GRAETZ PROBLEM WITH
FINITE FLUID AXIAL HEAT CONDUCTION (FROM SCHMIDT [252])

Pe = 1		Pe = 10		Pe = 25		Pe = 300	
x*	δ	x*	δ	x*	δ	x*	δ
.000986	.000383	.000329	.01587	.000329	.0822	.000329	.8959
.00202	.000528	.000675	.02032	.00142	.1465	.000675	.9032
.00312	.000786	.00226	.04697	.00321	.2273	.00104	.9211
.00427	.001045	.00427	.0728	.00549	.2964	.00183	.9415
.00549	.001364	.00690	.1002	.00855	.3595	.00321	.9572
.02198	.00336	.01049	.1304	.01282	.4201	.00427	.9636
.03147	.00573	.01567	.1650	.01923	.4808	.00549	.9685
.05769	.00925	.02381	.2062	.02991	.5444	.00690	.9724
.2038	.01827	.03846	.2576	.05128	.6141	.00855	.9755
.4744	.03024	.07265	.3257	.1154	.6878	.01049	.9782
.7308	.0329	.1581	.3865	.5000	.7196	.01282	.9806
1.5000	.0342	.2436	.4018			.01567	.9826
		.5000	.4091			.01923	.9845
						.02381	.9862

Pe = 50		Pe = 100		Pe = 200		.02991	.9878
						.03846	.9893
x*	δ	x*	δ	x*	δ	.05128	.9907
						.07265	.9920
.000329	.2277	.000329	.5059	.000347	.8053	.08974	.9926
.000675	.2577	.000675	.5366	.000711	.8183	.1154	.9932
.00226	.4167	.00142	.636	.00150	.8704	.1581	.9936
.00427	.5164	.00226	.6973	.00238	.8978	.2436	.9940
.00690	.5900	.00321	.7405	.00338	.9152	.5000	.9941
.01049	.6498	.00427	.7718	.00450	.9275		
.01567	.7019	.00549	.7974	.00579	.9368	Pe = 10000	
.02381	.7496	.00690	.8185	.00901	.9506		
.03846	.7954	.00855	.8366	.01652	.9646	x*	δ
.05128	.8180	.01049	.8524	.02510	.9718		
.07265	.8402	.01282	.8665	.04054	.9780	.000329	.99988
.1154	.8605	.01923	.8912	.07658	.9834	.00226	.99994
.1581	.8688	.02991	.9128	.1667	.9866	.00617	.99997
.2436	.8746	.05128	.9324	.5270	.9875	.02137	.99998
.5000	.8772	.1154	.9497			.1581	.99999

Kader [284] employed the Lévêque method to analyze the Graetz
problem with finite fluid axial heat conduction for the initial and boundary
conditions of Fig. 17a. He obtained the solution in terms of $\xi = (9x^*)^{1/3}$
(for $x^* < 0.06$) so that his solution is valid for large Pe. He derived, with
a rigorous mathematical analysis, the first three terms of the heat flux
which are identical to those of Eqs. (214) and (215). Thus, Kader's analysis
confirms the effect of fluid axial heat conduction being negligible for large
Pe flows.

Bodnarescu [285], Hennecke [68], Bes [286], Jones [287], and Verhoff
and Fisher [288] also extended the Graetz problem by considering axial
heat conduction within the fluid. They employed the initial and boundary
conditions of Fig. 17b for the analysis. Bodnarescu [285] obtained the
solution in terms of hypergeometric functions.

Hennecke [68] analyzed the problem numerically by a finite difference
method. He presented graphically $Nu_{x,T}$, $0.5\Phi_{x,T}$, $\Phi_{m,T}$, $(x/a)\Phi_{m,T}$, and
$(1 - \theta_m)$ as a function of $2x^*$ for Pe = 1, 2, 5, 10, 20, and 50. The Nu_T (comparable to the Table 6 values) and $2L_{th,T}^*$ were also graphically presented as

a function of Pe. The $Nu_{x,T}$ of Hennecke are presented in Table 15 and Fig. 19.

While Schmidt and Zeldin [66] employed the initial and boundary conditions of Fig. 17a, Hennecke [68] employed the conditions of Fig. 17b, for which case, as for Fig. 17a, the heat is conducted upstream of $x = 0$ ($t_w > t_f \geq t_e$). However, in Fig. 17b, the temperature of fluid is not imposed

TABLE 15

Circular Duct: $Nu_{x,T}$ as a Function of x^* and Pe for the Initial and Boundary Conditions of Fig. 17b for a Fully Developed Velocity Profile (from the Graphical Results of Hennecke [68])

x*	$Nu_{x,T}$						
	Pe=1	2	5	10	20	50	∞
0.0005	–	–	–	–	34.0	18.7	12.82
0.001	–	–	–	46.2	21.6	13.6	10.13
0.002	–	–	50.7	24.5	13.4	9.6	8.04
0.003	–	–	35.1	17.3	10.8	8.0	7.04
0.004	–	68.9	27.4	13.8	9.0	7.1	6.43
0.005	–	55.0	21.9	11.3	7.8	6.5	6.00
0.01	65.0	30.0	12.2	7.1	5.6	5.1	4.92
0.02	32.9	15.8	7.1	5.0	4.4	4.2	4.17
0.03	22.9	11.4	5.5	4.3	4.0	3.9	3.89
0.04	17.4	9.2	4.9	4.0	3.8	3.8	3.77
0.05	14.4	7.8	4.5	3.9	3.72	3.71	3.71
0.1	8.5	5.3	3.9	3.70	3.67	3.66	3.66
0.2	5.5	4.3	3.77	3.70	3.67	3.66	3.66
0.3	4.7	4.0	3.77	3.70	3.67	3.66	3.66
0.4	4.5	3.92	3.77	3.70	3.67	3.66	3.66
0.5	4.3	3.92	3.77	3.70	3.67	3.66	3.66
1.0	4.03	3.92	3.77	3.70	3.67	3.66	3.66
2.0	4.03	3.92	3.77	3.70	3.67	3.66	3.66

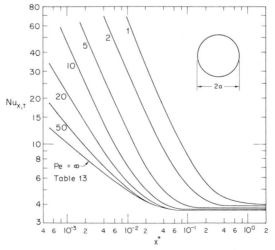

Fig. 19. Circular duct: $Nu_{x,T}$ as functions of x^* and Pe for the initial and boundary conditions of Fig. 17b for a fully developed velocity profile (from Table 15).

as constant at $x = 0$, and as a result of the upstream conduction, it would be higher than that in Fig. 17a. Consequently, t_m for specified Pe and x^* would also be higher for Fig. 17b compared to the Fig. 17a situation. Thus, for the specified Pe and x^*, the $(t_w - t_m)$ and the wall gradient for Fig. 17b would be lower and hence $Nu_{x,T}$ would be lower relative to the Fig. 17a situation. The difference between $Nu_{x,T}$ of Fig. 17a and 17b should increase with decreasing Pe. The results of Schmidt and Zeldin [66] and Hennecke [68] exhibit this trend. For the conditions of Fig. 17b, it should be emphasized that the effect of finite fluid axial heat conduction is to increase $Nu_{x,T}$ at a specified x^* as Pe decreases, as shown in Fig. 19. $Nu_{x,T}$ is infinite at $x^* = 0$ for all Péclet numbers, and has values corresponding to those of Table 6 at $x^* = \infty$. The thermal entrance length $L_{th,T}^*$ also increases as Pe decreases, ranging from 0.033 at Pe $= \infty$ to 0.5 at Pe $= 1$ [68].

Bes [286] solved the same problem (with Fig. 17b conditions) for the circular tube and parallel plates for both the Ⓣ and Ⓗ boundary conditions. He found that for Pe < 15, longitudinal conduction is significant for the circular tube. Unfortunately, he utilizes unfamiliar symbols and terminology, making an adequate evaluation difficult. Moreover, Bes's analysis appears to be in error, as mentioned on p. 134, because Nu_T does not approach those values in Table 6.

Jones [287] analyzed the same problem as Hennecke [68], Fig. 17b, by an eigenvalue approach. A closed-form solution was obtained by employing a two-sided Laplace transform. The eigenvalues were presented in the form of an asymptotic expansion in Péclet number, unlike Singh's approach [281], where the eigenvalues and eigenfunctions were needed for each Péclet number. Jones tabulated θ_m and $Nu_{x,T}$ for Pe $= 100, 200, 1000$, and 2000. For $x^* > 0.075$ and Pe > 200, the $Nu_{x,T}$ were approximated as

$$Nu_{x,T} = 3.657\{1 + 0.6072 \exp[4(p_1 - p_0)x^*]\} \qquad (226)$$

where the values of p_0 and p_1 are shown in the accompanying tabulation.

	Pe $= 200$	Pe $= 1000$	Pe $= 2000$
p_0	-3.65595	-3.65676	-3.65678
p_1	-22.2687	-22.3034	-22.3044

Jones' $Nu_{x,T}$ are lower than those of Singh [281], who considered the conditions of Fig. 17a. The difference is greater than 5% for $x^* < 0.025$. The results of Jones and those of Hennecke are both based on the conditions of Fig. 17b, but are not directly comparable because the results are computed for a quite different range of the Péclet number.

In the Jones and the Hennecke analyses, the solutions are complicated by the use of boundary conditions at minus infinity (Fig. 17b). Verhoff and

Fisher [288] alleviated the difficulty by employing an inverse-tangent transformation. As a result, the double infinite region of Fig. 17b was transformed into a region with finite boundaries. They then obtained a finite difference solution to the Graetz problem with finite fluid axial heat conduction for the boundary conditions of both Fig. 17b and 17c. Their results for the case of Fig. 17b are in close agreement with those Hennecke [68]. They found $Nu_{x,T}$ for the case of Fig. 17c lower than those for Fig. 17b for $x^* > 0$.

Tan and Hsu [289], Verhoff and Fisher [288], Newman [266], and Michelsen and Villadsen [227] employed the boundary condition of Fig. 17c to analyze the extended Graetz problem (with fluid axial heat conduction). Tan and Hsu employed the eigenvalue method, and determined the first 20 eigenvalues and the corresponding eigenfunctions by a Runge–Kutta technique. Their 20 eigenvalues and eigenfunctions are not sufficient to provide accurate Nusselt numbers for Pe < 10 with x^* approaching zero. This is because $Nu_{x,T}$ approached a finite value, in contradiction to approaching infinity as noted by other investigators [227,266,288].

Verhoff and Fisher [288] applied an inverse-tangent transformation to the extended Graetz problem (Fig. 17c), as mentioned above, and obtained the solution by a finite difference method. $Nu_{x,T}$ of Verhoff and Fisher compare favorably with those of Michelsen and Villadsen [227].

Newman [266] analyzed the same problem by the method of singular perturbation for high Péclet numbers. He found that axial diffusion is important only in a small region near $x^* = 0$ and $r = a$ and of spatial dimensions of the order of $a/Pe^{1/2}$. For the boundary conditions of Fig. 17c, the Nusselt number approaches infinity near $x^* = 0$ like $1/x^{*1/2}$; the Lévêque solution, which neglects axial conduction, predicts a behavior like $1/x^{*1/3}$.

Michelsen and Villadsen [227] analyzed this problem by a method that combines orthogonal collocation and matrix diagonalization techniques. They obtained J ($= 4x^* \Phi_{m,T}$) and $Nu_{x,T}$ for several Péclet numbers. For high values of x^*, their results are in excellent agreement with Tan and Hsu [289]. At low values of x^*, Michelsen and Villadsen showed that $Nu_{x,T}$ is proportional to $(x^*)^{-1/2}$, which is in agreement with the results of Newman [266]. According to Michelsen and Villadsen, the Fourier series of Tan and Hsu [289] is very slowly convergent; 20 eigenfunctions and eigenvalues are not sufficient to provide the accurate results and to show the trend of $(x^*)^{-1/2}$ dependence for low Péclet numbers.

f. Arbitrary Axial Temperature Distribution

When the effects of thermal energy sources, viscous dissipation, and flow work are neglected, the resulting energy equation is linear and homogeneous. It is then possible to construct a solution for any arbitrary variation in axial wall temperature by superposition using Duhamel's theorem [7,25,26,290]. Either the surface temperature is described by a large number of constant temperature steps [7,25], or a polynomial approximation is used [26], and the Graetz solution is superimposed accordingly using Duhamel's theorem.

Sellers et al. [25] obtained the formulas for the local Nusselt number, wall heat flux, and fluid bulk mean temperature when the wall temperature was varying linearly in the axial direction, $t_w - t_e = cx^*$ or ϑ_o of Eq. (229) as zero. The local Nusselt number for the linear t_w is expressed as [25]

$$\mathrm{Nu}_x = \frac{1 - 8\sum_{n=0}^{\infty}(G_n/\lambda_n^2)\exp(-2\lambda_n^2 x^*)}{\frac{11}{48} - 16\sum_{n=0}^{\infty}(G_n/\lambda_n^4)\exp(-2\lambda_n^2 x^*)} \tag{227}$$

For large x^*, Nu_x approaches 48/11, corresponding to the Ⓗ boundary condition. For small x^*, the Lévêque–type approximation yields [25]

$$\Phi_x = 1.615098(x^*)^{-1/3} \tag{228}$$

Grigull and Tratz [34] numerically obtained the solution for a linearly varying wall temperature. They graphically presented the local Nusselt number as a function of x^* with a dimensionless temperature ϑ_0

$$\vartheta_0 = \frac{32}{3}\frac{(t_w - t_e)_e}{(dt/dx^*)_w} \tag{229}$$

as a parameter, where the temperature jump at the wall at $x = 0$ is nondimensionalized by the wall temperature gradient. For small jumps ($0 < \vartheta_0 < 1$), the local Nusselt number asymptotically approached the value of 4.364, corresponding to the Ⓗ boundary condition. For large jumps, the local Nusselt number passed through a minimum and approached, for example, for $\vartheta_0 = 50$, a value of 3.66, corresponding to the Ⓣ boundary condition. Employing the superposition theorem, Kays [7] presented closed-form formulas for t_m and Nu_x for the same problem.

Kuga [268,291] obtained the solution for axially varying sinusoidal wall temperature. One sine wave temperature was imposed on the tube wall.

Shapovalov [292] considered the arbitrary wall temperature distribution with arbitrary initial conditions, and obtained a solution in terms of hypergeometric functions.

In the above analyses, the peripheral wall temperature distribution was assumed to be uniform. Bhattacharyya [293] investigated peripherally variable wall temperature with an axially constant wall temperature. The solution was obtained by the separation of variables and eigenvalue method. A specific example of cosine heat flux variation in the peripheral direction was considered. Bhattacharyya found that (1) thermal entrance length increases due to peripheral temperature variations, (2) the local Nusselt number $\mathrm{Nu}_{x,\mathrm{T}}$ depends upon the temperature jump $(t_{w,m} - t_e)$, and (3) usually

Nusselt numbers are higher along the minimum temperature line than those for the case of uniform peripheral heating. A reverse effect is observed for cooling. This happens up to a separating line where either one of two Nusselt numbers (corresponding to maximum and minimum temperature lines) becomes infinity because of $(t_w - t_m)$ approaching zero.

2. FINITE WALL THERMAL RESISTANCE, (T3)

For the thermal entrance problem with the (T3) boundary condition, the outside wall temperature is constant in the axial direction, while there is finite thermal resistance R_w in the normal direction. For $R_w = 0$, this problem corresponds to the (T) (Graetz) problem; and for $R_w = \infty$, it corresponds to the (H) problem. This (T3) boundary condition is further described on p. 23. A fully developed velocity profile is specified for this subsection.

Schenk and DuMoré [294] analyzed the (T3) thermal entrance problem theoretically, using the eigenvalue method. They tabulated the first three eigenvalues and related constants as well as the dimensionless local heat flux and fluid bulk mean temperature (as defined below) for the wall thermal resistance parameter $R_w = 0, 0.025, 0.25,$ and 1. Sideman et al. [230] supplemented and extended the work of Schenk and DuMoré [294]. They presented the first five eigenvalues and related constants for $R_w = 0, 0.025, 0.05, 0.10, 0.15, 0.25, 0.5,$ and 1.

Lock et al. [295] also employed the modified Graetz method to solve the (T3) thermal entrance problem while studying ice formation in a convectively cooled water pipe. They presented a transcendental equation for the eigenvalues and numerically evaluated the first five. However, they used the asymptotic eigenfunctions of the Graetz problem $(R_w = 0)$ [25] to arrive at the complete solution. Hwang and Yih [296] showed that it is inaccurate to use this asymptotic solution when R_w is large. The first few eigenvalues computed by Lock et al. [295] were quite inaccurate. Hwang and Yih [296] employed the conventional power series method to obtain the eigenvalues. The first four eigenvalues of Hwang and Yih are in excellent agreement with those of Hsu [297] for $R_w = 0, 0.1, 0.2, 0.4, 1, 2, 4,$ and ∞, except for λ_0 for $R_w = \infty$. Dzung [298] showed that Lock et al. employed only the first term of the asymptotic series to evaluate the eigenvalues, and hence their computed values are inaccurate. Dzung further showed that a significant improvement in accuracy can be made for all R_w, if the second terms are included.

Schenk and DuMoré, Sideman et al., and Lock et al. employed the modified Graetz method to obtain the solution for the (T3) thermal entrance problem. However, with this approach, the evaluation of higher eigenvalues becomes increasingly difficult. To alleviate this difficulty, Hsu [297] determined them by solving the Sturm–Liouville type equation using a Runge–Kutta successive approximation numerical procedure. He accurately obtained and tabulated the first ten eigenvalues and constants for $R_w = 0.005, 0.05, 0.25$, 0.5, and 1.0. He also presented implicit asymptotic formulas to determine higher eigenvalues and constants for the (T3) thermal entrance problem. In the analysis, Hsu also included the effect of uniform internal thermal energy generation within the fluid. He presented graphically $Nu_{x,T3}$ as a function of $2x^*$, (1) with the above-listed values of R_w as a parameter for zero thermal energy generation, and (2) with various values of thermal energy generation as a parameter for $R_w = 0.25$. $Nu_{x,T3}$ for zero thermal energy generation are presented in Table 16 and Fig. 20 based on the first ten eigenvalues and constants from Hsu [297] and the next 110 values calculated by the present authors from the implicit asymptotic equations of Hsu.

The infinite series solution for the temperature distribution and $Nu_{x,T3}$ converge very slowly near $x^* = 0$. Rosen and Scott [299] extended the Lévêque solution for the finite R_w case. They presented the dimensionless local heat flux $q_x''D_h/2k(t_{wo} - t_e)$ and the fluid bulk mean temperature $(t_m - t_e)/(t_{wo} - t_e)$ in tabular and graphical forms for $R_w = 0, 0.01, 0.025$, 0.05, and 0.25.

Fluid axial heat conduction has been neglected in the above analyses [230,294,297,299]. Hsu [233] investigated the finite wall resistance thermal entrance problem with finite fluid axial heat conduction. He assumed a uniform temperature profile at the entrance, $x^* = 0$. He solved the problem using an eigenvalue approach for the wall thermal resistance parameter $R_w = 0.005, 0.05$, and 0.25. For $Pe = 1$, the first 20 eigenvalues and related constants were presented, while for $Pe = 5, 10, 20, 30, 50$, and 100, the first 12 eigenvalues and related constants were presented. For $Pe = \infty$, Hsu listed the first 10 eigenvalues and related constants. He plotted $Nu_{x,T3}$ as a function of $2x^*$ for the above-listed Péclet numbers. Here Hsu used a solution method similar to the one he used for the (H) boundary condition [300]. As described on p. 132, his analysis in [300] was found to be in error; thus Hsu's constants and $Nu_{x,T3}$ in [233] may also be in error for low Péclet numbers.

McKillop et al. [231] analyzed the (T3) entrance problem by a finite difference method and presented $Nu_{x,T3}$ as a function of $4x^*$ for $R_w = 0$, 0.005, 0.025, 0.25, and ∞. His $Nu_{x,T3}$, calculated from Eq. (259) for $Pe = \infty$ and $R_w = 0.005$ and 0.25, are up to 2% higher than those of Table 16 for $10^{-4} \leq x^* \leq 0.05$.

Hickman [232] analyzed the (T3) thermal entrance problem by a Laplace transform technique for large x^*. He presented a formula for $Nu_{o,m}$, defined by Eq. (119), as functions of x^* and $1/R_w$.

Javeri [207] employed the Galerkin–Kantorowich method of variational calculus and analyzed the (T3) thermal entrance problem. He presented $Nu_{x,T3}$ as a function of $4x^*$ for $R_w = 0, 0.005, 0.05, 0.1$, and 0.25. When

TABLE 16

Circular Duct: $Nu_{x,T3}$ as a Function of x^* and R_w for a Fully Developed Velocity Profile (Based on the Results of Hsu [297])

x^*	$Nu_{x,T3}$				
	$R_w=.005$	$R_w=.05$	$R_w=.25$	$R_w=.05$	$R_w=1.0$
0.0001	22.936	25.176	26.437	26.684	26.815
0.00015	19.904	21.884	23.091	23.327	23.452
0.0002	18.001	19.801	20.962	21.190	21.312
0.0003	15.628	17.182	18.271	18.491	18.609
0.0004	14.139	15.527	16.564	16.778	16.894
0.0005	13.084	14.350	15.346	15.556	15.672
0.0006	12.283	13.453	14.415	14.623	14.738
0.0007	11.645	12.738	13.672	13.878	13.992
0.0008	11.121	12.149	13.059	13.263	13.377
0.0009	10.679	11.653	12.542	12.744	12.858
0.001	10.300	11.227	12.096	12.297	12.411
0.0015	8.971	9.732	10.531	10.727	10.841
0.002	8.145	8.804	9.552	9.745	9.859
0.003	7.126	7.662	8.342	8.530	8.643
0.004	6.499	6.959	7.591	7.776	7.889
0.005	6.061	6.471	7.067	7.249	7.362
0.006	5.734	6.106	6.673	6.853	6.966
0.007	5.478	5.820	6.364	6.542	6.654
0.008	5.271	5.589	6.113	6.289	6.401
0.009	5.099	5.397	5.904	6.078	6.190
0.01	4.954	5.236	5.727	5.899	6.012
0.015	4.469	4.695	5.129	5.295	5.407
0.02	4.196	4.388	4.784	4.945	5.056
0.03	3.912	4.064	4.408	4.561	4.671
0.04	3.783	3.913	4.222	4.368	4.476
0.05	3.722	3.838	4.122	4.263	4.368
0.06	3.693	3.800	4.068	4.203	4.307
0.07	3.678	3.781	4.038	4.170	4.272
0.08	3.671	3.771	4.021	4.150	4.252
0.09	3.667	3.766	4.012	4.139	4.239
0.1	3.666	3.763	4.006	4.133	4.232
∞	3.666	3.763	4.000	4.124	4.223

FIG. 20. Circular duct: $Nu_{x,T3}$ as a function of x^* and R_w for a fully developed velocity profile (from Table 16).

compared to $Nu_{x,T3}$ of Table 16, Javeri's $Nu_{x,T3}$ are about 3% higher at $x^* = 10^{-4}$ and are in excellent agreement for $x^* \geq 10^{-2}$.

3. RADIANT FLUX BOUNDARY CONDITION (T4)

For the thermal entrance problem with the (T4) boundary condition, the wall temperature is constant axially, while the radiant heat flux is specified along the periphery. This thermal entrance problem reduces to the (T) and (H) problems for the radiant flux parameter $\gamma = \infty$ and 0, respectively. The (T4) boundary condition is further described on p. 24. The velocity profile is specified as fully developed in this subsection.

Chen [235] obtained an approximate solution for the (T4) thermal entrance problem in terms of the Liouville–Neumann series. He also carried out an exact iterative numerical solution to the same problem for a wide range of the radiant flux parameter $\gamma = \varepsilon_w \sigma T_e^3 D_h / k$. He presented $Nu_{x,T4}$ in tabular and graphical forms for $\gamma = 0.2, 0.4, 1, 2, 4, 10, 20, 40$, and 100. For the range $0.001 < x^* \leq 0.1$ and $\gamma \leq 40$, his following approximation for $Nu_{x,T4}$ agrees within $\pm 2\%$ with his numerical solution:

$$Nu_{x,T4}/Nu_{x,H} = 0.928 - 0.023 \ln(\gamma/2) \qquad (230)$$

where $Nu_{x,H}$ are available in Table 18.

Dussan and Irvine [301] investigated the same problem analytically and experimentally. The problem was solved analytically by considering linearized radiation and an exponential kernel approximation. The wall heat flux q_x''/q_e'' was plotted against $2x^*$ with $\gamma/2$ as a parameter. The results are in good agreement with Chen's result for $x^* \geq 0.03$. Their experimental results agree satisfactorily with the analysis, except for the entrance region.

Kadaner et al. [236] also analyzed the same problem by an approximate method where the energy equation was first transformed into ordinary differential equations and then the solution was sought. They graphically presented $Nu_{x,T4}$ as a function of x^* with $\gamma/2$ as a parameter. The (H) and (T) local Nusselt numbers are shown as limiting cases for $\gamma = 0$ and ∞, respectively. Within the range $0.001 < x^* < 0.2$ and $0.2 < \gamma < 100$ with the ambient (radiation sink) temperature as zero, they presented the approximation

$$\frac{Nu_{x,T4}}{Nu_{x,H}} = 0.94 - \frac{0.0061 - 0.0053 \ln x^*}{1 + 0.0242 \ln x^*} \ln\left(\frac{\gamma}{2}\right) \qquad (231)$$

which agrees with their solution within 2%. This ratio $(Nu_{x,T4}/Nu_{x,H})$ increases monotonically with x^* for $\gamma > 2$ and decreases for $\gamma < 2$. In contradiction, Eq. (230) indicates a Nusselt number ratio that is independent of x^*. Equations (230) and (231) differ by about $\pm 3\%$ for $\gamma < 10$ and $0.001 < x^* < 0.2$. The difference increases for higher values of γ.

Salomatov and Puzyrev [302] also analyzed this problem by an approximate method using a Laplace transform. Their results are in satisfactory agreement with those of Chen [235] and Kadaner *et al.* [236].

Sikka and Iqbal [303] considered the steady radiant heat flux being incident on one-half of the tube circumference, while the fluid emanated heat through the wall on all sides to an absolute zero degree environment. This boundary condition may be approximated in some spacecraft nuclear reactor power plant components. A solution by a finite difference method was obtained in the thermal entrance region for a fully developed velocity profile. The local variations of the wall temperature and local Nusselt numbers were delineated as a function of $2x^*$ with $\gamma/2$ and ψ as parameters, where $\psi = \alpha_w(q_r''^3 a^4 \sigma/k^4)^{1/3}$. For $\gamma = 1$ and $\psi = 0$, their results agree with Chen's [235] for $x^* > 0.005$. For $x^* < 0.005$, their results are increasingly higher. This partial disagreement may be due to differences in methodologies.

4. SPECIFIED HEAT FLUX DISTRIBUTION

For this class of problems, the heat flux is specified along the tube wall. The wall thermal resistance normal to the fluid flow direction is assumed to be zero. The discussion is divided into the following subsections: constant axial wall heat flux, the Ⓗ problem; extensions to the Ⓗ problem; the Ⓗ problem with finite fluid axial heat conduction; the axially exponential wall heat flux, the Ⓗ5 problem; and problems with arbitrary axial wall heat flux distribution. The velocity profile is considered as fully developed in all of these subsections.

a. *Constant Wall Heat Flux,* Ⓗ[†]

After the Graetz problem, this is a fundamental and practically important thermal entrance problem. The heat flux along the tube wall is specified as constant. All other idealizations are the same as those for the Graetz problem. Specifically, an incompressible fluid with constant physical properties flows through a circular tube with a laminar developed velocity profile and a developing temperature profile. The temperature profile of the fluid at the inlet ($x = 0$) is uniform. Axial heat conduction, viscous dissipation, flow work, and thermal energy sources within the fluid are neglected. The resulting energy equation is the same as Eq. (196), the initial condition is the same as Eq. (197), and the wall boundary condition is $k(\partial t/\partial r)_w = q'' =$

[†] The Ⓗ1–Ⓗ4 thermal boundary conditions for the symmetrically heated circular duct are the same, and hence these are simply designated as the Ⓗ boundary condition, as described on pp. 25–26.

constant, instead of Eq. (198). The symmetry condition of Eq. (199) is still applicable.

Kays [40] first solved this problem by a finite difference method. He tabulated and graphically presented $Nu_{x,H}$ as a function of $4x^*$. Petukhov and Chzhen-Yun [304] showed that $Nu_{x,H}$ of Kays were about 20% higher than those of Siegel et al. [29]. Sellers et al. [25] analyzed this problem by an inversion method, knowing the solution to the Graetz problem. Siegel et al. [29] investigated the Ⓗ thermal entrance problem by the same method as Graetz [1], that is the method of separation of variables and the Sturm–Liouville theory, and obtained the following solution. It can be shown that $(t - t_w)_H$ calculated from Eqs. (232) and (233) for $x^* \to \infty$ is identical to that of Eq. (177) for $c_5 = c_6 = 0$:

$$\Theta = \frac{t - t_e}{(q''D_h/k)} - 4x^* + \frac{1}{2}\left(\frac{r}{a}\right)^2 - \frac{1}{8}\left(\frac{r}{a}\right)^4 - \frac{7}{48}$$

$$+ \frac{1}{2}\sum_{n=1}^{\infty} C_n R_n \exp(-2\beta_n^2 x^*) \tag{232}$$

$$\Theta_w = \frac{t_w - t_e}{(q''D_h/k)} = 4x^* + \frac{11}{48} + \frac{1}{2}\sum_{n=1}^{\infty} C_n R_n(1)\exp(-2\beta_n^2 x^*) \tag{233}$$

$$\Theta_m = \frac{t_m - t_e}{(q''D_h/k)} = 4x^* \tag{234}$$

$$Nu_{x,H} = \left[\frac{11}{48} + \frac{1}{2}\sum_{n=1}^{\infty} C_n R_n(1)\exp(-2\beta_n^2 x^*)\right]^{-1} \tag{235}$$

Here β_n, R_n, and C_n are eigenvalues, eigenfunctions, and constants, respectively. The first seven of these quantities were determined to six digit accuracy by Siegel et al. [29]. Hsu [305] extended the work of Siegel et al. and reported the first 20 values for each of β_n^2, $R_n(1)$, and C_n with eight to ten digit accuracy. These are listed in Table 17. He also presented approximate formulas for higher eigenvalues and constants. Of particular interest are

$$\beta_n = 4n + \tfrac{4}{3} \tag{236}$$

$$C_n R_n(1) = -2.401006045\beta_n^{-5/3} \tag{237}$$

Note that γ_m and A_m of Kays [7] are the same as β_n and $-1/[C_n R_n(1)\beta_n^4]$ in the above equations. The asymptotic formula for A_m presented by Kays [7] is incorrect; it should have been $0.416492\gamma_m^{-7/3}$ instead of $0.358\gamma_m^{-2.32}$.

This Ⓗ thermal entrance problem is quite useful for internal flow forced-convection heat transfer. Consequently, $Nu_{x,H}$ values have been accurately established by Shah [270] and are presented in Table 18. Equation (235) was

TABLE 17

CIRCULAR DUCT: INFINITE SERIES SOLUTION
FUNCTIONS FOR THE (H) THERMAL
ENTRANCE PROBLEM (FROM HSU [305])

n	β_n^2	$-C_n R_n(1)$
1	25.679611	0.19872216
2	83.861753	0.06925746
3	174.16674	0.03652138
4	296.53630	0.02301407
5	450.94720	0.01602945
6	637.38735	0.011906317
7	855.849532	0.009249488
8	1106.329035	0.007427222
9	1388.822594	0.006117477
10	1703.327852	0.005141193
11	2049.843045	0.004391938
12	2428.366825	0.003803024
13	2838.898142	0.003330824
14	3281.436173	0.002945767
15	3755.980271	0.002627194
16	4262.529926	0.002360296
17	4801.084748	0.002135757
18	5371.644444	0.001940852
19	5974.208812	0.001774030
20	6608.777727	0.001628990

solved for $x^* > 10^{-4}$ using the first 121 terms of the series; β_n^2 and $C_n R_n(1)$ up to $n = 20$ were taken from Table 17, and $n = 11$ to 121 were determined from Eqs. (236) and (237). For $x^* \le 10^{-4}$, the extended Lévêque-type solution of Worsøe-Schmidt [200] was used as described below. $\text{Nu}_{x,\text{H}}$ of Table 18 are plotted in Fig. 15. The thermal entrance length $L_{\text{th,H}}^*$ defined by Eq. (128), was computed by Shah [270] as

$$L_{\text{th,H}}^* = 0.0430527 \tag{238}$$

$L_{\text{th,H}}^*$ computed by the method of Sandall and Hanna [58], Eq. (129), is 0.0365, 15% lower than the value of Eq. (238).

Grigull and Tratz [34] investigated the same problem numerically by a finite difference method and presented $\text{Nu}_{x,\text{H}}$ graphically. Based on the numerical results, they approximated Eq. (242) for $\text{Nu}_{x,\text{H}}$.

The series of Eq. (235) converges uniformly for all nonzero x^*, but the convergence is very slow as x^* approaches zero. In this region, a similarity solution, similar to the Lévêque solution for the (T) problem, was provided by Worsøe-Schmidt [200]. Based on his results, the following formula for $\text{Nu}_{x,\text{H}}$ is derived.

$$\frac{1}{\text{Nu}_{x,\text{H}}} = 0.768058587x^{*1/3} + 0.625298952x^{*2/3} + 0.62172x^*$$

$$+ 0.67151x^{*4/3} + 0.77044x^{*5/3} + 0.93839x^{*2}$$

$$+ 1.21900x^{*7/3} - 4x^* \tag{239}$$

TABLE 18

x^*	$Nu_{x,H}$	$Nu_{m,H}$	$N_H(x)$
0.000001	129.203	194.2	0.000190
0.0000015	112.753	169.5	0.000248
0.000002	102.360	153.9	0.000299
0.000003	89.307	134.4	0.000390
0.000004	81.062	122.0	0.000471
0.000005	75.190	113.2	0.000544
0.000006	70.707	106.5	0.000613
0.000007	67.124	101.1	0.000677
0.000008	64.167	96.67	0.000738
0.000009	61.665	92.91	0.000797
0.000010	59.510	89.68	0.000853
0.000015	51.889	78.24	0.00111
0.00002	47.077	71.01	0.00133
0.00003	41.037	61.94	0.00173
0.00004	37.224	56.20	0.00207
0.00005	34.511	52.12	0.00239
0.00006	32.440	49.01	0.00268
0.00007	30.787	46.52	0.00295
0.00008	29.422	44.46	0.00321
0.00009	28.269	42.73	0.00345
0.00010	27.275	41.24	0.00369
0.00015	23.762	35.94	0.00474
0.0002	21.555	32.61	0.00565
0.0003	18.790	28.42	0.00722
0.0004	17.048	25.78	0.00857
0.0005	15.813	23.90	0.00977
0.0006	14.872	22.47	0.0109
0.0007	14.123	21.33	0.0119
0.0008	13.506	20.39	0.0128
0.0009	12.985	19.60	0.0137
0.0010	12.538	18.91	0.0145
0.0015	10.967	16.50	0.0182
0.002	9.9863	14.99	0.0212
0.003	8.7724	13.10	0.0262
0.004	8.0200	11.92	0.0302
0.005	7.4937	11.08	0.0336
0.006	7.0986	10.45	0.0365
0.007	6.7881	9.948	0.0391
0.008	6.5359	9.537	0.0414
0.009	6.3261	9.191	0.0434
0.010	6.1481	8.896	0.0453
0.015	5.5469	7.869	0.0526
0.02	5.1984	7.241	0.0576
0.03	4.8157	6.488	0.0637
0.04	4.6213	6.043	0.0672
0.05	4.5139	5.747	0.0692
0.06	4.4522	5.536	0.0704
0.07	4.4162	5.379	0.0711
0.08	4.3949	5.257	0.0715
0.09	4.3823	5.160	0.0717
0.10	4.3748	5.082	0.0719
0.15	4.3645	4.845	0.0721
0.2	4.3637	4.724	0.0722

$Nu_{x,H}$ were computed from this equation by Shah [270] and were found to be identical up to five or more digits with those of Eq. (235) for $0.0002 < x^* < 0.005$. The agreement between $Nu_{x,H}$ of Eqs. (235) and (239) was found to be within $\pm 0.27\%$ for the extended range $0.00005 \leq x^* \leq 0.02$. $Nu_{x,H}$ in Table 18 for $x^* \leq 10^{-4}$ are based on the above extended Lévêque-type solution, Eq. (239).

The approximate equations (240) and (241) are recommended by Shah [270] and Eq. (242) by Grigull and Tratz [34] for the range of x^* indicated:

$$\text{Nu}_{x,\text{H}} = \begin{cases} 1.302(x^*)^{-1/3} - 1 & \text{for } x^* \le 0.00005 \quad (240) \\ 1.302(x^*)^{-1/3} - 0.5 & \text{for } 0.00005 \le x^* \le 0.0015 \quad (241) \\ 4.364 + 8.68(10^3 x^*)^{-0.506} e^{-41x^*} & \text{for } x^* \ge 0.0015 \quad (242) \end{cases}$$

These equations provide $\text{Nu}_{x,\text{H}}$ within -0.5, ± 1.0, and $+0.4\%$, respectively, when compared with $\text{Nu}_{x,\text{H}}$ of Table 18.

The foregoing three equations cover a complete range of x^*. In constrast, Churchill and Ozoe [306] proposed the following single relation for the entire range of x^*.

$$\frac{\text{Nu}_{x,\text{H}} + 1}{5.364} = \left[1 + \left(\frac{220}{\pi} x^* \right)^{-10/9} \right]^{3/10} \quad (243)$$

They formulated this expression based on asymptotic formulas for small and large values of x^*. This equation predicts $\text{Nu}_{x,\text{H}}$ lower than those of Table 18 for $x^* \le 0.01$ with a maximum error of 3.4% at $x^* = 0.003$. It predicts $\text{Nu}_{x,\text{H}}$ higher than those of Table 18 for $x^* > 0.01$ with a maximum error of 4.9% at $x^* = 0.06$.

For the (H) boundary condition, heat transfer through a single duct in a one fluid heat exchanger is known at any location. As a result, there is no need to evaluate the mean Nusselt number $\text{Nu}_{m,\text{H}}$. This may be why it is generally not available in the literature. However, in a two-fluid heat exchanger, knowledge of the mean Nusselt number is necessary to determine the overall heat transfer coefficient. $\text{Nu}_{m,\text{H}}$ were determined by Shah [270] by integrating the 6 to 8 digits $\text{Nu}_{x,\text{H}}$ at x^* values of Table 18. The calculated $\text{Nu}_{m,\text{H}}$ and $N_{\text{H}}(x)$ are presented in Table 18. The incremental heat transfer number $N_{\text{H}}(x)$ for fully developed flow is found as

$$N_{\text{H}}(\infty) = 0.0722 \quad (244)$$

Shah [270] recommended the following approximate formulas for $\text{Nu}_{m,\text{H}}$:

$$\text{Nu}_{m,\text{H}} = \begin{cases} 1.953(x^*)^{-1/3} & \text{for } x^* \le 0.03 \quad (245) \\ 4.364 + \dfrac{0.0722}{x^*} & \text{for } x^* > 0.03 \quad (246) \end{cases}$$

The calculated $\text{Nu}_{m,\text{H}}$ by these equations agree with those of Table 18 within ± 3.1 and $+2.1\%$ respectively.

b. (H) *Problem with Nonuniform Fluid Inlet Temperature*

Hicken [274] investigated the effect of nonuniform fluid inlet temperature on the thermal entrance heat transfer. He considered five different (one-period) sinusoidal fluid inlet temperature profiles. All other idealizations made in the analysis were the same as those of Siegel *et al.* [29]. He tabulated eigenvalues and constants and graphically presented $1/\text{Nu}_{x,\text{H}}$ as a function of x^*. He found, as expected, that if the temperature is higher at the center for the same $t_{m,e}$, the $\text{Nu}_{x,\text{H}}$ are higher relative to the uniform entrance temperature profile situation; if the fluid temperature is lower at the center, $\text{Nu}_{x,\text{H}}$ are lower.

c. (H) *Problem with Internal Thermal Energy Generation*

Sparrow and Siegel [307] extended the analysis of Siegel *et al.* [29] by including the internal thermal energy generation. The thermal energy generation was allowed to vary in an arbitrary manner both longitudinally along the tube and radially across the cross section. Inman [308] experimentally studied the temperature distribution in laminar flow through an insulated circular tube with the internal thermal energy generation. His results are in excellent agreement with the theoretical prediction of Sparrow and Siegel [307].

d. (H) *Problem with Viscous Dissipation*

While the effect of viscous dissipation is negligible for low-velocity gas flows, it is important for liquids and high-velocity gas flows. The effect of viscous dissipation on the (H) thermal entrance heat transfer has been investigated by Brinkman [61], and Ou and Cheng [239], who treated viscosity and other fluid properties as constants and the velocity profile as fully developed.

The (H) thermal entrance solution, as outlined on p. 124, is valid for both situations, whether the fluid flowing through the tube is being heated or being cooled. However, when viscous dissipation is considered finite (Brinkman number, $\text{Br}' \neq 0$), the solutions are different for heating of the fluid ($q' > 0$) relative to cooling ($q' < 0$). The fluid heating and cooling situations result in positive and negative values of Br', respectively. The influence of viscous dissipation on t_m and t_w is discussed separately below.

Based on Eq. (24), the rate of thermal energy generated per unit volume of a circular tube, at a local point (r, θ) due to viscous dissipation is given by

$$\frac{\mu}{g_c J}\left(\frac{\partial u}{\partial r}\right)^2 \tag{247}$$

Integration of this term over the flow cross section results in $8\pi\mu u_m{}^2/g_c J$ for the energy rate per unit length for a fully developed velocity profile and constant fluid properties. This remains constant regardless of the wall heat transfer rate. Hence, from an energy balance, the axial change in the fluid bulk mean temperature for the thermally developing or developed (H) problem (q' = constant) is given by

$$Wc_{\mathrm{p}} \frac{dt_{\mathrm{m}}}{dx} = q' + \frac{8\pi\mu u_{\mathrm{m}}^{2}}{g_{\mathrm{c}}J} \tag{248}$$

Integration of this equation from $x = 0$ to x and rearrangement yields

$$\Theta_{\mathrm{m}} = \frac{t_{\mathrm{m}} - t_{\mathrm{e}}}{q'' D_{\mathrm{h}}/k} = 4x^{*} + 32x^{*}\,\mathrm{Br}' \tag{249}$$

Equation (249) shows that the axial distribution of t_{m} is linear and that its slope depends upon the value of Br'. The first and second terms on the right-hand side of Eq. (249) are due to wall heat transfer and viscous dissipation. Note that when Br' $= -1/8$, heat transfer from the fluid to the wall is compensated for by the thermal energy generated due to viscous dissipation; and there is no change of t_{m} from the inlet value of t_{e}. In this case, the temperature profile develops in the tube as shown in Fig. 21.

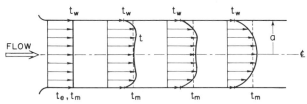

FIG. 21. Fluid temperature development in the thermal entrance region for the Ⓗ boundary condition for Br' $= -1/8$. The fluid is being cooled at the wall. The dotted line represents the fluid bulk mean temperature which remains constant at each cross section. The temperature profiles are drawn for increasing x^{*} from left to right.

The effect of viscous dissipation on the axial distribution of t_{w} cannot be presented by a simple equation. The closed-form expression for t_{w} in the presence of viscous dissipation involves eigenvalues and eigenfunctions as given by Ou and Cheng [239]. The influence of viscous dissipation on t_{w} and t_{m} is discussed qualitatively below.

As the rate of shear is higher near the wall, the thermal energy generated due to viscous dissipation [Eq. (247)] is more pronounced near the wall region compared to the central region of the tube. Hence, viscous dissipation affects the wall temperature t_{w} more significantly than it affects the local fluid temperature near the center of the tube. The fluid bulk mean temperature t_{m}, by its definition, depends upon the temperature and velocity profiles across the cross section. Since the velocity is low in the near wall region where viscous dissipation plays an important role, its influence on t_{m} is relatively smaller. Thus, viscous dissipation has a strong influence on t_{w} and a small influence on t_{m}. The analytical results of Brinkman [61] and Ou and Cheng [239] are discussed in further detail below.

Brinkman [61] considered the adiabatic wall case and investigated the influence of viscous dissipation on the fluid temperature profile when the entering temperature profile was uniform. He found the fluid temperature rising rapidly near the wall for increasing x^*. The direct effect of viscous dissipation is zero at the center of the tube, but the centerline fluid temperature rises slowly for increasing x^* due to heat conduction from the wall region. Thus, Brinkman found the fluid temperature profile becoming more nonuniform for increasing x^*.

Ou and Cheng [239] analyzed the (H) thermal entrance problem having finite viscous dissipation. For the fluid heating situation $Br' > 0$, t_w and t_m increase at a faster rate in the thermal entrance region compared to the $Br' = 0$ situation. As discussed above, viscous dissipation will increase t_w more rapidly than t_m. This results in $(t_w - t_m)$ increasing with higher values of Br' at any specified x^*. Consequently, Nu_x at any x^* decreases as Br' increases for the (H) boundary condition. The fully developed Nu_H for finite viscous dissipation is expressed by Eq. (179) with $S^* = 0$. Note that in the fully developed region, $(t_w - t_m)$ is a constant dependent upon Br'.

For the fluid cooling situation, $q'' < 0$ and $Br' < 0$, the problem is more complex and depends upon the magnitude of Br'. As seen from Eq. (249),[†] for $-1/8 < Br' < 0$, the fluid bulk mean temperature t_m is always less than the uniform entrance temperature t_e $(t_w < t_m < t_e)$, indicating the dominance of cooling over viscous heating. The fluid continues to cool down even in the asymptotic region for the foregoing range of Br'. Viscous dissipation increases t_w substantially and t_m slightly in comparison to the $Br' = 0$ situation. Hence, as expected, Ou and Cheng showed that both t_m and $(t_m - t_w)$ decrease and $Nu_{x,H}$ increases for decreasing Br' in the range $-1/8 < Br' < 0$.

Ou and Cheng showed that for $-11/48 < Br' \leq 0$, t_w is always lower than t_m. For $-11/48 < Br' < -1/8$ [from Eq. (249) with negative q''], t_m increases with increasing x^* because the thermal energy generated due to viscous dissipation exceeds the fluid-to-wall heat transfer rate. Hence, for this Br' range, t_m and t_w continue to rise with increasing x^*, and t_m is always higher than t_w. This results in a positive Nusselt number, which increases as Br' drops further below $-1/8$.

For $Br' < -11/48$, the thermal energy generated due to viscous dissipation is higher than the fluid-to-wall heat transfer rate. Hence, t_m continues to increase starting from $x^* = 0$, as can be seen from Eq. (249) for negative q''. However, near $x^* = 0$, because of the imposed constant heat transfer rate at the wall, t_w is lower than t_m. At some $x^* > 0$, viscous dissipation being strong enough causes t_w to rise above t_m. A typical fluid temperature development, t_m and t_w for $Br' < -11/48$, are shown in Fig. 22. It is interesting

---

[†] Ou and Cheng erroneously quoted the range as $-11/48 < Br' < 0$. All of their discussion for $Br' < 0$ is confusing, because they do not explain the variations of t_w and t_m in the entrance region for different ranges of Br' for $Br' < 0$.

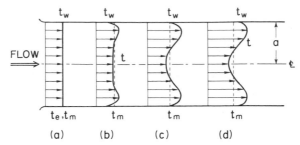

FIG. 22. Fluid temperature development in the thermal entrance region for the Ⓗ boundary condition for Br′ < −11/48. The fluid is being cooled at the wall. The dotted line represents the fluid bulk mean temperature. The temperature profiles are drawn for increasing x^* from left to right.

to note the three situations, $t_w < t_m$, $t_w = t_m$, and $t_w > t_m$ in Fig. 22b,c,d. They result in positive, infinite, and negative $Nu_{x,H}$, respectively. The location $t_m(x) = t_w(x)$ represents a point of singularity in the $Nu_{x,H}$ vs. x^* curve. Ou and Cheng showed that this singularity occurs at lower x^* for decreasing values of Br′.

The following additional points may be observed from Fig. 22: (1) Because of the imposed Ⓗ boundary condition, the temperature gradient at the wall is constant and the fluid is being cooled in the wall region. (2) Because of the dominance of viscous dissipation, the thermal energy generated is conducted in radial directions toward both the wall and the center, thereby increasing all three temperatures; $t_w > t_m > t_e$ indicates an overall increasing temperature or heating effect for the fluid, even though the imposed boundary condition says that the fluid is being cooled at the wall, in the sense of a constant q' from the fluid.

e. Ⓗ *Problem with Fluid Axial Heat Conduction*

The Ⓗ thermal entrance problem with fully developed laminar velocity profile has been analyzed for only two sets of initial and boundary conditions as shown in Fig. 23a,c. It will be shown that axial heat conduction within the fluid is negligible for Pe > 10, except for the immediate neighborhood of $x^* = 0$.

Hsu [300] considered a uniform entrance temperature profile at $x^* = 0$ (Fig. 23a) and analyzed the Ⓗ thermal entrance problem including fluid axial conduction. He tabulated the first 12 eigenvalues and constants for Pe = 5, 10, 20, 30, 50, and 100. He also presented graphically $Nu_{x,H}$ as a function of $2x^*$ for these Péclet numbers. Pirkle and Sigillito [309] pointed out two errors in Hsu's analysis. They presented correct formulas for the constants in the infinite series solutions for Θ and $Nu_{x,H}$. However, they did not present the numerical values.

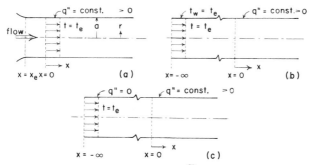

FIG. 23. Initial and boundary conditions for the Ⓗ thermal entrance problem with finite fluid axial heat conduction.

Petukhov and Tsvetkov [310] considered the initial and boundary conditions of Fig. 23c, namely, uniform entrance temperature profile at $x = -\infty$, with the wall region $-\infty < x < 0$ adiabatic ($q'' = 0$), and the wall region $0 < x < \infty$ having a constant heat flux q'' (>0) to the fluid. They numerically obtained the approximate temperature distributions and presented $Nu_{x,H}$ for Pe = 1, 10, and 45. While $Nu_{x,T}$ is infinite at $x = 0$ for all Péclet numbers for the initial and boundary conditions of Fig. 17b, Petukhov and Tsvetkov showed that $Nu_{x,H}$ for the above boundary condition (Fig. 23c) has a finite value at $x = 0$, for Pe $< \infty$, which decreases with decreasing Péclet numbers. Additionally, $Nu_{x,H}$ increases with decreasing Péclet numbers at any $x^* \gtrsim 0.0075$. This latter behavior is the same as that for $Nu_{x,T}$. Thus, the $Nu_{x,H}$ vs. x^* curves, for various Pe, have an inflection point at $x^* \simeq 0.0075$ for the case of uniform entrance temperature profile at $x = -\infty$. This characteristic (the inflection point), as shown in Fig. 24, is not found when the entrance temperature profile is uniform at $x = 0$, the case treated by Hsu [300]. Even though the results of Petukhov and Tsvetkov are qualitatively correct, Hsu [311] found their approximate method considerably in error.

Hennecke [68] independently investigated the same problem (Fig. 23c). He employed a finite difference method and matched the solutions point by point at $x^* = 0$. He presented graphically $Nu_{x,H}$ and Θ_m as a function of $2x^*$ for Pe = 1, 2, 5, 10, 20, and 50. Also presented were $2L_{th,H}^*$ and $Nu_{x,H}$ at $x^* = 0$ as a function of Pe. Table 19 and Fig. 24 are prepared from Hennecke's graphical results. As shown in Fig. 24, Hennecke found the behavior of $Nu_{x,H}$ vs. x^* to be the same as that observed by Petukhov and Tsvetkov [310]. Based on his results, Hennecke concluded that when $x^* \geq 0.005$, the effect of axial heat conduction may be neglected for Pe > 10 for the Ⓗ boundary condition with uniform entrance temperature at $x = -\infty$.

TABLE 19

Circular Duct: $Nu_{x,H}$ as a Function of x^* and Pe for the
Initial and Boundary Conditions of Fig. 23c for a Fully
Developed Velocity Profile (from the Graphical Results
of Hennecke [68])

x^*	$Nu_{x,H}$						
	Pe=1	2	5	10	20	50	∞
0.0005	7.06	7.51	8.35	9.67	11.67	14.41	15.81
0.001	7.05	7.47	8.14	9.21	10.62	11.85	12.54
0.002	6.99	7.26	7.81	8.53	9.30	9.71	9.99
0.003	6.93	7.15	7.55	8.02	8.47	8.75	8.77
0.004	6.87	7.03	7.31	7.66	7.90	8.01	8.02
0.005	6.83	6.93	7.10	7.33	7.45	7.48	7.49
0.01	6.61	6.58	6.41	6.31	6.26	6.20	6.15
0.02	6.28	5.98	5.64	5.41	5.27	5.22	5.20
0.03	6.03	5.62	5.25	5.00	4.88	4.84	4.82
0.04	5.80	5.40	4.99	4.76	4.70	4.66	4.62
0.05	5.65	5.22	4.81	4.63	4.58	4.54	4.51
0.1	5.12	4.67	4.47	4.40	4.39	4.37	4.37
0.2	4.65	4.40	4.37	4.36	4.36	4.36	4.36
0.3	4.48	4.37	4.36	4.36	4.36	4.36	4.36
0.4	4.39	4.36	4.36	4.36	4.36	4.36	4.36
0.5	4.36	4.36	4.36	4.36	4.36	4.36	4.36

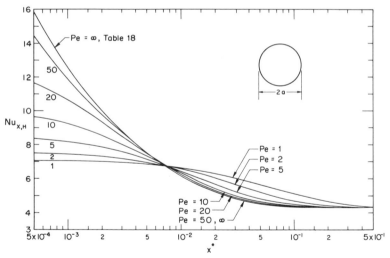

FIG. 24. Circular duct: $Nu_{x,H}$ as a function of x^* and Pe for the initial and boundary conditions of Fig. 23c for a fully developed velocity profile (from Table 19).

Bes [286] solved the same problem as Hennecke [68]. His analysis appears to be in error, since he did not find the above inflection point and his asymptotic Nusselt numbers approach different values for different Pe, instead of 4.364.

Hsu [311] considered the same problem as Hennecke [68], and solved it by constructing a series of product solutions, which led to the determination of a set of eigenfunctions expressing the radial variations of temperature. Hsu employed the Runge–Kutta integration scheme and determined the first 20 eigenvalues and eigenfunctions for the heated and adiabatic regions. $Nu_{x,H}$ obtained by Hsu are in excellent agreement with those of Hennecke.

Davis [312] also studied the same problem analyzed by Hsu [311], however, by a different method, so that his eigenvalues were presented by a transcendental equation in terms of the confluent hypergeometric function. The determination of eigenvalues by Davis's method was consequently simpler and more accurate. The eigenvalues of Davis and Hsu are identical up to first six digits.

Verhoff and Fisher [288] analyzed this problem (Fig. 23c) first by employing an inverse-tangent transformation to reduce the infinite region into a finite region, and then carrying out the solution by a finite difference method. Their results are in good agreement with those of Hennecke [68] and Hsu [311].

Smith et al. [313] considered a more general problem than that of Hsu [311]. They allowed the specified wall heat flux in Fig. 23c to have arbitrary circumferential variations, and also allowed for thermal energy sources and viscous dissipation. They proposed a solution in the form of a series of product solutions to the differential equation. Use was made of the orthogonality relationships among solution components, which yielded a simpler procedure to obtain the fluid temperature distribution in the entrance region. Hsu [311] did not make the use of these relationships.

f. Exponential Wall Heat Flux, (H5)

The exponential axial wall heat flux boundary condition is described by Eqs. (62) and (63). Spektor and Rassadkin [314] analyzed this (H5) thermal entrance problem for fully developed laminar flow through a circular tube. They employed the approximate method of Kadaner et al. [236] in which the partial differential energy equation was first transformed into ordinary differential equations. They graphically presented $Nu_{x,H5}$ as a function of x^* for the exponent $m = -20, -10, 0, 20,$ and 40 as a parameter. For all these exponents, $Nu_{x,H5}$ are about the same at $x^* = 10^{-3}$, and the fully developed values of Nu_{H5} are in good agreement with those of Table 9. Like the fully developed Nu_{H5}, the local Nusselt numbers are higher for positive values of m and lower for negative values of m compared to those for the case $m = 0$.

Based on the (H) thermal entrance solution of Siegel et al. [29], the (H5) thermal entrance solution can be obtained by the superposition method [7,29]. The local Nusselt number thus found is

$$Nu_{x,H5} = \left(\sum_{n=1}^{\infty} \frac{-C_n R_n(1)\beta_n^2}{2\beta_n^2 + m} \{1 - \exp[-(2\beta_n^2 + m)x^*]\} \right)^{-1} \quad (250)$$

where the constants β_n^2, $C_n R_n(1)$ are available in Table 17 for n up to 20, and are determined by Eqs. (236) and (237) for higher n. The fully developed Nu_{H5}, Eq. (186), is obtained from Eq. (250) for $x^* \to \infty$.

g. *Arbitrary Heat Flux Distribution*

Similar to the problem of arbitrary axial temperature distribution described on p. 118, the problem with arbitrary variations in axial heat flux can be handled by the superposition method, as first described by Siegel *et al.* [29]. For an arbitrary analytical axial wall heat flux distribution, Noyes [30] derived a closed-form integrated solution for the local Nusselt number. Hasegawa and Fujita [31] and Bankston and McEligot [32] refined the superposition method of Siegel *et al.* Shapovalov [315] considered an arbitrary wall heat flux distribution with arbitrary initial condition and obtained the temperature distribution in terms of hypergeometric functions. Dzung [316], Hsu [305], Kuga [317], Kays [7], and Reisman [318] considered the sinusoidal axial wall heat flux. They presented closed-form formulas for the wall temperature, and local and mean Nusselt numbers. Since it is difficult to experimentally generate a true sine wave axial wall heat flux, Reisman [318] approximated one sine wave heat flux imposed on the tube wall by three step-varying wall heat fluxes. Reisman derived analytical expressions for the wall temperature and local Nusselt numbers by Duhamel's superposition theorem. His experimentally measured t_w and Nu_x are in fair agreement with the theoretical predictions.

In the above cases, the peripheral wall heat flux distribution was assumed to be uniform. Bhattacharyya and Roy [319] investigated peripherally variable and axially constant wall heat flux distribution. The solution to this problem was obtained by the separation of variables technique with the related eigenvalues and constants reported in tabular form. As an example, they considered the following heat flux variations in the peripheral direction: (1) $q'' = q_m''(1 + 0.2 \cos \theta)$ and (2) $q'' = q_m''(1 + 0.2 \cos 2\theta)$. They showed that the first harmonic has a greater effect on wall temperature or Nusselt number than the higher harmonics. They also extended the above thermal entrance problem to the one with arbitrary wall heat flux in the peripheral as well as the axial direction by applying Duhamel's superposition theorem.

5. CONJUGATED PROBLEMS

In the four preceding subsections, a thermal boundary condition was specified at the wall for the forced convection problem, and the solution to the problem was obtained for the fluid medium only. In contrast, the conjugated problem is formulated for the entire solid body–fluid medium system, rather than using *a priori* fluid boundary conditions. Then a simultaneous solution to fluid and solid media is obtained as discussed on pp. 11–12.

Luikov et al. [18,19] solved the conjugated problem for the circular tube. The solution is presented in a closed-form expression involving complicated functions, integrals, and infinite series. Since no numerical results are presented, it is not possible to compare this solution to the conventional solution using an a priori boundary condition.

Mori et al. [20] solved the conjugated problem for the circular tube as formulated by Eqs. (14)–(20). Since by specification there was no internal thermal energy generation in the tube wall, and since the thermal boundary condition at the outside wall of the tube was also specified as uniform peripherally, the solution to their conjugated problem considered rigorously the effect of axial heat conduction in the tube wall. They considered two thermal boundary conditions specified at the outside wall of the circular tube: (1) constant heat flux, and (2) constant temperature. They also assumed the wall–fluid interface temperature distribution in the axial direction in a power series form with unknown coefficients. With this temperature distribution as the interface boundary condition, the solution to the energy equation for the fluid was obtained directly by superposing the Graetz solution as outlined by Sellers et al. [25] and Kays [7]. The solution to the energy equation for the wall was derived readily since all the boundary conditions were now known. Equating the temperature and heat fluxes across the solid and fluid media at the interface, they obtained the unknown coefficients for the power series, thus completing the solution to the conjugated problem. They presented the axially local Nusselt number in the form

$$
\text{Nu}_x = \text{Nu}\left(\frac{x}{L}, \frac{\text{Pe}\, D_\text{h}}{L}, \frac{a'}{L}, \frac{a'}{D_\text{h} R_\text{w}}, \frac{D_\text{h}}{L}\right) \tag{251}
$$

The detailed solutions for the interface temperature and Nu_x were presented graphically as a function of x/L for $0.001 \leq a'/L \leq 0.05$ and $1 \leq a'/D_\text{h} R_\text{w} \leq 5000$ when $\text{Pe}\, D_\text{h}/L = 50$ and $D_\text{h}/L = 0.1$. The following conclusions may be drawn from this study:

(1) The local Nusselt numbers for this conjugated problem falls between the Ⓗ and Ⓣ solutions of the conventional convection problem, for both boundary condition cases, constant heat flux and constant temperature specified at the outside wall.

(2) With constant heat flux specified at the outside wall: For a "thin" wall $(a'/L \to 0)$, the solution to the conjugated problem, in terms of Nu_x, approaches that for the Ⓗ conventional convection problem. However, for a "thick" wall, it approaches the Ⓣ solution because axial heat conduction tends to level out the temperature. The criteria for the "thin" wall

condition depend upon R_w; a wall may be considered thin when $a'/L = 0.0001$ for $R_w \simeq 2 \times 10^{-7}$ and when $a'/L = 0.001$ for $R_w > 10^{-5}$.

(3) With constant temperature specified at the outside wall: For a thin wall, the solution to the conjugated problem, in terms of Nu_x, approaches that for the Ⓣ conventional convection problem. However, for a thick wall, it approaches the Ⓗ solution. This is because the thick wall has a higher radial thermal resistance, which allows the temperature gradient to develop on the inside wall in the axial direction. A wall may be considered thin when $R_w < 0.01$.

(4) All of the foregoing conclusions were based on the results $Pe\,D_h/L = 50$. Mori *et al.* showed that for the case of constant heat flux at the outside wall, the results of the conjugated problem approach the Ⓗ solution for increasing values of $Pe\,D_h/L$, and they approach the Ⓣ solution for decreasing values of $Pe\,D_h/L$ for $a'/L = 0.03$ and $R_w = 0.03$.

Of the foregoing observations, the following points deserve special emphasis.

(1) For the case of specified constant heat flux at the outside wall, the axial heat conduction in the wall will reduce the interface Nu_x. The lowest Nu_x values will correspond to those for the conventional Ⓣ solution.

(2) For the case of specified constant wall temperatures at the outside wall, the finite resistance in the wall will increase the interface Nu_x. The largest Nu_x values will correspond to those for the conventional Ⓗ solution.

D. Simultaneously Developing Flow

The flow having both velocity and temperature profiles developing together is referred to as simultaneously developing flow. The associated forced convection heat transfer problem is referred to as the combined hydrodynamic and thermal entry length problem, or in short, as the combined entry length problem. The solutions for the circular tube are outlined separately below for the Ⓣ, Ⓗ, Ⓣ3, and Ⓓ𝑡 boundary conditions.

1. Constant Wall Temperature, Ⓣ

The combined entry length problem for the circular tube was first investigated numerically by Kays [40]. He considered the Ⓣ, Ⓗ, and Ⓓ𝑡 boundary conditions for the fluid with $Pr = 0.7$. The Langhaar velocity profile was employed for the axial velocity distribution. The radial velocity component and fluid axial heat conduction [in Eq. (31)] were neglected. The solution was obtained in the range $0.001 \leq x^* \leq 0.2$ by a finite difference method employing a relatively large grid size. The local and mean Nusselt numbers were tabulated and presented graphically. As shown later, the neglect of the radial velocity component overestimates the calculated Nusselt numbers for simultaneously developing flow in comparison to those by more accurate methods. Goldberg [320] extended Kays' solution to cover a Prandtl number range of 0.5 to 5. Goldberg's results are summarized by Rohsenow and Choi [5]. Tien and Pawelek [321] analyzed the $\widehat{T}$ combined entry length problem for fluids with high Prandtl number. A solution to simultaneously developing flow was obtained by using the Lévêque-type approximation and assuming a very thin thermal boundary layer. The solution is thus valid for small x^* and large Pe. They depicted $0.5Nu_{x,T}$ and $0.5Nu_{m,T}$ as a function of $4x^*$ (range 10^{-6} to 10^{-1}) for Pr = 5, 10, 15, 20, 40, and 100. Tien and Pawelek's results for Pr = 5 are in good agreement with those of Goldberg [5,320].

Stephan [322] employed an approximate series solution to solve the $\widehat{T}$ combined entry problem and correlated his results by an equation similar to Eq. (329). However, the predicted $Nu_{m,T}$ are as much as 30% high compared to Table 18 values; hence the equation is not presented here.

Kakaç and Özgü [215] employed Sparrow's velocity profile [162] and obtained a finite difference solution to the combined entry length problem for the $\widehat{T}$ and $\widehat{H}$ boundary conditions and Pr = 0.7. Their local Nusselt numbers are in good agreement with those of Ulrichson and Schmitz [323] for large x^*.

Ulrichson and Schmitz [323] refined Kays' solution [40] by utilizing the axial velocity component of Langhaar's solution and subsequently the radial component from the continuity equation. The solution to the combined entry length problem was obtained by a finite difference method for the fluid with Pr = 0.7. They found that the effect of radial velocity component (which was neglected by Kays) on $Nu_{x,T}$ was significant only for $x^* < 0.04$. In this region, $Nu_{x,T}$ obtained were lower than those determined by Kays [40]. The continuity equation is not satisfied when the radial velocity component is neglected in a developing velocity field. This neglect results in higher than actual $(\partial t/\partial r)_w$ and consequently higher $Nu_{x,T}$ as found by Kays. In the graph by Ulrichson and Schmitz of $Nu_{x,T}$ vs. x^*, the abscissa is misprinted as $16x^*$ instead of x^*.

Hornbeck [216] further refined the combined entry length problem by employing an all-numerical finite difference method for the $\widehat{T}$ and $\widehat{H}$ boundary conditions and Pr = 0.7, 2, and 5. Except for the linearization of the momentum equation at section $x = x_1$ by means of the velocity at $x = x_1 - \Delta x$, no additional idealizations were involved in the analysis. To obtain the velocity and subsequently the temperature profiles in the entrance region, variable grid sizes were used, with the fine grid size near the entrance and near the wall, and gradually a coarser grid as fully developed flow was approached. Relative to Ulrichson and Schmitz [323], his computed $\mathrm{Nu}_{x,\mathrm{T}}$ for $\mathrm{Pr} = 0.7$ is 4% lower at $x^* = 0.001$ and is in agreement for $x^* \geq 0.01$. Using the graphical results of Hornbeck, as no tabular results are available, Figs. 25 and 26 and Table 20 were prepared for $\mathrm{Nu}_{x,\mathrm{T}}$ and $\mathrm{Nu}_{m,\mathrm{T}}$ as functions of x^* and Pr. A comparison of Hornbeck's graphical results with those of Table 13 revealed that his Nusselt numbers asymptotically approached a value lower than 3.66. Consequently, $\mathrm{Nu}_{x,\mathrm{T}}$ and $\mathrm{Nu}_{m,\mathrm{T}}$ in Figs. 25 and 26

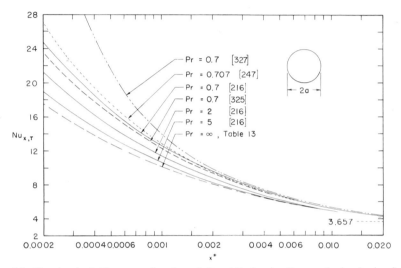

FIG. 25. Circular duct: $\mathrm{Nu}_{x,\mathrm{T}}$ as a function of x^* and Pr for simultaneously developing flow.

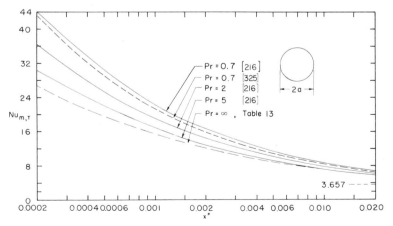

FIG. 26. Circular duct: $\mathrm{Nu}_{m,\mathrm{T}}$ as a function of x^* and Pr for simultaneously developing flow.

TABLE 20

Circular Ducts: $Nu_{x,T}$ and $Nu_{m,T}$ as Functions
of x^* and Pr for Simultaneously Developing Flow
(from the Graphical Results of Hornbeck [216])

x^*	$Nu_{x,T}$			$Nu_{m,T}$		
	Pr=.7	Pr=2	Pr=5	Pr=.7	Pr=2	Pr=5
0.0002	24.8	21.2	18.9	44.1	36.4	30.3
0.0003	21.0	18.2	16.4	37.8	31.1	26.6
0.0004	18.6	16.2	14.7	33.6	27.7	24.0
0.0005	16.8	14.8	13.6	30.5	25.2	22.1
0.0006	15.5	13.8	12.7	28.1	23.4	20.7
0.0008	13.7	12.4	11.5	24.6	20.8	18.4
0.0010	12.6	11.4	10.6	22.2	19.1	16.9
0.0015	10.8	9.8	9.2	18.7	16.2	14.4
0.002	9.6	8.8	8.2	16.7	14.4	12.8
0.003	8.2	7.5	7.2	14.1	12.4	11.1
0.004	7.3	6.8	6.5	12.4	11.1	9.9
0.005	6.7	6.2	6.05	11.3	10.2	9.2
0.006	6.25	5.8	5.70	10.6	9.5	8.6
0.008	5.60	5.3	5.27	9.5	8.5	7.7
0.010	5.25	4.93	4.92	8.7	7.8	7.2
0.015	4.60	4.44	4.44	7.5	6.8	6.3
0.020	4.28	4.17	4.17	6.8	6.2	5.8
∞	3.66	3.66	3.66	3.66	3.66	3.66

TABLE 21

Circular Duct: $Nu_{x,T}$ and $Nu_{m,T}$ as a
Function of x^* for $Pr = 0.7$ for
Simultaneously Developing Flow

x^*	Manohar [247]	Hwang [325]	
	$Nu_{x,T}$	$Nu_{x,T}$	$Nu_{m,T}$
0.00003571	62.25	60.58	85.37
0.00007143	42.98	41.63	67.38
0.0001071	35.27	33.07	57.28
0.0001429	30.77	28.09	50.52
0.0001786	27.72	24.98	45.73
0.0002143	26.13	22.91	42.06
0.0002857	22.78	20.26	36.91
0.0003571	20.50	18.53	33.40
0.0007143	14.89	14.03	24.84
0.001071	12.4	11.94	20.74
0.001429	11.0	10.65	18.46
0.001786	9.99	9.757	16.73
0.002143	9.26	9.086	15.57
0.002857	8.24	8.129	13.82
0.003571	7.54	7.469	12.61
0.007143	5.84	5.793	9.553
0.01071	5.11	5.081	8.165
0.01429	4.69	4.671	7.341
0.01786	4.42	4.409	6.777
0.02143	4.23	4.224	6.368
0.02857	3.998	3.993	5.801
0.03571	3.846	3.862	5.425
0.07143	3.641	3.674	4.580
0.07661	3.632	-	-
0.1071	-	3.655	4.273

and Table 20 for $Pr = 5$ have been slightly modified for $x^* > 0.008$ to asymptotically approach the $Pr = \infty$ curve[†] in these figures.

Hornbeck [216] also investigated the effect of inlet velocity profile and viscous dissipation on heat transfer. Using a modified Wang and Longwell velocity profile [191] at the inlet, he obtained $Nu_{m,T}$ higher than those for the uniform inlet velocity profile, 9% higher at $x^* = 0.00025$ and essentially no difference for $x^* \geq 0.01$. He also found $Nu_{m,T}$ to be higher when viscous dissipation was taken into account; about 5% higher for the range $0.00025 \leq x^* \leq 0.075$ with the Eckert number $Ec = -0.1$ and $Pr = 0.7$ ($Br = -0.07$).

Bender [217] numerically solved the combined entry length problem for the Ⓣ and Ⓗ boundary conditions. He graphically presented $(1 - \theta_m)$ as a function of x^* for $Pr = 0.1, 1.0, 2.4, 6.25, 10, 30$, and ∞.

Manohar [175] independently refined the work of Ulrichson and Schmitz [323] by solving the nonlinear momentum and linear energy equations numerically by an iterative procedure. Relative to Ulrichson and Schmitz, Manohar's $Nu_{x,T}$ is 2% lower at $x^* = 0.001$ and identical for $x^* \geq 0.01$. Manohar's $Nu_{x,T}$ [247] are compared with Hornbeck's results in Fig. 25. Manohar's $Nu_{x,T}$ approaches 3.63 asymptotically, a value 1% lower than the exact value of 3.66. This error may be due to the derivatives at the wall calculated from the first-order one-sided difference formulas [247]. Incidentally, curves 1 and 2 in Fig. 1 of Manohar [175] are drawn incorrectly [247].

Hwang and Sheu [324] also independently analyzed the Ⓣ and Ⓗ simultaneously developing flow for the circular tube. The problem was solved by two methods: (1) a method similar to that used by Ulrichson and Schmitz [323] as described on p. 139, and (2) a complete finite difference method. $Nu_{x,T}$ and $Nu_{m,T}$ were calculated for $Pr = 0.7, 1, 7$, and 10. Their results for $Pr = 0.7$ are presented in Table 21 and Figs. 25 and 26. $Nu_{x,T}$ of Hwang and Sheu is 2.5% lower than that of Hornbeck at $x^* = 0.001$, and $Nu_{m,T}$ is 3.6% lower at $x^* = 0.001$. $Nu_{x,T}$ of Hwang and Sheu for $Pr = 7$ and 10 are lower than those of Hornbeck for $Pr = 5$ and agree closely for $Pr = \infty$ (Table 13). Hwang and Sheu also plotted the ratio of the radial convective term to the axial convective term, $(v\,\partial t/\partial r)/(u\,\partial t/\partial x)$, for $Pr = 0.7$. They showed that the magnitude of this ratio decreases very slowly with increasing axial distances. Close to the fully developed region, $x^* = 0.036$, the magnitude of the above ratio is greater than 0.1 for $r/a \geq 0.7$.

Javeri [207] solved the Ⓣ combined entry length problem for the circular tube by the Galerkin–Kantorowich variational method. He also investigated the influence of the radial velocity component on $Nu_{x,T}$ and found the

---

† Refer to p. 16 for the meaning of the $Pr = \infty$ case.

results in close agreement with those of Ulrichson and Schmitz [323] and Hwang and Sheu [324]. He tabulated $Nu_{x,T}$ as a function of $4x^*$ for $Pr = 0$, 0.1, 0.7, 1, 10, and ∞. His $Nu_{x,T}$ for the $v \neq 0$ case are within 2% of Hornbeck [216] and Hwang [325], Tables 20 and 21, for $x^* > 0.0005$. Based on a comparison on Fig. 25, it appears that Javeri's $Nu_{x,T}$ for the above Prandtl numbers are accurate within 2% for $x^* > 0.0005$, except for possibly some typographical errors.

The thermal entrance length $L^*_{th,T}$ for simultaneously developing flow and $Pr = 0.7$ is found as

$$L^*_{th,T} = 0.037 \tag{252}$$

based on both sets of Table 21. $L^*_{th,T}$ for $Pr = \infty$ from Eq. (208) is 0.033. Thus, as expected, $L^*_{th,T}$ increases with decreasing Pr and is a weak function of Pr for $Pr \geq 0.7$.

For small x^*, Kays [40] derived an expression for $Nu_{m,T}$ based on the Pohlhausen solution for a flat plate, for simultaneously developing flow. The correct expression is presented by Kreith [326] as follows:

$$Nu_{m,T} = \frac{1}{4x^*} \ln\left(\frac{1}{1 - 2.654x^{*0.5}Pr^{-0.167}}\right) \tag{253}$$

Churchill and Ozoe [273] reviewed the literature related to the refinement of the Pohlhausen solution. They showed that the Pohlhausen solution is valid for high Pr fluids and proposed the following equation for simultaneously developing flow in a circular tube for small x^*:

$$Nu_{x,T} - \frac{1}{2} Nu_{m,T} = \frac{0.6366[(4/\pi)x^*]^{-1/2}}{[1 + (Pr/0.0468)^{2/3}]^{1/4}} \tag{254}$$

$Nu_{m,T}$ of Eqs. (253) and (254) agree with each other within $\pm 2\%$ for $Pr > 2$ for $10^{-7} \leq x^* \leq 10^{-3}$. A comparison with Hornbeck's results in Table 20 shows that $Nu_{x,T}$ predicted by Eq. (254) would represent an asymptote for $x^* \leq 10^{-3}$ and $0.7 \leq Pr \leq 5$. Churchill and Ozoe then combined Eqs. (225) and (254) to cover the complete Pr and x^* range, as follows:

$$\frac{Nu_{x,T} + 1.7}{5.357\{1 + [(388/\pi)x^*]^{-8/9}\}^{3/8}}$$
$$= \left[1 + \left(\frac{\pi/(284x^*)}{\{1 + (Pr/0.0468)^{2/3}\}^{1/2}\{1 + [(388/\pi)x^*]^{-8/9}\}^{3/4}}\right)^{4/3}\right]^{3/8} \tag{255}$$

This equation approaches Eq. (225) for $Pr \to \infty$. Churchill and Ozoe also showed that this equation approaches an asymptotic formula for $Pr = 0$ and low values of x^*. It approaches Eq. (254) for $x^* \to 0$ and 3.657 for $x^* \to \infty$. For $0.0002 \leq x^* \leq 0.004$, Eq. (255) predicts $Nu_{x,T}$ higher than those of

Table 20 within 6% for Pr = 0.7, 8% for Pr = 2, and 11% for Pr = 5. The agreement between $Nu_{x,T}$ of Eq. (255) and Table 20 degenerates further for higher x^*, up to 25% at $x^* = 0.020$.

Zeldin and Schmidt [327] employed a velocity profile based on the complete Navier–Stokes equations and solved the Ⓣ thermal entry length problem, Eq. (31). They stated that the flow work term $u(dp/dx)/J$ in Eq. (24) is not negligible for a gas; moreover, it is appropriate to consider the gas as having a constant density along the streamline and constant μ and k. With these idealizations, i.e., considering flow work and constant ρ, the resulting energy equation is of the same form as Eq. (31), but the thermal diffusivity α contains c_v instead of c_p. On this basis, with Pr = 0.505 for air based on c_v, instead of 0.707 based on c_p, they determined $Nu_{x,T}$ and $\Phi_{m,T}$ for simultaneously developing flow for Re = 500. They also obtained a thermal entry length solution for a fully developed velocity profile. Their results are tabulated in [13,328]. Their $Nu_{x,T}$ for the fully developed velocity profile include the effect of fluid axial heat conduction and agree with their earlier results in [66] for Pe = 252.5. However, as noted on p. 113, their earlier results for $x^* < 0.0025$ diverge from the theory for Pr = ∞ and hence are questionable for low x^* for different Pe. Their results for simultaneously developing flow [327] are compared in Fig. 25 with the boundary layer type solution. Here again the solution appears to diverge below $x^* = 0.0025$. Consequently, these numerical results of Zeldin and Schmidt also appear to be questionable for low x^*. For $x^* > 0.0025$, $Nu_{x,T}$ of Zeldin and Schmidt in Fig. 25 are about 2% higher than those of [175,216] for which the effect of flow work is neglected. Thus, it appears that the effect on heat transfer of the flow work term is negligible. This can also be inferred from the following discussion on the Ⓗ boundary condition results.

To assess the order of magnitude of the flow work term $u(dp/dx)/J$ in the energy equation [e.g., Eq. (24)], it is compared with the convective term $\rho c_p u(\partial t/\partial x)$ for the Ⓗ boundary condition (for which $\partial t/\partial x$ is constant) as a convenience. If it is assumed that (dp/dx) is approximately constant, and both terms are integrated over the flow cross-sectional area, the results are $u_m A_c(dp/dx)/J$ and $(\rho u_m A_c)c_p(\partial t/\partial x)$. Dividing these terms by the perimeter P, it can be shown that these terms are E_{std} [see Eq. (573)] and q'', respectively. For a gas turbine regenerator application, typical values are $E_{std} = 0.005$ hp/ft$^2 \simeq 12.5$ Btu/hr ft^2 and $q'' = 1500$ Btu/hr ft^2. Thus the flow work is less than 1% of the convective heat transfer; and it can be concluded to be negligible in its influence on the Nusselt number.

2. CONSTANT WALL HEAT FLUX, Ⓗ

The combined entry problem with the Ⓗ boundary condition was first investigated by Kays [40] for Pr = 0.7. He employed the Langhaar velocity profile [153] for the axial velocity distribution. The radial velocity component and fluid axial heat conduction were neglected. The solution in the range $0.001 \leq x^* \leq 0.2$ was numerically obtained by a finite difference method employing a relatively large grid size. $Nu_{x,H}$ were tabulated and presented graphically as a function of $4x^*$. As shown later, the neglect of the radial velocity component overestimates the calculated $Nu_{x,H}$ in comparison to those by more accurate methods.

Heaton et al. [44] used a different approach, but again employing the Langhaar velocity profile. They obtained a generalized entry region temperature profile that could be used in the energy integral equation. The solutions were then obtained in terms of Bessel and Thomson functions for the entire family of concentric annular ducts for the Ⓗ boundary condition and $Pr - 0.01$, 0.7, and 10. These $Nu_{x,H}$ are tabulated by Kays [7]. As mentioned on p. 71, Shumway and McEligot [45] pointed out an error in the analysis for the pressure drop by Heaton et al. [44]. However, $Nu_{x,H}$ of Heaton et al. are correct. For $x^* \leq 0.004$, the $Nu_{x,H}$ of Heaton et al. lie between those of Manohar [175] and Hornbeck [216]. For $x^* > 0.004$, Heaton's $Nu_{x,II}$ are about 3% higher.

Petukhov and Chzhen-Yun [304] obtained the Ⓗ combined entry length solution by an integral method. For the hydrodynamic entry length solution, they equated Langhaar's equation for the velocity of the potential core to that by Schiller [49] to arrive at the unknown velocity boundary layer thickness as a function of x^+. Subsequently, the axial velocity distribution in the boundary layer was obtained from Schiller's assumed parabolic velocity distribution in the boundary layer. Petukhov and Chzhen-Yun considered a logarithmic form for the thermal boundary layer thickness to subsequently solve the energy equation. They determined and presented graphically $Nu_{x,H}$ as a function of x^* for $Pr = 0.7$, 1, 10, 100, 1000, and ∞ for $10^{-7} \leq x^* \leq 10^{-1}$. Their results for the case of developed velocity profile are in excellent agreement with those of Siegel et al. [29]. They showed that $Nu_{x,H}$ of Kays [40] for $Pr = 0.7$ were high by about 20% for $10^{-3} \leq x^* \leq 0.04$.

Roy [329] considered three regions in the thermal entrance: simultaneous development of the velocity and temperature profiles, fully developed velocity and developing temperature profiles, and both profiles developed. The problem was handled by an integral method for the Ⓗ boundary condition for $Pr = 1$, 10, 100, and 1000.

Kakaç and Özgü [215] employed Sparrow's velocity profile [162] and obtained a finite difference solution for the combined entry length problem for $Pr = 0.7$. Their computed $Nu_{x,H}$ are in good agreement with those of Ulrichson and Schmitz [323] for large x^*.

Ulrichson and Schmitz [323] refined Kays' solution [40] by employing the Langhaar velocity profile as the axial component and subsequently utilizing the radial component from the continuity equation. The solution was obtained by a finite difference method for a fluid with $Pr = 0.7$. They found that the effect of the radial velocity component (which was neglected by Kays) on $Nu_{x,H}$ was significant only for $x^* < 0.04$. In this region, $Nu_{x,H}$ obtained were lower than those by Kays [40]. In the plot of $Nu_{x,H}$ vs. x^*, the abscissa is misprinted as $16x^*$ instead of x^* in [323].

Hwang and Sheu [324] independently analyzed the same problem by the method of Ulrichson and Schmitz [323] for $Pr = 0.1$, 0.7, 7, and 10 and presented graphically $Nu_{x,H}$ and $Nu_{m,H}$.

Hornbeck [216] further refined the combined entry problem by employing an all-numerical finite difference method for the Ⓗ boundary condition and $Pr = 0.7$, 2, and 5. All the pertinent idealizations in the solution are mentioned earlier on p. 139. His computed $Nu_{x,H}$ are in close agreement with the results of Ulrichson and Schmitz [323]. Based on the graphical results of Hornbeck, since tabular results are not available, Fig. 27 and Table 22 are prepared for $Nu_{x,H}$ as functions of x^* and Pr.

Hornbeck also investigated the effect of inlet velocity profile and viscous dissipation on $Nu_{x,H}$. These effects were opposite to those found for $Nu_{m,T}$ (as described on p. 142). Specifically, $Nu_{x,H}$ for the modified inlet velocity profile or viscous dissipation, is lower than $Nu_{x,H}$ for the uniform velocity profile and no viscous dissipation.

Bender [217] numerically solved the combined entry length problem for the Ⓗ boundary condition. He graphically presented $2/Nu_{x,H}$ as a function of x^* for $Pr = 0.1$, 1, 10, and ∞.

Manohar [175] independently further refined the solution of Ulrichson and Schmitz [323] by an iterative numerical solution. Manohar's $Nu_{x,H}$ is 3% higher at $x^* = 0.001$ than that of Ulrichson and Schmitz, and identical for $x^* > 0.002$. Manohar's $Nu_{x,H}$ are compared with Hornbeck's results in Fig. 27 and are reported in Table 22 [247].

Bankston and McEligot [330] employed a finite difference method and analyzed turbulent and laminar heat transfer to gases with large property variations in a circular tube. They considered the constant and variable axial heat flux boundary conditions. Their results for the constant axial heat flux boundary condition, constant properties, and a developed laminar entering velocity profile are in excellent agreement with those of Table 18. Their $Nu_{x,H}$ for simultaneously developing flow for $Pr = 0.7$ are presented in Table 22, and agree very well with those of Hornbeck [216] as shown in Fig. 27. The thermal entrance length $L_{th,H}^*$ for simultaneously developing flow and $Pr = 0.7$ is found as

$$L_{th,H}^* = 0.053 \qquad (256)$$

based on $Nu_{x,H}$ of Bankston and McEligot. $L_{th,H}^*$ for $Pr = \infty$ from Eq. (238)

TABLE 22

Circular Duct: $Nu_{x,H}$ for Simultaneously Developing Flow

x*	$Nu_{x,H}$ [216]			$Nu_{x,H}$ [330]	x*	$Nu_{x,H}$ [247]
	Pr=.7	Pr=2	Pr=5	Pr=.7		Pr=.7
0.0002	–	–	27.5	–	0.00003571	88.85
0.0003	29.8	26.3	23.7	–	0.00007143	63.16
0.0004	26.3	23.4	21.1	–	0.0001071	51.46
0.0005	23.7	21.2	19.2	24.18	0.0001429	44.56
0.0006	21.8	19.6	17.8	–	0.0001786	39.88
0.0008	19.2	17.3	15.8	–	0.0002143	37.52
0.0010	17.5	15.8	14.4	17.62	0.0002857	32.93
0.0015	14.7	13.4	12.2	–	0.0003571	29.63
0.002	13.0	11.8	10.9	–	0.0007143	21.25
0.0025	–	–	–	11.83	0.001071	17.55
0.003	11.0	10.1	9.4	–	0.001429	15.37
0.004	9.8	9.0	8.5	–	0.001786	13.90
0.005	9.0	8.2	7.9	8.936	0.002143	12.82
0.006	8.4	7.7	7.4	–	0.002857	11.33
0.008	7.5	6.9	6.7	–	0.003571	10.31
0.010	6.9	6.4	6.2	6.928	0.007143	7.854
0.015	6.0	5.7	5.6	–	0.01071	6.792
0.020	5.5	5.3	5.2	–	0.01429	6.179
0.025	–	–	–	5.267	0.01786	5.774
0.05	–	–	–	4.617	0.02143	5.486
0.10	–	–	–	4.385	0.02857	5.108
0.25	–	–	–	4.363	0.03571	4.879
∞	4.36	4.36	4.36	4.364	0.07143	4.460
					0.07661	4.439

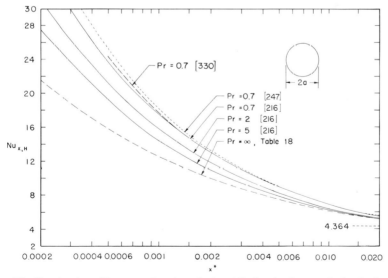

FIG. 27. Circular duct: $Nu_{x,H}$ as a function of x^* and Pr for simultaneously developing flow (from Table 22).

is 0.043. Thus, as expected, $L_{th,H}^*$ increases with decreasing Pr and is a weak function of Pr for Pr $\geq$ 0.7.

Based on the flat plate solution, Churchill and Ozoe [306] proposed the following equation for $Nu_{x,H}$ for small x^*:

$$Nu_{x,H} = \frac{[(4/\pi)x^*]^{-1/2}}{[1 + (Pr/0.0207)^{2/3}]^{1/4}} \tag{257}$$

This equation is an asymptotic formula for x^* approaching zero. According to Churchill and Ozoe, Kays' [7] flat plate solution expressed by his Eq. (10-41) is valid only for high Pr fluids. A comparison with Hornbeck's results in Table 22 shows that $Nu_{x,H}$ predicted by Eq. (257) would represent an asymptote for $x^* \lesssim 10^{-5}$ and $0.7 \leq Pr \leq 5$. Churchill and Ozoe then combined Eqs. (243) and (257) to cover the complete Pr and x^* range as

$$\frac{Nu_{x,H} + 1}{5.364\{1 + [(220/\pi)x^*]^{-10/9}\}^{3/10}}$$
$$= \left[1 + \left(\frac{\pi/(115.2x^*)}{\{1 + (Pr/0.0207)^{2/3}\}^{1/2}\{1 + [(220/\pi)x^*]^{-10/9}\}^{3/5}}\right)^{5/3}\right]^{3/10} \tag{258}$$

This equation approaches Eq. (243) for Pr $\to \infty$, Eq. (257) for $x^* \to 0$, and 4.364 for $x^* \to \infty$. Churchill and Ozoe also showed that this equation approaches an asymptotic formula for Pr = 0 and low values of x^*. This equation agrees with Hornbeck's $Nu_{x,H}$ of Table 22 within 7% for Pr = 0.7, 5% for Pr = 2, and 3% for Pr = 5. The agreement with Manohar's results is within 9% for Pr = 0.7.

Collins [331] analyzed the (H) combined entry length problem for the circular tube by a finite difference method. He included the effect of viscous dissipation and viscosity variation with temperature. His analytical predictions agreed well with the experimental results of Butterworth and Hazell [43]. He concluded that the effect of viscous dissipation is only minor and reduces the entrance region Nusselt numbers slightly.

McMordie and Emery [214] considered finite fluid axial heat conduction for the (H) simultaneously developing flow. They presented $Nu_{x,H1}$ as a function of $1/x^*$ for Pr = 0.005, 0.02, 0.03 and Re = 500, 1000, 1500, and 2000.

All the investigations described so far are concerned with either thermally developing flow alone or both the thermal and velocity boundary layers developing from the same location. However, heat transfer experiments are often conducted in an apparatus where heating starts after a short hydrodynamic entry section. Hence, the thermal entry length solutions with heating started at different locations in the hydrodynamic entry length region are useful. Such studies have been conducted by Butterworth and Hazell [43] and Bankston and McEligot [330].

Butterworth and Hazell analyzed the problem for Pr = 60 to 550, using Langhaar's axial velocity profile and neglecting the radial velocity component. They found that the effect of heating started at different locations in the hydrodynamic entry length was negligible for large Pr > 500. Their theoretical predictions did not take into account the effects of viscous dissipation and viscosity variations with temperature. Their theoretical Nusselt numbers were in good agreement with their experimental values for Pr $\lesssim$ 300, but were lower for Pr $\sim$ 500. Butterworth and Hazell attributed the measured high heat transfer coefficients to significant viscous dissipation and viscosity variation with temperature. Collins [331] found that the effect of viscosity variation with temperature increases Nusselt numbers as found by Butterworth and Hazell. However, viscous dissipation slightly reduces the entrance Nusselt numbers.

Bankston and McEligot [330] employed an all-numerical method and obtained $Nu_{x,H}$ for air (Pr = 0.7) with heating started at $x_{hy}^+ = x_{hy}/(D_h \, Re) = 0, 0.003, 0.006, 0.012$, and 0.024, where x_{hy} is the duct length from the entrance to the section where heating starts (see Fig. 28). The ratio of this $Nu_{x,H}$ to $Nu_{x,H}$ of Table 18 (parabolic entrance velocity profile), designated as ζ, is presented in Table 23 and Fig. 28. From these results, it is found that if the heating starts at $x_{hy}^+ > 0.006$, the effect of a partially developed velocity profile on the entrance region Nusselt number is less than 10%.

3. (T3) BOUNDARY CONDITION

McKillop et al. [231] analyzed the simultaneously developing flow of a non-Newtonian fluid through a circular tube with the (T3) boundary condition. They included the effect of temperature-dependent viscosity. The solution, obtained by a finite difference method, is presented graphically as $Nu_{x,T3}$ vs. $4x^*$ for $R_w = 0, 0.005, 0.025, 0.25, 0.1$, and ∞ for a Newtonian fluid with Pr = 0.7. Their fully developed Nu_{T3} agree well with those reported in Table 7. Like other solutions, their $Nu_{x,T3}$ increases with decreasing values of x^*. As x^* approaches zero, $Nu_{o,x}$ approaches the value of $1/R_w$.

McKillop et al. [231] correlated their numerical results by a least squares method in the following form:

$$\ln\left(\frac{Nu_{x,T3}}{Nu_{T3}}\right) = \sum_{j=1}^{3} a_j \left[\ln\left(\frac{1}{4x^*}\right)\right]^j \qquad (259)$$

with a maximum deviation of 2% for $1.25 \times 10^{-4} < x^* < 0.025$. The coefficients a_j for the case of a Newtonian fluid are presented in Table 24.

$Nu_{x,T}$ of McKillop et al. from Eq. (259) for $R_w = 0$ and Pr = ∞ agrees with the Graetz solution Table 13 within 1%, while $Nu_{x,T}$ magnitudes for $R_w = 0$ and Pr = 0.7 are up to 10% higher than those of Hornbeck reported in Table 20. $Nu_{x,H}$, corresponding to $R_w = \infty$ and Pr = 0.7, are up to 8% higher than those of Hornbeck in Table 22.

TABLE 23

CIRCULAR DUCT: ζ FOR $Pr = 0.7$ FOR A DEVELOPING
TEMPERATURE PROFILE AND A PARTIALLY DEVELOPED
VELOCITY PROFILE (FROM BANKSTON AND McELIGOT [330])[†]

x*	ζ				
	$x^+_{hy=0}$	0.003	0.006	0.012	0.024
0.0005	1.526	1.158	1.100	1.055	1.022
0.001	1.404	1.153	1.098	1.054	1.022
0.0025	1.272	1.137	1.092	1.051	1.021
0.005	1.192	1.115	1.082	1.047	1.019
0.01	1.126	1.086	1.064	1.038	1.016
0.025	1.058	1.042	1.033	1.021	1.008
0.05	1.021	1.015	1.012	1.007	1.002
0.1	1.002	1.002	1.002	1.001	1.000
0.25	1.000	1.000	1.000	1.000	1.000

[†] ζ is the ratio of $Nu_{x,H}$ for heating started at x_{hy}/D_h Re
to $Nu_{x,H}$ for the parabolic entering velocity profile (Table 18).

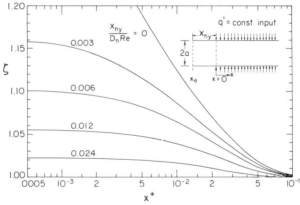

FIG. 28. Influence of partial flow development on the thermal entrance heat transfer for a circular duct for $Pr = 0.7$ (from Table 23).

TABLE 24

CIRCULAR DUCT: Ⓣ3 COMBINED ENTRY LENGTH SOLUTION BY
McKILLOP et al. [231][‡]

R_w	Pr = 0.7			Pr = 1		
	a_1	a_2	a_3	a_1	a_2	a_3
0	−0.125	0.0906	−0.00463	−0.130	0.0886	−0.00438
0.005	−0.138	0.0959	−0.00488	−0.136	0.0911	−0.00438
0.025	−0.150	0.105	−0.00558	−0.138	0.0966	−0.00482
0.25	−0.117	0.103	−0.00536	−0.121	0.0996	−0.00548
∞	−0.0977	0.0937	−0.00520	−0.106	0.0923	−0.00496
	Pr = 10			Pr = ∞		
0	−0.0932	0.0653	−0.00264	−0.102	0.0746	−0.00410
0.005	−0.0914	0.0647	−0.00240	−0.102	0.0756	−0.00416
0.025	−0.0949	0.0690	−0.00268	−0.105	0.0782	−0.00434
0.25	−0.0904	0.0741	−0.00338	−0.0899	0.0776	−0.00450
∞	−0.0783	0.0678	−0.00292	−	−	−

[‡] The constants a_1, a_2 and a_3 of Eq. (259).

$\mathrm{Nu}_{x,\mathrm{T3}}$ of McKillop *et al.* for $R_w = 0.005$ agree with those of Javeri [207] of Table 24a within 3, 5, and 5%, except for the endpoints, for Pr = 10, 1, and 0.7, respectively. $\mathrm{Nu}_{x,\mathrm{T3}}$ of McKillop *et al.* for $R_w = 0.250$ agree with those of Table 24a within 3 and 4%, except for the endpoints, for Pr = 10 and 1, respectively. $\mathrm{Nu}_{x,\mathrm{T3}}$ of McKillop *et al.* for $R_w = 0.25$ and Pr = 0.7 agree with those of Table 24a within 7% for $0.005 \le x^* \le 0.25$.

Javeri [207] analyzed the (T3) simultaneously developing flow by the Galerkin–Kantorowich variational method, and presented $\mathrm{Nu}_{x,\mathrm{T3}}$ as

TABLE 24a

Circular Duct: $\mathrm{Nu}_{x,\mathrm{T3}}$ as a Function of x^*, Pr, and R_w for
Simultaneously Developing Flow (from Javeri [207])

x^*	$\mathrm{Nu}_{x,\mathrm{T3}}$					
	Pr = ∞	Pr=10	Pr=1	Pr=.7	Pr=.1	Pr=0
			$R_w=0.005$			
0.00025	15.87	18.27	24.78	26.24	34.64	43.38
0.00050	12.98	14.04	17.05	17.81	22.93	30.17
0.00125	9.621	9.837	11.64	12.04	14.54	18.78
0.0025	7.558	7.501	8.597	8.874	10.64	13.71
0.0050	6.061	5.977	6.514	6.691	7.914	10.23
0.0125	4.672	4.677	4.776	4.872	5.649	7.385
0.025	4.026	4.027	4.037	4.090	4.666	6.259
0.050	3.722	3.723	3.733	3.740	4.176	5.869
0.125	3.669	3.669	3.669	3.669	3.892	5.832
0.25	3.666	3.666	3.667	3.667	3.750	5.832
			$R_w=0.05$			
0.00025	18.12	22.52	29.69	31.26	40.59	50.35
0.00050	14.31	16.41	20.75	21.72	28.14	37.00
0.00125	10.40	11.02	13.43	13.97	17.49	23.70
0.0025	8.154	8.276	9.738	10.10	12.47	17.00
0.0050	6.469	6.471	7.217	7.441	9.043	12.36
0.0125	4.924	4.927	5.126	5.247	6.204	8.518
0.025	4.193	4.194	4.244	4.307	4.974	6.942
0.050	3.838	3.838	3.853	3.866	4.348	6.310
0.125	3.763	3.763	3.764	3.764	4.008	6.224
0.25	3.763	3.763	3.763	3.763	3.855	6.224
			$R_w=0.1$			
0.00025	18.68	22.99	30.56	32.40	43.45	53.19
0.00050	14.75	16.81	21.58	22.89	30.11	38.89
0.00125	10.77	11.39	14.09	14.76	18.72	25.32
0.0025	8.468	8.590	10.26	10.70	13.34	18.35
0.0050	6.705	6.715	7.608	7.872	9.651	13.41
0.0125	5.110	5.113	5.368	5.504	6.576	9.255
0.025	4.329	4.331	4.407	4.479	5.223	7.473
0.050	3.935	3.937	3.965	3.974	4.495	6.684
0.125	3.845	3.846	3.846	3.846	4.105	6.547
0.25	3.844	3.844	3.844	3.844	3.941	6.547
			$R_w=0.25$			
0.00025	20.55	26.09	34.99	34.96	49.72	54.41
0.00050	15.82	18.69	23.87	24.62	33.26	39.32
0.00125	11.43	12.36	15.22	15.74	20.13	25.68
0.0025	8.938	9.287	11.02	11.43	14.32	19.02
0.0050	7.112	7.201	8.159	8.429	10.39	14.26
0.0125	5.389	5.391	5.760	5.884	7.118	10.09
0.025	4.574	4.578	4.684	4.756	5.621	8.181
0.050	4.120	4.129	4.165	4.171	4.759	7.289
0.125	4.001	4.003	4.003	4.003	4.285	7.088
0.25	4.000	4.000	4.000	4.000	4.105	7.087

functions of R_w and Pr as listed in Table 24a. His Nu_{T3} are in excellent agreement with those of Table 7. His $Nu_{x,T}$ for the Graetz problem and for the Ⓣ simultaneously developing flow problem are compared with the literature results on pp. 103 and 142, respectively.

4. Ⓐ̲ₜ BOUNDARY CONDITION

Simultaneously developing flow for the circular tube with constant wall-to-fluid bulk mean temperature difference Ⓐ̲ₜ boundary condition was investigated by Kays [40]. He employed the Langhaar velocity profile for the axial component and neglected the radial velocity component. The solution was obtained by a finite difference method. The local Nusselt numbers are tabulated and presented graphically as a function of $4x^*$. The local Nusselt numbers fall between the Ⓣ and Ⓗ solutions and approach those for the Ⓣ and Ⓗ boundary conditions for x^* approaching 0 and ∞, respectively.

Chapter VI

Parallel Plates

The parallel plates duct is the geometry amenable to the simplest mathematical treatment, even more so than the circular duct. Consequently, laminar flow and heat transfer for the parallel plates have been analyzed in great detail. The parallel plates duct geometry is a limiting geometry for the family of rectangular ducts and also for concentric annular ducts. For most cases, it forms an upper bound for fluid friction and heat transfer for the two duct classes. Since these geometries are widely used in fluid flow and heat transfer devices, even though parallel plates as such are not extensively used, the detailed analytical results for laminar flow and heat transfer for parallel plates are described below. Very few experimental results exist for duct geometries approaching the parallel plates geometry.

A. Fully Developed Flow

The fully developed laminar velocity profile and friction factor for a plate spacing of $2b$ with coordinate axes at the center are given by Rohsenow and Choi [5] and McCuen et al. [332] as

$$u = \tfrac{1}{2}c_1(y^2 - b^2) \tag{260}$$

$$u_m = -\tfrac{1}{3}c_1 b^2 \tag{261}$$

$$f\,\text{Re} = 24 \tag{262}$$

The heat transfer results are now described separately for each of the boundary conditions.

153

1. CONSTANT WALL TEMPERATURE AND WALL HEAT FLUX, Ⓣ AND Ⓗ

The Ⓣ boundary condition problem was first studied by Nusselt [333] in 1923. Later, Lévêque [199], Norris and Streid [334], and Hahnemann and Ehret [335] independently investigated the Ⓣ problem. Glaser [237] investigated fully developed laminar heat transfer for the Ⓗ boundary condition. The Ⓗ1–Ⓗ4 boundary conditions are all identical for parallel plates, as is the case for the circular tube, and are all designated as the Ⓗ boundary condition.

Depending upon the temperature or heat flux specified at either wall, there are four fundamental problems for heat transfer. These are shown in Fig. 8 and discussed on p. 32. As mentioned there, the fundamental boundary conditions of the fourth and fifth kinds are identical for the parallel plates.

McCuen *et al.* [332] obtained the four fundamental solutions for parallel plates. The fully developed temperature profiles corresponding to these four solutions are presented in Fig. 29 along with a thermal entrance profile.

The fully developed Nusselt numbers corresponding to these four solutions are

$$\text{First kind:} \quad \text{Nu}_1 = \text{Nu}_2 = 4 \qquad (263)$$

$$\text{Second kind:} \quad \text{Nu}_1 = 0, \qquad \text{Nu}_2 = 5.385 \qquad (264)$$

$$\text{Third kind:} \quad \text{Nu}_1 = 0, \qquad \text{Nu}_2 = 4.861 \qquad (265)$$

$$\text{Fourth kind:} \quad \text{Nu}_1 = \text{Nu}_2 = 4 \qquad (266)$$

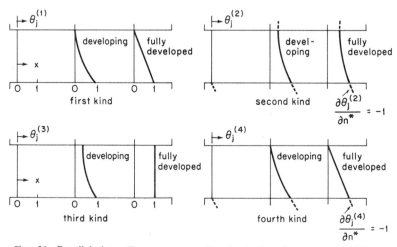

FIG. 29. Parallel plates: Temperature profiles for the four fundamental problems.

Here subscripts 1 and 2 refer to walls 1 and 2 in Fig. 8. The above Nusselt numbers are based on the difference between the wall temperature and the fluid bulk mean temperature.

As mentioned previously, if viscous dissipation and internal thermal energy sources are neglected in Eq. (24), the resulting equation is linear and homogeneous; then any complex problem can be handled by the superposition of these four fundamental solutions. Of particular interest are three problems described in Fig. 30. The Nusselt numbers for these special cases are outlined separately in the following subsection. They are defined as

$$\mathrm{Nu}_j = \frac{q_j'' D_h}{k(t_j - t_m)} \tag{267}$$

where j is wall 1 or 2, and t_j is the temperature of the jth wall.

a. *Constant Temperatures at Each Wall*

The temperature at each wall is specified as shown in Fig. 30a. The fully developed Nusselt number depends upon whether these temperatures are different or equal:
If $t_{w1} \neq t_{w2}$,

$$\mathrm{Nu}_1 = 4, \qquad \mathrm{Nu}_2 = 4 \tag{268}$$

If $t_{w1}, t_{w2} = t_w$,

$$\mathrm{Nu}_T = 7.54070087 \tag{269}$$

The special case of equal wall temperatures is the ⓣ boundary condition. The corresponding Nusselt number at either wall is the same and is designated as Nu_T in Eq. (269).

When t_{w1} approaches t_{w2}, the fully developed Nusselt number is given by Eq. (268), which is significantly lower than that of Eq. (269) for the case of $t_{w1} = t_{w2}$. This difference in the Nusselt numbers may be explained as follows. When the wall temperatures are unequal and $L/r_h = \infty$ (Fig. 30a), the heat transfer is into the fluid at one wall and out of the other, and this asymmetry is maintained in the limits as t_{w1} approaches t_{w2}. In contrast, when $t_{w1}, t_{w2} = t_w$, a common magnitude, and $L/r_h < \infty$, heat transfer is symmetrical (either in or out) and this is maintained in the limit as L/r_h goes

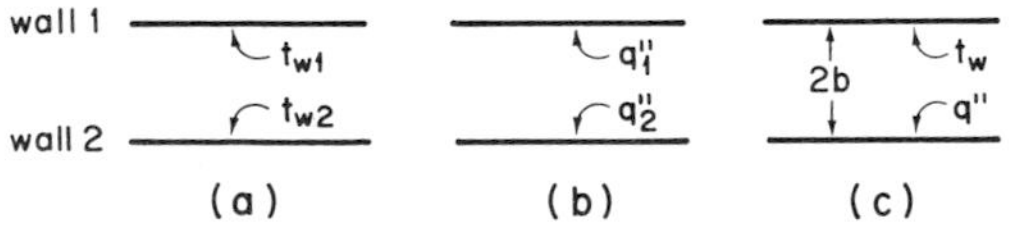

FIG. 30. Specification of wall temperatures and heat fluxes for three parallel plates problems.

to ∞. Thus the apparent paradox of $Nu_T \simeq 1.9\ Nu_1$ by Eqs. (268) and (269) is a result of comparing in the limit (and different limits) an asymmetrical heat transfer problem with a symmetrical one. In the limit for the symmetrical situation, the heat flux goes to zero because the temperature difference for the heat transfer goes to zero.

Cheng and Wu [336] considered finite viscous dissipation and obtained the fundamental solution of the first kind. They derived the fully developed Nusselt numbers with $t_{w1} > t_{w2}$ as

$$Nu_1 = \frac{4(1-6\ Br)}{1-(48/35)Br}, \qquad Nu_2 = \frac{4(1+6\ Br)}{1+(48/35)Br} \qquad (269a)$$

Note that the Brinkman number of Cheng and Wu is $-2\ Br$ in the present terminology.

Pahor and Strand [337] included the effect of axial heat conduction in the fluid for the Ⓣ heat transfer problem. The Ⓣ temperature distribution was expressed in terms of confluent hypergeometric functions. They presented graphically fully developed Nu_T as a function of Pe. They also formulated the following asymptotic formulas:

$$Nu_T = \begin{cases} 7.540\left(1 + \dfrac{3.79}{Pe^2} + \cdots\right) & \text{for} \quad Pe \gg 1 \qquad (270) \\[3mm] 8.118(1 - 0.031\ Pe + \cdots) & \text{for} \quad Pe \ll 1 \qquad (271) \end{cases}$$

Nu_T calculated from Eq. (270) are higher than those of Table 25, ranging from 0% at $Pe = \infty$ to 1% at $Pe = 20$. Nu_T calculated from Eq. (271) are lower than those of Table 25, ranging from 0% at $Pe = 0$ to 1% at $Pe = 0.6$.

Grosjean et al. [338] refined the work of Pahor and Strand [337] for small Péclet numbers ($Pe \ll 1$) and presented the following asymptotic formula:

$$Nu_T = 8.11742(1 - 0.030859\ Pe + 0.0069436\ Pe^2 - \cdots) \qquad (272)$$

TABLE 25

Parallel Plates: Nu_T as a Function of Pe
for Fully Developed Laminar Flow
(from Ash [226])

Pe	Nu_T	Pe	Nu_T
∞	7.5407	1.0572	7.9998
132.36	7.5408	0.8156	8.0242
92.24	7.5408	0.6508	8.0416
69.78	7.5408	0.5326	8.0546
28.86	7.5408	0.4444	8.0644
9.970	7.6310	0.02352	8.1144
3.548	7.8142	0.00616	8.1166
2.104	7.9084	0.000706	8.1176
1.4368	7.9640	0	8.118

Nu_T computed from Eq. (272) are lower than those of Table 25, ranging from 0% at Pe = 0 to 1% at Pe = 1.

Ash [226] also considered the effect of axial heat conduction in the fluid. Nu_T as a function of Pe of Ash are reported in Table 25.

The foregoing results are valid for the case when viscous dissipation and flow work within the fluid are identically zero. For very long ducts and liquid flows, viscous dissipation may not be neglected. Ou and Cheng [229] reported that the asymptotic Nusselt number for the finite viscous dissipation case is 17.5 and is independent of the Brinkman number. For gases, the thermal energy generated due to viscous dissipation is identical to the thermal energy absorbed due to the flow work effect, as shown on p. 184 for the fully developed velocity profile and constant properties. When the effect of viscous dissipation and flow work are considered in a gas, Ou and Cheng [229] showed that the temperature gradient at the wall and asymptotic Nusselt number approach zero and θ_m approaches (27/35)Br.

b. Constant but Different Heat Fluxes at Each Wall

The heat flux at each wall is specified as shown in Fig. 30b. The fully developed Nusselt numbers are obtained as follows:

If $q_1'' \neq q_2''$,

$$Nu_1 = \frac{140}{26 - 9(q_2''/q_1'')}, \qquad Nu_2 = \frac{140}{26 - 9(q_1''/q_2'')} \qquad (273)$$

If $q_1'' = q_2''$,

$$Nu_H = 140/17 = 8.2352941 \qquad (274)$$

If $q_1'' = 0$ (one wall adiabatic),

$$Nu_1 = 0, \qquad Nu_2 = 140/26 = 5.3846154 \qquad (275)$$

Note that when $q_1''/q_2'' = 26/9$, $Nu_2 = \infty$ from Eq. (273). In this case, it does not imply an infinite heat transfer rate but only means that $(t_{w2} - t_m) = 0$.

For the special case of heat fluxes equal at both walls, i.e., the Ⓗ boundary condition, the fluid local and bulk mean temperatures are

$$t_H = t_w - \frac{3}{2}\frac{q''}{bk}\left(\frac{5}{12}b^2 - \frac{y^2}{2} + \frac{y^4}{12b^2}\right) \qquad (276)$$

$$t_{m,H} = t_w - \frac{17}{35}\frac{u_m b^2}{\alpha}\left(\frac{dt_m}{dx}\right) = t_w - \frac{17}{140}\frac{q''D_h}{k} \qquad (277)$$

Tao [79] included the internal thermal energy generation for laminar flow between parallel plates with the Ⓗ boundary condition and solved the problem by a complex variables technique. Tyagi [63] further extended the results of Tao by including viscous dissipation. Tyagi presented Nu_H as functions of $c_3/c_4 b^2$ and $\mu c_1/k c_2$. His expression for Nu_H in terms of more familiar dimensionless groups S^* and Br' is

$$Nu_H = \frac{140}{17}\left[\frac{1}{1 + (3/68)S^* + (108/17)Br'}\right] \tag{278}$$

Perkins, in a discussion of Tyagi's results, incorrectly related $c_3/c_4 b^2$ and $\mu c_1/k c_2$ to S^* and Br'. They should be

$$\frac{c_3}{c_4 b^2} = -\frac{S^*}{3[S^* + 4(1 + 12\ Br')]} \tag{279}$$

$$\frac{\mu c_1}{k c_2} = -\frac{48\ Br'}{S^* + 4(1 + 12\ Br')} \tag{280}$$

Br' and S^* are both zero when viscous dissipation and internal thermal energy generation are neglected within the fluid. In the presence of thermal energy sources and viscous dissipation, it may be noted that the enthalpy rise (or drop) of the fluid, based on Tyagi's results, is given by

$$W c_p \frac{dt_m}{dx} = q' + SA_c + \frac{48\mu u_m^2 A_c}{g_c J D_h^2} \tag{281}$$

c. Temperature Specified at One Wall, Heat Flux at Other Wall

Constant temperature and heat flux are specified as shown in Fig. 30c. The Nusselt number is dependent on whether the specified heat flux is zero or nonzero:
If $q'' \neq 0$,

$$Nu_1 = 4, \qquad Nu_2 = 4 \tag{282}$$

If $q'' = 0$,

$$Nu_1 = 4.8608125, \qquad Nu_2 = 0 \tag{283}$$

2. FINITE WALL THERMAL RESISTANCE, Ⓣ3

Based on the results of Sideman et al. [230], for the Ⓣ3 thermal entry length problem, fully developed Nu_{T3} and Nu_o are calculated and presented in Table 26.

TABLE 26

Parallel Plates: Nu_{T3} and Nu_o for Fully Developed Laminar Flow

(from Sideman et al. [230])

R_w	Nu_{T3}	Nu_o
0	7.541	7.541
0.0025	7.553	7.413
0.0125	7.599	6.940
0.025	7.650	6.422
0.050	7.731	5.576
0.075	7.793	4.918
0.125	7.881	3.970
0.25	8.000	2.667
0.50	8.095	1.604
∞	8.235	0

TABLE 26a

Parallel Plates: Nu_{H5} for Fully Developed Laminar Flow [From Eq. (285)]

m	Nu_{H5}	m	Nu_{H5}
-80	6.15	10	8.45
-70	6.46	20	8.66
-60	6.75	30	8.85
-50	7.03	40	9.04
-40	7.30	50	9.23
-30.16	7.541	60	9.40
-30	7.55	70	9.57
-20	7.79	80	9.74
-10	8.02	90	9.90
0	8.235	100	10.06

Hickman [232] analyzed the Ⓣ3 thermal entrance problem by a Laplace transform technique and presented the following formula for the first approximation to the asymptotic Nusselt number:

$$Nu_{T3} = \frac{1 + (140/17)R_w}{(74/561) + R_w} \qquad (284)$$

Nu_{T3} calculated by Eq. (284) are higher than those of Table 26, ranging from 0% for $R_w = \infty$ to 0.53% for $R_w = 0$.

3. Exponential Axial Wall Heat Flux, Ⓗ5

Gräber [244] studied the axial exponential wall heat flux distribution for parallel plates by introducing a parameter F_o, defined on p. 85. He presented graphically Nu_{H5}/Nu_{H1} as a function of F_o (range from -2 to 8).

The fully developed Nu_{H5} can also be obtained in a closed form from the Ⓗ5 thermal entrance solution of Eq. (328) by letting $x^* \to \infty$:

$$Nu_{H5} = \frac{3}{8}\left[\sum_{n=1}^{\infty} \frac{-C_n Y_n(1)\beta_n^2}{(32/3)\beta_n^2 + m}\right]^{-1} \qquad (285)$$

where β_n and $-C_n Y_n(1)$ for n up to 10 are presented in Table 33 and higher values may be determined from Eqs. (313) and (314). Unfortunately, the series of Eq. (285) is very slowly convergent. The results of Table 26a are obtained by taking the first 10^5 terms of this series!

As noted on p. 24, the Ⓣ3 boundary condition is a special case of the Ⓗ5 boundary condition for fully developed flow for $m = -4 Nu_o$. The Nu_{H5} for parallel plates for $-30.16 \leq m \leq 0$, calculated from Eq. (285), are up to 0.1% higher than Nu_{T3} of Table 26 for $m = -4 Nu_o$.

B. Hydrodynamically Developing Flow

1. INTEGRAL SOLUTIONS

The hydrodynamic entry length problem for parallel plates was first investigated by Schiller [49] by the integral method. He considered two regions in the flow cross section: (1) a parabolic boundary layer developing near the wall with an impressed pressure gradient, and (2) a straight potential core in the remaining central cross section. This method provides good results at the entrance, but poor results downstream.

Naito and Hishida [145] and Naito [146] employed the Kármán–Pohlhausen integral method to analyze the entrance flow. The velocity profile in the boundary layer was approximated by a polynomial of the fourth degree. The resultant continuity and momentum equations represent a set of simultaneous, nonlinear ordinary differential equations. Naito and Hishida [145] integrated these equations numerically by the so-called Runge–Kutta–Gill method. Naito [146] solved these equations by the method of undetermined coefficients. The velocity profiles, u_{max}/u_m, $f_{app}\,Re$, and $K(x)$ of Naito and Hishida and those of Naito agreed with each other within $\pm 0.12\%$ for $x^+ < 0.004$. These results agreed with those of Bodoia and Liu of Table 28 within 1% for $0.000625 \le x^+ \le 0.004$.

Gupta [150] refined Schiller's solution by emplying the method of Campbell and Slattery [148], whereby he took into account the effect of viscous dissipation in the fluid. Ishizawa [339] also proposed an integral method in which the effect of viscous dissipation was also included in the fluid. Ishizawa's results are in good agreement with those of Bodoia and Osterle [172]. Williamson [151] also used the Campbell and Slattery approach, but with a different assumption for the velocity profile. His solution resulted in a more rapid flow development.

Bhatti and Savery [340] refined Schiller's solution by employing a parabolic velocity profile in the boundary layer, an integral form of the continuity equation, and the mechanical energy equation. They presented closed-form formulas for the boundary layer thickness, u, v, dp/dx, $f_x\,Re$, and $f_m\,Re$. Their results are in excellent agreement with those of Bodoia and Osterle [172].

2. AXIALLY MATCHED SOLUTIONS

Schlichting [65,137] applied a perturbation method to solve the hydrodynamic entry length problem. His method consisted of smoothly joining two asymptotic series solutions, one based on the perturbed Blasius solution for an external boundary layer development in the entrance region and the other for a perturbed Hagen–Poiseuille flow of a parabolic velocity distribution downstream. Collins and Schowalter [140] and Roidt and Cess [141] refined the solution by retaining more terms in Schlichting's upstream and downstream series velocity distribution. Their results approach the numerical results of Bodoia and Osterle [172] to be described.

In Schlichting's method, the flow model at the entrance consisted of a boundary layer near the walls together with a central uniform potential core. The boundary layer growth was calculated for an external accelerated stream by employing the perturbed Blasius solution. The fluid velocity in the central potential core was increased downstream to satisfy the continuity equation, but was assumed uniform with parallel streamlines. Van Dyke [142] showed that this uniform and parallel velocity profile in the central core represents a paradox. The displacement effect of boundary layers induces a change in the potential flow so that it does not resemble a uniform core. Van Dyke showed that Schlichting's series for small as well as large x^+ applies only to conditions far downstream and matches with another expansion (carried out by Van Dyke) valid near the inlet. Van Dyke considered two entry conditions: velocity profile uniform at $x = -\infty$ (Fig. 14c), and velocity profile uniform at $x = 0$ (Fig. 14a). The first condition is also referred to as the entry condition for "a cascade of plates in a uniform oncoming stream" or simply as "infinite cascade." Van Dyke's results for Re = 300 are in excellent agreement with numerical solutions of Wang and Longwell [191], Gillis and Brandt [192], and Bodoia and Osterle [172].

Wilson [143] independently investigated and reported complementary aspects of the same problem studied by Van Dyke. He presented the criticism of earlier boundary layer type investigations from a mathematical point of view, in contrast to Van Dyke's criticism from a physical viewpoint. Wilson also concluded that the earlier work was formally inconsistent, and the most satisfactory model is that of uniform flow into an infinite cascade of parallel plates (uniform at $x = -\infty$, Fig. 14c).

Kapila et al. [144] further extended the work of Van Dyke [142] and Wilson [143]. They considered the parallel plates placed in a uniform stream, which results in the boundary condition of Fig. 14c slightly modified; there is no frictionless wall, the flow has infinite extent in the transverse y direction, with $u_e = u_m$ at $x = -\infty$, and parallel plates are placed in this stream starting at $x = 0$. The solution to this problem, by employing the second-order boundary layer theory, provided more insight into the entry flow problem. Some effects left out in the solution to the problems of Figs. 14b and 14c were revealed, and new effects were discovered; these effects are of second-order magnitude.

TABLE 27

PARALLEL PLATES: u/u_m AND Q^* AS A FUNCTION OF $X^+(=10^3 x^+)$ FOR DEVELOPING
LAMINAR FLOW (FROM BODOIA [341])

y/b \ X^+	0.0625	0.125	0.250	0.375	0.500	0.625	0.750
0	1.0615	1.0751	1.1013	1.1244	1.1443	1.1615	1.1767
0.1	1.0615	1.0751	1.1013	1.1244	1.1443	1.1615	1.1767
0.2	1.0615	1.0751	1.1013	1.1244	1.1443	1.1615	1.1766
0.3	1.0615	1.0751	1.1013	1.1244	1.1442	1.1613	1.1763
0.4	1.0615	1.0751	1.1012	1.1243	1.1438	1.1604	1.1745
0.5	1.0615	1.0751	1.1010	1.1234	1.1414	1.1555	1.1665
0.6	1.0615	1.0750	1.0993	1.1176	1.1290	1.1351	1.1373
0.7	1.0612	1.0725	1.0863	1.0874	1.0788	1.0655	1.0501
0.8	1.0551	1.0485	1.0132	0.9665	0.9204	0.8798	0.8455
0.9	0.9587	0.8655	0.7194	0.6204	0.5567	0.5136	0.4832
1.0	0.0000	0.0000	0.0000	0.0000	0.0000	0.0000	0.0000
Q^*	1.011	1.009	1.005	1.003	1.002	1.002	1.001

y/b \ X^+	1.000	1.250	1.500	1.750	2.000	2.500	3.125
0	1.2031	1.2259	1.2463	1.2648	1.2818	1.3121	1.3441
0.1	1.2030	1.2258	1.2460	1.2643	1.2811	1.3105	1.3412
0.2	1.2028	1.2252	1.2448	1.2623	1.2778	1.3043	1.3306
0.3	1.2017	1.2228	1.2406	1.2556	1.2684	1.2887	1.3067
0.4	1.1972	1.2144	1.2275	1.2373	1.2447	1.2542	1.2601
0.5	1.1813	1.1893	1.1928	1.1935	1.1923	1.1871	1.1786
0.6	1.1339	1.1253	1.1144	1.1030	1.0918	1.0715	1.0504
0.7	1.0185	0.9896	0.9644	0.9429	0.9246	0.8950	0.8677
0.8	0.7922	0.7535	0.7241	0.7011	0.6825	0.6541	0.6291
0.9	0.4427	0.4162	0.3971	0.3825	0.3708	0.3534	0.3383
1.0	0.0000	0.0000	0.0000	0.0000	0.0000	0.0000	0.0000
Q^*	1.001	1.001	1.001	1.000	1.000	1.000	1.000

y/b \ X^+	3.750	5.000	6.250	9.375	12.5	62.5	∞
0	1.3707	1.4111	1.4388	1.4758	1.4903	1.499999	1.500
0.1	1.3663	1.4039	1.4292	1.4629	1.4762	1.485000	1.485
0.2	1.3511	1.3803	1.3993	1.4239	1.4336	1.440000	1.440
0.3	1.3195	1.3357	1.3454	1.3573	1.3619	1.364000	1.365
0.4	1.2626	1.2635	1.2628	1.2611	1.2604	1.260000	1.260
0.5	1.1703	1.1565	1.1467	1.1336	1.1284	1.124999	1.125
0.6	1.0337	1.0095	0.9938	0.9733	0.9653	0.959999	0.960
0.7	0.8475	0.8197	0.8022	0.7796	0.7708	0.765000	0.765
0.8	0.6112	0.5870	0.5720	0.5526	0.5451	0.540000	0.540
0.9	0.3275	0.3132	0.3042	0.2926	0.2880	0.250000	0.285
1.0	0.0000	0.0000	0.0000	0.0000	0.0000	0.000000	0.000
Q^*	1.000	1.000	1.000	1.000	1.000	0.999999	1.000

3. Linearized Solutions

Han [154] employed Langhaar's linearization method for the solution of hydrodynamic entry length problem for parallel plates. Sparrow et al. [162] used the stretched coordinate linearization method for the solution. The results of Sparrow et al. are in good agreement with the numerical results of Bodoia and Osterle [172].

4. Numerical Solutions

Bodoia and Osterle [172] solved the velocity problem numerically by linearizing the momentum equation at any cross section $x = x_1$ by means of velocity at $x = x_1 - \Delta x$. The dimensionless velocity and pressure drop were calculated by a finite difference method using nine grid points between the centerline of the channel and one wall. The dimensionless flow rate $Q^* = Q/A_c u_m$ was first determined by evaluating the velocity profile by a finite difference method and then numerically integrating the velocity profile. Theoretically Q^* should be 1. The computed flow was found to be up to 1.1% higher from $X^+ = 0$ to $X^+ = 1.75$ ($X^+ - 10^3 x^+$), thus establishing excellent accuracy for the solution. The convergence of the solution was checked by considering 14 instead of 9 grid points for X^+ up to 1. The error incurred was 0.2%, well within the inherent error of the finite difference formulation. The Bodoia and Osterle solutions for the parallel plates velocity problem are presented in Tables 27 and 28 [341].

TABLE 28

Parallel Plates: u_{max}/u_m, f_{app} Re, and $K(x)$ for Developing Laminar Flow

Bodoia [341]				Liu [182]			
x^+	$\dfrac{u_{max}}{u_m}$	$f_{app}Re$	$K(x)$	x^+	$\dfrac{u_{max}}{u_m}$	$f_{app}Re$	$K(x)$
0.0000625	1.062	496.8	0.1182	0.0001183	1.072	–	–
0.000125	1.075	306.6	0.1413	0.0001479	1.080	–	–
0.000250	1.101	210.0	0.1860	0.0002656	1.105	209.1	0.1967
0.000375	1.124	174.3	0.2255	0.0003839	1.126	174.5	0.2311
0.000500	1.144	153.3	0.2585	0.0005023	1.143	153.2	0.2596
0.000625	1.162	138.4	0.2861	0.0007391	1.173	127.1	0.3048
0.000750	1.177	127.2	0.3098	0.001209	1.219	100.4	0.3696
0.001000	1.203	111.1	0.3484	0.001683	1.256	86.00	0.4173
0.001250	1.226	99.96	0.3798	0.002156	1.289	76.76	0.4550
0.001500	1.246	91.73	0.4064	0.002630	1.317	70.16	0.4855
0.001750	1.265	85.36	0.4295	0.004985	1.411	53.04	0.5791
0.002000	1.282	80.21	0.4497	0.006882	1.450	46.44	0.6178
0.002500	1.312	72.38	0.4838	0.008772	1.472	42.44	0.6470
0.003125	1.344	65.42	0.5178	0.01067	1.484	39.29	0.6526
0.003750	1.371	60.33	0.5450	0.01256		37.13	0.6598
0.005000	1.411	53.26	0.5852	0.02011	1.499	32.34	0.6708
0.006250	1.439	48.50	0.6126	0.02768	1.500	30.07	0.6720
0.009375	1.476	41.34	0.6501	0.03525	1.500	28.77	0.6725
0.01250	1.490	37.31	0.6654	0.04283	1.500	27.93	0.6732
0.03125	–	29.41	0.6760	0.05040	1.500	27.34	0.6734
0.06250	1.500	26.70	0.6760	∞	1.500	24.00	–

To obtain the hydrodynamic entrance length solution, Bodoia and Osterle employed Eqs. (9), (13), and

$$\frac{1}{bu_{\mathrm{m}}} \int_0^b u \, dy = 1 \qquad (286)$$

for the three unknowns u, v, and p. Wendel and Whitaker [190] showed that Eq. (286) is simply an integrated form of Eq. (13) subject to the appropriate boundary condition. Hence, the three unknowns are determined from two equations. Even then the results are in good agreement with Schlichting [65,137] because of a "tight connection between u and p, with v having little influence."

Shah and Farnia [181] employed the finite difference method of Patankar and Spalding [248] to analyze the hydrodynamic entry length problem for concentric annuli. Their results for parallel plates, as tabulated by Liu [182], are reported in Table 28. The f_{app} Re factors of Bodoia and Osterle and of Shah and Farnia are in excellent agreement.

In all of the foregoing analyses, the velocity profile is considered as uniform at the entrance. Kiya et al. [342] considered the inlet flow with a constant velocity gradient normal to the duct axis. As a result, the inlet velocity profile was linear, minimum at one wall, maximum at the other wall and the mean velocity represented by the centerline. The solution to the hydrodynamic entry length problem was obtained by a finite difference method. They found that if the flow rate through the channel is constant, the incremental pressure drop in the entrance region is considerably decreased as the normal velocity gradient at the entrance section is increased. For example, for the dimensionless velocity gradient $\kappa = b(\partial u/\partial y)_{\mathrm{e}}/u_{\mathrm{m}} = 0.6$ at the inlet, the incremental pressure drop number $K(x)$ is about 62% of the value for the uniform entrance velocity profile, $\kappa = 0$. They found that the hydrodynamic entrance length was reduced significantly with an increase of the transverse velocity gradient κ at the inlet section.

5. SOLUTIONS OF NAVIER–STOKES EQUATIONS

In all of the above solutions, except for that of Kiya et al. [342], the velocity distribution at the entrance was considered to be uniform, and the idealizations were made that the effects of the $\mu(\partial^2 u/\partial x^2)$ and $\partial p/\partial y$ terms in the momentum equations were negligible (same as the boundary layer type idealizations). The effects of these terms are retained in the solution of the complete set of Navier–Stokes equations. The phenomenon of the velocity overshoots mentioned in the following solutions is discussed on p. 91.

The Navier–Stokes equations for parallel plates are solved for three initial conditions similar to those of Fig. 14a–c. Wendel and Whitaker [190], Gillis and Brandt [192], McDonald *et al.* [193], Morihara and Cheng [194], Chen [197], and Narang and Krishnamoorthy [198] obtained the solutions for the initial condition of Fig. 14a. Wang and Longwell [191], Schmidt and Zeldin [188,189], and McDonald *et al.* [193] analyzed the initial condition of Fig. 14b. Wang and Longwell [191] studied the stream tube condition of Fig. 14c.

Wendel and Whitaker [190] were the first investigators who solved two momentum equations and the continuity equation, although neglecting the axial momentum diffusion $\mu(\partial^2 u/\partial x^2)$. The solution was obtained by a finite difference method, using a coarse grid, for Re = 8 and a uniform entrance velocity profile at $x^+ = -0.125$. They concluded that the boundary layer type equation, as solved by Bodoia and Osterle [172], would tend to make the hydrodynamic entry length too long if the uniform velocity profile is assumed at $x^+ = 0$; however, the neglect of the pressure gradient $(\partial p/\partial y)$ across the duct has more than a compensating effect. Consequently, the development length calculated by boundary layer type equations is too small compared to that calculated by solving the complete Navier–Stokes equations. This conclusion is confirmed by more rigorous solutions of the Navier–Stokes equations for the initial condition corresponding to Fig. 14a as discussed below.

Gillis and Brandt [192] analyzed numerically the complete Navier–Stokes equations for parallel plates. They obtained solutions for several Reynolds numbers with a uniform velocity profile at the entrance (Fig. 14a). Their velocity profiles u and v for Re = 0, 40, and 400 are also tabulated by Morihara [343]. They also obtained the velocity overshoots as described on p. 91.

McDonald *et al.* [193] also solved the complete set of Navier–Stokes equations for two initial conditions, uniform and irrotational entries, Fig. 14a,b. They compared the numerical methods of Wang and Longwell [191], Gillis and Brandt [192], and others and showed that their own numerical method is better. They obtained a detailed solution for Re = 300. They concluded that the centerline velocity distribution of Gillis and Brandt has converged and is in excellent agreement with their results, while the results of Wang and Longwell appear to be high. They also noted that the centerline velocities are higher for the irrotational entry than that for the uniform entry; the difference between the two is greater for parallel plates compared to the circular tube. The centerline velocities for the initial conditions of Fig. 14a and Fig. 14b differ by a maximum of 8% at $x^+ \simeq 0.0006$ for parallel plates at Re = 300. McDonald *et al.* also found the above-mentioned velocity overshoots.

Morihara and Cheng [194] employed the method of quasi-linearization to solve the complete set of Navier–Stokes equations for parallel plates for the initial condition of Fig. 14a. The resultant equations were solved by a finite difference method for Re varying from 0 to 4000. They presented graphically the velocity profiles and the centerline velocity development for the above Reynolds numbers, and vorticity distribution for Re = 400 and 4000. Their velocity profiles in the entrance region are in excellent agreement with those of Gillis and Brandt [192] for $x^+ > 0.00625$; for small x^+, the agreement is within 2.5%. They also determined L_{hy}^+ for Re = 0.2, 4.0, 400, and 4000.

Morihara and Cheng also found the overshoots in the axial velocity profiles for all Reynolds numbers investigated, however, the magnitude of overshoots decreased with increasing Re and almost disappeared at Re = 4000. The physical phenomenon causing the velocity overshoots was explained by them as described earlier.

Chen [197] employed an approximate integral method to solve the complete Navier–Stokes equations for the initial condition corresponding to Fig. 14a. His results are in good agreement with the results discussed above for Re > 20.

Narang and Krisnamoorthy [198] linearized the inertia terms of x and y momentum equations, as described on p. 74, and obtained solutions for Re ranging from 2 to 4000. They presented graphically typical velocity profiles and average $\Delta p/(\mu u_m/b g_c)$ as a function of x/b for the above Reynolds numbers. They found that the maximum transverse velocity v is $0.302 u_m$, which occurs at $x/b = 0.2$ and $y/b = 0.684$ for Re = 2. It decreases to $0.0127 u_m$ for Re = 4000. They also determined L_{hy} for the above Reynolds number range and found them in excellent agreement with those of Morihara and Cheng [194].

Wang and Longwell [191] were the first investigators to rigorously solve the complete set of Navier–Stokes equations. These equations were written in terms of the stream function and vorticity. The subsequent solution was obtained by a finite difference iterative method. Two initial conditions were studied: irrotational flow at the entrance ($x = 0$), Fig. 14b, and uniform velocity distribution far upstream ($x = -\infty$), Fig. 14c. Numerical results were presented for the velocity distribution and pressure drop in the entrance region for Re = 300. One interesting result of the analysis was that Wang and Longwell obtained concave velocity profiles in the entrance region near the center of the plates. Normally, one would expect either a uniform or a convex profile. They concluded that if the velocity distribution and pressure gradients were required near the entrance region, the boundary layer type idealizations were not appropriate and full differential equations must be solved with realistic boundary conditions. Van Dyke [142], and Burke and Berman [249] pointed out that Wang and Longwell analyzed the hydrodynamic entrance problem for the initial condition of Fig. 14b and not of Fig. 14a as mentioned in [191].

Schmidt and Zeldin [188, 189] also obtained a solution to the complete Navier–Stokes equations, for the initial condition of Fig. 14b. They reported the nondimensional pressure distribution and cross-sectional area average $\bar{K}(x)$ for Re = 100, 500, and 10,000. Their f_{app} Re and $\bar{K}(x)$ values are presented in Table 29 [252] for the entry condition of Fig. 14b.

As discussed on p. 92, Abarbanel et al. [196] obtained an exact solution for the Stokes flow problem (Re = 0) (for the initial condition of Fig. 14a) and showed that the velocity overshoots are a realistic part of the solution. The overshoots are weaker for the entrance velocity profile deviating from a uniform one.

6. Apparent Friction Factors

f_{app} Re and $K(x)$ are presented in Table 28 for the hydrodynamic entrance region of parallel plates. For a "short duct," a comparison of f_{app} Re of Bodoia [341] and of Eq. (188) reveals that f_{app} Re of Bodoia are slightly different, ranging from 3.9% higher at $x^+ = 0.00025$ to 2.1% lower at $x^+ = 0.001$. The f_{app} Re factors of Liu [182] are also in excellent agreement with Eq. (188), varying from 1% higher at $x^+ = 0.0002$ to 1% lower at $x^+ = 0.001$.

TABLE 29

Parallel Plates: f_{app} Re and $\bar{K}(x)$ as Functions of x^+ and Re for Developing Laminar Flow, Based on the Solution to the Complete Naviers–Stokes Equations (from Schmidt [252])

x^+	Re = 10000		Re = 500		Re = 100	
	f_{app}Re	$\bar{K}(x)$	f_{app}Re	$\bar{K}(x)$	f_{app}Re	$\bar{K}(x)$
0.0000307	626.1	0.0739	1024.3	0.1228	–	–
0.0000473	510.7	0.0921	812.9	0.1493	236.0	0.04011
0.0000648	433.8	0.1063	677.8	0.1695	324.5	0.07790
0.0000833	380.9	0.1190	584.1	0.1867	355.8	0.1106
0.0001029	341.6	0.1308	514.7	0.2020	361.5	0.1390
0.0001468	285.9	0.1527	416.7	0.2291	342.7	0.1859
0.0002213	231.4	0.1836	322.1	0.2639	297.2	0.2418
0.0003141	194.3	0.2139	258.4	0.2945	255.2	0.2905
0.0004312	166.2	0.2453	211.2	0.3229	219.8	0.3376
0.0005833	143.5	0.2789	174.4	0.3509	189.3	0.3858
0.0007892	124.3	0.3166	144.7	0.3810	162.4	0.4368
0.001083	107.2	0.3605	120.2	0.4169	137.5	0.4917
0.001538	91.39	0.4146	99.33	0.4634	113.6	0.5514
0.002009	81.14	0.4592	87.63	0.5034	97.86	0.5936
0.002750	70.68	0.5135	74.28	0.5531	81.97	0.6376
0.004083	59.38	0.5780	61.53	0.6130	65.90	0.6843
0.005250	53.12	0.6115	54.69	0.6444	57.70	0.7078
0.007194	46.29	0.6413	47.37	0.6724	49.31	0.7284
0.01108	38.92	0.6615	39.60	0.6914	40.75	0.7426
0.02275	31.35	0.6688	31.67	0.6983	32.22	0.7481

For a long duct with $x^+ \gtrsim 0.01$, f_{app} Re can be determined by Eq. (92) for the known values of f Re $(=24)$ and $K(\infty)$. The theoretically determined values of $K(\infty)$ vary from 0.60 to 0.74, as compared by Lakshmana Rao and Sridharan [135], Schmidt and Zeldin [189], Ishizawa [339], Kiya et al. [342], and Bhatti and Savery [340], except for Han [154], who found it to be 0.85. No experimental value of $K(\infty)$ exists for truly parallel plates. The numerical results of Liu [182] are believed to be the most accurate; hence $K(\infty)$ is taken as 0.674 based on his results. The f_{app} Re factors then can be determined from

$$f_{app} \text{ Re} = \begin{cases} \dfrac{3.44}{(x^+)^{1/2}} & \text{for} \quad x^+ \lesssim 0.001 \qquad (287) \\[3mm] 24 + \dfrac{0.674}{4x^+} & \text{for} \quad x^+ \gtrsim 0.01 \qquad (288) \end{cases}$$

Shah [259] combined these equations in a manner similar to Eq. (192) to cover the complete range x^+.

$$f_{app} \text{ Re} = \frac{3.44}{(x^+)^{1/2}} + \frac{24 + 0.674/(4x^+) - 3.44/(x^+)^{1/2}}{1 + 0.000029(x^+)^{-2}} \qquad (289)$$

f_{app} Re of Eq. (289) agree with those of Liu within $\pm 2.4\%$ (within $\pm 1.5\%$ for all but three points). f_{app} Re of Eq. (289) agree with those of Bodoia within $\pm 2.5\%$ for $x^+ > 0.0004$.

$K(\infty)$ and hence f_{app} Re are dependent upon Re for low Re, as shown by the solutions of the Navier–Stokes equations. For this case, Chen [197] presented the following equation for $K(\infty)$, based on his results by an integral method:

$$K(\infty) = 0.64 + (38/\text{Re}) \qquad (290)$$

It should be emphasized that in all of the above equations, the hydraulic diameter $D_h = 4b$ is used in the definitions of x^+ and Re.

7. HYDRODYNAMIC ENTRANCE LENGTH

The hydrodynamic entrance length L_{hy}^+, defined on p. 41 (u_{max} approaching $0.99u_{max,fd}$), is constant when based on a boundary layer type momentum equation (9). Bodoia [341] and Liu [182] both obtained L_{hy}^+ as 0.011. The values of L_{hy}^+ obtained by different investigators are compared by Lakshmana Rao and Sridharan [135], Kiya et al. [342], and Bhatti and Savery [340].

L_{hy}^+ is a function of Re for low Reynolds number flows when the complete set of Navier–Stokes equations is solved. Atkinson et al. [257] presented the following equation for L_{hy} when the entrance profile is uniform at $x = 0$, Fig. 14a:

$$L_{hy}/D_h = 0.3125 + 0.011\,Re \qquad (291)$$

This equation was obtained by a linear combination of creeping flow and boundary layer type solutions. Instead of Eq. (291), Chen [197] proposed an equation similar to Eq. (195):

$$\frac{L_{hy}}{D_h} = \frac{0.315}{0.0175\,Re + 1} + 0.011\,Re. \qquad (292)$$

L_{hy}/D_h of this equation are within 3% of those by Morihara and Cheng [194] for $Re = 2,400$, and 4000. L_{hy}/D_h of this equation is 12% higher than that of Morihara and Chen for $Re = 40$.

C. Thermally Developing and Hydrodynamically Developed Flow

Thermal entrance solutions with a fully developed laminar velocity profile for flow between parallel plates are divided into six categories: (1) specified wall temperature distribution, (2) specified wall heat flux distribution, (3) fundamental solutions of third and fourth kinds, (4) finite wall thermal resistance, (5) exponential axial wall heat flux, and (6) solutions to conjugated problems. These are discussed below separately. All the solutions obtained up to 1961 are summarized by McCuen et al. [332].

1. SPECIFIED WALL TEMPERATURE DISTRIBUTION

Once the fundamental solution of the first kind is obtained, the thermal entrance solution for parallel plates can be obtained for any arbitrary or equal temperature distribution at each wall. However, the particular case of equal wall temperatures is practically more important when considering parallel plates as a limiting rectangular and/or concentric annular duct geometry. Hence, the $\textcircled{T}$ problem is discussed first before the more general problem of the first kind.

a. The $\textcircled{T}$ Problem

This thermal entrance problem with a fully developed laminar velocity profile was first solved by Nusselt [333] in 1923. He employed the same idealizations and an approach similar to that of Graetz [1] for the solution. This problem for parallel plates is well known as the *Graetz–Nusselt problem*.

As mentioned in the circular duct section, the closed-form solution has been obtained primarily by two methods (the Graetz and Lévêque). The solution by the Lévêque method is described after Nusselt's solution by the Graetz method.

Nusselt [333] obtained the first three terms of the following infinite series solution:

$$\theta = \frac{t - t_w}{t_e - t_w} = \sum_{n=0}^{\infty} C_n Y_n \exp\left(-\frac{32}{3}\lambda_n^2 x^*\right) \tag{293}$$

$$\theta_m = \frac{t_m - t_w}{t_e - t_w} = 3 \sum_{n=0}^{\infty} \frac{G_n}{\lambda_n^2} \exp\left(-\frac{32}{3}\lambda_n^2 x^*\right) \tag{294}$$

$$\mathrm{Nu}_{x,\mathrm{T}} = \frac{8}{3} \frac{\sum_{n=0}^{\infty} G_n \exp(-\frac{32}{3}\lambda_n^2 x^*)}{\sum_{n=0}^{\infty} (G_n/\lambda_n^2) \exp(-\frac{32}{3}\lambda_n^2 x^*)} \tag{295}$$

$$\mathrm{Nu}_{m,\mathrm{T}} = \frac{1}{4x^*} \ln\left(\frac{1}{\theta_m}\right) \tag{296}$$

where λ_n, Y_n, and C_n are eigenvalues, eigenfunctions, and constants, respectively, and $G_n = -(C_n/2)Y_n'(1)$. These are listed in Table 30.

TABLE 30

PARALLEL PLATES: INFINITE SERIES SOLUTION
FUNCTIONS FOR THE GRAETZ–NUSSELT PROBLEM
(BASED ON THE RESULTS OF BROWN [260])

n	λ_n	G_n
0	1.68159 53222	0.85808 6674
1	5.66985 73459	0.56946 2850
2	9.66824 24625	0.47606 5463
3	13.66766 14426	0.42397 3730
4	17.66737 35653	0.38910 8706
5	21.66720 53243	0.36346 5044
6	25.66709 64863	0.34347 5506
7	29.66702 10447	0.32726 5745
8	33.66696 60687	0.31373 9318
9	37.66692 44563	0.30220 4200

An independent verification of Nusselt's solution was accomplished by Norris and Streid [334], Purday [124], Prins et al. [344], and Yih and Cermak [345]. Thus the determination of the first three eigenvalues and eigenfunctions of a series solution for $\mathrm{Nu}_{x,\mathrm{T}}$ was exhaustively accomplished. For higher eigenvalues and eigenfunctions, Sellers et al. [25] extended the Graetz–Nusselt problem for the circular tube and parallel plates by employing the WKBJ approximate method.[†] They derived the asymptotic expressions for the higher eigenvalues and constants, and presented the first ten eigenvalues and related constants with 0 to 4 decimal point accuracy.

Brown [260] refined the work of Sellers *et al.* and reported the first ten eigenvalues and constants to ten decimal point accuracy. Based on Brown's tabulation, λ_n and G_n magnitudes are given in Table 30. For higher values of n, the asymptotic expressions of Sellers *et al.* [25] are

$$\lambda_n = 4n + \tfrac{5}{3} \tag{297}$$

$$G_n = 1.01278729\lambda_n^{-1/3} \tag{298}$$

Chandler *et al.* [212] solved the Graetz–Nusselt problem by the Monte Carlo method.

As mentioned in the introduction to this section, the parallel plates duct geometry forms an upper bound for the families of rectangular and concentric annular ducts. As these geometries are widely used, numerical results for parallel plates, in terms of θ_m, $Nu_{x,T}$, $Nu_{m,T}$, and $N_T(x)$ as a function of x^*, have been accurately calculated by Shah [270] and are presented in Table 31. These calculations were based on Eqs. (294)–(296), using the first 121 terms of the series; G_n and λ_n, up to $n = 9$, were taken from Table 30; and for $n > 9$, G_n and λ_n were determined from Eqs. (297) and (298). The computed values of these variables are listed in Table 31 for $x^* > 10^{-4}$. The values for $x^* \leq 10^{-4}$ are determined from the extended Lévêque solution to be described below. $Nu_{x,T}$ and $Nu_{m,T}$ from Table 31 are plotted in Fig. 31.

The related variables $\Phi_{x,T}$, $\Phi_{m,T}$, and $Nu_{a,T}$ for this Graetz–Nusselt problem can be evaluated, using the results of Table 31, from Eqs. (123), (124), and (116), respectively.

$L_{th,T}^*$ and $N_T(x)$ defined by Eqs. (128) and (130) at $x^* = \infty$ were calculated by Shah [270] as follows:

$$L_{th,T}^* = 0.00797350 \tag{299}$$

$$N_T(\infty) = 0.02348094. \tag{300}$$

The foregoing solution by the Graetz method converges very slowly as x^* approaches zero. To resolve this difficulty, Lévêque [3,199] formulated a method, the idealizations and details of which have already been described in the circular duct section on p. 105. The extended Lévêque solution for

---

[†] The WKBJ method reduces a singular perturbation problem of an ordinary differential equation to a regular perturbation problem. The asymptotic solution is then obtained in terms of the perturbed parameter (e.g., large eigenvalue). Refer to Froman and Froman [346] and Kumar [347] for further details.

TABLE 31

PARALLEL PLATES: θ_m, $Nu_{x,T}$, $Nu_{m,T}$, AND $N_T(x)$ FOR A FULLY
DEVELOPED VELOCITY PROFILE (FROM SHAH [270])

x^*	θ_m	$Nu_{x,T}$	$Nu_{m,T}$	$N_T(x)$
0.000001	0.99926	122.943	184.548	0.0001770
0.0000015	0.99903	107.374	161.185	0.0002305
0.000002	0.99883	97.538	146.425	0.0002778
0.000003	0.99847	85.187	127.886	0.0003610
0.000004	0.99814	77.385	116.173	0.0004345
0.000005	0.99785	71.830	107.833	0.0005015
0.000006	0.99757	67.589	101.465	0.0005635
0.000007	0.99731	64.200	96.375	0.0006218
0.000008	0.99705	61.403	92.173	0.0006771
0.000009	0.99681	59.037	88.619	0.0007297
0.00001	0.99658	56.999	85.557	0.0007802
0.000015	0.99553	49.793	74.728	0.001008
0.00002	0.99458	45.245	67.890	0.001207
0.00003	0.99291	39.539	59.305	0.001553
0.00004	0.99142	35.939	53.885	0.001854
0.00005	0.99004	33.379	50.027	0.002124
0.00006	0.98876	31.426	47.083	0.002373
0.00007	0.98755	29.867	44.731	0.002603
0.00008	0.98640	28.582	42.790	0.002820
0.00009	0.98530	27.495	41.149	0.003025
0.0001	0.98423	26.560	39.736	0.003220
0.00015	0.97937	23.262	34.745	0.004081
0.0002	0.97504	21.188	31.598	0.004811
0.0003	0.96736	18.600	27.657	0.006035
0.0004	0.96052	16.977	25.177	0.007054
0.0005	0.95425	15.830	23.416	0.007938
0.0006	0.94840	14.960	22.077	0.008722
0.0007	0.94287	14.270	21.009	0.009428
0.0008	0.93761	13.704	20.130	0.01007
0.0009	0.93258	13.229	19.389	0.01066
0.001	0.92774	12.822	18.752	0.01121
0.0015	0.90565	11.408	16.517	0.01346
0.002	0.88604	10.545	15.125	0.01517
0.003	0.85137	9.5132	13.409	0.01760
0.004	0.82065	8.9100	12.354	0.01925
0.005	0.79258	8.5166	11.623	0.02041
0.006	0.76648	8.2456	11.081	0.02124
0.007	0.74191	8.0532	10.662	0.02185
0.008	0.71860	7.9146	10.326	0.02229
0.009	0.69636	7.8139	10.053	0.02261
0.01	0.67503	7.7405	9.8249	0.02284
0.015	0.57936	7.5826	9.0972	0.02335
0.02	0.49804	7.5495	8.7133	0.02345
0.03	0.36632	7.5411	8.3234	0.02348
0.04	0.27241	7.5407	8.1277	0.02348
0.05	0.20148	7.5407	8.0103	0.02348
0.06	0.14902	7.5407	7.9320	0.02348
0.07	0.11022	7.5407	7.8761	0.02348
0.08	0.08152	7.5407	7.8342	0.02348
0.09	0.06029	7.5407	7.8016	0.02348
0.1	0.04459	7.5407	7.7755	0.02348
0.15	0.00987	7.5407	7.6972	0.02348
0.2	0.00218	7.5407	7.6581	0.02348

parallel plates can also be presented by Eq. (212), except that the variables ξ and η include $6x^*$ instead of $9x^*$ in their definitions.

Lévêque [3,199] and Lundberg et al. [42] obtained the first term of the Lévêque series. Mercer [348] evaluated θ_1, θ_2, and θ_3 of Eq. (212) numerically and tabulated as a function of η. He did not compute $\Phi_{x,T}$ or θ_m. Worsøe-Schmidt [200] analyzed the perturbation solution by a finite difference method and tabulated $-\theta_n'(0)$ for n up to 6. Krishnamurty and Venkata Rao

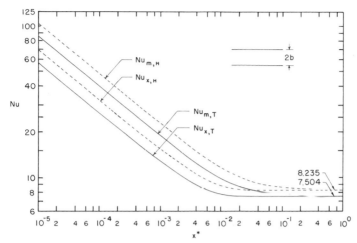

Fig. 31. Parallel plates: $Nu_{x,T}$, $Nu_{m,T}$, $Nu_{x,H}$, and $Nu_{m,H}$ for a fully developed velocity profile (from Tables 31 and 34).

[349] also obtained the first term of the Lévêque series and presented formulas for $Nu_{x,T}$ with one or both sides of parallel plates heated.

The dimensionless wall heat fluxes $\Phi_{x,T}$ and $\Phi_{m,T}$ based on the Worsøe-Schmidt solution [200] are

$$\Phi_{x,T} = 1.232550580(x^*)^{-1/3} - 0.4 - 0.267383252(x^*)^{1/3}$$
$$- 0.30675(x^*)^{2/3} - 0.43668x^* - 0.72132(x^*)^{4/3} - 1.29647(x^*)^{5/3}$$

$$(301)$$

$$\Phi_{m,T} = 1.848825871(x^*)^{-1/3} - 0.4 - 0.200537439(x^*)^{1/3}$$
$$- 0.18405(x^*)^{2/3} - 0.21834x^* - 0.30914(x^*)^{4/3} - 0.48618(x^*)^{5/3}$$

$$(302)$$

$Nu_{x,T}$ and $Nu_{m,T}$ may then be determined from the relationships of Eqs. (123) and (124).

Based on Eqs. (301) and (302), θ_m, $Nu_{x,T}$, $Nu_{m,T}$, and $N_T(x)$ were calculated by Shah [270] for the Graetz–Nusselt problem for $10^{-6} \leq x^* \leq 5 \times 10^{-2}$. $Nu_{x,T}$ calculated from Eqs. (123), (124), and (302) were identical up to five or more digits with those of Eq. (295) for $0.0005 \leq x^* \leq 0.001$ and were at most 0.004% higher for $5 \times 10^{-5} < x^* < 5 \times 10^{-2}$. The results for the ⓉT problem in Table 31 for $x^* \leq 10^{-4}$ are based on the foregoing extended Lévêque solution.

The following operationally more convenient approximate equations are recommended by Shah [270] for $Nu_{x,T}$ and $Nu_{m,T}$ for the indicated ranges of x^*:

$$Nu_{x,T} = \begin{cases} 1.233(x^*)^{-1/3} + 0.4 & \text{for} \quad x^* \leq 0.001 \quad (303) \\ 7.541 + 6.874(10^3 x^*)^{-0.488} e^{-245x^*} & \text{for} \quad x^* > 0.001 \quad (304) \end{cases}$$

$$Nu_{m,T} = \begin{cases} 1.849(x^*)^{-1/3} & \text{for} \quad x^* \leq 0.0005 \quad (305) \\ 1.849(x^*)^{-1/3} + 0.6 & \text{for } 0.0005 < x^* \leq 0.006 \quad (306) \\ 7.541 + \dfrac{0.0235}{x^*} & \text{for} \quad x^* > 0.006 \quad (307) \end{cases}$$

Values from these approximations are only slightly in error relative to those in Table 31 by at most ± 1.1, ± 0.6, ± 0.5, ± 2.8, and $+3.0\%$, respectively.

The Graetz–Nusselt problem with finite axial heat conduction in the fluid has been analyzed by Schmidt and Zeldin [66] for the boundary conditions of Fig. 17a and by Bodnarescu [285], Agrawal [350], Bes [286], and Deavours [351,352] for the boundary conditions of Fig. 17b.

Schmidt and Zeldin [66] employed a finite difference method for the semi-infinite region ($0 \leq x < \infty$) with uniform temperature at the inlet (Fig. 17a). They graphically presented $Nu_{x,T}$ and $\Phi_{m,T}$ as a function of $2x^*$ with Pe as a parameter. Similar to the circular tube case, they also presented the energy content of the fluid as functions of $2x^*$ and Pe. Their results [252] for the energy content of the fluid are given in Table 32 and Fig. 32.

Bodnarescu [285] solved the more general problem with finite fluid axial heat conduction. He employed boundary conditions similar to those of Fig. 17b, but with one wall at temperature t_1 and the other at t_2 for $-\infty < x < 0$ and at t_3 and t_4, respectively, for $0 \leq x \leq \infty$. The solution was obtained in terms of hypergeometric functions. Agrawal [350] analyzed the problem for the boundary conditions of Fig. 17b. He presented the eigenfunctions as an infinite Fourier-sine series, derived expressions for the local temperature distribution and $Nu_{x,T}/4$, and outlined a detailed solution for Pe $= 4$. Bes [286] also analyzed the Fig. 17b problem; however, his results are questionable, as discussed in the circular tube section, p. 117. Deavours [351] obtained a temperature distribution for the Graetz–Nusselt problem with the effect of axial heat conduction and viscous dissipation included. He employed the boundary conditions of Fig. 17b and an arbitrary velocity profile at the entrance. The temperature distribution was obtained in terms of a nonorthogonal eigenfunction expansion. The expansion coefficients were determined by a vector system of equations. After presenting the same solution, Deavours [352] obtained numerical results for Pe $= 1$ and 10 for the temperature distribution upstream and downstream of the cross section of the step change in

TABLE 32

PARALLEL PLATES: ENERGY CONTENT OF THE FLUID FOR THE GRAETZ–NUSSELT PROBLEM WITH FINITE FLUID AXIAL HEAT CONDUCTION (FROM SCHMIDT [252])

Pe = 10		Pe = 25		x^*	δ		
x^*	δ	x^*	δ		Pe=200	Pe=300	Pe=400
.000164	.00840	.000164	.0502	.000164	.6791	.8229	.8914
.000337	.0111	.000337	.06185	.000337	.7024	.8362	.8998
.000712	.01998	.000520	.07996	.000520	–	.8654	.9188
.001603	.03696	.001131	.1289	.000712	.7816	.8853	.9314
.002747	.05533	.002137	.1890	.001131	.8260	.9108	.9475
.004274	.07636	.003452	.2473	.001603	.8551	.9269	.9571
.006410	.1014	.005245	.3064	.002137	.8761	.9382	.9639
.009615	.1326	.007835	.3687	.002747	.8924	.9468	.9690
.01496	.1732	.01190	.4369	.003452	.9056	.9537	.9731
.02564	.2293	.01923	.5140	.004274	.9167	.9593	.9765
.03632	.2659	.03632	.6014	.005245	.9262	.9642	.9793
.07906	.3302	.07906	.6627	.006410	.9346	.9684	.9817
.1218	.3476	.1218	.6757	.007835	.9420	.9721	.9859
.2500	.3562	.2500	.6815	.009615	.9489	.9754	.9877
				.01190	.9551	.9785	.9893
				.01496	.9609	.9813	.9908

Pe = 50		Pe = 100		x^*	δ		
x^*	δ	x^*	δ				
				.01923	.9663	.9840	.9923
				.02564	.9712	.9864	.9934
				.03632	.9758	.9885	.9940
.000164	.1496	.000164	.3687	.05769	.9796	.9903	.9949
.000337	.1747	.000337	.4046				
.000520	.2141	.000520	.4660	.07906	.9810	.9910	.9951
.001360	.3384	.000916	.5510	.1218	.9819	.9915	.9953
.002431	.4314	.001861	.6597	.2500	.9824	.9917	–
.003846	.5084	.003086	.7283				
.005800	.5765	.004738	.7808				

Pe = 1		Pe = 10000	
x^*	δ	x^*	δ
.009726	.00237	.0001	.99964
.04762	.008	.0346	.99998
.2308	.023		
1.0000	.0301		

Additional Pe = 50 / Pe = 100 rows:

Pe = 50		Pe = 100	
x^*	δ	x^*	δ
.008673	.6401	.007085	.8224
.01331	.7020	.01068	.8581
.02208	.7636	.01690	.8900
.04487	.8216	.03220	.9186
.07906	.8443	.04487	.9310
.2500	.8545	.07881	.9406
		.2500	.9458

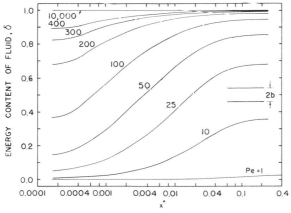

FIG. 32. Parallel plates: Energy content of the fluid for the Graetz–Nusselt problem with finite fluid axial heat conduction (from Table 32).

the temperature. He considered the velocity profile fully developed, Eq. (260). Using asymptotic expansions for the functions involved, he also made an estimate for the number of terms required to achieve a given level of accuracy in the temperature determination for a given Péclet number.

The Graetz–Nusselt problem with thermal energy generation within the fluid has been investigated by Sparrow et al. [353]. They obtained the thermal entry solution with both surfaces at equal and uniform temperatures as well as with an arbitrary temperature distribution. The thermal energy sources were assumed to have arbitrary longitudinal as well as cross-sectionally symmetric transverse variations. The solution was obtained by the Graetz method. The first ten eigenvalues and constants for both walls at equal and uniform temperatures are tabulated by Sparrow et al.

The effect of viscous dissipation and flow work on the Ⓣ thermal entrance problem has been investigated by Ou and Cheng [229]. They obtained the solution by an eigenvalue method and tabulated the first ten constants associated with the problem. They presented graphically the fluid temperature profiles, θ_m, and $Nu_{x,T}$ as functions of x^* and Br. Note that the combined effect of viscous dissipation and flow work is represented by the Brinkman number Br (refer to p. 14). It should be emphasized that the following discussion is applicable to gas flows only, because the flow work generally is zero for liquids, as mentioned on p. 13.

As discussed on pp. 183–184, the viscous dissipation term represents a nonuniformly distributed heat source, and the flow work term represents a nonuniformly distributed heat sink. The thermal energy is generated mainly near the wall region due to the viscous dissipation effect; it is absorbed mainly near the channel central region due to the flow work effect. The amount of thermal energy generated due to the viscous dissipation effect at a cross section, Eq. (324), is the same as the amount of thermal energy absorbed by the flow work effect, Eq. (326). However, a part of the thermal energy generated due to viscous dissipation may be transfered away at the wall. Hence θ_m depends upon Br.

For the fluid heating situation (Br > 0, $t_w > t_e$), the wall temperature gradient is reduced due to the strong viscous dissipation effect near the wall region. This results in $Nu_{x,T}$ lower for Br > 0 in comparison to $Nu_{x,T}$ for Br = 0. At a given x^*, the temperature gradient at the wall decreases for increasing Br (more viscous dissipation near the wall). This results in a lower amount of heat transfer through the wall for increasing Br, which also results in a slower rise in t_m for increasing x^*. Thus, $Nu_{x,T}$ is reduced for Br > 0 due to two effects: a decrease in the wall temperature gradient, and an increase in $(t_w - t_m)$. Ou and Cheng showed that $Nu_{x,T}$ approaches zero, and θ_m approaches (27/35) Br for $x^* \to \infty$.

For the fluid cooling situation (Br < 0, $t_w < t_e$), the temperature gradient at the wall is increased due to the strong viscous dissipation effect near the wall region. This results in both higher wall heat transfer rate and higher $Nu_{x,T}$ compared to the Br $= 0$ situation. The fluid bulk mean temperature also drops more rapidly due to higher wall heat transfer. At some x^*, $t_m = t_w$, producing a singularity ($Nu_{x,T} = \infty$) in the $Nu_{x,T}$ vs. x^* curve. Beyond this location, while heat transfer at the wall is still from the fluid to the wall, t_m is lower than t_w (a temperature profile similar to that of Fig. 22d), which results in negative Nusselt numbers. Ou and Cheng showed that the formulation of $Nu_{x,T}$ singularity occurs at lower x^* for decreasing Br for Br < 0.

b. *Fundamental Solution of the First Kind and Arbitrary Axial Wall Temperature Distribution*

Fundamental solution of the first kind has been analyzed in great detail by McCuen et al. [332]. They extensively reported the results in tabular and graphical forms. Their thermal entrance solutions are reported in Chapter XII.

Yih and Cermak [345] solved the same problem. Utilizing the superposition method, they arrived at solutions for variable and unequal wall temperatures and insulated boundary conditions. Their method has been outlined by Klein and Tribus [354].

A fundamental solution of the first kind obtained by the modified Graetz method requires an evaluation of what is called even and odd eigenvalues and constants. Cess and Shaffer [355] summarized the first three even and odd (6 total) eigenvalues and constants for the uniform and unequal wall temperatures, and presented asymptotic expressions for higher eigenvalues and constants. McCuen et al. [332] also solved the unequal wall temperature problem by the superposition method and reported the first four even and the first three odd eigenvalues and constants, with asymptotic formulas for higher values. Hatton and Turton [356] independently also solved the problem of uniform and unequal wall temperatures. They reported the first eight even and odd (16 total) eigenvalues and constants with asymptotic formulas for higher values. Only the even eigenvalues and constants are required for the equal wall temperature case. The first ten such even eigenvalues and constants are reported in Table 30.

In all of the foregoing analyses, axial heat conduction in the fluid was neglected. Hwang and Cheng [357] analyzed the fundamental problem of the first kind by including fluid axial heat conduction. They considered a uniform fluid temperature distribution at the entrance ($x^* = 0$). Thus the initial and boundary conditions were similar to those of Fig. 17a, except that the other wall of the parallel plates was maintained at $t = t_e$ for the fundamental solution of the first kind. They obtained an eigenvalue solution and tabulated up to the first 17 eigenvalues for Pe = 20,40, 60, 100 200, and ∞.

Wu *et al.* [358] extended the fundamental problem of the first kind of Hwang and Cheng [357] by considering the inlet fluid temperature uniform at $x = -\infty$, and both walls being adiabatic for $-\infty < x \leq 0$. Thus initial and boundary conditions were similar to those of Fig. 17c, except that for $0 \leq x < \infty$ one wall was at t_w, the other at t_e, and $t_w > t_e$. The solution was obtained by the eigenfunction expansion method utilizing the Gram–Schmidt orthonormalization procedure. Numerical results were obtained for Pe = 1, 5, 10, and 50. They concluded that the effect of fluid axial heat conduction is appreciable on bulk temperature and local Nusselt numbers for Pe < 50. For low Péclet number (Pe < 50), axial heat conduction can cause a considerable increase in thermal entry lengths L_{th}^* for both upstream and downstream regions.

Cheng and Wu [336] included the effect of viscous dissipation in the fluid and analyzed the fundamental problem of the first kind. They considered only the fluid heating situation (Br > 0). They obtained the eigenvalues and eigenfunctions by applying the fourth-order Runge–Kutta method, and presented numerical values for the first eight eigenvalues and constants. They presented graphically the developing temperature profiles, bulk mean temperature, and local Nusselt numbers as functions of $32x^*/3$ and -2 Br. For the fluid heating situation, the fluid bulk mean temperature rises axially near the entrance due to two effects: heat transfer from the wall to the fluid, and viscous heating effect. The typical temperature profile development is similar to the one shown in Fig. 16, except that one should consider the region between the lower wall and the tube centerline (which would be the upper wall for parallel plates), and the temperature at the centerline should be fixed at t_e. Subsequently, the local Nusselt number behavior obtained by Cheng and Wu can be explained in a straightforward manner.

Sadikov used the simplified energy Eq. (168) and solved the thermal entry length problem for parallel plates with the entering fluid at a uniform [208] and also a nonuniform temperature [209]. In both cases, the wall temperature was assumed linearly varying with the axial distance. He presented graphically the local temperature distribution and the Nusselt numbers as a function of the axial distance.

In the classical Graetz–Nusselt problem, the temperature of both walls is kept uniform axially. If these walls contain a series of line thermal energy sources or discrete heating strips, the wall temperature may be approximated by an axially cyclic sinusoidal temperature distribution. Even though this problem can be analyzed directly by the superposition method, it has not been solved in this manner. Savkar [206] solved it by a variational method.

The dimensionless frequency (wave number) ω was varied from 10 to 10,000 where $\omega = (3\pi/16)D_h \, \text{Pe}/\lambda$ and λ is the wavelength of the sine wave. Savkar found that the effect of a sinusoidal wall temperature distribution is negligible for $\omega > 1000$. For this case, the solution approaches that of the Graetz–Nusselt problem.

Instead of line thermal energy sources, Povarnitsyn and Yurlova [359] considered the continuous but arbitrary thermal energy sources $S(x)$, dependent on the axial direction. They treated the wall thickness a' as very thin, so that the following axial boundary condition is obtained at the inner wall surfaces, the outer wall surfaces being insulated:

$$k_w \frac{\partial^2 t}{\partial x^2} + S(x) - \frac{k}{a'}\frac{\partial t}{\partial y} = 0 \qquad \text{at} \quad y = \pm b \qquad (308)$$

In addition, they considered viscous dissipation and flow work effects in the gas. This problem for the fully developed velocity profile was solved by the Laplace transform method. The wall temperature distribution was presented graphically for various thermal energy source functions, viscous dissipation numbers, and uniform as well as parabolic inlet fluid temperatures.

2. SPECIFIED WALL HEAT FLUX DISTRIBUTION

Once the fundamental solution of the second kind is obtained, the thermal entrance solution for parallel plates can be obtained for any arbitrary equal or different heat fluxes at each wall. However, the equal and constant wall heat flux case, the Ⓗ problem, is treated separately because of its importance as a limiting solution for the rectangular and concentric duct family. The fundamental solution of the second kind and arbitrary variation in wall heat fluxes are described after the Ⓗ problem.

a. The Ⓗ Thermal Entrance Problem

The thermal entry length problem for parallel plates with constant and equal wall heat fluxes was first investigated by Cess and Shaffer [360] employing idealizations the same as those for the circular tube. They used the method of Siegel et al. [29] and obtained the first three terms of the following infinite series solution:

$$\Theta = \frac{t - t_e}{q'' D_h / k} = 4x^* + \frac{3}{16}\left(\frac{y}{b}\right)^2 - \frac{1}{32}\left(\frac{y}{b}\right)^4 - \frac{39}{1120}$$

$$+ \frac{1}{4}\sum_{n=1}^{\infty} C_n Y_n \exp\left(-\frac{32}{3}\beta_n^2 x^*\right) \qquad (309)$$

$$\Theta_w = \frac{t_w - t_e}{q''D_h/k} = 4x^* + \frac{17}{140} + \frac{1}{4} \sum_{n=1}^{\infty} C_n Y_n(1) \exp\left(-\frac{32}{3} \beta_n^2 x^*\right) \quad (310)$$

$$\Theta_m = \frac{t_m - t_e}{q''D_h/k} = 4x^* \quad (311)$$

$$Nu_{x,H} = \left[\frac{17}{140} + \frac{1}{4} \sum_{n=1}^{\infty} C_n Y_n(1) \exp\left(-\frac{32}{3} \beta_n^2 x^*\right)\right]^{-1} \quad (312)$$

Here β_n, Y_n, and C_n are eigenvalues, eigenfunctions, and constants, respectively, and are reported in Table 33.

TABLE 33

PARALLEL PLATES: INFINITE SERIES
SOLUTION FUNCTIONS FOR THE (H)
THERMAL ENTRANCE PROBLEM
(BASED ON THE RESULTS OF
SPARROW *et al.* [353])

n	β_n	$-C_n Y_n(1)$
1	4.287224	0.2222280
2	8.30372	0.0725316
3	12.3106	0.0373691
4	16.3145	0.0232829
5	20.3171	0.0161112
6	24.3189	0.0119190
7	28.3203	0.0092342
8	32.3214	0.0074013
9	36.3223	0.0060881
10	40.3231	0.0051116

Sparrow *et al.* [353] extended the analysis of Cess and Shaffer [360] for uniform and also arbitrary but equal wall heat fluxes with internal thermal energy generation. They reported the first ten eigenvalues and constants in tabular form, from which Table 33 is prepared. For higher values of n, Cess and Shaffer presented the asymptotic formulas as follows:

$$\beta_n = 4n + \tfrac{1}{3} \quad (313)$$

$$C_n Y_n(1) = -2.401006045 \beta_n^{-5/3} \quad (314)$$

Sparrow and Siegel [204] applied a variational method to solve the same (H) thermal entrance problem. Their first eigenvalue and constants are in good agreement with those of Cess and Shaffer.

Accurate numerical results for this problem in terms of $Nu_{x,H}$ have been evaluated by Shah [270] using the first 121 terms of Eq. (312). The first ten eigenvalues and constants were taken from Table 33; the rest were evaluated from the asymptotic formulas of Eqs. (313) and (314). The calculated $Nu_{x,H}$ are reported in Table 34 for $x^* > 10^{-4}$. The values for $x^* \leq 10^{-4}$ in Table 34 were established using the extended Léveque-type solution to be described

TABLE 34

PARALLEL PLATES: $Nu_{x,H}$, $Nu_{m,H}$, AND $N_H(x)$ FOR A
FULLY DEVELOPED VELOCITY PROFILE (FROM SHAH [270])

x^*	$Nu_{x,H}$	$Nu_{m,H}$	$N_H(x)$
0.000001	148.773	223.2	0.000215
0.0000015	129.944	195.0	0.000280
0.000002	118.049	177.1	0.000338
0.000003	103.110	154.7	0.000439
0.000004	93.673	140.6	0.000529
0.000005	86.954	130.5	0.000611
0.000006	81.824	122.8	0.000687
0.000007	77.724	116.6	0.000759
0.000008	74.339	111.6	0.000827
0.000009	71.477	107.3	0.000891
0.00001	69.011	103.55	0.000953
0.000015	60.292	90.45	0.00123
0.00002	54.787	82.18	0.00148
0.00003	47.880	71.79	0.00191
0.00004	43.521	65.24	0.00228
0.00005	40.419	60.57	0.00262
0.00006	38.054	57.00	0.00293
0.00007	36.165	54.16	0.00321
0.00008	34.607	51.81	0.00349
0.00009	33.290	49.82	0.00374
0.0001	32.153	48.11	0.00399
0.00015	28.154	42.06	0.00507
0.0002	25.636	38.25	0.00600
0.0003	22.488	33.47	0.00757
0.0004	20.512	30.46	0.00889
0.0005	19.113	28.33	0.0100
0.0006	18.050	26.70	0.0111
0.0007	17.205	25.40	0.0120
0.0008	16.511	24.33	0.0129
0.0009	15.928	23.43	0.0137
0.001	15.427	22.65	0.0144
0.0015	13.681	19.93	0.0175
0.002	12.604	18.22	0.0200
0.003	11.299	16.11	0.0236
0.004	10.516	14.80	0.0263
0.005	9.9878	13.89	0.0283
0.006	9.6085	13.21	0.0298
0.007	9.3249	12.67	0.0310
0.008	9.1073	12.24	0.0320
0.009	8.9374	11.88	0.0328
0.01	8.8031	11.58	0.0334
0.015	8.4393	10.59	0.0353
0.02	8.3107	10.03	0.0359
0.03	8.2458	9.446	0.0363
0.04	8.2368	9.145	0.0364
0.05	8.2355	8.963	0.0364
0.06	8.2353	8.842	0.0364
0.07	8.2353	8.755	0.0364
0.08	8.2353	8.690	0.0364
0.09	8.2353	8.640	0.0364
0.1	8.2353	8.599	0.0364
0.15	8.2353	8.478	0.0364
0.2	8.2353	8.417	0.0364

next. $Nu_{x,H}$ are plotted in Fig. 31. The thermal entrance length was evaluated as [270]

$$L^*_{th,H} = 0.0115439 \tag{315}$$

$L^*_{th,H}$ computed by Hanna et al. [59] is 0.00871, a magnitude 33% lower than the value of Eq. (315).

The series of Eq. (312) converges uniformly for all nonzero x^*, but the convergence is very slow as x^* approaches zero. In this region, a Lévêque-type similarity solution was provided by Worsøe-Schmidt [200]. Based on his results, the following formula is derived for $Nu_{x,H}$:

$$\frac{1}{Nu_{x,H}} = 0.670960978x^{*1/3} + 0.159064137x^{*2/3}$$

$$+ 0.12012x^* + 0.12495x^{*4/3} + 0.15602x^{*5/3}$$

$$+ 0.22176x^{*2} + 0.34932x^{*7/3} - 4x^* \qquad (316)$$

$Nu_{x,H}$ were computed from this equation by Shah [270] for $10^{-6} \leq x^* \leq 10^{-2}$. $Nu_{x,H}$ of Eq. (316) are identical up to five or more digits of Eq. (312) for $0.0005 \leq x^* \leq 0.005$ and are at most 0.08% higher for the extended range $5 \times 10^{-5} \leq x^* \leq 10^{-2}$. $Nu_{x,H}$ in Table 34 for $x^* \leq 10^{-4}$ are based on Eq. (316).

The following operationally more convenient approximate equations are recommended by Shah [270] for Eqs. (312) and (316) for the indicated range of x^*.

$$Nu_{x,H} = \begin{cases} 1.490(x^*)^{-1/3} & \text{for } x^* \leq 0.0002 & (317) \\ 1.490(x^*)^{-1/3} - 0.4 & \text{for } 0.0002 < x^* \leq 0.001 & (318) \\ 8.235 + 8.68(10^3 x^*)^{-0.506}e^{-164x^*} & \text{for } x^* > 0.001 & (319) \end{cases}$$

Equation (317) predicts $Nu_{x,H}$ higher than those in Table 34 by at most 0.2% for $x^* \leq 0.00006$ and lower by at most 0.6% for $x^* > 0.00006$. Equations (318) and (319) predict $Nu_{x,H}$ within ± 0.8 and $\pm 0.6\%$, respectively, when compared to those of Table 34.

The mean Nusselt numbers $Nu_{m,H}$ were also determined for the parallel plates duct by integration of $Nu_{x,H}$ of Table 34 [270]. $Nu_{m,H}$ and $N_H(x)$ are presented in Table 34. $Nu_{m,H}$ are plotted in Fig. 31. The incremental heat transfer number $N_H(x)$ for fully developed flow is

$$N_H(\infty) = 0.0364 \qquad (320)$$

The following approximate equations are recommended by Shah [270] for $Nu_{m,H}$

$$Nu_{m,H} = \begin{cases} 2.236(x^*)^{-1/3} & \text{for } x^* \leq 0.001 & (321) \\ 2.236(x^*)^{-1/3} + 0.9 & \text{for } 0.001 < x^* < 0.01 & (322) \\ 8.235 + \dfrac{0.0364}{x^*} & \text{for } x^* \geq 0.01 & (323) \end{cases}$$

The calculated $Nu_{m,H}$ by these equations agree with those of Table 34 within ± 1.3, ± 2.7, and $+2.6\%$, respectively.

The Ⓗ thermal entrance problem with finite fluid axial heat conduction has been solved by Hsu [311] for the initial and boundary conditions of Fig. 23c, and by Jones [361] for those of Fig. 23b. Hsu [311] obtained a closed-form eigenvalue solution. He reported the first 20 eigenvalues and constants for $Pe = 1$ and 5. Additionally, he presented graphically $Nu_{x,H}$ as a function of $32x^*/3$ for $Pe = 1, 2.5, 5, 10, 20, 30, 40$, and ∞. In this plot, similar to the circular tube case, $Nu_{x,H}$ is finite at $x^* = 0$ for finite values of Pe and has an inflection point at $x^* \simeq 0.0015$. Jones [361] analyzed the problem for the initial and boundary conditions of Fig. 23b. The solution was obtained using a double-sided Laplace transform. The detailed temperature distribution 4Θ was obtained and graphically presented for $Pe = 0.125$ and 0.25.

The effect of viscous dissipation on the Ⓗ thermal entrance problem for parallel plates has been investigated by Hwang et al. [362] using a finite difference method. They presented the thermal entrance temperature profiles, $4\Theta_w$, $4\Theta_m$, and $Nu_{x,H}$ for $Br' = -0.25, -0.125, 0, 0.125$, and 0.25. The effect of viscous dissipation (Br') on thermal entrance t_w, t_m, and $Nu_{x,H}$ for parallel plates is very similar to that for the circular tube, as discussed on p. 129, except for the different range of Br' as described below.

For parallel plates, by integrating $\mu(\partial u/\partial r)^2/g_c J$ over the flow cross section area, it can be shown that, per unit duct length,

$$\frac{\text{thermal energy generated}}{\text{due to viscous dissipation}} = \frac{48\mu u_m{}^2 A_c}{g_c J D_h{}^2} \tag{324}$$

Following the development of Eq. (249), the fluid bulk mean temperature distribution for parallel plates is given by

$$\Theta_m = \frac{t_m - t_e}{q'' D_h/k} = 4x^* + 48x^* \, Br' \tag{325}$$

For the case of positive q'', i.e., heat transfer from the wall to the fluid, $Br' > 0$, fluid is heated and t_m always increases with increasing x^*. For the case of negative q'', i.e., heat transfer from the fluid to the wall, $Br' < 0$, fluid cools down and t_m decreases with increasing x^* for $-1/12 < Br' < 0$. However, for $Br' < -1/12$, the thermal energy generated due to viscous dissipation is more than the fluid-to-wall heat transfer rate, and t_m increases with increasing x^* in spite of negative q''. Based on Eq. (278) with $S^* = 0$, it is found that for $Br' > -17/108$, t_w is always less than t_m even when t_m increases with x^*. For $Br' < -17/108$, t_m is greater than t_w for low x^*, and t_w is greater than t_m for high x^*, a situation similar to one described by Fig. 22. The effect of viscous dissipation on $Nu_{x,H}$ is qualitatively similar to that for the circular tube described on p. 131 except for the point of singularity being $Br' = -17/108$.

For a perfect gas, the effect of the flow work term $u(dp/dx)/J$ on heat transfer is of the same order of magnitude as that of the viscous dissipation term. For a fully developed velocity profile and constant properties, integration of this flow work term yields, per unit duct length,

$$\frac{\text{thermal energy generated}}{\text{due to the flow work effect}} = -\frac{48\mu u_m^2 A_c}{g_c J D_h^2} \tag{326}$$

The negative sign on the right-hand side of this equation shows that the flow work term acts as a distributed heat sink. The thermal energy absorbed due to the flow work effect is identical to the thermal energy generated due to viscous dissipation, Eq. (324). The net result is that the fluid bulk mean temperature is unaffected due to these effects, so that

$$\Theta_m = 4x^* \tag{327}$$

Thus, t_m increases with increasing x^* for the fluid heating situation; t_m decreases with increasing x^* for the fluid cooling situation, a trend different from when the effect of viscous dissipation is considered alone. The local effect of flow work and viscous dissipation is of an opposite nature. This is because these two effects are proportional to u and $(\partial u/\partial r)^2$, respectively, so that the heat sinks prevail in the central region and heat sources dominate in the wall region.

The combined effect of flow work and viscous dissipation on the wall temperature distribution is complex. The solution to the thermal entrance problem has been obtained by Ou and Cheng [229]. They showed that for $Br' > 0$, t_w is always higher than t_m. For $-17/108 < Br' < 0$, t_w is always lower than t_m. For $Br' < -17/108$, t_w is lower than t_m near $x^* = 0$, and t_w is higher than t_m for higher x^*. The location for $t_w(x) = t_m(x)$ represents a point of singularity in the $Nu_{x,H}$ vs. x^* curve. It must be emphasized that the combined qualitative effect of flow work and viscous dissipation on $Nu_{x,H}$ vs. x^* curve is similar to that for viscous dissipation alone; however, the quantitative effects are different. The temperature profile development is significantly different from that of Fig. 22. t_m is unaffected due to the combined effect of flow work and viscous dissipation; it depends only upon the amount and direction of heat flow at the wall. As shown in Eq. (325), t_m is dependent upon Br' when the effect of viscous dissipation is considered alone. Ou and Cheng [229] presented the fluid temperature development, t_m, t_w, and $Nu_{x,H}$, as a function of x^*. The fully developed Nusselt number for the combined effect is the same as that for the effect of viscous dissipation alone for all Br', as rationalized by Ou and Cheng, and is given by Eq. (278) with $S^* = 0$.

b. *Fundamental Solution of the Second Kind and Arbitrary Axial Wall Heat Flux Distribution*

The fundamental solution of the second kind obtained by the modified Graetz method also requires the determination of what is called odd and even eigenvalues and constants. Cess and Shaffer reported the first four odd [363] and the first three even [360] eigenvalues and constants, and presented asymptotic formulas for higher values. Cess and Shaffer then generalized the solution to the arbitrary prescribed wall heat flux case. McCuen *et al.* [332] also presented the first three even and four odd eigenvalues and constants, and generalized the solution for the prescribed arbitrary axial wall heat flux distribution. McCuen *et al.* then presented Nusselt numbers and dimensionless temperatures in graphical and tabular forms. Based on their results, the fundamental solution of the second kind is reported in Chapter XII.

Dzung [364] obtained the fundamental solution of the second kind and also analyzed the Ⓗ thermal entrance and the axially sinusoidal wall heat flux problems. He tabulated related eigenvalues and constants.

3. FUNDAMENTAL SOLUTIONS OF THE THIRD AND FOURTH KINDS

McCuen *et al.* [332] obtained the fundamental solutions of the third and fourth kinds (Fig. 8). They summarized the detailed solutions in tabular and graphical forms for all four fundamental boundary conditions (their results are reported in Chapter XII). With these results, the thermal entry length solution for any arbitrary combinations of wall heat fluxes and temperatures can be worked out, provided that axial heat conduction, viscous dissipation, and thermal energy sources within the fluid are negligible.

Tay and De Vahl Davis [213] also obtained the fundamental solutions of all four kinds for parallel plates by a finite element method. Additionally, they obtained solutions for both walls at uniform as well as linearly varying temperatures. However, they reported solutions (1) for both walls with linearly varying temperatures and (2) one wall at uniform heat flux, the other being insulated. Their results are in excellent agreement with those of McCuen *et al.* [332].

4. FINITE WALL THERMAL RESISTANCE, Ⓣ3

The Ⓣ3 thermal entrance problem for parallel plates (outside temperatures of both walls are equal and uniform in the axial direction with a finite thermal resistance in the normal direction) has been analyzed in considerable detail. van der Does de Bye and Schenk [365] solved it for $R_w = 0.025$ and 0.25 using the eigenvalue method (the modified Graetz method). Berry [366] also considered finite wall resistance for the circular tube and parallel plates, but examples were worked out only for slug flow. Schenk [367] extended Berry's work considering fully developed laminar flow. Schenk and Beckers [368] dealt with the case of finite wall resistance and nonuniform inlet temperature profile. The calculations were made for a linear transverse temperature distribution at the inlet and the wall resistance parameter $R_w = 0, 0.125$, and ∞. Dennis and Poots [369] used the Rayleigh approximate method to solve the (T3) problem with $R_w = 0, 0.025$, and 0.25.

Sideman et al. [230] supplemented and extended the work of Schenk et al. [365,368] for the case of finite wall resistance. They determined and reported the first five eigenvalues and constants for $R_w = 0, 0.0125, 0.025, 0.05, 0.075, 0.125, 0.25$, and 0.5. All of the above thermal entrance solutions are valid for fully developed laminar flow between parallel plates when the effects of axial heat conduction, viscous dissipation and internal thermal energy generation are negligible.

To obtain the solution, Schenk et al. and Sideman et al. employed the modified Graetz method for which the evaluation of higher eigenvalues and constants becomes increasingly difficult. Hsu [297] alleviated this difficulty by numerically solving the Sturm–Liouville type differential equation. He determined and tabulated the first ten eigenvalues and constants for $R_w = 0.0025, 0.025, 0.125, 0.25$, and 0.5. He also presented implicit asymptotic formulas for the determination of higher eigenvalues and constants. In the analysis, Hsu also included the effect of uniform internal thermal energy generation within the fluid. He presented graphically $Nu_{x,T3}$ as a function of $32x^*/3$: (1) with the above-listed values of R_w as a parameter for zero thermal energy generation, and (2) with various values of thermal energy generation as a parameter for $R_w = 0.5$. The $Nu_{x,T3}$ for zero thermal energy generation are presented in Table 35 and Fig. 33, based on the first ten eigenvalues and constants from Hsu [297] and the next 110 values calculated by the present authors from the implicit asymptotic equations of Hsu.

Lakin [370] derived the explicit asymptotic formulas for higher eigenvalues and eigenfuctions for the (T3) problem by the so-called turning-point approximations. However, he did not obtain the asymptotic formulas for the constants needed to determine $Nu_{x,T3}$. Lakin's asymptotic formula for higher eigenvalues is only first order and hence does not have the parameter n for the nth eigenvalue. Consequently, Lakin's very approximate results are not comparable with those of Hsu [297].

Hickman [232] analyzed the (T3) thermal entrance problem by a Laplace transform technique for large x^*. He presented a formula for $Nu_{o,m}$, defined by Eq. (119), as a function of x^* and $1/2R_w$.

In all of the foregoing work, the thermal resistance of each wall was assumed to be the same. Schenk [371] considered one wall insulated ($R_w = \infty$) with the other wall having a finite thermal resistance, $R_w = 0, 0.15$, and 0.25.

TABLE 35

Parallel Plates: $Nu_{x,T3}$ as a Function of x^* and R_w for a
Fully Developed Velocity Profile (Based on the Results
of Hsu [297])

x^*	$Nu_{x,T3}$				
	$R_w=.0025$	$R_w=.025$	$R_w=.125$	$R_w=.25$	$R_w=0.5$
0.0001	27.071	28.245	31.325	31.710	31.922
0.00015	23.656	24.930	27.343	27.716	27.925
0.0002	21.516	22.813	24.837	25.202	25.409
0.0003	18.851	20.093	21.706	22.059	22.262
0.0004	17.185	18.329	19.743	20.086	20.287
0.0005	16.010	17.056	18.354	18.690	18.888
0.0006	15.120	16.081	17.301	17.630	17.826
0.0007	14.414	15.303	16.464	16.788	16.982
0.0008	13.836	14.663	15.778	16.096	16.289
0.0009	13.350	14.126	15.201	15.514	15.706
0.001	12.935	13.667	14.706	15.015	15.206
0.0015	11.495	12.075	12.985	13.277	13.461
0.002	10.616	11.105	11.928	12.207	12.387
0.003	9.566	9.948	10.654	10.913	11.085
0.004	8.953	9.269	9.896	10.139	10.305
0.005	8.553	0.824	9.391	9.621	9.781
0.006	8.277	8.514	9.032	9.251	9.406
0.007	8.081	8.191	8.768	8.976	9.126
0.008	7.939	8.127	8.569	8.768	8.913
0.009	7.836	8.006	8.417	8.607	8.747
0.01	7.760	7.915	8.300	8.481	8.617
0.015	7.597	7.708	8.006	8.155	8.272
0.02	7.562	7.658	7.919	8.051	8.157
0.03	7.56	7.641	7.885	8.006	8.103
0.04	7.55	7.637	7.881	8.001	8.096
0.05	7.55	7.634	7.881	8.000	8.095
∞	7.553	7.650	7.881	8.000	8.095

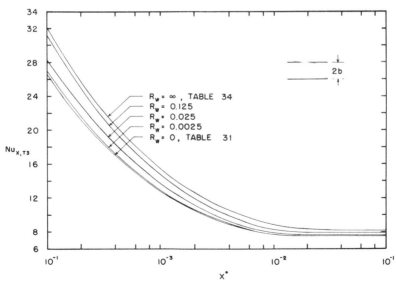

FIG. 33. Parallel plates: $Nu_{x,T3}$ as a function of x^* and R_w for a fully developed velocity profile (from Table 35).

5. EXPONENTIAL AXIAL WALL HEAT FLUX, (H5)

Based on the (H) thermal entrance solution described on p. 179, the (H5) thermal entrance solution can be obtained by the superposition method [7,29]. The local Nusselt number is thus found as

$$\text{Nu}_{x,\text{H5}} = \frac{3}{8}\left(\sum_{n=1}^{\infty} \frac{-C_n Y_n(1)\beta_n^2}{(32/3)\beta_n^2 + m}\left\{1 - \exp\left[-\left(\frac{32}{3}\beta_n^2 + m\right)x^*\right]\right\}\right)^{-1} \quad (328)$$

where the constants β_n and $-C_n Y_n(1)$ are available in Table 33 for values of $n \le 10$ and can be determine by Eqs. (313) and (314) for $n > 10$.

6. CONJUGATED PROBLEM

In the five preceding subsections, a thermal boundary condition was specified at inside walls, and the solution was sought for the fluid medium only. In contrast, a conjugated problem, as described on p. 11, is formulated for the entire solid body–fluid medium system and the solution is obtained considering a continuous temperature and heat flux at the solid–fluid interface.

Shelyag [21] analyzed the conjugated problem (similar to the one formulated on p. 12) for parallel plates when the outside walls were subjected to a constant heat flux. The solution was obtained by an eigenvalue method. The axial wall temperature distribution was presented graphically in dimensional form. They found that the interface wall temperature at the entrance ($x = 0$) was significantly different from the gas inlet temperature. They also reported an experimental investigation of the same problem and found good agreement with the theoretical predictions. Since the results of Shelyag are presented for specific cases in dimensional form only, they are not compared with the dimensionless results of Mori et al. [23] described later.

Davis and Gill [22] analyzed a conjugated problem for parallel plates, with laminar Poiseuille–Couette flow, with one wall at the temperature of the entering fluid and the outside of the other wall at constant heat flux. The Poiseuille flow[†] and Couette flow[‡] are special cases of the Poiseuille–Couette flow. They showed that the solution depended upon four dimensionless groups: x/L, Pe, a'/L, a'/LR_w. These dimensionless groups are in a slightly different form than those of Eq. (163). Davis and Gill presented

---

[†] Refer to p. 78 for the definition of Poiseuille flow.

[‡] Fully developed, steady state, laminar flow of an incompressible fluid between two parallel plates (one of which is at rest, the other moving at a constant velocity parallel to itself) is referred to as *Couette flow*. The viscosity of the fluid is considered as constant, and there are no body forces. The invariant velocity profile obtained for Couette flow is linear at any cross section of the duct.

graphically solid–fluid interface temperatures as a function of $4x^*$, where Pe in x^* is based on maximum velocity. They found that the effect of wall axial heat conduction (as included in the conjugated problem) can significantly affect the fluid temperature field and lower the Nusselt number relative to those predicted for a uniform heat flux at the solid–fluid interface. They found fair agreement between the experimental results and their analysis.

Luikov et al. [18,19] also analyzed the conjugated problem for parallel plates. However, their results are too complex to interpret for comparison with the conventional convection problem with specified boundary conditions.

Mori et al. [23] analyzed a conjugated problem for parallel plates that was similar to that formulated by Eqs. (14)–(20) for the circular tube. They employed the method of analysis similar to the one they used for the circular tube [20], and derived the same conclusions as for the circular tube as summarized on pp. 137–138. In particular, for the case of specified constant heat flux at the outside wall, axial heat conduction in the wall reduces the interface Nu_x. The lowest magnitudes of Nu_x correspond to those for the Ⓣ solution. For the case of specified constant wall temperature at the outside wall, the finite resistance in the wall increases the interface Nu_x. The highest values of Nu_x correspond to those for the Ⓗ solution. The effect of wall heat conduction is negligible for $R_w \leq 0.0005$ and $a'/L \leq 0.001$. In addition, Mori et al. conducted an experimental investigation of the measurement of the interface Nu_x, when the outside wall was maintained at a constant temperature. Their measured Nu_x at the interface for a rectangular duct of $\alpha^* = 1/6$ agreed with their theoretical results within $\pm 10\%$, thus providing a reasonable verification.

D. Simultaneously Developing Flow

The combined entry length problem, where both the velocity and temperature profiles are developing together, has been analyzed primarily for equal and constant wall temperatures or heat fluxes. No systematic attempt has been made to obtain all fundamental solutions. The reason for this is that they are of less practical use unless they can be employed with the superposition theorem to approximate the arbitrary variations in wall temperatures or heat fluxes. Unfortunately, a large number of such solutions with heating started at different locations would be required for this purpose (refer to p. 34 for further discussion). For this reason only several combined entry length solutions are available. They are described next.

1. SPECIFIED WALL TEMPERATURE DISTRIBUTION

Simultaneous development of velocity and temperature profiles for parallel plates was first investigated by Sparrow [372] for equal wall temperatures, the Ⓣ boundary condition. Sparrow used Schiller's velocity profile and employed the Kármán–Pohlhausen integral method. Rohsenow and Choi [5] have graphically presented Sparrow's $Nu_{m,T}$ as a function of x^* for $Pr = 0, 0.01, 0.1, 0.72, 1, 2, 10, 50$, and ∞. For $Pr = 0.1$, Sparrow's results agree with those of Table 36 within 5% for $10^{-4} \leq x^* \leq 0.1$.

Slezkin [373] and Murakawa [374] considered the Ⓣ combined entry length problem for the circular tube and parallel plates, but they did not present exact solutions of the momentum and energy equations.

Stephan [322] employed an approximate series solution for the Ⓣ combined entry length problem and approximated his $Nu_{m,T}$ by the following empirical equation for a Pr range of 0.1 to 1000:

$$Nu_{m,T} = 7.55 + \frac{0.024(x^*)^{-1.14}}{1 + 0.0358(x^*)^{-0.64}Pr^{0.17}} \tag{329}$$

Hwang and Fan [218] obtained an all-numerical finite difference solution for the Ⓣ combined entry problem. They used the same method as that of Bodoia and Osterle [172] for the velocity problem. They obtained the solution for the Prandtl number range of 0.01 to 50. Their $Nu_{m,T}$ are reported in Table 36 and are plotted in Fig. 34 [375].

Mercer et al. [376] also solved the same problem by a finite difference method, but they first transformed the differential equations in terms of the stream function. They also reported an experimental investigation with results that are in good agreement with their theoretical predictions. In addition, they proposed an equation similar to Eq. (329) for $Nu_{m,T}$.

In practical applications, the velocity profile at the duct inlet may be asymmetric, which would result in the developing velocity profiles asymmetric in the entrance region. To investigate the effect of such a velocity profile at the duct inlet on pressure drop and heat transfer, Kiya et al. [342] analyzed the Ⓣ simultaneously developing flow problem for parallel plates. For the convenience of analysis, they considered a linear inlet velocity profile (having a constant velocity gradient normal to the duct axis) and the fluid Prandtl number of 0.72. The results for the velocity problem are described on p. 164. They obtained results for the thermal entrance problem by a finite difference method for various values of the dimensionless inlet velocity gradients κ. The inlet velocity profile is uniform when $\kappa = 0$. They found that the mean heat transfer rate ($Nu_{x,T}$ and $Nu_{m,T}$) is much higher for the wall at the high inlet velocity side, and the difference relative to the low-velocity side increases with higher values of κ. However, an average value of the heat

TABLE 36

PARALLEL PLATES: $Nu_{m,T}$ AS A FUNCTION OF x^* AND Pr FOR SIMULTANEOUSLY
DEVELOPING FLOW (FROM HWANG [375])

Pr = 0.1		Pr = 0.72		Pr = 10		Pr = 50	
x^*	$Nu_{m,T}$	x^*	$Nu_{m,T}$	x^*	$Nu_{m,T}$	x^*	$Nu_{m,T}$
0.0000625	115.7	0.0000434	116.1	0.0000125	130.7	0.0000075	150.2
0.000188	60.50	0.0000868	72.87	0.0000188	114.3	0.0000138	108.4
0.000313	47.77	0.000260	44.14	0.000025	98.65	0.0000200	89.93
0.000469	38.53	0.000434	35.09	0.000050	70.69	0.0000250	80.79
0.000625	33.88	0.000608	30.23	0.000075	58.66	0.0000563	54.95
0.00125	25.16	0.000955	24.91	0.000100	51.53	0.0000875	45.19
0.00188	21.25	0.00130	21.90	0.000125	46.68	0.000119	39.77
0.00250	18.92	0.00174	19.52	0.000188	38.99	0.000150	36.20
0.00313	17.33	0.00260	16.63	0.000250	34.38	0.000275	28.73
0.00375	16.17	0.00347	14.90	0.000500	25.73	0.000400	25.15
0.00438	15.27	0.00434	13.74	0.00075	21.94	0.000525	22.91
0.00500	14.55	0.00608	12.24	0.00106	19.25	0.000625	21.62
0.00563	13.96	0.00868	10.96	0.00200	15.44	0.000938	18.92
0.00625	13.47	0.0148	9.593	0.00313	13.40	0.00125	17.26
0.00938	11.84	0.0234	8.827	0.00625	11.01	0.00156	16.10
0.0125	10.94	0.0321	8.474	0.00938	9.978	0.00188	15.23
0.0281	9.240	0.0434	8.225	0.0125	9.402	0.00313	13.11
0.0438	8.691	0.0651	7.986	0.0250	8.474	0.00438	11.94
0.0594	8.396	0.0942	7.792	0.0406	8.106	0.00563	11.17
0.0750	8.218	0.1519	7.707	0.0656	7.878	0.00688	10.63

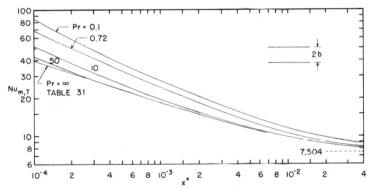

FIG. 34. Parallel plates: $Nu_{m,T}$ as a function of x^* and Pr for simultaneously developing flow
(from Table 36).

transfer rate for the two walls is smaller only by a few percent than that for
the case of a uniform inlet velocity profile, $\kappa = 0$.

Miller and Lundberg [377] extended the work of [218,322,376] for the
boundary condition of constant but unequal wall temperatures. They employed Bodoia's velocity distribution [172]. For the case of equal wall
temperatures, their solution agrees with that of Mercer et al. [376] for $x^* >$
10^{-3}. For smaller values of x^*, their results are about 10% higher due to
their use of a finer grid size at the entrance. Their results for unequal wall
temperatures are presented in terms of dimensionless heat flux functions for
$0.5 \leq Pr \leq 10$.

2. CONSTANT AND EQUAL WALL HEAT FLUXES, (H)

Siegel and Sparrow [378] solved the case of equal and constant wall heat fluxes by the integral method, employing Schiller's velocity profile, used by Sparrow [372]. They presented $Nu_{x,H}$ as functions of x^* and Prandtl number (from 0.01 to 50). Their results agree reasonably well with those of Table 37.

Using Langhaar's velocity profile, Han [155] also solved the equal and constant wall heat flux case for the thermal entrance region. The rate of approach of his local Nusselt number to its asymptotic value was quite different from the results of Siegel and Sparrow [378]. Miller [379] showed that Han's solution was in poor agreement with a similar solution employing the Schlichting velocity profile [137].

Naito [146] first obtained the solution for the hydrodynamic entrance region by the Kármán–Pohlhausen integral method as briefly discussed on p. 160. He subsequently obtained the solution for the (H) combined entry length problem by the modified Kármán–Pohlhausen method. The temperature profile in the thermal boundary layer was approximated by a polynomial of the fourth degree. He presented closed-form formulas for the thermal boundary layer thickness, $Nu_{x,H}$, and $Nu_{m,H}$ for Pr = 0.01 to 1000. He also provided the limiting value of x^* for each Pr below which his solution is valid and accurate. His $Nu_{x,H}$ are in excellent agreement with those of Siegel and Sparrow [378] at low x^*, and a maximum of 5% higher for high x^*. The $Nu_{x,H}$ of Naito are lower than those of Table 37. For $0.7 \le Pr \le 1.0$, his $Nu_{x,H}$ are within 6% of those of Table 37 for $x^* < 0.03$. For Pr = 10, his $Nu_{x,H}$ are within 2% for $x^* < 0.0008$. For higher x^*, his results are either negative or too low compared to those of Table 37.

Bhatti and Savery [340] analyzed the (H) combined entry length problem by an integral method employing their hydrodynamic entry length solution reported on p. 160. They presented graphically $Nu_{x,H}/4$ as a function of $16x^*$ Pr for Pr ranging from 0.01 to 10,000. Their $Nu_{x,H}$ are lower than those of Hwang and Fan [218], Table 37. The difference ranged from 16% at $x^* = 10^{-3}$ to 6% at $x^* = 10^{-2}$ for Pr = 0.01, from 15% at $x^* = 10^{-3}$ to 10% at $x^* = 10^{-2}$ for Pr = 0.7, and about 6% for $0.0006 \le x^* \le 0.003$ for Pr = 10. Their fully developed Nu_H are lower than 8.235 for $Pr \ge 0.7$, thus indicating $Nu_{x,H}$ obtained by their method are too low.

Hwang and Fan [218] obtained an all-numerical finite difference solution for the (H) combined entry length problem for Pr = 0.1, 0.72, 1, 2, 10, and 50. Their $Nu_{x,H}$ are presented in Table 37 and Fig. 35 [375]. There appears to be an error in $Nu_{x,H}$ of Hwang and Fan. What they designate $Nu_{x,H}$ for Pr = 0.1 should be $Nu_{x,H}$ for Pr = 0.01. For $x^* < 0.005$, $Nu_{x,H}$ for the latter case agree within 6% with those of Naito for Pr = 0.01. Accordingly, Pr = 0.01 is reported in Table 37 and Fig. 35 instead of Pr = 0.1 as used by Hwang and Fan.

TABLE 37

PARALLEL PLATES: $Nu_{x,H}$ AS A FUNCTION OF x^* AND Pr FOR SIMULTANEOUSLY
DEVELOPING FLOW (FROM HWANG AND FAN [218])

Pr = 0.01		Pr = 0.7		Pr = 1		Pr = 10	
x^*	$Nu_{x,H}$	x^*	$Nu_{x,H}$	$-x^*$	$Nu_{x,H}$	x^*	$Nu_{x,H}$
0.001	24.5	0.000714	21.98	0.00050	24.34	0.000050	50.74
0.002	18.5	0.00179	15.11	0.00125	16.62	0.000125	34.07
0.004	13.7	0.00625	10.03	0.00438	10.79	0.000438	20.66
0.007	11.1	0.0107	8.90	0.0075	9.31	0.00075	17.03
0.01	10.0	0.0286	8.24	0.0200	8.31	0.00200	12.60
0.02	9.0	0.0893	8.22	0.0625	8.23	0.00625	9.50
0.04	8.5	0.143	8.22	0.100	8.23	0.0100	8.80
0.07	8.3						
0.20	8.23						

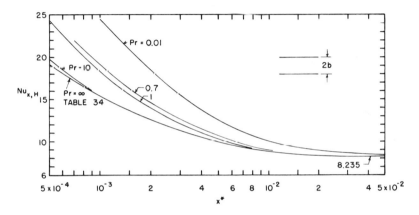

FIG. 35. Parallel plates: $Nu_{x,H}$ as a function of x^* and Pr for simultaneously developing flow (from Table 37).

Gavrilov and Dul'nev [380] combined in a particular way $Nu_{x,H}$ of simultaneously developing flow over a flat plate and Nu_H of fully developed flow through parallel plates, and derived approximate formulas for $Nu_{x,H}$ that could be used throughout the entrance and fully developed regions of parallel plates. $Nu_{x,H}$ determined from their formula are in satisfactory agreement with those reported by Petukhov [8] for $0.01 \le Pr \le 50$.

The thermal entrance length $L_{th,H}^*$ for simultaneously developing flow through parallel plates, based on $Nu_{x,H}$ of Tables 37 and 34, is found as 0.030, 0.017, 0.014, 0.012, and 0.0115, respectively, for Pr = 0.01, 0.7, 1, 10, and ∞.

3. FUNDAMENTAL SOLUTIONS OF THE SECOND KIND

Heaton et al. [44] used the Kármán–Pohlhausen integral method and Langhaar's velocity profile to solve the combined entry problem for parallel plates with one wall having a constant (nonzero) flux and the other wall being insulated (zero flux). Their Nusselt numbers for the heated wall and the influence coefficient $\theta^* = -(\theta_{j0}^{(2)} - \theta_{m0}^{(2)})/(\theta_{00}^{(2)} - \theta_{m0}^{(2)})$ are presented in Table 38. Other pertinent results may be obtained from [7, 44].

TABLE 38

PARALLEL PLATES: LOCAL NUSSELT NUMBERS AT THE HEATED WALL
FOR FUNDAMENTAL SOLUTIONS OF THE SECOND KIND FOR
SIMULTANEOUSLY DEVELOPING FLOW (FROM HEATON et al. [44])

x^*	$Pr = 0.01$		$Pr = 0.70$		$Pr = 10.0$	
	$Nu_{x,jj}^{(2)}$	θ^*	$Nu_{x,jj}^{(2)}$	θ^*	$Nu_{x,jj}^{(2)}$	θ^*
0.0001	–	–	52.8	0.0116	40.4	0.0081
0.0010	24.2	0.0484	18.50	0.0370	15.56	0.0311
0.0025	15.8	0.0790	12.60	0.0630	11.46	0.0573
0.005	11.7	0.117	9.62	0.0962	9.20	0.0920
0.010	8.80	0.176	7.68	0.154	7.49	0.149
0.025	6.48	0.311	6.13	0.265	6.09	0.264
0.05	5.77	0.378	5.55	0.327	5.55	0.327
0.10	5.53	0.376	5.40	0.345	5.40	0.345
0.25	5.47	0.361	5.39	0.346	5.39	0.346
∞	5.39	0.346	5.39	0.346	5.39	0.346

Naito [146] also analyzed the combined entry length problem for the fundamental boundary condition of the second kind. He employed the Kármán–Pohlhausen integral method for both the velocity and temperature problems. He presented formulas for $Nu_{x,jj}^{(2)}$ as a function of x^* for Pr ranging from 0.01 to 1000. His $Nu_{x,jj}^{(2)}$ are lower than those of Table 38 and agree with them within 5 and 7% for $x^* < 0.025$ for $Pr = 0.01$ and 0.70, respectively. There appears to be an error in Naito's equation for $Nu_{x,jj}^{(2)}$ for $Pr = 10.0$ since his results are too low or negative compared to those of Table 38.

4. FUNDAMENTAL SOLUTIONS OF THE THIRD KIND

Sparrow [372] employed Schiller's velocity profile and the Kármán–Pohlhausen integral method to solve the combined entry length problem for parallel plates with one wall at a uniform temperature and the other wall insulated.

Stephan [381] obtained an approximate series solution for the same problem. The mean Nusselt number for the heated wall was correlated by a rather complicated equation.

Mercer et al. [376] also analyzed the same problem by employing a finite difference method. They correlated the results by the following equation:

$$\mathrm{Nu_{m,T}} = 4.86 + \frac{0.32(4x^*)^{-1.2}}{1 + 0.24(4x^*)^{-0.7}\,\mathrm{Pr}^{0.17}} \tag{330}$$

This equation agrees with the results of Sparrow [372] and Stephan [381] within 7% for $0.1 \leq \text{Pr} \leq 10$, and it reduces to the following "flat plate" equation for small values of x^*:

$$\Phi_{m,T} = 0.667 \left(\frac{D_h}{x} \right) \text{Re}_x^{0.5} \, \text{Pr}^{0.33} \tag{331}$$

where the coefficient 0.667 is slightly different from the well-recognized value of 0.664 [7]. Note that $\text{Nu}_{m,T}$ and $\Phi_{m,T}$ are approximately the same for low values of x^*. Using the exact relationship of Eq. (124) between $\text{Nu}_{m,T}$ and $\Phi_{m,T}$, one can arrive at the $\text{Nu}_{m,T}$ expression of Eq. (253) from $\Phi_{m,T}$ of Eq. (331) with a constant of 0.664.

Lombardi and Sparrow [382] experimentally determined local Sherwood numbers for parallel plates using the naphthalene sublimation technique. Employing the heat and mass transfer analogy, they arrived at $\text{Nu}_{x,T}$ for simultaneously developing flow. Their experimental results agree with Sparrow's theoretical results [372] within 5%.

Chapter VII

Rectangular Ducts

Fluid flow and heat transfer study in rectangular ducts requires a two-dimensional analysis, in contrast to the one-dimensional analysis for circular tube and parallel plates geometries. Since the analysis is more complex, rectangular ducts have not been analyzed in as much detail. For fully developed flow, f Re, Nu_T, Nu_{H1}, and Nu_{H2}, as well as the Nusselt numbers for other thermal boundary conditions, have been determined. The hydrodynamic entrance problem, as well as the thermal entrance problem with fully developed laminar flow, has also been investigated. However, there remains a need for the refinement of the thermal entry length solutions for the simultaneously developing flow. So far these have been obtained by neglecting the transverse velocity components v and w. Experience with the circular duct, as discussed on pp. 139 and 146, indicates that this neglect introduces significant errors into the solution for the entrance region close to inlet ($x^* = 0$).

A. Fully Developed Flow

Initially, a review is made of the fully developed fluid flow problem for rectangular ducts. Then the fully developed heat transfer results are presented for different thermal boundary conditions.

1. FLUID FLOW

The fully developed velocity profile for rectangular ducts has been determined, using an analogy with the stress function of the theory of elasticity

(Timoshenko and Goodier [72]) by Dryden *et al.* [4], and Marco and Han [70]. Consider the cross section of a rectangular duct, characterized by its aspect ratio $\alpha^* = 2b/2a$, as shown in Fig. 36 with the flow direction along the x axis (perpendicular to the plane of paper).

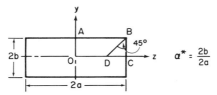

FIG. 36. A rectangular duct.

The velocity profile, provided by the solution of Eq. (4) with the boundary condition of Eq. (6), is

$$u - = -\frac{16 c_1 a^2}{\pi^3} \sum_{n=1,3,\dots}^{\infty} \frac{1}{n^3}(-1)^{(n-1)/2}\left[1 - \frac{\cosh(n\pi y/2a)}{\cosh(n\pi b/2a)}\right]\cos\left(\frac{n\pi z}{2a}\right) \quad (332)$$

$$u_m = -\frac{c_1 a^2}{3}\left[1 - \frac{192}{\pi^5}\left(\frac{a}{b}\right)\sum_{n=1,3,\dots}^{\infty}\frac{1}{n^5}\tanh\left(\frac{n\pi b}{2a}\right)\right] \quad (333)$$

This velocity profile is in excellent agreement with the experimental results of Holmes and Vermeulen [383] and Muchnik *et al.* [384].

Since Eq. (332) involves considerable computational complexity, Purday [124] proposed a simple approximation in the following form for the aspect ratio $\alpha^* \le 0.5$:

$$\frac{u}{u_{\text{max}}} = \left[1 - \left(\frac{y}{b}\right)^n\right]\left[1 - \left(\frac{z}{a}\right)^m\right] \quad (334)$$

where $n = 2$ and m is obtained, by applying the principle of minimum energy dissipation, as 2.37, 3.78, 5.19, 6.60, 13.6, and ∞ for $\alpha^* = 2b/2a = 0.5$, 1/3, 0.25, 0.20, 0.10, and 0, respectively. Holmes and Vermeulen [383] presented values of m and n based on the experimental measurements of the velocity gradient at the wall and the area under the velocity profile curve. Natarajan and Lakshmanan [385] solved the momentum equation (4) by a finite difference method, and matched the velocity profile to the empirical equation (334) to arrive at m and n as

$$m = 1.7 + 0.5(\alpha^*)^{-1.4} \quad (335)$$

$$n = \begin{cases} 2 & \text{for} \quad \alpha^* \le 1/3 \quad (336) \\ 2 + 0.3(\alpha^* - 1/3) & \text{for} \quad \alpha^* \ge 1/3 \quad (337) \end{cases}$$

The integration of Eq. (334) over the duct cross section yields

$$\frac{u}{u_m} = \left(\frac{m+1}{m}\right)\left(\frac{n+1}{n}\right)\left[1 - \left(\frac{y}{b}\right)^n\right]\left[1 - \left(\frac{z}{a}\right)^m\right] \tag{338}$$

$$\frac{u_{max}}{u_m} = \left(\frac{m+1}{m}\right)\left(\frac{n+1}{n}\right) \tag{339}$$

The values of m and n by Natarajan and Lakshmanan yield profiles that are in good agreement with the experimental results of Holmes and Vermeulen [383]. The approximate u_{max}/u_m of Eq. (339) are within 0.9% of those exact values in Table 39. Natarajan and Lakshmanan also presented a closed-form formula for f Re as a function of α^*, which is within $\pm 4.5\%$ of the exact values in Table 42. A more accurate formula for f Re is presented by Eq. (341).

Lundgren et al. [51] determined u_{max}/u_m, $K_d(\infty)$, and $K_e(\infty)$, which are presented in Table 39.

TABLE 39

RECTANGULAR DUCTS: u_{max}/u_m,
$K_d(\infty)$ AND $K_e(\infty)$ FOR FULLY
DEVELOPED LAMINAR FLOW
(FROM LUNDGREN et al. [51])

α^*	$\dfrac{u_{max}}{u_m}$	$K_d(\infty)$	$K_e(\infty)$
1.000	2.0962	1.3785	2.1541
0.750	2.0774	1.3727	2.1329
0.500	1.9918	1.3474	2.0389
0.400	1.9236	1.3283	1.9690
0.250	1.7737	1.2876	1.8256
1/6	1.6758	1.2600	1.7325
0.125	1.6283	1.2454	1.6848
0.100	1.6009	1.2365	1.6560
0.050	1.5488	1.2183	1.5990
0	1.5000	1.2000	1.5429

Eckert and Irvine [386,387] were among the first investigators who determined the value of $[(K(\infty) + 1]$ for rectangular and triangular ducts. Their graphical values of $K(\infty)$ for rectangular ducts ranged from 2.0 for $\alpha^* = 1$ to 0.66 for $\alpha^* = 0$. The values of $K(\infty)$ obtained by different investigators are compared in Table 40. The analytical values of $K(\infty)$ of Miller and Han [158] are in the closest agreement with the experimental values of Beavers et al. [388]. $K(\infty)$ of Lundgren et al. [51] are higher than the experimental values, as mentioned on p. 43. Han [154] predicts rapid flow development, and his $K(\infty)$ are also too high.

Hydrodynamic entrance lengths L_{hy}^+ for rectangular ducts of various investigators are compared in Table 41. L_{hy}^+ for all data in Table 41 is defined as the dimensionless duct length required to achieve the centerline velocity

TABLE 40

Rectangular Ducts: $K(\infty)$ for Fully Developed Laminar Flow

α^*	Miller Han [158]	Fleming Sparrow [166]	Wiginton Dalton [168]	Lundgren [51]	Han [154]	Beavers et al. [388]
1	1.433	–	1.63	1.552	2.02	1.31
0.750	–	–	–	1.520	2.00	1.21
0.500	1.281	1.46	1.44	1.383	1.80	1.18
0.250	–	–	1.13	1.076	1.36	–
0.200	0.931	0.96	1.01	–	–	0.88
0.125	–	–	–	0.879	1.10	0.72
0	0.658	0.65	–	0.686	0.85	0.60

TABLE 41

Rectangular Ducts: L_{hy}^{+} for Fully Developed Laminar Flow

α^*	Wiginton Dalton [168]	Fleming Sparrow [166]	Han [154]	McComas [48]
1	0.09	–	0.0752	0.0324
0.750	–	–	0.0735	0.0310
0.500	0.085	0.095	0.0660	0.0255
0.250	0.075	–	0.0427	0.0147
0.200	0.08	0.08	–	–
0.125	–	–	0.0227	0.00938
0	–	–	0.0099	0.00588

as 99% of the corresponding fully developed value. L_{hy}^{+} of Wiginton and Dalton [168] are the most accurate; those of Fleming and Sparrow [166] are based on the graphical results; those of Han [154] are low because of his calculated rapid flow development; and those of McComas [48] are too low, as mentioned on p. 42.

With u_m of Eq. (333), f Re for the rectangular ducts can be expressed in a closed form as

$$f\,\mathrm{Re} = -\frac{8c_1 a^2}{u_m[1 + (a/b)]^2} \tag{340}$$

The friction factors were calculated accurately from Eq. (340) by Shah and London [13]. Since the series of Eq. (333) converges rapidly, a seven-digit accuracy was established for f Re by taking up to 30 terms. The results are presented in Table 42 and Fig. 37. The f Re factors of Shih [108], determined by a point-matching method, are in excellent agreement with those of Table 42. The f Re factors of Table 42 may be approximated by the following equation:

$$f\,\mathrm{Re} = 24[1 - 1.3553\alpha^* + 1.9467\alpha^{*2} - 1.7012\alpha^{*3}$$
$$+ 0.9564\alpha^{*4} - 0.2537\alpha^{*5}] \tag{341}$$

TABLE 42

RECTANGULAR DUCTS: f Re, Nu_T, Nu_{H1}, AND
Nu_{H2} FOR FULLY DEVELOPED LAMINAR FLOW,
FOR ALL FOUR WALLS TRANSFERRING HEAT
(FROM SHAH AND LONDON [13])

α^*	fRe	Nu_T	Nu_{H1}	Nu_{H2}
1.000	14.22708	2.976	3.60795	3.091
0.900	14.26098	–	3.62045	–
1/1.2	14.32808	–	3.64531	–
0.800	14.37780	–	3.66382	–
0.750	14.47570	–	3.70052	–
1/1.4	14.56482	3.077	3.73419	–
0.700	14.60538	–	3.74961	–
2/3	14.71184	3.117	3.79033	–
0.600	14.97996	–	3.89456	–
0.500	15.54806	3.391	4.12330	3.02
0.400	16.36810	–	4.47185	–
1/3	17.08967	3.956	4.79480	2.97
0.300	17.51209	–	4.98989	–
0.250	18.23278	4.439	5.33106	2.94
0.200	19.07050	–	5.73769	2.93
1/6	19.70220	5.137	6.04946	2.93
1/7	20.19310	–	6.29404	2.94
0.125	20.58464	5.597	6.49033	2.94
1/9	20.90385	–	6.65106	2.94
1/10	21.16888	–	6.78495	2.95
1/12	21.58327	–	6.99507	–
1/15	22.01891	–	7.21683	–
1/20	22.47701	–	7.45083	–
1/50	23.36253	–	7.90589	–
0	24.00000	7.541	8.23529	8.235

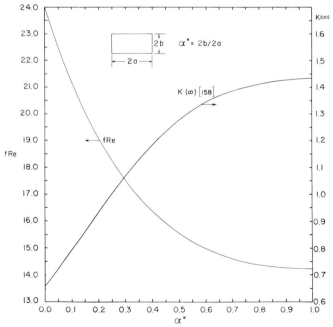

FIG. 37. Rectangular ducts: f Re and $K(\infty)$ for fully developed laminar flow (from Tables 40–42).

This equation predicts f Re higher by a maximum of only 0.05% compared to those of Table 42.

Rothfus *et al.* [389] presented the fully developed friction factor as a function of the aspect ratio and Reynolds number for laminar, transition, and turbulent flow regimes. For laminar flow, f Re factors presented are the same as those of Table 42. The associated constants of the functional relationships for the transition and turbulent flow regimes were derived from the experimental data and are presented graphically.

For fully developed flow, the friction factor is not only a function of Reynolds number (such that f Re $= K_f$, a constant), but it also depends upon the shape of the duct cross section (refer to Table 138). Tirunarayanan and Ramachandran introduced a concept of a shape factor, as described below, to correlate the f Re factors for rectangular ducts [390] and isosceles triangular ducts [391].

The flow field in a noncircular duct having corners is divided into as many numbers of flow regions as there are corners. It is hypothesized that the flow in each such region is influenced by its own corner. Each region is then treated as a separate flow passage with its characteristic dimension as the length of the path of least shear resistance in the flow field connecting the corner and the point of maximum velocity. The average characteristic dimension for the noncircular duct, $\bar{B}$, is taken as the arithmetic average of the characteristic dimensions of all flow regions. The *shape factor* characterizing the duct geometry is defined as the ratio of the foregoing average dimension to the duct perimeter.

A rectangular duct, such as shown in Fig. 36, has four identical flow regions, one of which is $OABC$. The characteristic dimension for the region $OABC$ is the line ODB. From the geometry,

$$ODB = a + b(\sqrt{2} - 1) = \bar{B} \tag{342}$$

$$\frac{\bar{B}}{P} = \frac{a + b(\sqrt{2} - 1)}{4(a + b)} = \frac{1 + \alpha^*(\sqrt{2} - 1)}{4(1 + \alpha^*)} \tag{343}$$

Tirunarayanan and Ramachandran [390] derived the following correlation for rectangular ducts:

$$f \, \text{Re} = 14.227 + 1402.5\left(\frac{\bar{B}}{P} - \frac{\sqrt{2}}{8}\right)^{1.90} \tag{344}$$

The f Re factors calculated from this equation are slightly lower than those of Table 42, with a maximum difference of only 1.7% for the rectangular duct of $\alpha^* = 0.2$.

The fully developed pressure gradient in a duct depends upon the surface area of contact. The ratio of fully developed pressure gradient in a rectangular duct to that in a circular duct of the same flow area would then depend upon the perimeter ratio. For this purpose Natarajan and Lakshmanan [392] proposed the equation

$$\frac{(dp/dL)_r}{(dp/dL)_c} = 0.861\left(\frac{P_r}{P_c}\right)^{2.75} \tag{345}$$

where suffixes r and c stand for rectangular and circular ducts, respectively. The experimental f Re values of several investigators, as referenced by Rothfus *et al.* [389], agree with Eq. (345) within 4%. It is evident from this equation that the pressure drop in a rectangular duct of any aspect ratio is higher than that in a circular duct of the same cross-sectional area; the pressure drop ratio is 1.20 and 4.30 for $\alpha^* = 1$ and 0.125, respectively. Equation (345) could be expressed in terms of f Re and α^*, using $(f\text{ Re})_c = 16$, as

$$(f\text{ Re})_r = 8.968\left(\frac{1 + \alpha^*}{\sqrt{\alpha^*}}\right)^{0.75} \tag{346}$$

This equation agrees with the f Re factors of Table 42 within 6% for $\alpha^* \geq 0.125$.

2. SPECIFIED WALL TEMPERATURE AND Ⓣ

In 1953, Clark and Kays [94], the first investigators to analyze rectangular ducts in detail, numerically evaluated fully developed Nu_T for $\alpha^* = 0$, 0.5, and 1. Miles and Shih [393] refined this treatment by employing a finer grid for the finite difference solution, a $40 \times 40\alpha^*$ instead of $10 \times 10\alpha^*$, where α^* is the aspect ratio. They reported Nu_T for $\alpha^* = 0.125$, 1/6, 0.25, 1/3, 0.5, 1/1.4, and 1. Along with the thermal entrance solution, Lyczkowski *et al.* [38] determined Nu_T for $\alpha^* = 0.25$, 0.5, 2/3, and 1 by a finite difference method. Nu_T of Lyczkowski *et al.* differed from Nu_T of Miles and Shih by 0.36, 0.12, and 0.03%, respectively for $\alpha^* = 0.25$, 0.5, and 1. Nu_T of Miles and Shih are presented in Table 42 and Fig. 38. They may be approximated, within $\pm 0.1\%$, by the formula

$$Nu_T = 7.541(1 - 2.610\alpha^* + 4.970\alpha^{*2} - 5.119\alpha^{*3} + 2.702\alpha^{*4} - 0.548\alpha^{*5}) \tag{347}$$

Schmidt and Newell [95] considered one or more walls being heated at a constant surface temperature, the other walls being adiabatic. They solved the fully developed heat transfer problem by a finite difference method. The rectangular duct was divided into: a 20×20 grid when symmetry was present about both axes, and a 20×10 grid when the symmetry was present

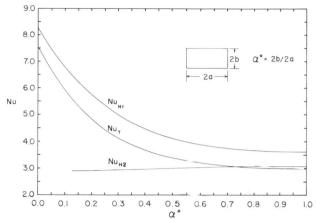

FIG. 38. Rectangular ducts: Nu_T, Nu_{H1}, and Nu_{H2} for fully developed laminar flow (from Table 42).

about one axis. A 10×10 grid was used when symmetry was not present about any axis, requiring that the complete duct be considered. According to Miles and Shih [393], the influence of grid size on the calculated Nusselt numbers is virtually eliminated if a $40 \times 40\alpha^*$ grid is employed. Consequently, the results in Table 43 for nonsymmetric heating are not as accurate as those for symmetric heating. Shah and London [13] defined the Nusselt number with the hydraulic diameter based on the wetted perimeter, instead of the heated perimeter, as done in [95]. They recomputed Nu_T, as provided by Schmidt [252], and are reported in Table 43 and Fig. 39.

Ikryannikov [394] obtained the Ⓣ temperature distribution in a rectangular duct by employing a finite integral Fourier sine transformation. He considered viscous dissipation in the energy Eq. (24), but neglected the convective term $\rho c_p u(\partial t/\partial x)$, thermal energy sources, flow work, and fluid axial heat conduction. He presented graphically the maximum temperature at the midpoint of the channel as a function of α^* of the duct. The temperature distribution in the midplane of the square duct was also presented graphically.

3. SPECIFIED AXIAL WALL HEAT FLUX AND Ⓗ1

Glaser [237] employed a finite difference method and obtained the fully developed Nu_{H1} for a square duct. Clark and Kays [94] analyzed this problem for rectangular ducts in detail, by numerically evaluating Nu_{H1} for $\alpha^* = 0$, 0.25, 1/3, 0.5, 1/1.4, and 1 using a $10 \times 10\alpha^*$ grid. Miles and Shih [393] refined the calculations by employing a finer grid of $40 \times 40\alpha^*$. They

TABLE 43

Rectangular Ducts: Nu_T for Fully Developed Laminar Flow,
for One or More Walls Transferring Heat (from Schmidt [252])

$\dfrac{2b}{2a}$	Nu_T Table 42 ▭	Nu_T ⊢2a⊣ 2b				
0.0	7.541	7.541	7.541	7.541	0	4.861
0.1	–	5.858	6.095	6.399	0.457	3.823
0.2	–	4.803	5.195	5.703	0.833	3.330
0.3	–	4.114	4.579	5.224	1.148	2.996
0.4	–	3.670	4.153	4.884	1.416	2.768
0.5	3.391	3.383	3.842	4.619	1.647	2.613
0.6	–	3.198	–	–	–	2.509
0.7	–	3.083	3.408	4.192	2.023	2.442
0.8	–	3.014	–	–	–	2.401
0.9	–	2.980	–	–	–	2.381
1.0	2.976	2.970	3.018	3.703	2.437	2.375
1.43	–	3.083	2.734	3.173	2.838	2.442
2.0	3.391	3.383	2.602	2.657	3.185	2.613
2.5	–	3.670	2.603	2.333	3.390	2.768
3.33	–	4.114	2.703	1.946	3.626	2.996
5.0	–	4.803	2.982	1.467	3.909	3.330
10.0	–	5.858	3.590	0.843	4.270	3.823
∞	7.541	7.541	4.861	0	4.861	4.861

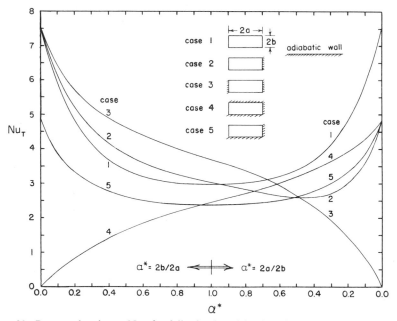

FIG. 39. Rectangular ducts: Nu_T for fully developed laminar flow, for one or more walls transferring heat (from Table 43).

determined Nu_{H1} for $\alpha^* = 0.125, 1/6, 0.25, 1/3, 0.5, 1/1.4,$ and 1. Their Nu_{H1} are up to 0.05% lower than those in Table 42.

Marco and Han [70] derived the (H1) temperature distribution for the rectangular ducts by an analogy method using an existing solution for the small deflection of a thin plate under a uniform lateral load, with the plate simply supported along all edges. Nu_{H1} from [70] can be expressed as

$$Nu_{H1} = \frac{64}{(1 + \alpha^*)^2 \pi^2} \frac{\left[\sum_{m=1,3,\ldots}^{\infty} \sum_{n=1,3,\ldots}^{\infty} \frac{1}{m^2 n^2 (m^2 + n^2 \alpha^{*2})} \right]^2}{\left[\sum_{m=1,3,\ldots}^{\infty} \sum_{n=1,3,\ldots}^{\infty} \frac{1}{m^2 n^2 (m^2 + n^2 \alpha^{*2})^3} \right]} \quad (348)$$

The present authors, by taking up to the first 250 terms in each series, established six-digit accuracy for Nu_{H1} as presented in Table 42 and Fig. 38. Nu_{H1} is approximated, within $\pm 0.03\%$, by the following:

$$Nu_{H1} = 8.235[1 - 2.0421\alpha^* + 3.0853\alpha^{*2}$$
$$- 2.4765\alpha^{*3} + 1.0578\alpha^{*4} - 0.1861\alpha^{*5}] \quad (349)$$

Savino and Siegel [395] investigated the effect of unequal heat fluxes. They found that poor convection due to low velocities in the corner and along the narrow walls always caused peak temperatures to occur at the corners. Also, the lowest peak temperatures were found when all heating took place at only the broad sides rather than when heating was partially distributed on the short sides.

Schmidt and Newell [95] determined the Nusselt number by a finite difference method, when one or more walls had the (H1) boundary condition, the other walls being adiabatic. Shah and London [13], after redefining the Nusselt number with the hydraulic diameter based on the wetted perimeter (instead of the heated perimeter as was done in [95]), recomputed the Nusselt numbers of Schmidt [252] and are reported in Table 44 and depicted in Fig. 40.

4. SPECIFIED AXIAL WALL HEAT FLUX AND (H2)

Cheng [396] analyzed the (H2) boundary condition with all four walls being heated. He arrived at closed-form expressions for the temperature distribution and Nu_{H2}, and calculated Nu_{H2} for $\alpha^* = 0.1, 0.25, 0.5,$ and 1. However, there appears to be an error in the computed Nu_{H2} by Cheng for the square duct as pointed out by Sparrow and Siegel [119] and confirmed by Cheng [107]. Sparrow and Siegel [119] developed a variational approach for the (H2) boundary condition. Shah [116] employed a discrete least squares method and determined Nu_{H2} for rectangular ducts. Nu_{H2} of Shah

TABLE 44

RECTANGULAR DUCTS: Nu_{H1} FOR FULLY DEVELOPED LAMINAR FLOW,
FOR ONE OR MORE WALLS TRANSFERRING HEAT (FROM SCHMIDT [252])

$\frac{2b}{2a}$	Nu_{H1} Table 42	Nu_{H1}				
0.0	8.235	8.235	8.235	8.235	0	5.385
0.1	6.785	6.700	6.939	7.248	0.538	4.410
0.2	5.738	5.704	6.072	6.561	0.964	3.914
0.3	4.990	4.969	5.393	5.997	1.312	3.538
0.4	4.472	4.457	4.885	5.555	1.604	3.279
0.5	4.123	4.111	4.505	5.203	1.854	3.104
0.6	3.895	3.884	-	-	-	2.987
0.7	3.750	3.740	3.991	4.662	2.263	2.911
0.8	3.664	3.655	-	-	-	2.866
0.9	3.620	3.612	-	-	-	2.843
1.0	3.608	3.599	3.556	4.094	2.712	2.836
1.43	3.750	3.740	3.195	3.508	3.149	2.911
2.0	4.123	4.111	3.146	2.947	3.539	3.104
2.5	4.472	4.457	3.169	2.598	3.777	3.279
3.33	4.990	4.969	3.306	2.182	4.060	3.538
5.0	5.738	5.704	3.636	1.664	4.411	3.914
10.0	6.785	6.700	4.252	0.975	4.851	4.410
∞	8.235	8.235	5.385	0	5.385	5.385

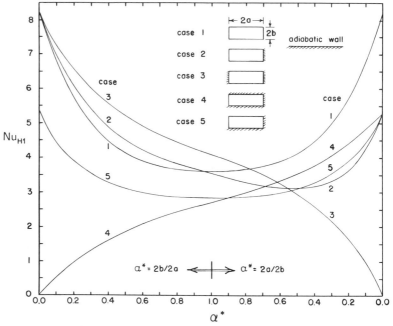

FIG. 40. Rectangular ducts: Nu_{H1} for fully developed laminar flow, for one or more walls transferring heat (from Table 44).

are in good agreement with Cheng's results except for $\alpha^* = 1$. Iqbal et al. [92] also obtained Nu_{H2} for the rectangular ducts by a variational method. The three-digit, truncated values of Nu_{H2} of both Shah and Iqbal et al. are in excellent agreement. Hence, since Iqbal et al. covered a wider range of α^*, their truncated Nu_{H2} are presented in Table 42 and Fig. 38.

It should be mentioned that Nu_{H2} for a rectangular duct with an aspect ratio approaching zero ($\alpha^* \to 0$) *cannot* approach $Nu_{H2} = 8.235$ of parallel plates. This is because the imposed constant heat flux on the short sides of the rectangular duct affects Nu_{H2} even when α^* approaches zero.

The temperature distribution around the duct periphery is not uniform for the (H2) boundary condition. Shah [116] obtained the normalized maximum and minimum wall temperatures, defined as follows, for rectangular ducts:

$$t^*_{w,max} = \frac{t_{w,max} - t_c}{t_{w,m} - t_c}, \qquad t^*_{w,min} = \frac{t_{w,min} - t_c}{t_{w,m} - t_c} \tag{350}$$

where t_c is the fluid temperature at the duct centerline. These temperatures are presented in Table 45. The maximum temperature occurs at the corners and the minimum temperature occurs on the midpoint of the long side. This information may be useful for the proper design of hot and cold regions of the duct.

TABLE 45

Rectangular Ducts: $t^*_{w,max}$
and $t^*_{w,min}$ for the (H2)
Fully Developed Flow
(from Shah [116])

α^*	$t^*_{w,max}$	$t^*_{w,min}$
1.000	1.39	0.769
0.750	1.41	0.649
0.500	1.50	0.499
0.250	1.76	0.311
0.125	2.11	0.192

Ikryannikov [397] studied laminar heat transfer through rectangular ducts having a constant axial wall heat flux, and arbitrary but the same peripheral heat flux distribution on each pair of opposite walls. Particularly, he investigated the peripheral temperature distribution for the following two problems: (1) On one pair of opposite walls, a sinusoidal wall heat flux distribution was specified in the peripheral direction, the other pair of opposite walls being adiabatic. He considered $\alpha^* = 2a/2b = 7, 5, 2, 1$, and 0.2. (2) On one pair of opposite walls, a parabolic wall heat flux distribution was specified in the peripheral direction, the other pair of opposite walls being adiabatic. Here, he considered $\alpha^* = 2a/2b = 3.3, 2, 1, 0.2$, and 0.1. In both problems, the "$2b$" walls were adiabatic.

5. SPECIFIED AXIAL WALL HEAT FLUX AND (H4)

Han [398] generalized the results of Marco and Han [70] by considering one pair of opposite walls as primary surfaces having the (H1) boundary condition, and the other pair of opposite walls as extended surfaces or fins having the (H4) boundary condition. On the extended surfaces, the heat is transferred from the primary surface to the fluid via fins. Thus, on those fins the temperature is not constant in the peripheral direction, but the axial heat flux is constant. Han obtained closed-form expressions for the temperature distribution and fully developed Nusselt numbers for a rectangular duct with these combined boundary conditions. The Nusselt number is expressed functionally as

$$
\mathrm{Nu} = \mathrm{Nu}_{\mathrm{H1}} F_c\!\left(\frac{\alpha^* K_p}{1 + \alpha^*}, \frac{1}{\alpha^*}\right) \tag{351}
$$

where $\mathrm{Nu}_{\mathrm{H1}}$ is given by Eq. (348), and F_c is the correction factor dependent upon the fin parameter $\alpha^* K_p/(1 + \alpha^*)$ and the aspect ratio $\alpha^* = 2a/2b$. The width of the primary surface of the rectangular channel is $2a$ and the height of the secondary surface (fins) of rectangular channel is $2b$; K_p is the peripheral heat conduction parameter. Han tabulated the correction factor for $\alpha^* = 1$, 0.667, and 0.5 with the fin parameter as 0, 1, 2, 4, and 10. Han found that the correction factor can be significantly lower than unity for small values of the fin parameter, as might occur in a liquid metal heat exchanger. The model analyzed by Han would be approximated in a counterflow heat exchanger with equal thermal capacity rates for the two fluids.

Siegel and Savino [399] considered the effect of peripheral wall heat conduction in the broad walls with nonconducting insulated narrow walls. They also extended their analysis to include different boundary conditions at the corners so that the effect of heat conduction in the insulated side walls could be investigated [400]. They found that the peripheral wall conduction substantially lowered the peak wall temperatures at the corners and temperature gradients along the broad walls.

Lyczkowski et al. [38] analyzed the (H4) boundary condition for the square duct and determined $\mathrm{Nu}_{\mathrm{H4}}$ by a finite difference method for $K_p = 0$, 0.5, 1.0, and 2.0 as presented in Table 46. It appears that $\mathrm{Nu}_{\mathrm{H4}} = 3.23$ for $K_p = 0$ by [38] has not converged to the expected[†] limiting value of $\mathrm{Nu}_{\mathrm{H2}} = 3.091$ [107]. However, the convergence is improved for higher values of K_p.

Iqbal et al. [39] analyzed the (H4) problem of combined free and forced convection through vertical noncircular ducts. They employed a variational

---

[†] See p. 28 for the limiting cases of $\mathrm{Nu}_{\mathrm{H4}}$.

TABLE 46

RECTANGULAR DUCTS: Nu_{H4} FOR FULLY
DEVELOPED LAMINAR FLOW (FROM
LYCZKOWSKI et al. [38] AND THE GRAPHICAL
RESULTS OF IQBAL et al. [39])

K_p	[38]	Iqbal et al. [39]		
		Nu_{H4}		
	$\alpha^*=1$	$\alpha^*=1$	$\alpha^*=\frac{1}{2}$	$\alpha^*=\frac{1}{3}$
0	3.23	3.08	3.04	2.95
0.5	3.41	–	–	–
1	3.47	3.50	3.87	4.24
2	3.52	–	–	–
100	–	3.606	–	–
∞	–	3.608	4.12	4.79

method for the analysis and obtained specific results for rectangular ducts of $\alpha^* = 1$, $1/2$, and $1/3$. Their Nu_{H4} for pure forced convection are presented in Table 46.

B. Hydrodynamically Developing Flow

The hydrodynamically developing flow for rectangular ducts has been studied experimentally by Sparrow et al. [50], Goldstein and Kreid [401], Beavers et al. [388], and Muchnik et al. [384]. Sparrow et al. [50] measured axial velocity profiles and pressure drops for rectangular ducts of $\alpha^* = 1$, 0.5, and 0.2. Their axial velocity profiles for $\alpha^* = 0.2$ are in good agreement with the analytical values of Curr et al. [177]. Their Δp^* in the entrance region for $\alpha^* = 0.2$ are in good agreement with those of Beavers et al. [388] and Curr et al. [177]. Their Δp^* for $\alpha^* = 0.5$ have uncertain validity owing to the possible presence of secondary flow [388]. Goldstein and Kreid [401] measured the velocity profiles and the centerline velocity development for a square duct using a laser-Doppler flowmeter. Their velocity profiles and the u_c/u_m development are in excellent agreement with the corresponding analytical results of Curr et al. [177]. Beavers et al. [388] made pressure drop measurements in the entrance region of rectangular ducts of $0.0196 \leq \alpha^* \leq 1$. These experiments are most comprehensive and accurate. The f_{app} Re factors of Beavers et al. for $\alpha^* = 1$, 0.5 and 0.2 are in excellent agreement with those of Table 47. Muchnik et al. [384] measured the u_c/u_m development for $\alpha^* = 0.5$, 0.25, and 0.1 and found to be in good agreement with that of Sparrow et al. [50].

The hydrodynamic entry length problem for rectangular ducts has been studied analytically by two methods: (1) the linearization of the momentum equation, and (2) finite difference methods. In all of the solutions described next, the velocity profile is considered uniform at the duct entrance, $x = 0$.

1. SOLUTIONS BY LINEARIZED MOMENTUM EQUATION

Han [154] solved the x momentum equation by Langhaar's linearization technique, and determined the pressure drop in the entrance region from the same x momentum equation evaluated at the duct centerline and integrated from $x = 0$ to x. Han reported u_c/u_m and Δp^* in tabular and graphical forms for rectangular ducts of $\alpha^* = 0, 0.125, 0.25, 0.5, 0.75$, and 1. Han's analysis predicts somewhat more rapid flow development and hence increased pressure drop than was observed experimentally by Sparrow et al. [50], Goldstein and Kreid [401], and Muchnik et al. [384].

Fleming and Sparrow [166] linearized the x momentum equation by introducing a stretched coordinate in the flow direction [refer to Eq. (167)]. At any cross section, the pressure gradient was then determined by two ways: integrating the momentum equation, and integrating the mechanical energy equation across the flow cross section. The mean velocity weighing factor $\varepsilon(x)$ in Eq. (167) was then determined by equating the pressure gradients of the foregoing two methods. Fleming and Sparrow reported hydrodynamic entrance solutions for rectangular ducts with $\alpha^* = 0.2$ and 0.5. The Δp^* of Fleming and Sparrow for $\alpha^* = 0.2$ are in excellent agreement with the experimental results of Sparrow et al. [50]. The centerline velocity development of Fleming and Sparrow for the $\alpha^* = 0.5$ duct is about 2% slower than the experimental measurements of Muchnik et al. [384].

In the analysis, Fleming and Sparrow [166] expressed the dimensionless axial velocity as the sum of the fully developed velocity and an entrance length correction or velocity difference. This velocity difference was obtained by an eigenvalue method. Wiginton and Dalton [168][†] employed a slightly different exponential function in the eigenfunction expansion of this velocity difference. A comparison of the two methods is provided on p. 72. They reported hydrodynamic entrance solutions for $\alpha^* = 0.2, 0.5$, and 1. For $\alpha^* = 0.2$, the pressure drop predicted by Wiginton and Dalton is about 2–3% higher than that of Fleming and Sparrow, and Beavers et al. For $\alpha^* = 0.5$, the Δp^* of these two works are in excellent agreement. For $\alpha^* = 1$, Wiginton and Dalton's Δp^* are about 4% higher than the experimental values of Beavers et al. [388].

Miller and Han [158] refined the solution of Han [154]. They employed Langhaar's linearization technique, but determined Δp^* by first integrating the mechanical energy equation across a flow cross section and then integrating the resulting equation from $x = 0$ to x. They presented graphically $K(x)$ for $\alpha^* = 0, 0.2, 0.5$, and 1. Their f_{app} Re calculated from $K(x)$ are up to 2% lower than those of Table 47 for $x^+ \geq 0.01$.

---

† The ordinates of Figs. 2, 4, and 6 of [168] are in error; they should start from 0 instead of 1.

2. FINITE DIFFERENCE SOLUTIONS

In the foregoing investigations, because of the linearization of the momentum equation, the transverse velocity components v and w were not involved. Patankar and Spalding [176] proposed a general numerical marching procedure for the calculation of the transport processes in a three-dimensional flow. To determine u, v, w, and p for a duct flow, the continuity and the Navier–Stokes equations (three momentum equations) are put into a finite difference form. The solutions are obtained by a marching procedure that incorporates a guess-and-correct feature for the pressure field. Patankar and Spalding obtained the hydrodynamic entry length solution for a square duct. Their f_{app} Re factors are in excellent agreement with those of Table 47 for $\alpha^* = 1$.

Caretto et al. [402] proposed two finite difference marching procedures to analyze three-dimensional internal flows. One of these methods employs the same set of equations as those of Patankar and Spalding [176], but solves them in a different method. The other method considers the x direction vorticity instead of the pressure as a main variable. Curr et al. [177] employed these methods to analyze the hydrodynamic entry length problem for rectangular ducts. They presented graphically Δp^*, u_c/u_m, and u/u_m for rectangular ducts of $\alpha^* = 1, 0.5$, and 0.2. Their Δp^* are in excellent agreement with the experimental results of Beavers et al. [388]. Their u_c/u_m (where u_c is the centerline velocity) development for the $\alpha^* = 1$ duct is in excellent agreement with the experimental measurements of Goldstein and Kreid [401]. Their velocity profiles u/u_m as a function of x^+ for $\alpha^* = 0.2$ are in excellent agreement with those of Sparrow et al. [50]. The f_{app} Re factors, based on the graphical Δp^* of Curr et al., are presented in Table 47 and Fig. 41. These f_{app} Re factors are in excellent agreement with those of Eq. (188) for $x^+ \lesssim 0.003$. Shah [270] presented these f_{app} Re factors by Eq. (576).

Carlson and Hornbeck [178] analyzed the three-dimensional entrance flow in a square duct by a finite difference method. They employed two mathematical models for the analysis. In the first model, they employed the x momentum equation, the continuity equation, and the following relationship between the transverse velocities:

$$(a - y)w = (a - z)v \tag{352}$$

which was arrived at by considering the velocity at any point as inversely proportional to its distance from the center. The integral form of the continuity equation was used to determine the pressure distribution. In the second model, Navier–Stokes equations (three momentum equations) and a continuity equation were solved at Re = 2000. They presented graphically the detailed results, including the transverse velocity components, for the

TABLE 47

Rectangular Ducts: f_{app} Re for
Developing Laminar Flow (from the
Graphical Results of Curr et al. [177])

x^+	f_{app} Re		
	$\alpha^* = 1$	$\alpha^* = .5$	$\alpha^* = .2$
0.001	111.0	111.0	111.0
0.002	80.2	80.2	80.2
0.003	66.0	66.0	66.1
0.004	57.6	57.6	57.9
0.005	51.8	51.8	52.5
0.006	47.6	47.6	48.4
0.007	44.6	44.6	45.3
0.008	41.8	41.8	42.7
0.009	39.9	40.0	40.6
0.010	38.0	38.2	38.9
0.015	32.1	32.5	33.3
0.020	28.6	29.1	30.2
0.030	24.6	25.3	26.7
0.040	22.4	23.2	24.9
0.050	21.0	21.8	23.7
0.060	20.0	20.8	22.9
0.070	19.3	20.1	22.4
0.080	18.7	19.6	22.0
0.090	18.2	19.1	21.7
0.100	17.8	18.8	21.4

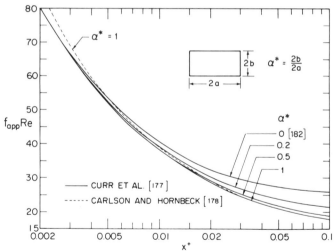

FIG. 41. Rectangular ducts: f_{app} Re for developing laminar flow.

square duct. Their f_{app} Re of the first model obtained from Carlson [179] are compared in Fig. 41, and are found to be 7 and 2% higher at $x^+ = 0.002$ and 0.01, respectively. Their f_{app} Re of the second model are about 4% lower than f_{app} Re of the first model for $x^+ \gtrsim 0.01$.

Rubin et al. [403] divided the *entry region* into three subregions: (1) Navier–Stokes region at the leading edge where the boundary layers are very thin; (2) entry flow region having thin boundary layers and a large potential core; and (3) fully viscous flow region in which the boundary layers are so thick that they fill a significant portion of the duct cross section. Rubin *et al.* analyzed subregion (2) by the matched boundary layer/potential core procedure, and analyzed subregion (3) by a numerical procedure for a square duct. They presented graphically the centerline velocity development, axial and secondary velocity profiles at different cross sections, and axial shear stress. They found reasonably good agreement between the above theoretical and numerical solutions. They showed that the axial flow behavior is virtually insensitive to the treatment of the secondary flow, although the inverse is not true.

Miller [185] proposed a method of transforming an initial value problem to a boundary value problem, and analyzed the hydrodynamic entrance problem for the square duct. For his analysis, he used the fully developed velocity profile of Eq. (332) with $2b = 2a$, $f \, \mathrm{Re} = 14.06$, and $L_{hy}^+ = 0.0328$ from McComas [48]. The subsequent boundary layer problem was solved by a finite difference method for $\mathrm{Re} = 1000$, and $f_{app} \, \mathrm{Re}$ factors were presented graphically. Because Miller also used the lower value of L_{hy}^+ by McComas (refer to p. 42 for the reason) and $f_{app} \, \mathrm{Re} = 14.06$ at $L_{hy}^+ = 0.0328$ (instead of 24 from Fig. 41), $f_{app} \, \mathrm{Re}$ determined in the entrance region are substantially lower than those by other investigators [158,166,168]. For example, the maximum value of $K(x)$ determined from $f_{app} \, \mathrm{Re}$ of Miller was 0.16 at $x^+ \simeq 0.012$; for higher or lower values of x^+, the $K(x)$ decreased and approached zero. This is contrary to the trend based on the results of Table 47, which indicate $K(x)$ monotonically increasing from 0 at $x^+ = 0$ to 1.43 at $x^+ = \infty$ (Table 40).

3. SOLUTION BY A FINITE ELEMENT METHOD

Godbole [404] employed a finite element method to analyze creeping flow ($\mathrm{Re} = 0$) through rectangular ducts. The solution was obtained by making use of a standard computer program for elastic body mechanics using the analogy between the creeping flow and incompressible flow problems. Godbole presented u/u_m along the centerline and center of the duct quadrant of rectangular ducts of $\alpha^* = 1, 0.75, 0.5$, and 0.25.

C. Thermally Developing and Hydrodynamically Developed Flow

1. SPECIFIED WALL TEMPERATURE AND Ⓣ

Dennis *et al.* [405] considered the case of constant temperature on all four walls of the rectangular duct and solved the thermal entry length problem by an eigenvalue method for a fully developed laminar velocity profile. They presented the first three eigenvalues and constants for $\alpha^* =$ 0.125, 0.25, 0.5, 2/3, and 1. But these values, except for the first term for rectangular ducts other than for the square duct, are in error as reported by Kays [7] and Lyczkowski [406].

Montgomery and Wibulswas [210] used an explicit finite difference method and solved the thermal entry length problem for rectangular ducts with $\alpha^* = 1/6$, 0.2, 0.25, 1/3, 0.5, and 1. They neglected the effect of axial heat conduction, viscous dissipation, and thermal energy sources within the fluid. Their $Nu_{x,T}$ and $Nu_{m,T}$ are presented in Table 48 [220]. The effect of α^* on $Nu_{x,T}$ for the developed velocity profile case is shown in Fig. 42.

TABLE 48

RECTANGULAR DUCTS: $Nu_{x,T}$ AND $Nu_{m,T}$ AS FUNCTIONS OF x^* AND α^* FOR A FULLY DEVELOPED VELOCITY PROFILE (FROM WILBULSWAS [220])

$\dfrac{1}{x^*}$	$Nu_{x,T}$						$Nu_{m,T}$					
	Aspect Ratio, α^*						Aspect Ratio, α^*					
	1.0	0.5	1/3	0.25	0.2	1/6	1.0	0.5	1/3	0.25	0.2	1/6
0	2.65	3.39	3.96	4.51	4.92	5.22	2.65	3.39	3.96	4.51	4.92	5.22
10	2.86	3.43	4.02	4.53	4.94	5.24	3.50	3.95	4.54	5.00	5.36	5.66
20	3.08	3.54	4.17	4.65	5.04	5.34	4.03	4.46	5.00	5.44	5.77	6.04
30	3.24	3.70	4.29	4.76	5.31	5.41	4.47	4.86	5.39	5.81	6.13	6.37
40	3.43	3.85	4.42	4.87	5.22	5.48	4.85	5.24	5.74	6.16	6.45	6.70
60	3.78	4.16	4.67	5.08	5.40	5.64	5.50	5.85	6.35	6.73	7.03	7.26
80	4.10	4.46	4.94	5.32	5.62	5.86	6.03	6.37	6.89	7.24	7.53	7.77
100	4.35	4.72	5.17	5.55	5.83	6.07	6.46	6.84	7.33	7.71	7.99	8.17
120	4.62	4.93	5.42	5.77	6.06	6.27	6.86	7.24	7.74	8.13	8.39	8.63
140	4.85	5.15	5.62	5.98	6.26	6.47	7.22	7.62	8.11	8.50	8.77	9.00
160	5.03	5.34	5.80	6.18	6.45	6.66	7.56	7.97	8.45	8.86	9.14	9.35
180	5.24	5.54	5.99	6.37	6.63	6.86	7.87	8.29	8.77	9.17	9.46	9.67
200	5.41	5.72	6.18	6.57	6.80	7.02	8.15	8.58	9.07	9.47	9.79	10.01

TABLE 49

RECTANGULAR DUCTS: $Nu_{x,T}$ AS A FUNCTION OF x^* AND α^* FOR A FULLY DEVELOPED VELOCITY PROFILE (FROM LYCZKOWSKI et al. [38])

$\alpha^* = 1$		$\alpha^* = 0.5$		$\alpha^* = 0.25$	
x^*	$Nu_{x,T}$	x^*	$Nu_{x,T}$	x^*	$Nu_{x,T}$
.00750	4.458	.004219	5.869	.002930	7.405
.01125	4.029	.006328	5.236	.004395	6.662
.01500	3.782	.008438	4.852	.005859	6.204
.01875	3.594	.01055	4.587	.007324	5.888
.02250	3.457	.01266	4.391	.008789	5.658
.02625	3.353	.01477	4.241	.01025	5.485
.03000	3.273	.01688	4.123	.01172	5.351
.03375	3.211	.01898	4.028	.01318	5.245
.03750	3.162	.02109	3.956	.01465	5.162
.05625	3.034	.04219	3.604	.02930	4.813
.07500	2.993	.06328	3.501	.05859	4.649
.09375	2.981	.08438	3.451	.1025	4.554
.1125	2.977	.1266	3.409	.1465	4.508
.1500	2.975	.1477	3.401	.2197	4.470
.1875	2.975	.1688	3.396	.2564	4.460
.2625	2.975	.1793	3.395	.2813	4.455

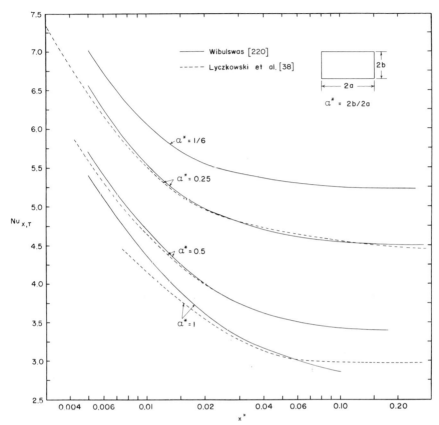

FIG. 42. Rectangular ducts; $Nu_{x,T}$ for a fully developed velocity profile; the influence of α^*. Similar behavior is obtained for $Nu_{m,T}$, $Nu_{x,H1}$, and $Nu_{m,H1}$ of Tables 48 and 51.

Lyczkowski et al. [38] considered a variety of thermal boundary conditions, such as zero thermal resistance on all four walls, and one wall, two walls, and three walls with various finite wall thermal resistances together with the Ⓣ condition on the remaining walls. They used the modified DuFort and Frankal explicit finite difference method and numerically evaluated thermal entrance Nusselt numbers for $\alpha^* = 0$, 0.125, 0.25, 0.5, and 1. Presented in Table 49 are $Nu_{x,T}$ vs. x^* for $\alpha^* = 0.25$, 0.5, and 1, when all four walls are uniformly at Ⓣ. The results of Tables 48 and 49 for $\alpha^* = 0.25$ and 0.5 are generally in agreement within about 2%, except for $\alpha^* = 1$, as can be seen from Fig. 42. The disagreement is due to the two different finite difference numerical methods employed. The more accurate results of Lyczkowski et al. [38] are recommended for design purposes, particularly for $\alpha^* = 0.5$, and $\alpha^* = 1$ for $x^* \geq 0.04$. $Nu_{x,T}$ of Chandrupatla and Sastri as described below are recommended for the square duct.

Chandrupatla and Sastri [407] and Chandrupatla [407a] analyzed the Ⓣ, Ⓗ1, and Ⓗ2 thermal entrance problem for laminar flow of Newtonian and non-Newtonian fluids through a square duct. The finite difference solutions were obtained by using the iterative extrapolated Liebmann method. Their results for a Newtonian fluid are presented in Table 49a. Their $Nu_{x,T}$ are in excellent agreement with those of Wibulswas [220] for $x^* \leq 0.05$ and with those of Lyczkowski et al. [38] for $x^* > 0.5$.

TABLE 49a

Square Duct: Nusselt Numbers for a Fully Developed
Velocity Profile and a Developing Temperature Profile
(from Chandrupatla and Sastri [407])

$\dfrac{1}{x^*}$	$Nu_{x,T}$	$Nu_{m,T}$	$Nu_{x,H1}$	$Nu_{m,H1}$	$Nu_{x,H2}$	$Nu_{m,H2}$
0	2.975	2.975	3.612	3.612	3.095	3.095
10	2.976	3.514	3.686	4.549	3.160	3.915
20	3.074	4.024	3.907	5.301	3.359	4.602
25	3.157	4.253	4.048	5.633	3.481	4.898
40	3.432	4.841	4.465	6.476	3.843	5.656
50	3.611	5.173	4.720	6.949	4.067	6.083
80	4.084	5.989	5.387	8.111	4.654	7.138
100	4.357	6.435	5.769	8.747	4.993	7.719
133.3	4.755	7.068	6.331	9.653	5.492	8.551
160	–	–	6.730	10.279	5.848	9.128
200	5.412	8.084	7.269	11.103	6.330	9.891

The thermal entrance lengths $L_{th,T}^*$, based on $Nu_{x,T}$ of Table 49 have been determined by the present authors and are reported in Table 50. An inspection of $L_{th,T}^*$ vs. α^* shows that $L_{th,T}^*$ does not monotonically decrease with decreasing values of α^*, but passes through a maximum. This trend is different from the dependence of $L_{th,H1}^*$ on α^* as shown in Table 50. Thus the dependence of $L_{th,T}^*$ on α^* requires confirmation by more refined solutions.

Krishnamurty and Sambasiva Rao [408] employed a Lévêque-type approximation and arrived at an expression for local and mean Nusselt numbers for rectangular ducts for the two cases when one and when all four walls transfer heat. They included empirically the effect of temperature-dependent viscosity.

DeWitt and Snyder [409] considered fluid axial heat conduction and viscous dissipation, and solved the Ⓣ thermal entry length problem for rectangular ducts of $\alpha^* = 0.1, 0.2, 0.5$, and 1. The approximate series solution was obtained by the Galerkin method. The fluid bulk mean temperature and $Nu_{x,T}$ were presented as a function of the dimensionless axial distance with the Péclet number as a parameter. Since the definitions used for the

TABLE 50

RECTANGULAR DUCTS: $L_{th,T}^*$ AND $L_{th,H1}^*$ FOR THE
DEVELOPED AND DEVELOPING VELOCITY PROFILES
(DETERMINED BY THE PRESENT AUTHORS)[†]

α^*	Developed velocity profile (Pr = ∞)		Simul. dev. flow Pr = 0.7
	$L_{th,T}^*$	$L_{th,H1}^*$	$L_{th,H1}^*$
0	0.0080	0.0115	0.017
1/4	0.054	0.042	0.136
1/3	—	0.048	0.17
1/2	0.049	0.057	0.23
1	0.041	0.066	0.34
circular tube	0.0335	0.0431	0.053

[†] The rectangular duct $L_{th,T}^*$ for simultaneously developing flow are not presented, because no tabulation is available for $Nu_{x,T}$. For simultaneously developing flow through a circular duct, $L_{th,T}^* = 0.037$ for Pr = 0.7.

dimensionless variables and parameters are unconventional, it is difficult to interpret their thermal entry length solutions.

Chiranjivi and Parabrahmachary [410] experimentally studied laminar and turbulent heat transfer in rectangular ducts with $\alpha^* = 1$, 1/2, and 1/3. Only the 3-in. base (bottom side) was electrically heated and maintained at constant temperature; the other three sides were insulated. Water and water–glycerol mixture were used as the test fluids. The experimental data were correlated as the Colburn factor j (including the Sieder–Tate correction for viscosity) vs. the Reynolds number Re. The isothermal duct length upstream of the test section was too short to ensure a fully developed laminar velocity profile at the entrance of the test section. Hence their experimental correlation for laminar flows applies to thermally developing and partially hydrodynamically developed flow.

2. SPECIFIED AXIAL HEAT FLUX DISTRIBUTION

Sparrow and Siegel [204] developed a variational method and calculated the first two eigenvalues and constants for the square duct with the (H1) boundary condition.

Sadikov [15] studied the thermal entry length problem, based on his simplified energy equation (168), for rectangular ducts with prescribed wall heat flux boundary condition. He presented graphically the local peripheral temperature and Nusselt number, with dimensionless axial distance as a parameter, for the (H2) boundary condition for $\alpha^* = 1$ and 0.2 with Pr = 0.7 and Re = 10^4.

Introducing the same idealizations as they employed for their (T) solution, Montgomery and Wibulswas [210] solved the thermal entry length problem

TABLE 51

Rectangular Ducts: $Nu_{x,H1}$ and $Nu_{m,H1}$ as Functions of x^* and α^*
for a Fully Developed Velocity Profile (from Wibulswas [220])

$\dfrac{1}{x^*}$	$Nu_{x,H1}$ Aspect Ratio, α^*				$Nu_{m,H1}$ Aspect Ratio, α^*			
	1.0	0.5	1/3	0.25	1.0	0.5	1/3	0.25
0	3.60	4.11	4.77	5.35	3.60	4.11	4.77	5.35
10	3.71	4.22	4.85	5.45	4.48	4.94	5.45	6.03
20	3.91	4.38	5.00	5.62	5.19	5.60	6.06	6.57
30	4.18	4.61	5.17	5.77	5.76	6.16	6.60	7.07
40	4.45	4.84	5.39	5.87	6.24	6.64	7.09	7.51
60	4.91	5.28	5.82	6.26	7.02	7.45	7.85	8.25
80	5.33	5.70	6.21	6.63	7.66	8.10	8.48	8.87
100	5.69	6.05	6.57	7.00	8.22	8.66	9.02	9.39
120	6.02	6.37	6.92	7.32	8.69	9.13	9.52	9.83
140	6.32	6.68	7.22	7.63	9.09	9.57	9.93	10.24
160	6.60	6.96	7.50	7.92	9.50	9.96	10.31	10.61
180	6.86	7.23	7.76	8.18	9.85	10.31	10.67	10.92
200	7.10	7.46	8.02	8.44	10.18	10.64	10.97	11.23

for rectangular ducts for the (H1) boundary condition. Their $Nu_{x,H1}$ and $Nu_{m,H1}$ for $\alpha^* = 1, 0.5, 1/3$, and 0.25 are reported in Table 51 [220].

Perkins *et al.* [411] experimentally determined $Nu_{x,H1}$ for a square duct and found good agreement with $Nu_{x,H1}$ reported in Table 51. They provided the following best fit equation for $Nu_{x,H1}$ for a square duct, $\alpha^* = 1$:

$$Nu_{x,H1} = \frac{1}{0.277 - 0.152 \exp(-38.6x^*)} \qquad (353)$$

which agrees with Table 51 within $\pm 3\%$ for $x^* > 0.005$ (the last point of Table 51). Note that Eq. (353) has been slightly modified by the present authors so that $Nu_{H1} = 3.61$ for $x^* = \infty$, in contrast to 3.63 for the original equation.

Chandrupatla and Sastri [407] employed a finite difference method and analyzed the (H1) thermal entrance problem for a square duct. Their $Nu_{x,H1}$ and $Nu_{m,H1}$ are presented in Table 49a. Their $Nu_{x,H1}$ for $x^* < 0.04$ are higher than those of Wibulswas [220] with a maximum difference of 2% at $x^* = 0.005$.

The thermal entrance lengths $L^*_{th,H1}$, based on $Nu_{x,H1}$ of Table 51, have been determined by the present authors as 0.066, 0.057, 0.048, and 0.050, respectively, for $\alpha^* = 1, 0.5, 1/3$, and 0.25. $L^*_{th,H1}$ for $\alpha^* = 1$ is 0.067 based on $Nu_{x,H1}$ of Table 49a. A plot of $L^*_{th,H1}$ vs. α^* reveals that the foregoing $L^*_{th,H1} = 0.050$ for $\alpha^* = 0.25$ is in error; it should be 0.042. Hence it appears that $Nu_{x,H1}$ for $\alpha^* = 0.25$ in Table 51 may be about 1.5% too high for $x^* \gtrsim 0.03$. These $L^*_{th,H1}$ values are presented in Table 50 for the developed velocity profile.

Based on the same method employed for the specified wall temperature problem, Lyczkowski *et al.* [38] analyzed the thermal entry length problem for a variety of thermal boundary conditions, such as zero thermal resistance on all four walls, and one wall, two walls, and three walls with various finite wall thermal resistances together with the (H1) condition on the remaining walls. The results were presented graphically.

Chandrupatla and Sastri [407] analyzed the (H2) thermal entrance problem for the square duct. Their results are presented in Table 49a.

Hicken [412] determined the temperature distribution in the thermal entrance region of rectangular ducts with different but axially constant heat flux specified on each wall. He presented graphically the dimensional variation of the wall to fluid bulk mean temperature difference $(4D_h/P)\Theta_{w-m}$ over the periphery of a square duct when one, two, three, or four walls were equally heated, the others being adiabatic. He also presented $(4D_h/P)\Theta_{w-m}$ for $\alpha^* = 0.1$ for all four walls equally heated. The eigenvalues and constants for $\alpha^* = 0.1, 0.2, 0.5, 1, 2, 5$, and 10 are reported in [413].

3. LINEARLY VARYING WALL TEMPERATURE

Javeri [35] employed the Kantorowich variational method to analyze the thermal entrance problem for rectangular ducts. He considered the wall temperature uniform peripherally and varying linearly in the flow direction. He tabulated Nu_x as a function of $x^*(D_h/b)^2$ for $\alpha^* = 0, 0.125, 0.25, 0.5$, and 1.

D. Simultaneously Developing Flow

Montgomery and Wibulswas [219] obtained numerically the combined hydrodynamic and thermal entry length solutions for rectangular ducts of $\alpha^* = 1, 0.5, 1/3, 0.25$, and 1/6. In their analysis, they specified the transverse velocity components v and w to be zero and also neglected the effects of axial momentum and thermal diffusions. Their $Nu_{m,T}$ are presented in Table 52, and $Nu_{x,H1}$ and $Nu_{m,H1}$ are presented in Table 53 for $Pr = 0.72$ [220].

The thermal entrance lengths $L_{th,H1}^*$, based on $Nu_{x,H1}$ of Table 53 for $Pr = 0.72$, have been determined by the present authors and are reported in Table 50. Thus $L_{th,H1}^*$ for simultaneously developing flow decreases monotonically with decreasing α^*. An interpolation of the foregoing results provides $L_{th,H1}^* = 0.079$ for $\alpha^* = 0.125$. It is interesting to note that the thermal entrance lengths for rectangular ducts with $\alpha^* > 0$ are significantly higher than those for parallel plates. $L_{th,H1}^*$ for a square duct is one order of magnitude larger than that for a circular tube. These tentative conclusions should be confirmed with more refined solutions for the combined entry length problem.

TABLE 52

Rectangular Ducts: $Nu_{m,T}$ as a Function
of x^* and α^* for Simultaneously
Developing Flow, Pr = 0.72
(from Wibulswas [220])

$\frac{1}{x^*}$	$Nu_{m,T}$				
	Aspect Ratio, α^*				
	1.0	0.5	1/3	0.25	1/6
10	3.75	4.20	4.67	5.11	5.72
20	4.39	4.79	5.17	5.56	6.13
30	4.88	5.23	5.60	5.93	6.47
40	5.27	5.61	5.96	6.27	6.78
50	5.63	5.95	6.28	6.61	7.07
60	5.95	6.27	6.60	6.90	7.35
80	6.57	6.88	7.17	7.47	7.90
100	7.10	7.42	7.70	7.98	8.38
120	7.61	7.91	8.18	8.48	8.85
140	8.06	8.37	8.66	8.93	9.28
160	8.50	8.80	9.10	9.36	9.72
180	8.91	9.20	9.50	9.77	10.12
200	9.30	9.60	9.91	10.18	10.51
220	9.70	10.00	10.30	10.58	10.90

TABLE 53

Rectangular Ducts: $Nu_{x,H1}$ and $Nu_{m,H1}$ as Functions of x^* and α^* for
Simultaneously Developing Flow, Pr = 0.72 (from Wilbulswas [220])

$\frac{1}{x^*}$	$Nu_{x,H1}$				$Nu_{m,H1}$			
	Aspect Ratio, α^*				Aspect Ratio, α^*			
	1.0	0.5	1/3	0.25	1.0	0.5	1/3	0.25
5	—	—	—	—	4.60	5.00	5.57	6.06
10	4.18	4.60	5.18	5.66	5.43	5.77	6.27	6.65
20	4.66	5.01	5.50	5.92	6.60	6.94	7.31	7.58
30	5.07	5.40	5.82	6.17	7.52	7.83	8.13	8.37
40	5.47	5.75	6.13	6.43	8.25	8.54	8.85	9.07
50	5.83	6.09	6.44	6.70	8.90	9.17	9.48	9.70
60	6.14	6.42	6.74	7.00	9.49	9.77	10.07	10.32
80	6.80	7.02	7.32	7.55	10.53	10.83	11.13	11.35
100	7.38	7.59	7.86	8.08	11.43	11.70	12.00	12.23
120	7.90	8.11	8.37	8.58	12.19	12.48	12.78	13.03
140	8.38	8.61	8.84	9.05	12.87	13.15	13.47	13.73
160	8.84	9.05	9.38	9.59	13.50	13.79	14.10	14.48
180	9.28	9.47	9.70	9.87	14.05	14.35	14.70	14.95
200	9.69	9.88	10.06	10.24	14.55	14.88	15.21	15.49
220	—	—	—	—	15.03	15.36	15.83	16.02

Montgomery and Wibulswas [219] also investigated the effect of the
Prandtl number on heat transfer for the duct with $\alpha^* = 0.5$. Their thermal
entry $Nu_{m,H1}$ values as a function of the Prandtl number are presented in
Table 54 and Fig. 43 [220]. It should be emphasized that the parameters
Pr = ∞ and 0 in Table 54 and Fig. 43 have the following meaning: Pr = ∞
identifies the case of hydrodynamically fully developed laminar flow at
$x = 0$, and Pr = 0 identifies the case of slug flow. $Nu_{m,H1}$ for both cases
are valid for any fluid Prandtl number. Refer to p. 16 for further discussion.

TABLE 54

RECTANGULAR DUCT ($\alpha^* = 0.5$): $Nu_{m,H1}$ AS
A FUNCTION OF x^* AND Pr FOR SIMULTANEOUSLY
DEVELOPING FLOW (FROM WILBULSWAS [220])

$\dfrac{1}{x^*}$	$Nu_{m,H1}$				
	$Pr=\infty$	10	0.72	0.1	0
20	5.60	6.15	6.94	7.90	8.65
40	6.64	7.50	8.54	9.75	10.40
60	7.45	8.40	9.77	11.10	11.65
80	8.10	9.20	10.83	12.15	12.65
100	8.66	9.90	11.70	13.05	13.50
140	9.57	11.05	13.15	14.50	14.95
180	10.31	11.95	14.35	15.65	16.15
220	10.95	12.75	15.35	16.70	17.20
260	11.50	13.45	16.25	17.60	18.10
300	12.00	14.05	17.00	18.30	18.90
350	12.55	14.75	17.75	19.10	19.80
400	13.00	15.40	18.50	19.90	20.65

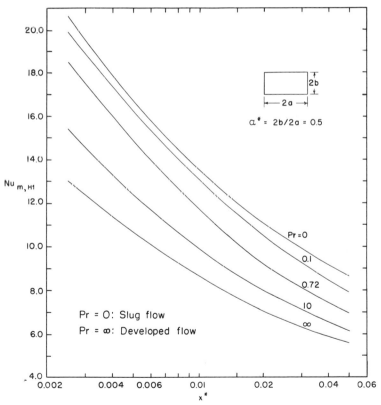

FIG. 43. Rectangular duct ($\alpha^* = 0.5$): $Nu_{m,H1}$ as functions of x^* and Pr for simultaneously developing flow (from Table 54).

Chandrupatla [407a] analyzed numerically the Ⓣ, Ⓗ①, and Ⓗ② combined entry length problems for Newtonian and non-Newtonian fluid flows through a square duct. He investigated the effect of Prandtl number on heat transfer, and tabulated local and mean Nusselt numbers for the three boundary conditions as a function of $1/x^*$ for $\text{Pr} = 0$, 0.1, 1, 10, and ∞.

Chapter VIII

Triangular Ducts

The triangular duct geometry is of considerable importance for fluid flow and heat transfer devices. As a result, it has been analyzed in great detail for fully developed flows. Because of mathematical complexities, it has not been analyzed thoroughly for hydrodynamic and thermally developing flows. The hydrodynamic entry length solution is available for isosceles triangular ducts with apex angles $2\phi = 30$, 60, and 90°. The thermal entry length solutions are available for isosceles triangular ducts with $2\phi = 60$ and 90° for the case of transverse velocity components v and w as zero for the simultaneously developing flow.

A. Fully Developed Flow

The hydrodynamically and thermally developed laminar flow solutions are described below in detail for the following duct geometries: the equilateral triangular duct, the equilateral triangular duct with rounded corners, isosceles triangular ducts, right triangular ducts, and arbitrary triangular ducts.

1. EQUILATERAL TRIANGULAR DUCT

The fully developed velocity profile and friction factor for the equilateral triangular duct of Fig. 44 have been well established as follows [4,70]:

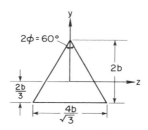

FIG. 44. An equilateral triangular duct.

$$u = \frac{c_1}{8b}\left[-y^3 + 3yz^2 + 2b(y^2 + z^2) - \frac{32}{27}b^3\right] \qquad (354)$$

$$u_m = -\frac{c_1}{15}b^2 \qquad (355)$$

$$f\,\mathrm{Re} = 40/3 = 13.333, \qquad D_h = 4b/3 \qquad (356)$$

Nusselt numbers for the equilateral triangular duct have been obtained for the Ⓣ, Ⓗ1, and Ⓗ2 boundary conditions. Kays and London [6,7] report $\mathrm{Nu_T}$ as 2.35 and Kutateladze [414] reports it as 2.70. From Schmidt and Newell's graphical results [95], $\mathrm{Nu_T}$ is found to be 2.47. By extrapolation of $\mathrm{Nu_{x,T}}$ of Table 63 for three Prandtl numbers to $1/x^*$ of zero, the average $\mathrm{Nu_T}$ is obtained as 2.47. The finite difference solution of Nakamura *et al.* [97] yielded $\mathrm{Nu_T}$ of 2.46. In light of these results, $\mathrm{Nu_T}$ for the equilateral triangular duct may be taken as 2.47 until a more refined magnitude is available. This value is 5.1% higher than that presented in [6,7].

$\mathrm{Nu_{H1}}$ for the equilateral triangular duct was first determined by Clark and Kays [94] by employing a finite difference method. Marco and Han [70] provided exact solutions for the velocity and temperature distributions for the Ⓗ1 boundary condition. By employing complex variable technique, Tao [79] arrived at $\mathrm{Nu_{H1}}$, which included thermal energy generation within the fluid. Tyagi [63] extended Tao's analysis by adding viscous dissipation effects. Based on Tyagi's results,

$$\mathrm{Nu_{H1}} = \frac{28}{9}\left(\frac{1}{1 + (1/12)S^* + (40/11)\mathrm{Br}'}\right) \qquad (357)$$

Tyagi presented the formula for $\mathrm{Nu_{H1}}$ in terms of the parameters c_3/c_4a^2 and $\mu c_1/kc_2$, which are dependent upon S^* and Br'. Perkins, in a discussion of [63], incorrectly related these parameters to S^* and Br'. They should have been

$$\frac{c_3}{c_4 a^2} = -\frac{1}{(20/3)\{1 + (4/S^*)[1 + (20/3)Br']\}}$$ (358)

$$\frac{\mu c_1}{k c_2} = -\frac{(80/3)Br'}{S^* + 4[1 + (20/3)Br']}$$ (359)

In the presence of thermal energy sources and viscous dissipation, the enthalpy rise (or drop) of the fluid, based on Tyagi's results [63], is

$$W c_p \frac{dt_m}{dx} = q' + SA_c + \frac{20\sqrt{3}\mu u_m^2}{g_c J}$$ (360)

Tyagi [415] extended his analysis to include the effect of flow work and presented closed-form formulas for u, u_m, t, t_m, and Nu_{H1}. For no internal thermal energy sources and viscous dissipation effects, Eq. (357) reduces to

$$Nu_{H1} = 28/9 = 3.111$$ (361)

a result 3.7% higher than first reported by Clark and Kays [94] and used in [6,7].

Lu and Miller [416] analyzed the same problem by employing an integral transform to obtain infinite series solutions for Nu_{H1} and other results of interest that are identical to those of Tao [79]. Aggarwala and Iqbal [78] employed the membrane analogy and obtained Nu_{H1} and $f\,Re$ that are in excellent agreement with the above values.

Nu_{H2} of the equilateral triangular duct was determined by Cheng [107], by employing a point-matching method, as

$$Nu_{H2} = 1.892$$ (362)

Shah [116] also obtained Nu_{H2} as 1.889 by applying a discrete least squares method.

2. EQUILATERAL TRIANGULAR DUCTS WITH ROUNDED CORNERS

Because of the manufacturing processes employed in a heat exchanger matrix with triangular flow passages, some of the passages will have one, two, or three rounded corners instead of sharp corners. This corner rounding effect has been investigated by Shah [116] for the equilateral triangular duct with fully developed laminar flow, by employing a discrete least squares method. The radius of each rounded corner is considered as $a/3$, as shown in Fig. 45, with $2a$ as the side length of the base triangle.

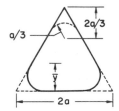

FIG. 45. An equilateral triangular duct with one, two, or three rounded corners. The solid lines represent an equilateral triangular duct with two rounded corners.

The flow friction and heat transfer characteristics together with the geometries of one, two, and three rounded corners are presented in Table 55. For comparison, the results for the base equilateral triangular duct are also in Table 55. As explained on p. 229, $\bar{y}$ and $\bar{y}_{max}$ in Table 55 refer to the distances, measured from the base, of the centroid and a point where the maximum fluid velocity occurs.

It may be noted that f Re and Nu_{H1} of the three rounded corners duct approach the corresponding values for the circular tube within 0.04 and 3.65%, respectively.

TABLE 55

EQUILATERAL TRIANGULAR DUCTS WITH NO, ONE, TWO, AND THREE ROUNDED CORNERS: GEOMETRICAL, FLOW, AND HEAT TRANSFER CHARACTERISTICS FOR FULLY DEVELOPED LAMINAR FLOW (FROM SHAH [116])

	No rounded corners	One rounded corner	Two rounded corners	Three rounded corners
$P/2a$	3.00000	2.77172	2.54343	2.31515
$A_c/(2a)^2$	0.43301	0.41399	0.39497	0.37594
$D_h/2a$	0.57735	0.59745	0.62115	0.64953
$\bar{y}/2a$	0.28868	0.26778	0.30957	0.28868
$\bar{y}_{max}/2a$	0.28868	0.28627	0.29117	0.28868
u_{max}/u_m	2.222	2.172	2.115	2.064
$K_d(\infty)$	1.429	1.406	1.379	1.353
$K_e(\infty)$	2.338	2.254	2.163	2.074
$K(\infty)$	1.818	1.698	1.567	1.441
L_{hy}^+	0.0398	0.0359	0.0319	0.0284
f Re	13.333	14.057	14.899	15.993
Nu_{H1}	3.111	3.401	3.756	4.205
Nu_{H2}	1.892	2.196	2.715	3.780
$t_{w,max}^*$	1.79	2.03	2.42	1.22
$t_{w,min}^*$	0.515	0.512	0.550	0.757

3. ISOSCELES TRIANGULAR DUCTS

a. *Fluid Flow*

The fully developed laminar velocity profile for isosceles triangular ducts has been determined by the variational, point-matching, discrete least squares, and other approximate methods.

Nuttall [417] analyzed laminar flow through isosceles triangular ducts utilizing the Rayleigh–Ritz variational method. He determined a constant, related to $f\,Re$ and $(2b/2a)$, for each of the 33 duct geometries he analyzed. As determined from Nuttall's constants, the $f\,Re$ factors are in good agreement with those of Table 57 for $0.5 \leq 2b/2a \leq 0.866$. For the values of $2b/2a$ outside this range, the agreement degenerates.

Sparrow [105] employed a point-matching method and presented the velocity profile in an infinite series form with a tabulation for the first 17 coefficients. The velocity profile and friction factors were presented graphically for the apex angle 2ϕ varying from 10 to 80°.

Migay [418] employed the same technique of solution as that for the circular sector duct [419] and obtained the following closed-form solutions for the velocity profile and friction factors for the isosceles triangular duct of Fig. 46 (Migay's results are also summarized by Petukhov [8]):

$$u = -\frac{c_1}{2}\frac{y^2 - z^2 \tan^2 \phi}{1 - \tan^2 \phi}\left[\left(\frac{z}{2b}\right)^{B-2} - 1\right] \tag{363}$$

$$u_m = -\frac{2c_1 b^2}{3}\frac{(B-2)\tan^2 \phi}{(B+2)(1 - \tan^2 \phi)} \tag{364}$$

$$f\,Re = \frac{12(B+2)(1 - \tan^2 \phi)}{(B-2)[\tan \phi + (1 + \tan^2 \phi)^{1/2}]^2} \tag{365}$$

where

$$B = \left[4 + \frac{5}{2}\left(\frac{1}{\tan^2 \phi} - 1\right)\right]^{1/2} \tag{366}$$

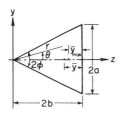

FIG. 46. An isosceles triangular duct.

The approximations employed by Migay to derive the above equations are not clear-cut to the authors. The f Re factors computed by Eq. (365) are higher than those in Table 57; they are only about 0.1% higher for $\infty \leq 2b/2a \leq 0.75$, but for lower values of $2b/2a$, the difference between f Re computed by Eq. (365) and that of Table 57 increases from 1.1% for $2b/2a = 0.5$ to 4.0% for $2b/2a = 0$. The limiting value of f Re at $2b/2a = 0$ ($2\phi = 180°$) is 12.479 by Eq. (365), in contrast to the exact value of 12. The f Re factors computed from Eq. (365) for $2\phi = 0$, 45, and 60° are identical to the entries in Table 57.

Lundgren et al. [51] determined f_D Re by employing a finite difference method and computed $K(\infty)$, $K_d(\infty)$, $K_e(\infty)$, and other flow results. McComas [48] evaluated the hydrodynamic entrance length L_{hy}^+. Their results for 2ϕ ranging from 10 to 80° are in excellent agreement with those of Tables 56 and 57.

Tirunarayanan and Ramachandran [391] introduced a concept of a shape factor, as described on p. 201, to correlate the f Re factors for the isosceles triangular ducts. Three lines drawn from the center of the inscribed circle perpendicular to each side of the isosceles triangle divide the flow field into three regions. The path of a least shear resistance for each flow region is the line connecting the corner to the center of the inscribed circle.[†] The shape factor for the isosceles triangular duct $\bar{B}/P$ is the arithmetic average of the above three paths divided by the duct perimeter:

$$\frac{\bar{B}}{P} = \frac{\cos \phi - (R_i/a)\sin \phi + 2(R_i/a)\sin \phi \csc(45° - \phi/2)}{6(1 + \sin \phi)} \tag{367}$$

Here the radius of the inscribed circle R_i is

$$R_i = a \tan(45° - \phi/2) \tag{368}$$

and 2ϕ is the apex angle of the isosceles triangular duct. Tirunarayanan and Ramachandran derived the following correlation for isosceles triangular ducts:

$$f \, \text{Re} = 13.33\left[3.878\left(\frac{\bar{B}}{P} - \frac{1}{3\sqrt{3}}\right) + 1 \right] \tag{369}$$

The f Re factors calculated from Eq. (369) are slightly higher, up to 0.4%, than those of Table 57 for isosceles triangular ducts with $2b/2a \geq 0.5$. The f Re factors from Eq. (369) for $2b/2a = 0.25$ and 0.125 are higher than those of Table 57 by 1.5 and 1.9%, respectively.

[†] Tirunarayanan and Ramachandran considered the point of maximum velocity for isosceles triangular ducts as the center of the inscribed circle. However, the maximum velocity occurs at a point between the centroid and the center of the inscribed circle as noted after Eq. (370).

Chiranjivi [420] employed the analogy between the stress function in torsion theory and the velocity field in laminar flow, and derived $f\,\mathrm{Re}$ factors for 44, 60, and 90° isosceles triangular ducts.

Shah [116] employed a discrete least squares method and obtained precise values for the flow and heat transfer characteristics of isosceles triangular ducts. The velocity profile was presented in closed form as

$$u = -c_1 \left[-\frac{r^2}{4} + \sum_{j=0}^{N} r^j (a_j \cos j\theta + b_j \sin j\theta) \right] \tag{370}$$

The coefficients[†] a_j and b_j for N up to 41 were determined by a least squares method with 91 points around the boundary. Subsequently, u_m was computed by numerical integration, and fluid flow results were determined from their definitions. These results[‡] are presented in Tables 56 and 57 and Fig. 47. For isosceles triangular ducts, except for the equilateral geometry, the point of maximum velocity does not occur either at the centroid of the cross section ($\bar{y} = 2b/3$ in Fig. 46) or at the center of the inscribed circle [R_i given by Eq. (368)]. However, it is located between these two points on the axis of symmetry. The distance $\bar{y}$ of the centroid and $\bar{y}_{max}$ where the maximum velocity occurs are both listed in Table 56.

The tabulated $f\,\mathrm{Re}$ of Shah and graphical $f\,\mathrm{Re}$ of Sparrow [105] are in good agreement. The experimental $f\,\mathrm{Re}$ determined by Carlson and Irvine [421] and Eckert and Irvine [422] for 2ϕ varying from 7.96 to 38.8° are within 2% of the corresponding values in Table 57.

In Tables 56 and 57, the results for $2b/2a = \infty$ and 0 are determined by employing the method of Chapter XI. The reason for $f\,\mathrm{Re} = 12$ for these special cases, relative to 24 for parallel plates, may be explained alternately as follows. The foregoing two special cases are the limiting cases: (1) $2\phi = 0°$ when $2a$ is kept finite, and (2) $2\phi = 180°$ when $2b$ is kept finite. The corresponding geometries appear to approximate parallel plates with $2a$ and $2b$ as maximum plate spacings, respectively. For parallel plates with these spacings, the hydraulic diameters are $4a$ and $4b$, respectively, and $f\,\mathrm{Re} = 24$. However, the hydraulic diameter of the isosceles triangular duct of Fig. 46 is

$$D_h = \frac{4b \sin \phi}{1 + \sin \phi} = \frac{2a \cos \phi}{1 + \sin \phi} \tag{371}$$

[†] The coefficients b_j of Eq. (370) and d_j of Eq. (372) are zero for a duct symmetrical about one axis and when θ is measured from the axis of symmetry. The coefficients a_j for the velocity problem and c_j for the (H1) and (H2) temperature problems have been determined for rectangular, isosceles triangular, rounded corner triangular, sine, rhombic, and trapezoidal ducts. They are deposited as Document No. NAPS 02464 with the National Auxiliary Publications Service, c/o Microfiche Publications, 305 E. 46 St., New York, N.Y. 10017.

[‡] Since the isosceles triangular ducts for $0 \le 2b/2a \le 1$ are not symmetrical with $0 \le 2a/2b \le 1$ ducts, one would not expect the curves in Fig. 47 to be symmetrical about the $\alpha^* = 1$ line.

TABLE 56

Isosceles Triangular Ducts: Geometrical and Flow Characteristics for Fully
Developed Laminar Flow (from Shah [116])

$\dfrac{2b}{2a}$	2ϕ	$\dfrac{\bar{P}}{2a}$	$\dfrac{D_h}{2a}$	$\dfrac{\bar{y}}{2a}$	$\dfrac{\bar{y}_{max}}{2a}$	$\dfrac{u_{max}}{u_m}$	$K_d(\infty)$	$K_e(\infty)$
∞	0	–	–	–	–	3.000	1.600	3.086
8.000	7.15	17.0312	.93945	2.66667	.83592	2.593	1.545	2.805
5.715	10.00	12.4737	.91633	1.90501	.74009	2.521	1.526	2.731
4.000	14.25	9.0623	.88278	1.33333	.64240	2.442	1.505	2.640
2.836	20.00	6.7588	.83910	0.94521	.55282	2.368	1.482	2.546
2.000	28.07	5.1231	.78078	0.66667	.46729	2.302	1.459	2.454
1.866	30.00	4.8637	.76733	0.62201	.45102	2.290	1.455	2.438
1.500	36.87	4.1623	.72076	0.50000	.40140	2.259	1.443	2.392
1.374	40.00	3.9238	.70021	0.45791	.38215	2.249	1.439	2.377
1.072	50.00	3.3662	.63707	0.35742	.33033	2.228	1.431	2.347
1.000	53.13	3.2361	.61803	0.33333	.31641	2.225	1.430	2.342
0.866	60.00	3.0000	.57735	0.28868	.28868	2.222	1.429	2.338
0.750	67.38	2.8028	.53518	0.25000	.26231	2.225	1.430	2.342
0.714	70.00	2.7434	.52057	0.23803	.25364	2.227	1.431	2.345
0.596	80.00	2.5557	.46631	0.19863	.22313	2.241	1.436	2.366
0.500	90.00	2.4142	.41421	0.16667	.19584	2.264	1.445	2.400
0.289	120.00	2.1547	.26795	0.09623	.12552	2.380	1.489	2.571
0.250	126.87	2.1180	.23607	0.08333	.11085	2.416	1.499	2.617
0.134	150.00	2.0353	.13165	0.04466	.06301	2.587	1.543	2.815
0.125	151.93	2.0308	.12311	0.04167	.05907	2.605	1.548	2.835
0	180.00	2.0000	.00000	0.00000	.00000	3.000	1.600	3.086

TABLE 57

Isosceles Triangular Ducts: Flow and Heat Transfer Characteristics for
Fully Developed Laminar Flow (from Shah [116]), and Nu_T (from Graphical
Results of Schmidt and Newell [95])

$\dfrac{2b}{2a}$	2ϕ	$K(\infty)$	L_{hy}^{+}	fRe	Nu_T	Nu_{H1}	Nu_{H2}	$t_{w,max}^{*}$	$t_{w,min}^{*}$
∞	0	2.971	.1048	12.000	0.943	2.059	0	–	–
8.000	7.15	2.521	.0648	12.352	1.46	2.348	0.039	–	–
5.715	10.00	2.409	.0590	12.474	1.61	2.446	0.080	–	–
4.000	14.25	2.271	.0533	12.636	1.81	2.575	0.173	4.14	–
2.836	20.00	2.128	.0484	12.822	2.00	2.722	0.366	4.07	–
2.000	28.07	1.991	.0443	13.026	2.22	2.880	0.747	3.83	.073
1.866	30.00	1.966	.0436	13.065	2.26	2.910	0.851	3.73	.127
1.500	36.87	1.898	.0418	13.181	2.36	2.998	1.22	3.38	.287
1.374	40.00	1.876	.0412	13.222	2.39	3.029	1.38	3.17	.347
1.072	50.00	1.831	.0401	13.307	2.45	3.092	1.76	2.53	.469
1.000	53.13	1.824	.0399	13.321	2.46	3.102	1.82	2.41	.483
0.866	60.00	1.818	.0398	13.333	2.47	3.111	1.892	1.79	.515
0.750	67.38	1.824	.0399	13.321	2.45	3.102	1.84	1.99	.499
0.714	70.00	1.829	.0400	13.311	2.45	3.095	1.80	2.04	.488
0.596	80.00	1.860	.0408	13.248	2.40	3.050	1.59	2.19	.432
0.500	90.00	1.909	.0421	13.153	2.34	2.982	1.34	2.30	.364
0.289	120.00	2.165	.0490	12.744	2.00	2.680	0.62	2.45	.167
0.250	126.87	2.235	.0515	12.622	1.90	2.603	0.490	2.47	.131
0.134	150.00	2.543	.0644	12.226	1.50	2.325	0.156	2.56	.045
0.125	151.93	2.574	.0659	12.196	1.47	2.302	0.136	2.65	.038
0	180.00	2.971	.1048	12.000	0.943	2.059	0	–	–

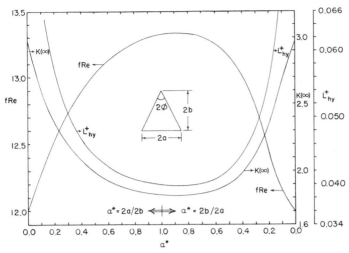

FIG. 47. Isosceles triangular ducts: $f\,Re$, $K(\infty)$, and L_{hy}^{+} for fully developed laminar flow (from Table 57).

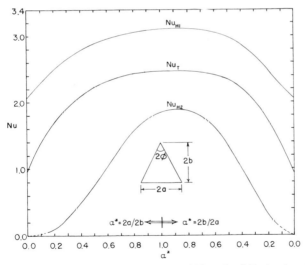

FIG. 48. Isosceles triangular ducts: Nu_T, Nu_{H1}, and Nu_{H2} for fully developed laminar flow (from Table 57).

yielding $2a$ for $2\phi = 0°$, and $2b$ for $2\phi = 180°$. These are one-half of D_h for parallel plates, and so the resulting f Re factors are 12 for these two limiting cases of isosceles triangular ducts.

b. *Heat Transfer*

The fully developed laminar Nusselt numbers for isosceles triangular ducts have been determined for the Ⓗ1, Ⓗ2, and Ⓣ boundary conditions.

Sparrow and Haji-Sheikh [423] employed a finite difference method and obtained Nu_{H1} for ducts of apex angle 2ϕ varying from 0 to 90°. They reported their results graphically, and these are in good agreement with those of Table 57. Iqbal *et al.* [76,92] employed the variational method and determined $2f$ Re, Nu_{H1}, and Nu_{H2}. Their graphical results are also in good agreement with those of Table 57.

Shah [116] presented the Ⓗ1 and Ⓗ2 temperature profiles in closed form as follows:

$$t = \frac{8f \operatorname{Re} q''}{kD_h{}^3}\left[-\frac{r^4}{64} + \sum_{j=0}^{N} \frac{r^{j+2}}{4(j+1)}(a_j \cos j\theta + b_j \sin j\theta)\right.$$
$$\left. + \sum_{j=0}^{N} r^j(c_j \cos j\theta + d_j \sin j\theta)\right] \tag{372}$$

where a_j and b_j are obtained from the solution of the velocity problem, Eq. (370). The unknown coefficients c_j and d_j were determined by a method of least squares by employing 91 points around the duct boundary for N up to 41. Subsequently, t_m was evaluated by numerical integration, and Nu_{H1} and Nu_{H2} from their definitions. These latter results are presented in Table 57 and Fig. 48. The wall temperature around the duct periphery is nonuniform for the Ⓗ2 boundary condition. The hot spot occurs at the sharpest corner of the isosceles triangle and the cold spot occurs centered on the shortest side. The corresponding dimensionless maximum and minimum wall temperatures $t^*_{w,max}$ and $t^*_{w,min}$ [defined by Eq. (350)] are presented in Table 57, where $t_{w,m}$ and t_c are the peripheral average wall temperature and the fluid temperature at the centroid of the duct, respectively.

In Table 57, Nu_T and N_{H1} for $2b/2a = \infty$ and 0 are determined by the method of Chapter XI. The reason that the values of Nu_T and Nu_{H1} are one-fourth that of the parallel plates may be explained alternatively as follows. In the absence of fluid axial heat conduction, viscous dissipation, and thermal energy sources, an energy balance on the duct length δx yields

$$q''P \, \delta x = (\rho A_c u_m)c_p \frac{dt_m}{dx} \delta x \tag{373}$$

Combining Eq. (373) with Eq. (104), Nu $= hD_h/k$ and $D_h = 4A_c/P$ provides

$$Nu = \frac{D_h{}^2 u_m}{4\alpha(t_{w,m} - t_m)} \frac{dt_m}{dx} \qquad (374)$$

As discussed above, D_h for the aforementioned limiting cases of isosceles triangular ducts is one-half that of the parallel plates, and hence the Nusselt number Nu is one-fourth that of the parallel plates.

Schmidt and Newell [95] considered the more general heat transfer problem, with one or more walls transferring heat and the others being adiabatic. They evaluated Nu_{H1} and Nu_T for these cases by a finite difference method and reported the results graphically as a function of the half apex angle ϕ. The Nu_T, presented in Table 57 and Fig. 48 for all three walls heated, are obtained from the graphical results of Schmidt and Newell [95]. Nu_T and Nu_{H1} with one or two walls transferring heat, are provided in Tables 58 and 59 [252] and Figs. 49 and 50. In the definition, Nu_T and Nu_{H1} have a hydraulic diameter D_h based on the wetted perimeter, instead of D_h based on the heated perimeter, as originally presented by Schmidt and Newell.

4. RIGHT TRIANGULAR DUCTS

a. *Fluid Flow*

Laminar flow through right triangular ducts has been analyzed by the elastic torsion analogy, finite difference, and variational methods. First the results are presented for a right-angled isosceles triangular duct; next the results are summarized for the more general right triangular duct.

The fully developed laminar velocity profile for the right-angled isosceles triangular duct of Fig. 51a is expressed as [4,70]

$$u = \frac{c_1}{2} \left\{ \frac{1}{2}(y + z)^2 - a(y + z) \right.$$

$$\left. + \frac{2}{a} \sum_{n=0}^{\infty} \frac{(-1)^n[\sinh(Nz)\cos(Ny) + \sinh(Ny)\cos(Nz)]}{N^3 \sinh(Na)} \right\} \qquad (375)$$

$$u_m = -\frac{c_1}{8a^3} \left[\frac{4}{3} a^5 - \sum_{n=0}^{\infty} \frac{1}{N^5 \tanh(Na)} \right] \qquad (376)$$

$$f \, Re = -\frac{4}{3 + 2\sqrt{2}} \frac{c_1 a^2}{u_m} = \frac{230}{3(3 + 2\sqrt{2})} = 13.154 \qquad (377)$$

where

$$N = \frac{(2n + 1)\pi}{2a} \qquad (378)$$

TABLE 58

Isosceles Triangular Ducts: Nu_T for Fully Developed
Laminar Flow, for One or More Walls Transferring Heat
(from Schmidt [252])

$\dfrac{2b}{2a}$	ϕ degrees	Nu_T			
∞	0	1.885	0.000	1.215	1.215
5.000	5.71	—	0.822	1.416	1.312
2.500	11.31	2.058	1.268	1.849	1.573
1.667	16.70	2.227	1.525	2.099	1.724
1.250	21.80	2.312	1.675	2.237	1.802
1.000	26.56	2.344	1.758	2.301	1.831
0.833	30.96	—	—	2.319	1.822
0.714	34.99	2.311	1.812	2.306	1.787
0.625	38.66	—	—	2.274	1.735
0.556	41.99	—	—	2.232	1.673
0.500	45.00	2.162	1.765	2.183	1.606
0.450	48.01	—	—	2.127	1.529
0.400	51.34	—	—	2.055	1.433
0.350	55.01	1.923	1.633	1.968	1.315
0.300	59.04	—	—	1.861	1.173
0.250	63.43	1.671	1.471	1.733	1.004
0.200	68.20	1.512	1.361	1.581	0.805
0.150	73.30	1.330	1.229	1.401	0.578
0.100	78.69	1.126	1.071	1.182	0.332
0.050	84.29	0.895	0.878	0.893	0.106
0	90.00	0.6076	0.6076	—	—

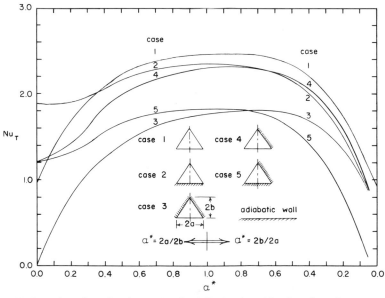

FIG. 49. Isosceles triangular ducts: Nu_T for fully developed laminar flow, for one or more walls transferring heat (from Table 58).

TABLE 59

ISOSCELES TRIANGULAR DUCTS: Nu_{H1} FOR FULLY DEVELOPED LAMINAR
FLOW, FOR ONE OR MORE WALLS TRANSFERRING HEAT
(FROM SCHMIDT [252])

$\frac{2b}{2a}$	ϕ degrees	Nu_{H1}			
∞	0	2.059	0	1.346	1.346
5.000	5.71	2.465	1.003	1.824	1.739
2.500	11.31	2.683	1.515	2.274	1.946
1.667	16.70	2.796	1.807	2.541	2.074
1.250	21.80	2.845	1.978	2.695	2.141
1.000	26.56	2.849	2.076	2.773	2.161
0.833	30.96	–	–	2.801	2.146
0.714	34.99	2.778	2.146	2.792	2.107
0.625	38.66	–	–	2.774	2.053
0.556	41.99	–	–	2.738	1.989
0.500	45.00	2.594	2.111	2.696	1.921
0.450	48.01	–	–	2.646	1.843
0.400	51.34	–	–	2.583	1.746
0.350	55.01	2.332	1.991	2.505	1.628
0.300	59.04	–	–	2.412	1.486
0.250	63.43	2.073	1.843	2.301	1.316
0.200	68.20	1.917	1.746	2.174	1.114
0.150	73.30	1.748	1.635	2.032	0.874
0.100	78.69	1.576	1.515	1.881	0.587
0.050	84.29	1.418	1.398	1.737	0.244
0	90.00	1.346	1.346	–	–

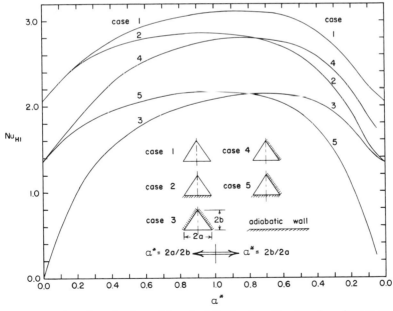

FIG. 50. Isosceles triangular ducts: Nu_{H1} for fully developed laminar flow, for one or more walls transferring heat (from Table 59).

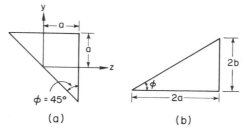

FIG. 51. (a) A right-angled isosceles triangular duct, (b) a right triangular duct.

The right triangular ducts with the apex angle ϕ varying from 0 to 90° were first analyzed by Sparrow and Haji-Sheikh [423] by a finite difference method. They presented graphically $f\,\mathrm{Re}$ and $K(\infty)$ from which Table 60 and Fig. 52 are prepared. Aggarwala and Iqbal [78] obtained $f\,\mathrm{Re}$ for two right triangular ducts ($\phi = 30$ and 45°), employing the membrane analogy. Iqbal *et al.* [76] employed a variational method and graphically presented $2f\,\mathrm{Re}$ for the complete family of right triangular ducts. The results of Sparrow and Haji-Sheikh and Iqbal *et al.* are in excellent agreement. $f\,\mathrm{Re}$ and other flow results for the right triangular duct family for $\phi = 0$ to 90° are symmetrical about $\phi = 45°$.

b. *Heat Transfer*

Marco and Han [70] determined the velocity and temperature distributions for the (H1) boundary condition for the right-angled isosceles triangular duct. Aggarwala and Iqbal [78] employed the analogy of membrane vibration to the (H1) problem and arrived at the solution for fully developed laminar flow through the right-angled isosceles triangular duct. They obtained

$$\mathrm{Nu_{H1}} = 2.982 \tag{379}$$

Sparrow and Haji-Sheikh [423] obtained $\mathrm{Nu_{H1}}$ for right triangular ducts using a finite difference method and presented them graphically for the apex angle varying from 0 to 90°. The $\mathrm{Nu_{H1}}$ values presented in Table 60 and Fig. 52 are obtained from these graphical results. Aggarwala and Iqbal [78] determined $f\,\mathrm{Re}$ and $\mathrm{Nu_{H1}}$ for two right triangular ducts ($\phi = 30$ and 45°) by employing the membrane analogy. Iqbal *et al.* [76] analyzed the right triangular ducts by a variational method. They presented graphically $2f\,\mathrm{Re}$ and $\mathrm{Nu_{H1}}$ for ϕ varying from 5 to 85°. Their results are in good agreement with those of Sparrow and Haji-Sheikh [423].

Iqbal *et al.* [92] also analyzed the (H2) boundary condition for right triangular ducts using a variational method. They presented graphically $\mathrm{Nu_{H2}}$ for ϕ varying from 5 to 80°. Their results are also presented in Table 60 and Fig. 52.

TABLE 60

RIGHT TRIANGULAR DUCTS: f Re, $K(\infty)$, Nu_{H1}, AND
Nu_{H2} FOR FULLY DEVELOPED LAMINAR FLOW
(FROM THE GRAPHICAL RESULTS OF SPARROW AND
HAJI-SHEIKH [23], AND IQBAL et al. [92])

ϕ deg.	$\dfrac{2b}{2a}$	f Re	$K(\infty)$	Nu_{H1}	Nu_{H2}
0	0	12.000	2.971	2.059	0
5	0.0875	12.27	2.65	2.26	0.02
10	0.1763	12.49	2.40	2.43	0.08
15	0.2679	12.68	2.21	2.57	0.20
20	0.3640	12.83	2.10	2.69	0.39
25	0.4663	12.94	2.01	2.80	0.62
30	0.5774	13.034	1.95	2.888	0.89
35	0.7002	13.09	1.91	2.94	1.14
40	0.8391	13.13	1.88	2.97	1.23
45	1.000	13.154	1.88	2.982	1.34
50	1.192	13.13	1.88	2.97	1.23
60	1.732	13.034	1.95	2.888	0.89
70	2.747	12.83	2.10	2.69	0.39
80	5.671	12.49	2.40	2.43	0.08
90	∞	12.000	2.971	2.059	0

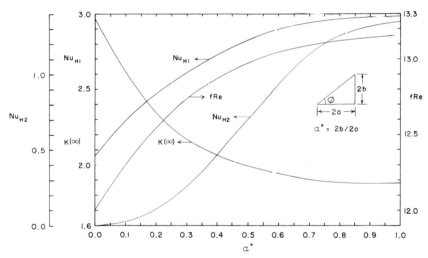

FIG. 52. Right triangular ducts: f Re, $K(\infty)$, Nu_{H1}, and Nu_{H2} for fully developed laminar flow (from Table 60).

5. ARBITRARY TRIANGULAR DUCTS

In the previous subsections, isosceles and right triangular ducts were considered. Nakamura et al. [97] analyzed the arbitrary triangular duct geometry by a finite difference method using a uniform triangular grid of arbitrary shape. A sketch of this duct is shown in Figs. 53 and 54. They employed the extrapolated Liebmann method to solve the system of linear algebraic equations resulting from a finite difference formulation of the momentum and energy equations. f Re, $K(\infty)$, Nu_{H1}, and Nu_T of Nakamura

TABLE 61

ARBITRARY TRIANGULAR DUCTS: f Re, $K(\infty)$, Nu_{H1}, AND Nu_T FOR FULLY
DEVELOPED LAMINAR FLOW (FROM NAKAMURA [424])

2ϕ	fRe	$K(\infty)$	Nu_{H1}	Nu_T	fRe	$K(\infty)$	Nu_{H1}	Nu_T
	$\delta = 0.9$				$\delta = 0.7$			
0	12.00	2.97	2.06	0.943	12.00	2.97	2.06	0.943
10	12.78	2.65	2.34	1.64	12.57	2.63	2.31	1.63
20	12.99	2.26	2.63	2.00	12.82	2.33	2.54	1.90
30	13.18	2.06	2.83	2.22	13.04	2.13	2.74	2.12
40	13.31	1.95	2.95	2.36	13.19	2.01	2.87	2.27
50	13.38	1.89	3.02	2.43	13.28	1.94	2.95	2.35
60	13.40	1.87	3.04	2.45	13.32	1.91	2.99	2.39
70	13.37	1.87	3.03	2.44	13.30	1.91	2.98	2.38
80	13.31	1.90	2.99	2.39	13.25	1.93	2.95	2.34
90	13.22	1.95	2.93	2.32	13.18	1.98	2.89	2.28
2ϕ	$\delta = 0.5$				$\delta = 0.4$			
0	12.00	2.97	2.06	0.943	12.00	2.97	2.06	0.943
10	12.44	2.65	2.27	1.58	12.41	2.67	2.26	1.55
20	12.60	2.46	2.42	1.76	12.53	2.53	2.36	1.68
30	12.80	2.28	2.58	1.94	12.68	2.38	2.49	1.82
40	12.96	2.15	2.71	2.08	12.81	2.25	2.60	1.95
50	13.06	2.07	2.79	2.17	12.91	2.17	2.68	2.04
60	13.12	2.02	2.84	2.22	12.98	2.12	2.73	2.09
70	13.13	2.01	2.85	2.23	13.00	2.10	2.75	2.11
80	13.11	2.02	2.83	2.21	12.99	2.10	2.74	2.10
90	13.05	2.06	2.79	2.17	12.95	2.13	2.71	2.06
2ϕ	$\delta = 0.3$				$\delta = 0.2$			
0	12.00	2.97	2.06	0.943	12.00	2.97	2.06	0.943
10	12.39	2.70	2.23	1.50	12.40	2.73	2.20	1.44
20	12.47	2.60	2.31	1.58	12.43	2.67	2.25	1.48
30	12.57	2.48	2.39	1.69	12.45	2.60	2.30	1.54
40	12.67	2.38	2.47	1.79	12.54	2.53	2.35	1.61
50	12.75	2.30	2.54	1.87	12.59	2.47	2.39	1.66
60	12.81	2.25	2.89	1.92	12.63	2.43	2.42	1.71
70	12.84	2.22	2.61	1.94	12.66	2.40	2.44	1.74
80	12.84	2.22	2.61	1.94	12.66	2.39	2.45	1.74
90	12.82	2.23	2.60	1.92	12.56	2.40	2.44	1.73

et al. from [424] are presented in Table 61; f Re and Nu_{H1} are also shown
in Figs. 53 and 54.

The results for isosceles triangular ducts, corresponding to $\delta = 1.0$ of
Nakamura *et al.*, are not presented here because of the more accurate and
complete results presented in Tables 56 and 57 and Figs. 47 and 48. A comparison reveals that f Re values of Nakamura *et al.* for $\delta = 1.0$ are higher
than those of Table 57, ranging from 3% for $2\phi = 10°$ to 0.6% for $2\phi = 60°$.
The Nu_{H1} are lower than those of Table 57, ranging from 4% for $2\phi = 10°$
to 2% for $2\phi = 60°$. The Nu_T are higher for $2\phi < 20°$ and lower for $2\phi > 20°$
than those of Table 57; $K(\infty)$ values are higher than those of Table 57,
ranging from 11% for $2\phi = 10°$ to 2% for $2\phi = 60°$.

The limiting values for $2\phi = 0°$ in Table 61, obtained by the method of
Chapter XI, are included to make the table more complete. f Re and Nu_{H1}

curves in Figs. 53 and 54 are smoothened between $2\phi = 0$ and $30°$. In this range, the values of Table 61 are higher than the smoothened values from the figures; similarly, Nu_T values of Table 61 are higher (although not shown here) between $2\phi = 0$ and $30°$. $K(\infty)$ values of Table 61 for $2\phi = 10°$ also

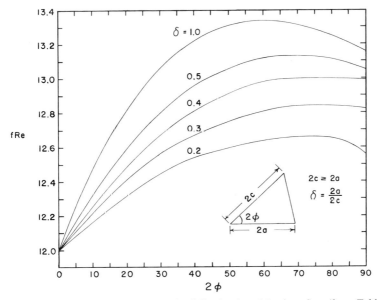

FIG. 53. Arbitrary triangular ducts: f Re for fully developed laminar flow (from Table 61).

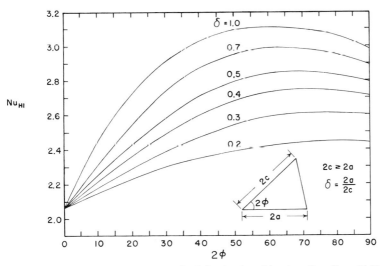

FIG. 54. Arbitrary triangular ducts: Nu_{H1} for fully developed laminar flow (from Table 61).

appear to have appreciable errors granting the validity of graphical interpolation. The deviations from the expected (smoothened) results for lower values of 2ϕ may be due to increased numerical erros associated with the finite difference method in this sharp angle range.

Semilet [425] reviewed the Russian and English language literatures for laminar heat transfer and pressure drop for *gas flow* in triangular ducts. He plotted the experimental Nusselt numbers from nine different sources. These results were all for different triangular geometries (the range of 2ϕ not specified) under possibly different thermal boundary conditions. Semilet arrived at the following best-fit curve for Re $\sim$ 50 to 8000:

$$Nu = 2.15 + (2.31)(10^{-3})\,Re + 1.25(10^{-7})\,Re^2 - (9.6)(10^{-12})\,Re^3 \quad (380)$$

This equation agrees with the experimental results within $\pm 10\%$. Semilet further simplified Eq. (380) to

$$Nu = 2.15 + 0.00245\,Re \quad (381)$$

for Re $\sim$ 50 to 2200 and $L/D_h > 50$. This equation agrees with the experimental results within $\pm 15\%$ in the laminar region.

Based on the experimental results, Semilet also presented the following equation for the friction factors for a triangular duct for Re $\sim$ 50 to 1000 and $L/D_h > 50$:

$$f = 0.007 + 12.5/Re \quad (382)$$

B. Hydrodynamically Developing Flow

The hydrodynamic entry length problem has been analyzed for three ducts: an equilateral triangular, and a 30° and a 90° isosceles triangular.

The hydrodynamic entry length problem for the equilateral triangular duct was first analyzed by Han and Cooper [156]. They employed Langhaar's linearization technique and von Kármán's integral procedure (with some modifications) to arrive at the solution. The results were presented in tabular and graphical forms. The analysis of Han and Cooper predicted somewhat more rapid flow development (i.e., high pressure drop at any x) than predicted by other investigators [158,166]. Han and Cooper obtained $K(\infty)$ as 2.382; a value considerably higher than 1.69 obtained from the results of [426] in Table 62.

Wibulswas [220], in his solution to the combined entry length problem for the equilateral triangular duct, solved the hydrodynamic entry length problem by a finite difference method. However, he did not report his results for the hydrodynamic entrance region.

Fleming and Sparrow [166] linearized the x momentum equation by introducing a stretched coordinate in the flow direction [refer to Eq. (167)]. They presented graphically the velocity profiles, Δp^*, and $K(x)$ for two duct geometries: the equilateral triangular, and the 30° isosceles triangular ducts. Their centerline velocity development for the equilateral triangular duct is in fair agreement with that of Han and Cooper. The f_{app} Re factors, based on their graphical $K(x)$, are reported in Table 62 and Fig. 55.

TABLE 62

ISOSCELES TRIANGULAR DUCTS: f_{app} Re FOR
DEVELOPING LAMINAR FLOW

x^+	f_{app}Re [166]		f_{app} Re [158]	f_{app} Re [426]	
	$2\phi=30°$	$2\phi=60°$	$2\phi=60°$	$2\phi=60°$	$2\phi=90°$
0.001	–	–	88.58	–	–
0.002	–	–	65.33	–	–
0.003	–	–	55.42	–	–
0.004	–	–	49.40	–	–
0.005	51.1	48.8	45.18	51.08	46.90
0.006	–	–	41.96	–	–
0.008	–	–	37.68	–	–
0.010	37.3	36.3	34.78	37.66	35.75
0.015	31.7	31.2	30.22	31.88	30.60
0.020	28.4	28.0	27.42	28.52	27.50
0.025	26.4	25.8	25.40	26.27	25.40
0.030	24.7	24.2	24.05	24.64	23.87
0.040	22.5	22.1	22.03	22.40	21.75
0.050	21.0	20.7	20.64	20.91	20.34
0.060	19.9	19.7	19.63	19.85	19.34
0.070	19.0	18.9	18.85	19.04	18.57
0.080	18.4	18.3	18.42	18.40	17.97
0.090	17.8	17.8	17.74	17.89	17.49
0.100	17.4	17.4	17.33	17.47	17.10
0.120	–	–	16.68	–	–
$K(\infty)$	1.85	1.67	1.61	1.692	1.634

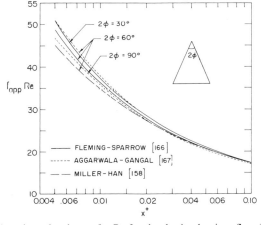

FIG. 55. Isosceles triangular ducts: f_{app} Re for developing laminar flow (from Table 62).

Aggarwala and Gangal [167] also linearized the x momentum equation by introducing a stretched coordinate. Subsequently, they obtained solution for the equilateral and right-isosceles triangular ducts, as mentioned on p. 72. Their f_{app} Re factors are presented in Table 62 [426] and Fig. 55. These factors for the equilateral triangular duct are believed to be more accurate and are recommended for the design.

Miller [185] employed a method of transforming an initial value probelm to a boundary value problem and analyzed the hydrodynamic entrance problem for the equilateral triangular duct. For reasons similar to those for the rectangular ducts discussed on p. 213, Miller's computed f_{app} Re are substantially low compared to those from Table 62.

Miller and Han [158] refined the solution of Han and Cooper [156] for the equilateral triangular duct. These investigators employed Langhaar's linearization technique. Han and Cooper determined Δp^* from the x momentum equation evaluated at the duct centerline and integrated from $x = 0$ to x. Miller and Han [158] determined Δp^* by first integrating the mechanical energy equation at x across the flow cross section and then integrating the resulting equation from $x = 0$ to x. The f_{app} Re factors of Miller and Han from [427] are about 3.6% lower than those of Fleming and Sparrow [166] of Table 62.

C. Thermally Developing Flow

1. EQUILATERAL TRIANGULAR DUCT

Kutateladze [414] presented graphically the thermal entrance $Nu_{m,T}$ for hydrodynamically developed and thermally developing laminar flow through an equilateral triangular duct. His $Nu_{m,T}$ are 12–20% lower when compared to $Nu_{m,T}$ of Wibulswas [220].

Wibulswas [220] obtained $Nu_{x,T}$, $Nu_{m,T}$, $Nu_{x,H1}$, and $Nu_{m,H1}$ for an equilateral triangular duct with slug flow (Pr = 0), with fully developed flow (Pr = ∞)[†] and with simultaneously developing flow for a fluid with Pr = 0.72. He used a finite difference method and neglected the effect of the transverse velocity components v and w, as well as the axial momentum and thermal diffusions, $\mu(\partial^2 u/\partial x^2)$ and $k(\partial^2 u/\partial x^2)$, respectively. His results are presented in Table 63 and Fig. 56.

Krishnamurty [428], employing the Lévêque-type approximation, derived expressions for the Ⓣ local and mean Nusselt numbers for the equilateral triangular duct.

[†] Refer to p. 16 for further clarification of these special cases.

TABLE 63

EQUILATERAL TRIANGULAR DUCT: $Nu_{x,T}$, $Nu_{m,T}$, $Nu_{x,H1}$, AND $Nu_{m,H1}$ AS FUNCTIONS OF x^* AND Pr (FROM WIBULSWAS [220])

$\dfrac{1}{x^*}$	$Nu_{x,T}$			$Nu_{m,T}$			$Nu_{x,H1}$			$Nu_{m,H1}$		
	Pr=∞	0.72	0	Pr=∞	0.72	0	Pr=∞	0.72	0	Pr=∞	0.72	0
10	2.57	2.80	3.27	3.10	3.52	4.65	3.27	3.58	4.34	4.02	4.76	6.67
20	2.73	3.11	3.93	3.66	4.27	5.79	3.48	4.01	5.35	4.76	5.87	8.04
30	2.90	3.40	4.46	4.07	4.88	6.64	3.74	4.41	6.14	5.32	6.80	9.08
40	3.08	3.67	4.89	4.43	5.35	7.32	4.00	4.80	6.77	5.82	7.57	9.96
50	3.26	3.93	5.25	4.75	5.73	7.89	4.26	5.13	7.27	6.25	8.20	10.65
60	3.44	4.15	5.56	5.02	6.08	8.36	4.49	5.43	7.66	6.63	8.75	11.27
80	3.73	4.50	6.10	5.49	6.68	9.23	4.85	6.03	8.26	7.27	9.73	12.35
100	4.00	4.76	6.60	5.93	7.21	9.98	5.20	6.56	8.81	7.87	10.60	13.15
120	4.24	4.98	7.03	6.29	7.68	10.59	5.50	7.04	9.30	8.38	11.38	13.82
140	4.47	5.20	7.47	6.61	8.09	11.14	5.77	7.50	9.74	8.84	12.05	14.46
160	4.67	5.40	7.88	6.92	8.50	11.66	6.01	7.93	10.17	9.25	12.68	15.02
180	4.85	5.60	8.20	7.18	8.88	12.10	6.22	8.33	10.53	9.63	13.27	15.50
200	5.03	5.80	8.54	7.42	9.21	12.50	6.45	8.71	10.87	10.02	13.80	16.00

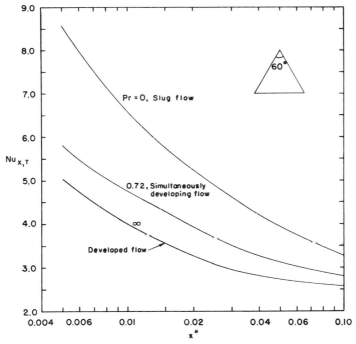

FIG. 56. Equilateral triangular duct: $Nu_{x,T}$ as a function of x^* and Pr (from Table 63). Similar behavior is obtained for $Nu_{m,T}$, $Nu_{x,H1}$, and $Nu_{m,H1}$ of Table 63.

2. RIGHT-ANGLED ISOSCELES TRIANGULAR DUCT

Wibulswas [220] solved the thermal entry length problem for the right-angled isosceles triangular duct ($\phi = 45°$). He obtained $\mathrm{Nu}_{x,T}$, $\mathrm{Nu}_{m,T}$, $\mathrm{Nu}_{x,H1}$, and $\mathrm{Nu}_{m,H1}$ for slug flow ($\mathrm{Pr} = 0$), fully developed flow ($\mathrm{Pr} = \infty$), and simultaneously developing flow for a fluid with $\mathrm{Pr} = 0.72$. His finite difference numerical solution employed the same idealizations mentioned previously for his equilateral triangular duct analysis. His results are presented in Table 64 and Fig. 57.

3. ISOSCELES TRIANGULAR DUCTS

Narayan Rao [429] studied experimentally the laminar and turbulent flow heat transfer and pressure drop behavior through isosceles triangular ducts with $2b/2a = 1.236$, 0.866, and 0.441. Heat transfer was allowed only from the bottom side of the isosceles triangular duct, the other two sides being insulated. Water and glycerol–water mixtures were used as the test fluids. The experimental results were correlated in terms of j and f factors as functions of Re, L/D_h, and a Grashof number correction factor that takes into account free convection effect. Since the isosceles triangular duct section upstream of the test section was not long enough to establish fully developed laminar flow, the velocity profile at $x = 0$ was only partially developed. Hence, the experimental results of Narayan Rao are for thermally developing flow with partially developed velocity profiles. This may be the reason for the unexpected exponents on Re and L/D_h in his correlations for j and f.

Chiranjivi and Balakameswar [430] experimentally studied laminar heat transfer to water from a two-ft long $3 \times 4 \times 4$ in. isosceles triangular duct, with the three-in. base electrically heated and the other two sides insulated. The experimental results were correlated in terms of the j factor as a function of the Reynolds number. Since the unheated isosceles triangular duct section upstream of the test section was not long enough, the velocity profile at $x = 0$ was only partially developed. Hence their correlation for the j factors is for the thermally developing and partially hydrodynamically developed laminar flow through the isosceles triangular duct.

Lebed' and Lobov [431] suggested that the turbulence upstream of a triangular heat exchanger affects the pressure drop and heat transfer in triangular channels of small D_h at low Re. They argued that the effect of the turbulence of the flow stream will be much more marked with the short channels. They obtained experimental pressure drop and heat transfer results for equilateral triangular ducts of L/D_h of 10, 20, 30, 50, and 100 for varying degree of upstream turbulence intensity. Nu and f for short triangular ducts were correlated to Nu and f for a long triangular duct of

TABLE 64

Right-Angled Isosceles Triangular Duct: $Nu_{x,T}$, $Nu_{m,T}$, $Nu_{x,H1}$, and $Nu_{m,H1}$ as Functions of x^* and Pr (from Wibulswas [220])

$\dfrac{1}{x^*}$	$Nu_{x,T}$			$Nu_{m,T}$			$Nu_{x,H1}$			$Nu_{m,H1}$		
	Pr=∞	0.72	0	Pr=∞	0.72	0	Pr=∞	0.72	0	Pr=∞	0.72	0
10	2.40	2.52	3.75	2.87	3.12	4.81	3.29	4.00	5.31	4.22	5.36	6.86
20	2.53	2.76	4.41	3.33	3.73	5.85	3.58	4.73	6.27	4.98	6.51	7.97
30	2.70	2.98	4.82	3.70	4.20	6.48	3.84	5.23	6.85	5.50	7.32	8.68
40	2.90	3.18	5.17	4.01	4.58	6.97	4.07	5.63	7.23	5.91	7.95	9.20
50	3.05	3.37	5.48	4.28	4.90	7.38	4.28	5.97	7.55	6.25	8.50	9.67
60	3.20	3.54	5.77	4.52	5.17	7.73	4.47	6.30	7.85	6.57	8.99	10.07
80	3.50	3.85	6.30	4.91	5.69	8.31	4.84	6.92	8.37	7.14	9.80	10.75
100	3.77	4.15	6.75	5.23	6.10	8.80	5.17	7.45	8.85	7.60	10.42	11.32
120	4.01	4.43	7.13	5.52	6.50	9.18	5.46	7.95	9.22	8.03	10.90	11.77
140	4.21	4.70	7.51	5.78	6.82	9.47	5.71	8.39	9.58	8.40	11.31	12.14
160	4.40	4.96	7.84	6.00	7.10	9.70	5.95	8.80	9.90	8.73	11.67	12.47
180	4.57	5.22	8.10	6.17	7.33	9.94	6.16	9.14	10.17	9.04	12.00	12.75
200	4.74	5.49	8.38	6.33	7.57	10.13	6.36	9.50	10.43	9.33	12.29	13.04

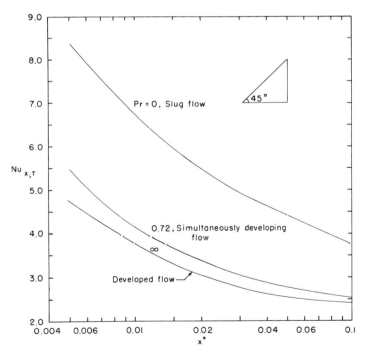

FIG. 57. Right-angled isosceles triangular duct: $Nu_{x,T}$ as a function of x^* and Pr (from Table 64). Similar behavior is obtained for $Nu_{m,T}$, $Nu_{x,H1}$, and $Nu_{m,H1}$ of Table 64.

$L/D_h = 100$ as a function of the ratio of upstream turbulence intensity of the corresponding ducts. The Nusselt number and friction factor for the duct of $L/D_h = 100$ were derived from Eqs. (381) and (382). The experimental results agreed with their correlations within $\pm 10\%$. The present authors believe that the increase in Nu and f for a short duct is not mainly due to upstream turbulence intensity, but is primarily associated with the developing boundary layers in the entrance section.

Chapter IX

Elliptical Ducts

Fully developed laminar fluid flow and heat transfer problems have been analyzed for elliptical ducts. Also, the thermal entry length solution for the case of hydrodynamically developed flow has been investigated to a limited extent. However, no hydrodynamic entry length solution has been reported for elliptical ducts.

A. Fully Developed Flow

1. Fluid Flow

The fully developed laminar velocity profile for the elliptical duct of Fig. 58, with $2a$ and $2b$ as major and minor axes, is given by Dryden *et al.* [4], Tao [79], and Shih [108], and f Re factors are presented by Lundgren *et al.* [51], as

$$u = \frac{c_1}{2(1 + \alpha^{*2})} (\alpha^{*2}z^2 + y^2 - b^2) \tag{383}$$

$$u_m = -\frac{c_1}{4}\left(\frac{b^2}{1 + \alpha^{*2}}\right) \tag{384}$$

$$f\,\mathrm{Re} = 2(1 + \alpha^{*2})\left[\frac{\pi}{E(m)}\right]^2 \tag{385}$$

$$D_h = \frac{\pi b}{E(m)}, \qquad A_c = \pi ab \tag{386}$$

247

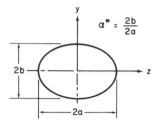

FIG. 58. An elliptical duct.

where $m = (1 - \alpha^{*2})$ and $E(m)$ is the complete elliptical integral of the second kind.

The f Re factors were computed from Eq. (385) by the present authors and are presented in Table 65 and Fig. 59. Lundgren et al. [51] determined $K(\infty)$ as 4/3. McComas [48] obtained L_{hy}^+ for elliptical ducts as presented in Table 65.

It is interesting to note that while one limiting case of elliptical ducts corresponds to the circular tube ($\alpha^* = 1$), the other limiting case, corresponding to $\alpha^* = 0$, does not approach the parallel plate geometry. All of the values in Table 65 for $\alpha^* = 0$ are different from those for parallel plates.

TABLE 65

ELLIPTICAL DUCTS: f Re FROM EQ. (385), L_{hy}^+ FROM
McCOMAS [48], Nu_T FROM DUNWOODY [434],
Nu_{H1} FROM EQ. (387), AND Nu_{H2} FROM IQBAL et al.
[92], FOR FULLY DEVELOPED LAMINAR FLOW[†]

α^*	fRe	L_{hy}^+	Nu_T	Nu_{H1}	Nu_{H2}
1.00	16.000	0.0260	3.658	4.364	4.364
0.90	16.022	0.0260	–	4.369	–
0.80	16.098	0.0259	3.669	4.387	–
0.75	16.161	0.0258	–	4.402	–
0.70	16.244	0.0257	–	4.422	–
2/3	16.311	0.0255	–	4.438	–
0.60	16.479	0.0253	–	4.477	–
0.50	16.823	0.0248	3.742	4.558	3.802
0.40	17.294	0.0241	–	4.666	–
1/3	17.681	0.0236	–	4.754	3.006
0.30	17.896	0.0233	–	4.803	–
0.25	18.240	0.0228	3.792	4.880	2.333
0.20	18.602	0.0224	–	4.962	1.820
1/6	18.847	0.0221	–	5.017	1.439
1/7	19.020	0.0219	–	5.056	1.156
1/8	19.146	0.0218	3.725	5.085	0.9433
1/9	19.241	0.0217	–	5.107	0.7812
1/10	19.314	0.0216	–	5.124	0.6562
1/16	19.536	0.0213	3.647	5.176	–
1/20	19.598	0.0213	–	5.191	–
0.00	19.739	0.0211	3.488	5.225	–

[†]For all α^*, $u_{max}/u_m = 2$, $K_d(\infty) = 4/3$, $K_e(\infty) = 2$, and $K(\infty) = 4/3$.

The values for $\alpha^* = 0$ in Table 65 are obtained either directly or by the method of Chapter XI.

Chiranjivi and Ravi Prasad [432] measured friction factors in three elliptical ducts with α^* of 0.25, 0.5, and 0.75. Unfortunately, instead of comparing the measured friction factors with the theoretical values, they derived correlations that included an L/D_h parameter for supposedly fully developed flow. They present two different correlations for elliptical ducts, one of which includes the parameter A/A_c, and the other includes the parameter L/D_h. Since these two parameters are related to each other from the definition of D_h, two different correlations presented by Chiranjivi and Ravi Prasad are, in fact, simply related by a constant. The present authors believe that the straight duct ahead of the test section was of insufficient length to ensure fully developed flow for elliptical ducts, and the test results of Chiranjivi and Ravi Prasad are valid neither for developing flow from the entrance nor for the fully developed flow.

Someswara Rao et al. [433] also derived Eq. (385) and presented f Re factors for elliptical ducts of α^* from 0 to 1. They measured friction factors for an elliptical duct of $\alpha^* = 0.25$, which agreed with the theoretical values within $\pm 10\%$.

2. HEAT TRANSFER

a. Constant Surface Temperature, ⓉⓉ

Dunwoody [434] determined Nu_T for elliptical ducts with $\alpha^* = 1/16$, 1/8, 1/4, 1/2, and 0.80. These were obtained as the asymptotic values for his thermal entry length solution. Later Schenk and Han [435] confirmed his results for $\alpha^* = 0.25$ and 0.80. James [126] presented Nu_T for $\alpha^* = 0$. All these results are presented in Table 65 and Fig. 60.

b. Constant Axial Wall Heat Flux, Ⓗ1 and Ⓗ2

The fully developed laminar Ⓗ1 heat transfer problem for elliptical ducts with internal thermal energy sources was first investigated by Tao [79]. He analyzed the problem by employing the method of complex variables. Tyagi [63] extended Tao's work by including viscous dissipation. The closed-form formulas were presented for u, u_m, t, t_m, and Nu_{H1}. In the absence of viscous dissipation and thermal energy sources, Nu_{H1} is expressed by Tyagi as

$$Nu_{H1} = \left[\frac{3\pi}{E(m)} \right]^2 \left\{ \frac{(1 + \alpha^{*2})[(1 + \alpha^{*4}) + 6\alpha^{*2}]}{17(1 + \alpha^{*4}) + 98\alpha^{*2}} \right\} \tag{387}$$

where $E(m)$ and m are as defined on p. 248.

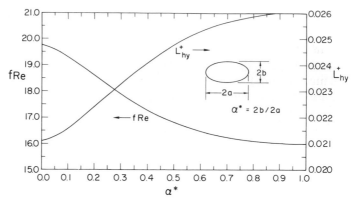

FIG. 59. Elliptical ducts: f Re and L_{hy}^+ for fully developed laminar flow (from Table 65).

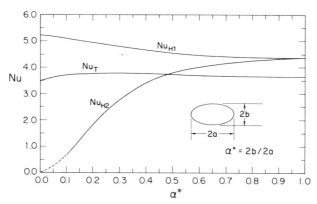

FIG. 60. Elliptical ducts: Nu_T, Nu_{H1}, and Nu_{H2} for fully developed laminar flow (from Table 65).

TABLE 66

ELLIPTICAL DUCT ($\alpha^* = 0.8$):
Nu_{T3} AND Nu_o FOR FULLY
DEVELOPED LAMINAR FLOW
(BASED ON THE RESULTS OF
SCHENK AND HAN [435])

R_w	Nu_{T3}	Nu_o
0.0	3.669	3.669
0.2	3.948	2.206
1.0	4.175	0.807
10.0	4.344	0.0978
∞	4.35	0

Nu_{H1} were evaluated from this equation by the present authors and are presented in Table 65 and Fig. 60.

Iqbal et al. [92] analyzed the (H2) problem for elliptical ducts by employing a conformal mapping method and also a variational method. Their conformal mapping results for Nu_{H2} for α^* from $1/10$ to $1/2$ are presented in Table 65 and Fig. 60.

c. Finite Wall Thermal Resistance, (T3)

Based on Schenk and Han's results for the (T3) thermal entry length solution [435], Nu_{T3} and Nu_o were determined by the present authors for an elliptical duct with $\alpha^* = 0.8$. These are presented in Table 66.

B. Thermally Developing Flow

The (T) thermal entrance problem for elliptical ducts has been analyzed by Dunwoody [434], Schenk and Han [435], Tao [205], Someswara Rao et al. [436], and James [126].

Based on his previous work with the free vibrations of membranes with elliptical boundaries, Dunwoody [434] arrived at an expression for $Nu_{x,T}$ in a double infinite series form. He evaluated numerically and tabulated necessary coefficients and eigenvalues for elliptical ducts of $\alpha^* = 1/16, 1/8, 1/4, 1/2,$ and 0.80. Schenk and Han [435] extended the work of Dunwoody for $\alpha^* = 0.25$ and 0.80 in order to check the accuracy of the results and obtain more insight into the physical aspects. The results of Schenk and Han, and Dunwoody are in excellent agreement. They then considered the finite wall thermal resistance (T3) ($R_w = 0, 0.2, 1,$ and 10) for the elliptical duct of $\alpha^* = 0.8$, and tabulated eigenvalues and constants. Tao [205] applied a variational method to solve the (T) thermal entrance problem. He presented a closed-form expression for $t_{m,T}$ in the thermal entrance region.

Someswara Rao et al. [436] solved the thermal entry length problem for short elliptical ducts with the (T) and (H1) boundary conditions. They employed the extended Lévêque method, which accounted for the circumferential variation of the slope of the linear velocity profile, to arrive at a solution that is thus not applicable to long ducts. The results do not asymptotically approach the fully developed values. Near the entrance, their results are in fair agreement with the results of Schenk and Han [435]. Someswara Rao et al. included a correction factor for the temperature-dependent viscosity in the empirical formulas for the local and average Nusselt numbers. James [126] also presented the (T) thermal entrance solution, based on the Lévêque theory, for short ducts with α^* from 1 to 0. His tabulated values for $Nu_{x,T}$ are in excellent agreement with the values of Someswara Rao et al. when the effect of temperature-dependent viscosity is eliminated.

Javeri [35] employed the Kantorowich variational method to analyze the thermal entrance problem for elliptical ducts. He considered the wall temperature uniform peripherally and varying linearly in the flow direction. He tabulated Nu_x as a function of $x^*(D_h/b)^2$ for $\alpha^* = 0.05$, 0.125, 0.25, 0.5, and 1.

Chapter X

Other Singly Connected Ducts

The remaining singly connected duct geometries analyzed in the literature are described in this chapter. Only the fully developed laminar fluid flow and heat transfer problems have been investigated for most of these geometries. The hydrodynamic entry length solution is available only for a circular sector duct. Thermal entry length solutions, based on the Lévêque method, have been obtained for the circular sector and annular sector ducts. The thermal entrance solutions for developed velocity profiles are reported for the semicircular duct only.

A. Sine Ducts

A sine duct geometry is bounded by one full wavelength sine wave curve and a straight line, as shown in Fig. 61.

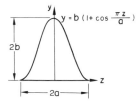

$$y = b \left(1 + \cos \frac{\pi z}{a}\right)$$

FIG. 61. A sine duct.

253

Sherony and Solbrig [36] analyzed fluid flow and heat transfer problems for sine ducts by a finite difference method. They presented f Re, $K(\infty)$, L_{hy}^{+}, Nu_{H1}, and Nu_{T} for the aspect ratios $2b/2a = 1.50$, 1.00, 0.75, 0.50, and 0.25. Their results for Nu_{T} are presented in Table 68 and Fig. 63.

Shah [116] employed a discrete least squares method to determine the coefficients[†] a_j and c_j of Eqs. (370) and (372) for sine ducts with $2 \le 2b/2a \le 1/8$. Subsequently, he obtained the fully developed flow and heat transfer characteristics of sine ducts. These results and the geometrical properties of sine ducts are presented in Tables 67 and 68 and Figs. 62 and 63. In these tables, the results for $2b/2a = 0$ and ∞ are obtained by the method of Chapter XI. For further clarification on $\bar{y}$, $\bar{y}_{max}$, $t_{w,max}^{*}$, etc., refer to pp. 229 and 207. The results for the velocity and (H1) temperature problems by Sherony and Solbrig [36] and Shah [116] are in good agreement.

Sherony and Solbrig [36] also analyzed the (T3) boundary condition for

TABLE 67

SINE DUCTS: GEOMETRICAL AND FLOW CHARACTERISTICS FOR
FULLY DEVELOPED LAMINAR FLOW (FROM SHAH [116])

$\dfrac{2b}{2a}$	$\dfrac{P}{2a}$	$\dfrac{D_h}{2a}$	$\dfrac{\bar{y}}{2a}$	$\dfrac{\bar{y}_{max}}{2a}$	$\dfrac{u_{max}}{u_m}$	$K_d(\infty)$	$K_e(\infty)$
∞	—	—	—	—	3.825	1.604	3.213
2	5.1898	.77074	0.75000	.46494	2.288	1.435	2.376
3/2	4.2315	.70897	0.56250	.40964	2.239	1.423	2.326
1	3.3049	.60516	0.37500	.33390	2.197	1.414	2.286
$\sqrt{3/2}$	3.0667	.56479	0.32476	.30773	2.191	1.414	2.283
3/4	2.8663	.52332	0.28125	.28205	2.190	1.415	2.287
1/2	2.4637	.40589	0.18750	.21347	2.211	1.429	2.334
1/4	2.1398	.23366	0.09375	.11926	2.291	1.467	2.474
1/8	2.0375	.12270	0.04688	.06173	2.357	1.496	2.582
0	2.0000	.00000	0.00000	.00000	2.400	1.512	2.648

TABLE 68

SINE DUCTS: FLOW AND HEAT TRANSFER CHARACTERISTICS FOR
FULLY DEVELOPED LAMINAR FLOW (FROM SHAH [116])
AND Nu_{T} (FROM SHERONY AND SOLBRIG [36])

$\dfrac{2b}{2a}$	$K(\infty)$	L_{hy}^{+}	fRe	Nu_{T}	Nu_{H1}	Nu_{H2}	$t_{w,max}^{*}$	$t_{w,min}^{*}$
∞	3.218	.1701	15.303	0.739	2.521	0	—	—
2	1.884	.0403	14.553	—	3.311	0.95	2.92	.002
3/2	1.806	.0394	14.022	2.60	3.267	1.38	2.93	.257
1	1.744	.0400	13.023	2.45	3.102	1.55	2.17	.398
$\sqrt{3/2}$	1.739	.0408	12.630	—	3.014	1.47	2.58	.396
3/4	1.744	.0419	12.234	2.33	2.916	1.34	2.93	.379
1/2	1.810	.0464	11.207	2.12	2.617	0.90	3.65	.266
1/4	2.013	.0553	10.123	1.80	2.213	0.33	4.16	.099
1/8	2.173	.0612	9.743	—	2.017	0.095	4.31	.030
0	2.271	.0648	9.600	1.178	1.920	0	—	—

[†] Refer to the footnote on p. 229.

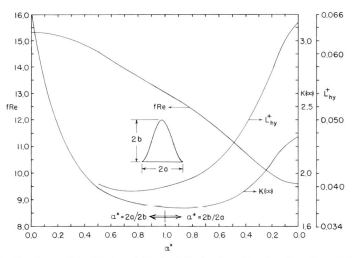

FIG. 62. Sine ducts: $f\,Re$, $K(\infty)$, and L_{hy}^+ for fully developed laminar flow (from Table 68).

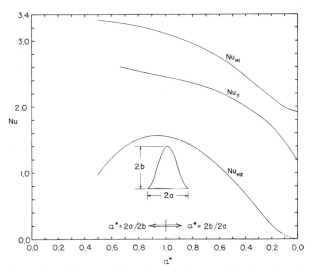

FIG. 63. Sine ducts: Nu_T, Nu_{H1}, and Nu_{H2} for fully developed laminar flow (from Table 68).[†]

a sine duct of $2b/2a = 0.5$. Their Nu_{T3} and Nu_o are presented in Table 69.

As discussed on p. 24, the (T3) boundary condition reduces to the (T) and (H2) boundary conditions for $R_w = 0$ and ∞, respectively. Hence Nu_{T3} for

---

[†] Some question remains for $2a/2b$ values below 0.5; further results are needed.

TABLE 69

Sine Duct ($2b/2a = 0.5$):
Nu_{T3} and Nu_o for Fully
Developed Laminar Flow
(from Sherony and Solbrig [36])

R_w	Nu_{T3}	Nu_o
0	2.12	2.12
0.01236	2.094	2.041
0.1236	1.974	1.587
1.236	1.524	0.529
12.36	1.508	0.0768
123.6	1.415	0.00804
∞	0.894	0

$R_w = \infty$ in Table 69 is taken as Nu_{H2} of Table 68 for $2b/2a = 0.5$. Since Nu_{H2} is lower than Nu_T for this sine duct geometry, Nu_{T3} decreases with increasing values of R_w. This trend is different from that for the circular tube, parallel plates, and elliptical duct as noted in the footnote on p. 24.

B. Trapezoidal Ducts

Shah [116] analyzed the trapezoidal duct of Fig. 64 by employing a discrete least squares method.[†] The fully developed flow and heat transfer results are presented in Tables 70 and 71 and Figs. 65 and 66. The results for for $2b/2a = 0$ (and $\phi \neq 0°$) are obtained by the method of Chapter XI. When $2a$ approaches zero (so that $2b/2a \to \infty$), the trapezoidal duct reduces to an isosceles triangular duct. When $\phi = 90°$, the trapezoidal duct reduces to a rectangular duct. The results for these limiting geometries are also included in Tables 70 and 71. It is interesting to note that the flow and heat transfer results for trapezoidal ducts with $2a/2b < 0.5$ are exceedingly sensitive to small variations in angle ϕ.

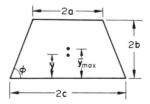

Fig. 64. A trapezoidal duct.

Chiranjivi and Sankara Rao [437] studied experimentally laminar and turbulent heat transfer through a trapezoidal duct with $2b/2a = 3.57$, $\phi = 70.1°$, and the length of base 7.94 cm. The base was heated electrically, the other three sides being insulated. Sankara Rao [438] investigated experimentally laminar fluid friction and heat transfer through three tra-

---

[†] Refer to the footnote on p. 229.

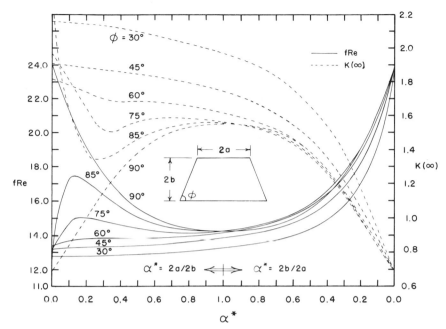

FIG. 65. Trapezoidal ducts: f Re and $K(\infty)$ for fully developed laminar flow (from Tables 70 and 71).

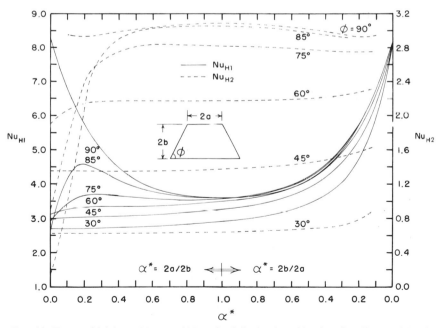

FIG. 66. Trapezoidal ducts: Nu_{H1} and Nu_{H2} for fully developed laminar flow (from Table 71).

TABLE 70: Trapezoidal Ducts: Geometrical and Flow Characteristics for Fully Developed Laminar Flow (from Shah [116])

$\dfrac{2b}{2a}$	$\dfrac{\bar{y}}{2a}$	$\dfrac{\bar{y}_{max}}{2a}$	$\dfrac{u_{max}}{u_m}$	$K_d(\infty)$	$K_e(\infty)$	$K(\infty)$	L_{hy}^+
\multicolumn{8}{c}{$\phi = 85°$}							
∞	1.90501	.74009	2.521	1.526	2.731	2.409	.0590
8	1.43804	.74010	2.186	1.368	2.147	1.557	.0318
4	1.07487	.73895	2.025	1.336	2.008	1.343	.0262
2	.70399	.66375	2.045	1.362	2.092	1.461	.0287
4/3	.52173	.52482	2.091	1.377	2.149	1.543	.0320
1	.41413	.42131	2.096	1.379	2.155	1.552	.0324
3/4	.32469	.33057	2.070	1.371	2.125	1.508	.0305
1/2	.22668	.22982	1.981	1.345	2.028	1.367	.0248
1/4	.11891	.11976	1.768	1.286	1.821	1.069	.0144
1/8	.06094	.06116	1.627	1.245	1.683	0.875	.00936
0	–	–	1.500	1.200	1.543	0.686	.00588
\multicolumn{8}{c}{$\phi = 75°$}							
∞	.62201	.45102	2.290	1.455	2.438	1.966	.0436
8	.58459	.45100	2.219	1.415	2.300	1.770	.0362
4	.52650	.44990	2.141	1.385	2.185	1.560	.0331
2	.42654	.42537	2.107	1.382	2.166	1.569	.0327
4/3	.35472	.37226	2.112	1.385	2.177	1.584	.0332
1	.30261	.32029	2.099	1.381	2.163	1.563	.0323
3/4	.25257	.26619	2.063	1.370	2.121	1.502	.0298
1/2	.18940	.19707	1.968	1.343	2.020	1.354	.0240
1/4	.10793	.11023	1.763	1.286	1.819	1.066	.0142
1/8	.05794	.05858	1.627	1.246	1.684	0.877	.00936
0	–	–	1.500	1.200	1.543	0.686	.00588
\multicolumn{8}{c}{$\phi = 60°$}							
∞	.28868	.28868	2.222	1.429	2.338	1.818	.0398
8	.28366	.28861	2.205	1.419	2.303	1.770	.0377
4	.27315	.28749	2.181	1.409	2.267	1.716	.0367
2	.24819	.27568	2.162	1.404	2.248	1.687	.0360
4/3	.22445	.25304	2.146	1.400	2.232	1.664	.0350
1	.20374	.22860	2.119	1.392	2.201	1.618	.0331
3/4	.18072	.20001	2.071	1.378	2.148	1.539	.0299
1/2	.14666	.15841	1.969	1.349	2.039	1.379	.0239
1/4	.09292	.09700	1.766	1.291	1.833	1.084	.0143
1/8	.05339	.05462	1.634	1.251	1.700	0.897	.00950
0	–	–	1.500	1.200	1.543	0.686	.00588
\multicolumn{8}{c}{$\phi = 45°$}							
∞	.16667	.19584	2.264	1.445	2.400	1.909	.0421
8	.16558	.19575	2.258	1.442	2.388	1.893	.0414
4	.16296	.19480	2.250	1.439	2.377	1.876	.0410
2	.15556	.18829	2.232	1.434	2.358	1.847	.0400
4/3	.14719	.17716	2.206	1.427	2.328	1.803	.0381
1	.13889	.16483	2.169	1.418	2.290	1.744	.0355
3/4	.12857	.14945	2.109	1.398	2.218	1.639	.0317
1/2	.11111	.12495	1.998	1.367	2.099	1.464	.0251
1/4	.07778	.08333	1.787	1.303	1.875	1.142	.0151
1/8	.04815	.05000	1.652	1.266	1.737	0.943	.00996
0	–	–	1.500	1.200	1.543	0.686	.00588
\multicolumn{8}{c}{$\phi = 30°$}							
∞	.09623	.12552	2.380	1.489	2.571	2.165	.0490
8	.09600	.12543	2.375	1.484	2.558	2.146	.0489
4	.09541	.12482	2.371	1.484	2.555	2.141	.0485
2	.09351	.12170	2.352	1.482	2.540	2.115	.0469
4/3	.09105	.11676	2.310	1.468	2.486	2.038	.0442
1	.08834	.11121	2.266	1.457	2.442	1.969	.0408
3/4	.08459	.10397	2.200	1.441	2.373	1.864	.0363
1/2	.07735	.09148	2.073	1.404	2.229	1.651	.0287
1/4	.06024	.06699	1.850	1.344'	1.988	1.288	.0174
1/8	.04103	.04361	1.678	1.272	1.768	0.992	.0112
0	–	–	1.500	1.200	1.543	0.686	.00588

pezoidal ducts with $2b/2a = 2.45$ ($\phi = 80°$), 3.09 ($\phi = 70°$), and 5.46 ($\phi = 70°$). He maintained the Ⓣ boundary condition during the experiments. In both of the foregoing studies [437,438], water and glycerol–water mixtures were used as test fluids. The experimental results were correlated as Colburn factor j vs. Reynolds number Re, with the effect of temperature-dependent viscosity included with the j factor. The isothermal duct section upstream of the test section was not sufficiently long to ensure a fully developed laminar velocity profile at the entrance of the test section. Hence their experimental correlation applies to thermally developing laminar flow with partially developed velocity profiles.

TABLE 71: TRAPEZOIDAL DUCTS: FLOW AND HEAT TRANSFER CHARACTERISTICS FOR
FULLY DEVELOPED LAMINAR FLOW (FROM SHAH [116])

$\frac{2b}{2a}$	fRe	Nu_{H1}	Nu_{H2}	$t^*_{w,max}$	$t^*_{w,min}$
		$\phi = 85°$			
∞	12.474	2.446	0.08	–	–
8	17.474	4.366	1.22	–	–
4	16.740	4.483	2.54	2.24	.297
2	15.015	3.896	3.01	1.55	.554
4/3	14.312	3.636	3.05	1.52	.687
1	14.235	3.608	3.05	1.55	.716
3/4	14.576	3.736	3.04	1.56	.611
1/2	15.676	4.175	2.98	1.64	.478
1/4	18.297	5.363	2.91	1.87	.305
1/8	20.599	6.501	2.89	2.18	.190
0	24.000	8.235	–	–	–
		$\phi = 75°$			
∞	13.065	2.910	0.851	3.73	.127
8	14.907	3.520	1.90	2.43	.369
4	14.959	3.720	2.57	1.61	.511
2	14.340	3.610	2.82	1.73	.611
4/3	14.118	3.542	2.83	1.76	.704
1	14.252	3.594	2.82	1.78	.612
3/4	14.697	3.766	2.81	1.79	.533
1/2	15.804	4.219	2.78	1.85	.432
1/4	18.313	5.371	2.75	2.03	.286
1/8	20.556	6.482	2.76	2.30	.181
0	24.000	8.235	–	–	–
		$\phi = 60°$			
∞	13.333	3.111	1.89	1.79	.515
8	13.867	3.284	2.09	1.96	.520
4	13.916	3.348	2.16	2.03	.514
2	13.804	3.350	2.17	2.07	.540
4/3	13.888	3.390	2.17	2.07	.490
1	14.151	3.495	2.17	2.07	.442
3/4	14.637	3.691	2.18	2.09	.401
1/2	15.693	4.140	2.20	2.13	.342
1/4	18.053	5.247	2.26	2.26	.242
1/8	20.304	6.341	2.31	2.47	.158
0	24.000	8.235	–	–	–
		$\phi = 45°$			
∞	13.153	2.982	1.35	2.29	.364
8	13.301	3.030	1.35	2.29	.364
4	13.323	3.048	1.35	2.30	.344
2	13.364	3.081	1.35	2.30	.326
4/3	13.541	3.155	1.36	2.30	.294
1	13.827	3.268	1.37	2.31	.277
3/4	14.260	3.469	1.40	2.32	.263
1/2	15.206	3.888	1.44	2.35	.236
1/4	17.397	4.943	1.55	2.46	.180
1/8	19.743	6.034	1.61	2.65	.122
0	24.000	8.235	–	–	–
		$\phi = 30°$			
∞	12.744	2.680	.621	2.45	.167
8	12.760	2.697	.625	2.45	.164
4	12.782	2.704	.624	2.44	.163
2	12.875	2.736	.622	2.44	.146
4/3	13.012	2.821	.635	2.44	.141
1	13.246	2.919	.640	2.46	.137
3/4	13.599	3.077	.652	2.47	.133
1/2	14.323	3.436	.685	2.55	.126
1/4	16.284	4.349	.744	2.83	.103
1/8	18.479	5.569	.888	2.89	.080
0	24.000	8.235	–	–	–

C. Rhombic Ducts

Iqbal *et al.* [76,92] analyzed laminar flow through rhombic ducts by
employing both finite difference and variational methods. They presented
graphically $2f$ Re and tabulated Nu_{H1} for $20° \leq \phi \leq 90°$ [76]. They also
tabulated Nu_{H2} for $60° \leq \phi \leq 90°$ [92].
Shah [116] studied fully developed laminar flow through rhombic ducts
by employing a discrete least squares method.[†] Nu_{H1} and Nu_{H2} of Iqbal

---

[†] Refer to the footnote on p. 229.

TABLE 72

RHOMBIC DUCTS: FLOW CHARACTERISTICS FOR
FULLY DEVELOPED LAMINAR FLOW (FROM SHAH [116])

ϕ	$\dfrac{u_{max}}{u_m}$	$K_d(\infty)$	$K_e(\infty)$	$K(\infty)$	L_{hy}^+
90	2.096	1.378	2.154	1.551	0.0324
80	2.102	1.381	2.163	1.564	0.0327
70	2.120	1.389	2.190	1.603	0.0336
60	2.151	1.402	2.239	1.673	0.0353
50	2.199	1.422	2.311	1.778	0.0380
45	2.230	1.436	2.361	1.850	0.0397
40	2.266	1.448	2.411	1.925	0.0419
30	2.359	1.481	2.541	2.120	0.0477
20	2.493	1.521	2.713	2.384	0.0570
10	2.689	1.562	2.908	2.693	0.0732
0	3.000	1.600	3.086	2.971	0.1048

TABLE 73

RHOMBIC DUCTS: FLOW AND HEAT TRANSFER
CHARACTERISTICS FOR FULLY DEVELOPED LAMINAR FLOW
(FROM SHAH [116])

ϕ	fRe	Nu_{H1}	Nu_{H2}	$t_{w,max}^*$	$t_{w,min}^*$
90	14.227	3.608	3.09	1.39	0.769
80	14.181	3.581	2.97	1.65	0.743
70	14.046	3.500	2.64	1.86	0.671
60	13.830	3.367	2.16	2.05	0.565
50	13.542	3.188	1.62	2.21	0.439
45	13.381	3.080	1.34	2.27	0.372
40	13.193	2.969	1.09	2.32	0.307
30	12.803	2.722	0.624	2.40	0.185
20	12.416	2.457	0.279	2.45	0.089
10	12.073	2.216	0.070	2.46	0.023
0	12.000	2.059	0	-	-

et al. and Shah are in excellent agreement. The flow and heat transfer results of Shah are presented in Tables 72 and 73 and Figs. 67 and 68. In this table, the results for $\phi = 0°$ were obtained by the method of Chapter XI.

D. Quadrilateral Ducts

Nakamura *et al.* [98] analyzed fully developed laminar fluid flow and heat transfer problems for arbitrary polygonal ducts. They employed a finite difference method with a uniform triangular grid of arbitrary shape. This method is the same as the one used by Nakamura *et al.* in analyzing arbitrary triangular ducts [97]. Specific examples were worked out for the quadrilateral duct of Fig. 69. The rectangular, trapezoidal, and rhombic ducts are the special cases of the quadrilateral duct. The numerical results obtained by Nakamura *et al.* for some quadrilateral ducts are presented in Table 74, where all angles are in degrees.

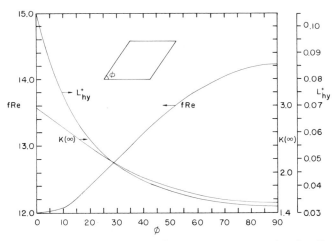

FIG. 67. Rhombic ducts: f Re, $K(\infty)$, and L_{hy}^+ for fully developed laminar flow (from Tables 72 and 73).

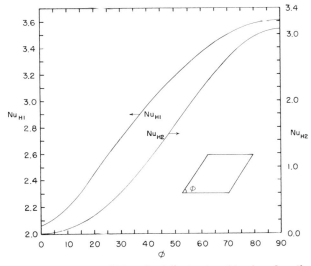

FIG. 68. Rhombic ducts: Nu_{H1} and Nu_{H2} for fully developed laminar flow (from Table 73).

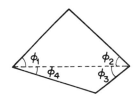

FIG. 69. A quadrilateral duct.

TABLE 74

QUADRILATERAL DUCTS: f Re, $K(\infty)$, Nu_{H1},
AND Nu_T FOR FULLY DEVELOPED LAMINAR FLOW
(FROM NAKAMURA et al. [98])

ϕ_1	ϕ_2	ϕ_3	ϕ_4	fRe	$K(\infty)$	Nu_{H1}	Nu_T
60	70	45	32.23	14.16	1.654	3.45	2.80
50	60	30	21.67	14.36	1.612	3.55	2.90
60	30	45	71.57	14.69	1.522	3.72	3.05
60	30	60	79.11	14.01	1.707	3.35	2.68

Nakamura et al. [98] also computed f Re factors for the foregoing four geometries by employing a finite difference method with a rectangular grid and a square grid and also by employing the discrete least squares method of Sparrow and Haji-Sheikh [114]. After making a comparison of the results obtained by various methods, they discussed the limitations of each method. They also computed f Re and Nu_{H1} for rectangular ducts of $\alpha^* = 1, 0.5$, and 0.25. These results are in good agreement with those in Table 42.

E. Regular Polygonal Ducts

Tao [82] and Hsu [439] analyzed fully developed laminar fluid flow and heat transfer problems for a hexagonal duct, while Cheng [69,106,107] and Shih [108] analyzed the general n-sided regular polygonal ducts.

Tao [82] employed the conformal mapping method to analyze fully developed laminar flow through a hexagonal duct with linearly varying wall temperature and arbitrary thermal energy generation. He presented the dimensionless heat flux as a function of thermal energy generation within the fluid.

Hsu [439] investigated fully developed laminar heat transfer for the hexagonal duct with and without thermal energy generation within the fluid. He considered two sets of boundary conditions: (1) heat fluxes from three pairs of mutually opposite side walls being pairwise the same, but for each pair, the heat flux magnitude being different, and (2) the heat flux on all six walls being uniform and equal. The velocity and temperature fields were determined by a finite difference method. The local Nusselt numbers along the walls were presented graphically, and the average Nusselt numbers, with and without thermal energy generation, were tabulated. Nu_{H2} for the hexagonal duct was reported as 3.795, which is 1.7% lower than that listed in Table 75.

Cheng [106] employed the 9-point matching method to determine u, τ, f Re, t, and Nu_{H1} for the n-sided regular polygonal ducts. He presented

f Re, τ_{max}, τ_{min}, and Nu_{H1} in tabular form for n varying from 3 to 20. His partial results are listed in Table 75. Cheng [69] extended his work by including the effect of uniform thermal energy generation and viscous dissipation within the fluid. The dimensionless heat flux and Nusselt numbers for the (H1) boundary condition were determined by the 10-point matching method and were presented graphically. Shih [108] independently evaluated f Re for regular polygonal ducts ($3 \leq n \leq 8$) using the 12-point matching method. His results are presented in Table 75. Cheng also employed the 12-point matching method to evaluate fully developed Nu_{H2} for regular polygonal ducts [107]. These too are presented in Table 75. The results of Table 75 are also shown in Fig. 70.

TABLE 75

Regular Polygonal Ducts: f Re, Nu_{H1},
Nu_{H2}, and Nu_T for Fully Developed
Laminar Flow (from Cheng [106,107]
and Shih [108])

n	fRe	Nu_{H1}	Nu_{H2}	Nu_T
3	13.333	3.111	1.892	2.47
4	14.227	3.608	3.091	2.976
5	14.737	3.859	3.605	–
6	15.054	4.002	3.862	–
7	15.31	4.102	4.009	–
8	15.412	4.153	4.100	–
9	15.52	4.196	4.159	–
10	15.60	4.227	4.201	–
20	15.88	4.329	4.328	–
∞	16.000	4.364	4.364	3.657

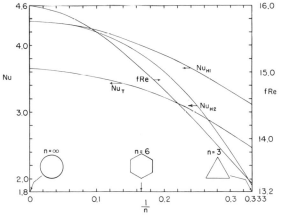

Fig. 70. Regular (n-sided) polygonal ducts: f Re, Nu_{H1}, Nu_{H2}, and Nu_T for fully developed laminar flow (from Table 75).

F. Circular Sector Ducts

1. Fully Developed Flow

Eckert and Irvine [419] and Eckert *et al.* [27] analyzed fully developed laminar flow through the circular sector duct (also referred to as a wedge-shaped duct) of Fig. 71. They derived the velocity profile, based on the torsion theory [72], as

$$u = \frac{c_1}{4}\left\{ r^2\left(1 - \frac{\cos 2\theta}{\cos 2\phi}\right) - \frac{16a^2(2\phi)^2}{\pi^3} \right.$$

$$\left. \times \sum_{n=1,3,\ldots}^{\infty} (-1)^{(n+1)/2}\left(\frac{r}{a}\right)^{n\pi/2\phi} \frac{\cos(n\pi\theta/2\phi)}{n[n+(4\phi/\pi)][n-(4\phi/\pi)]} \right\} \quad (388)$$

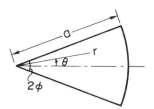

Fig. 71. A circular sector duct.

In addition to the velocity profile, Chiranjivi and Vidyanidhi [440] presented a formula for the mean velocity u_m that was found to be in error. The correct form is

$$u_m = -\frac{c_1}{16}\frac{a^2}{\phi}\left\{ (\tan 2\phi - 2\phi) - \frac{2048\phi^4}{\pi^5} \right.$$

$$\left. \times \sum_{n=1,3,\ldots}^{\infty} \frac{1}{n^2[n+(4\phi/\pi)]^2[n-(4\phi/\pi)]} \right\} \quad (389)$$

Eckert *et al.* [27,419] graphically presented the velocity profile u/u_m along the straight walls and centerline for the apex angle 2ϕ varying from 11 to 60°. They also graphically presented f_D Re for $11° \leq 2\phi \leq 60°$. Sparrow and Haji-Sheikh [423] extended this work and plotted f Re and $K(\infty)$ for $0° \leq 2\phi \leq 180°$. Their $K(\infty)$ are presented in Table 76 and Fig. 72 [441]. With u_m of Eq. (389), f Re for circular sector ducts can be expressed in closed form as

$$f\,\mathrm{Re} = -2c_1 a^2 \phi^2 / u_m(1 + \phi)^2 \quad (390)$$

TABLE 76

CIRCULAR SECTOR DUCTS: f Re FROM Eq. (390),
$K(\infty)$ AND Nu_{H1} FROM HAJI-SHEIKH [441],
AND Nu_{H2} FROM HU [443], FOR
FULLY DEVELOPED LAMINAR FLOW

2ϕ	fRe	$K(\infty)$	Nu_{H1}	Nu_{H2}
0	12.000	2.971	2.059	–
8	12.411	2.480	2.384	–
10	12.504	–	–	0.081
15	12.728	2.235	2.619	0.195
20	12.936	–	–	0.362
30	13.310	1.855	3.005	0.838
36	13.510	–	–	1.174
40	13.635	–	–	1.400
45	13.782	1.657	3.27	1.667
60	14.171	1.580	3.479	–
72	14.435	–	–	2.608
80	14.592	1.530	3.671	–
100	14.929	1.504	3.806	–
120	15.200	1.488	3.906	2.898
160	15.611	1.468	4.04	–
180	15.767	1.463	4.089	2.923

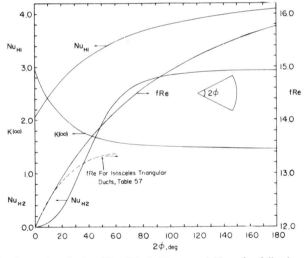

FIG. 72. Circular sector ducts: f Re, $K(\infty)$, Nu_{H1}, and Nu_{H2} for fully developed laminar flow (from Table 76).

The f Re factors were calculated by the present authors from Eq. (390) and are presented in Table 76 and Fig. 72.

It is interesting to note that the velocity profile and friction factors for the circular sector duct and the isosceles triangular duct agree very well for the apex angle 2ϕ of less than about 20°.

Based on the analogy with the deflection of a uniformly loaded thin plate simply supported around the rim, Eckert *et al.* [27] arrived at the following (H1) temperature distribution for circular sector ducts:

$$(t - t_w)_{H1} = c_1 c_2 a^4 \sum_{n=1,3,\ldots}^{\infty} (-1)^{(n+1)/2} \left\{ \frac{4r^4}{n\pi[16 - (n\pi/2\phi)^2][4 - (n\pi/2\phi)^2]a^4} \right.$$

$$+ \left(\frac{r}{a}\right)^{n\pi/2\phi} \frac{[6 + (n\pi/2\phi)]}{n\pi[16 - (n\pi/2\phi)^2][2 + (n\pi/2\phi)][1 + (n\pi/2\phi)]}$$

$$\left. - \left(\frac{r}{a}\right)^{(n\pi/2\phi)+2} \frac{1}{n\pi[4 - (n\pi/2\phi)^2][1 + (n\pi/2\phi)]} \right\} \cos\left(\frac{n\pi}{2\phi}\theta\right) \quad (391)$$

Eckert *et al.* used the above equation to present graphically the centerline ($\theta = 0$) temperature distribution and local wall heat flux distribution for circular sector ducts. They also presented graphically Nu_{H1} for $10° \le 2\phi \le 60°$. Sparrow and Haji-Sheikh [423] extended the results of Eckert *et al.* and plotted Nu_{H1} for $0° \le 2\phi \le 180°$. Their Nu_{H1} are presented in Table 76 and Fig. 72 [441].

Sparrow and Siegel [119] formulated a variational method to analyze laminar flow and heat transfer in a noncircular duct. As an illustrative example, they determined f Re and Nu_{H1} for a circular sector duct with $2\phi = 33.4°$. The computed results are in good agreement with those of Eckert and Irvine [419].

Eckert *et al.* [27] also analyzed the (H2) boundary condition for circular sector ducts. They presented a generalized solution for the temperature distribution with the constants determined to satisfy the boundary condition. The local fluid and wall temperature distributions and Nu_{H2} were presented graphically. Their Nu_{H2} agree within 6% of those in Table 76.

Hu and Chang [129,442] analyzed the fully developed laminar (H2) problem for a circular tube having n longitudinal fins, as summarized on p. 367. Circular tubes with n such full fins ($l = a$) have n circular sector passages having an apex angle $2\phi = 2\pi/n$. The Nu_{H2} of Hu [129,443], based on his full-fin geometries, are presented in Table 76.

2. HYDRODYNAMICALLY DEVELOPING FLOW

Wendt and Wiginton [170] employed a stretched coordinate linearization technique of Wiginton and Wendt [165] and Wiginton [169] to analyze the hydrodynamic entry length problem for circular sector ducts. As an example, they numerically evaluated Δp^* for a circular sector duct of $2\phi =$

$\pi/4$ (45°) and presented it graphically as a function of $4\phi x^+/(1 + \phi)^2$. However, they erroneously evaluated two integrals of their Eq. (1) in [170] and hence their Δp^* values are incorrect [444].

3. THERMALLY DEVELOPING AND HYDRODYNAMICALLY DEVELOPED FLOW

Chiranjivi and Vidyanidhi [440] employed the Lévêque method to obtain the thermal entrance solution for the circular sector duct. They presented $h_{m,T}$ and j factors for the circular sector duct when the curved wall was maintained at a constant temperature, different from the temperature of the entering fluid, and when the straight walls were adiabatic.

G. Circular Segment Ducts

1. FULLY DEVELOPED FLOW

Sparrow and Haji-Sheikh [114] employed a least squares point-matching method and analyzed fully developed laminar flow through circular segment ducts. The velocity and temperature profiles were presented in a series form, with the first 17 related constants tabulated. Their f Re, $K(\infty)$, Nu_{H1}, and Nu_{H2} are presented in Table 77 and Fig. 73.

Subrahmanyam [445] investigated experimentally laminar flow through three different circular segment ducts. The bottom side or the flat boundary was heated electrically. Water and glycerol–water mixture were used as heat transfer fluids. The experimental results were correlated as Colburn factor j (including the Sieder–Tate viscosity correction) vs. Reynolds number Re. Subrahmanyam also measured the pressure drops and correlated the experimental results as Fanning friction factor vs. Reynolds number. The isothermal duct section upstream of the test section was not long enough to ensure a fully developed velocity profile at the entrance of the test section. This may in part be the reason for the experimental friction factors being two to three times higher than the theoretical values for fully developed flow.

2. THERMALLY DEVELOPING AND HYDRODYNAMICALLY DEVELOPED FLOW

The limiting case of a circular tube with a twisted tape, when the pitch of the twisted tape approaches infinity, is a semicircular duct. The thermal entry length solution for a semicircular duct thus provides a lower bound for the circular tube with a twisted tape. Hong and Bergles [211,446] obtained the thermal entry length solutions for the semicircular duct for

TABLE 77

CIRCULAR SEGMENT DUCTS: f Re, $K(\infty)$, Nu_{H1},
AND Nu_{H2} FOR FULLY DEVELOPED LAMINAR
FLOW (FROM SPARROW AND HAJI-SHEIKH [114])

2ϕ	fRe	$K(\infty)$	Nu_{H1}	Nu_{H2}
0	15.555	1.740	3.580	0
10	15.558	1.739	3.608	0.013
20	15.560	1.734	3.616	0.052
40	15.575	1.715	3.648	0.207
60	15.598	1.686	3.696	0.456
80	15.627	1.650	3.756	0.785
120	15.690	1.571	3.894	1.608
180	15.767	1.463	4.089	2.923
240	15.840	1.385	4.228	3.882
300	15.915	1.341	4.328	4.296
360	16.000	1.333	4.364	4.364

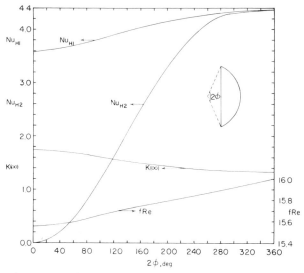

FIG. 73. Circular segment ducts: f Re, $K(\infty)$, Nu_{H1}, and Nu_{H2} for fully developed laminar flow (from Table 77).

two boundary conditions. In the axial direction, they considered constant wall heat flux; in the peripheral direction, they considered (1) constant wall temperature all along the periphery, and (2) constant wall temperature along the semicircular arc and zero heat flux (adiabatic) along the diameter. They analyzed the momentum equation by employing a successive overrelaxation finite difference method. They solved the energy equation by the DuFort–Frankel method. The local Nusselt numbers obtained for these two boundary conditions are presented in Table 78.

TABLE 78

Semicircular Duct: $Nu_{x,H1}$ as a Function of x^* for a Fully Developed Velocity Profile (from Hong and Bergles [446])

x^*	$Nu_{x,H1}$		x^*	$Nu_{x,H1}$	
	⌓	⌓		⌓	⌓
0.000458	17.71	17.43	0.0279	4.767	4.339
0.000954	13.72	13.41	0.0351	4.562	4.037
0.00149	11.80	11.37	0.0442	4.429	3.830
0.00208	10.55	10.08	0.0552	4.276	3.686
0.00271	9.605	9.141	0.0686	4.217	3.543
0.00375	8.475	8.127	0.0849	4.156	3.425
0.00493	7.723	7.375	0.105	4.124	3.330
0.00627	7.137	6.788	0.130	4.118	3.265
0.00777	6.556	6.312	0.159	4.108	3.208
0.00946	6.300	5.912	0.196	–	3.171
0.0128	5.821	5.368	0.241	–	3.161
0.0168	5.396	4.935	0.261	–	3.160
0.0217	5.077	4.579	∞	4.089	3.160

H. Annular Sector Ducts

Sparrow et al. [447] analyzed laminar flow through the annular sector duct (also referred to as a truncated sectorial duct) of Fig. 74. They determined the velocity profile by the method of separation of variables and the use of linear superposition. The local velocity u and the mean velocity u_m were expressed as

$$-\frac{u}{-\frac{1}{4}(c_1 r_i^2)} = -\left(\frac{r}{r_i}\right)^2 + \frac{2}{\pi}\sum_{n=1}^{\infty}\frac{1/n}{1+(2\beta^*/n\pi)^2}\left[1-\frac{(-1)^n}{r^{*2}}\right]$$

$$\times \frac{\cosh[(n\pi/\beta^*)(\phi-\theta)]}{\cosh(n\pi/\beta^*)\phi}\sin\frac{n\pi X}{\beta^*} + \frac{2}{\pi}\sum_{n=1}^{\infty}\frac{1}{n}[1-(-1)^n]$$

$$\times \left[\frac{\sinh[(n\pi/2\phi)(\beta^*-X)]+(1/r^{*2})\sinh(n\pi X/2\phi)}{\sinh(n\pi\beta^*/2\phi)}\right]\sin\frac{n\pi\theta}{2\phi}$$

(392)

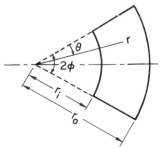

FIG. 74. An annular sector duct.

where

$$X = \ln \frac{r}{r_i}, \qquad \beta^* = \ln \frac{1}{r^*} \tag{393}$$

$$\frac{u_m}{-[c_1 r_i{}^4 / 4\phi(r_o{}^2 - r_i{}^2)]} = \frac{\phi}{2}\left[1 - \frac{1}{r^{*4}}\right] + \frac{4}{\pi}\sum_{n=1}^{\infty}\frac{(n\pi/\beta^*)^2\{1 - [(-1)^n/r^{*2}]\}^2}{n[4 + (n\pi/\beta^*)^2]^2}$$

$$\times \tanh\frac{n\pi\phi}{\beta^*} + \frac{2}{\pi}\sum_{n=1}^{\infty}\frac{(1/n)[1 - (-1)^n]^2}{[4 - (n\pi/2\phi)^2]\sinh(n\pi\beta^*/2\phi)}$$

$$\times \left\{\frac{4\phi}{n\pi}\left[\frac{1}{r^{*4}} - 1\right]\sinh\frac{n\pi\beta^*}{2\phi}\right.$$

$$\left. - \left[\frac{1}{r^{*4}} + 1\right]\cosh\frac{n\pi\beta^*}{2\phi} + \frac{2}{r^{*2}}\right\} \tag{394}$$

TABLE 79

ANNULAR SECTOR DUCTS: fRe FOR FULLY DEVELOPED LAMINAR FLOW
(FROM SHAH AND LONDON [13])

r^*	fRe								
	$2\phi=5°$	$10°$	$15°$	$20°$	$30°$	$40°$	$50°$	$60°$	$90°$
0.01	12.50	12.735	12.939	13.124	13.466	13.765	14.027	14.256	14.800
0.05	13.476	13.537	13.670	13.796	14.020	14.212	14.376	14.518	14.849
0.10	14.490	14.470	14.500	14.527	14.576	14.616	14.650	14.682	14.804
0.15	15.454	15.305	15.207	15.118	14.969	14.850	14.761	14.704	14.729
0.20	16.335	16.023	15.775	15.557	15.200	14.932	14.748	14.641	14.687
0.25	17.114	16.608	16.194	15.840	15.281	14.895	14.658	14.546	14.712
0.30	17.779	17.052	16.462	15.971	15.236	14.778	14.540	14.467	14.821
0.35	18.318	17.349	16.579	15.960	15.095	14.623	14.434	14.438	15.020
0.40	18.754	17.502	16.554	15.828	14.894	14.471	14.378	14.486	15.306
0.45	19.017	17.502	16.396	15.594	14.671	14.359	14.398	14.626	15.678
0.50	19.139	17.358	16.124	15.292	14.467	14.322	14.519	14.873	16.129
0.55	19.114	17.077	15.761	14.955	14.323	14.392	14.758	15.233	16.654
0.60	18.935	16.669	15.338	14.630	14.283	14.596	15.131	15.711	17.247
0.65	18.592	16.149	14.899	14.372	14.393	14.962	15.648	16.312	17.904
0.70	18.069	15.549	14.510	14.252	14.703	15.515	16.321	17.034	18.619
0.75	17.346	14.929	14.267	14.364	15.266	16.277	17.154	17.879	19.389
0.80	16.411	14.412	14.319	14.829	16.141	17.271	18.155	18.847	20.213
0.85	15.306	14.251	14.898	15.808	17.384	18.514	19.330	19.940	21.090
0.90	14.342	14.971	16.352	17.496	19.058	20.034	20.694	21.169	22.031
0.95	15.055	17.619	19.179	20.153	21.288	21.932	22.351	22.56	–

r^*	120°	150°	180°	210°	240°	270°	300°	330°	350°
0.01	15.190	15.479	15.704	15.888	16.047	16.191	16.324	16.451	16.532
0.05	15.105	15.339	15.575	15.818	16.063	16.306	16.541	16.765	16.908
0.10	15.007	15.288	15.616	15.960	16.301	16.627	16.932	17.215	17.391
0.15	14.988	15.374	15.806	16.237	16.645	17.021	17.364	17.675	17.866
0.20	15.066	15.572	16.095	16.591	17.043	17.449	17.811	18.133	18.329
0.25	15.242	15.859	16.455	16.994	17.472	17.891	18.259	18.583	18.779
0.30	15.505	16.216	16.863	17.429	17.917	18.338	18.704	19.023	19.214
0.35	15.845	16.628	17.307	17.883	18.371	18.786	19.143	19.452	19.635
0.40	16.251	17.082	17.776	18.350	18.828	19.231	19.574	19.868	20.043
0.45	16.713	17.571	18.264	18.825	19.286	19.671	19.995	20.273	20.437
0.50	17.225	18.088	18.764	19.304	19.742	20.104	20.408	20.666	20.818
0.55	17.778	18.626	19.275	19.785	20.195	20.530	20.810	21.047	21.186
0.60	18.367	19.181	19.793	20.266	20.643	20.949	21.203	21.417	21.542
0.65	18.988	19.752	20.315	20.746	21.085	21.360	21.587	21.776	21.887
0.70	19.637	20.335	20.841	21.223	21.523	21.763	21.960	22.125	22.221
0.75	20.312	20.929	21.369	21.699	21.955	22.159	22.326	22.465	22.545
0.80	21.012	21.533	21.900	22.172	22.382	22.549	22.684	22.797	22.862
0.85	21.736	22.149	22.436	22.648	22.810	22.939	23.043	23.130	23.180
0.90	22.499	22.795	23.000	23.152	23.269	23.363	23.440	–	–

Sparrow *et al.* subsequently determined the f Re factors from Eq. (87), employing u_m of Eq. (394) and D_h from its definition as

$$\frac{D_h}{r_i} = \frac{4\phi[(r_o/r_i)^2 - 1]}{2\phi(r_o/r_i + 1) + 2(r_o/r_i - 1)} \tag{395}$$

Sparrow *et al.* presented graphically f Re factors for r_i/r_o varying from 0.05 to 0.95, and 2ϕ varying from 0 to 360°. They also tabulated f Re factors for smaller values of 2ϕ (5–60°) for $0.05 \le r^* \le 0.95$.

The present authors computed the f Re factors for annular sector ducts, by employing up to 1200 terms in the series of Eq. (394), for a wide range of 2ϕ and r^*. These results are presented in Table 79 and Fig. 75. Note that Eq. (394) assumes an indeterminate form for $n\pi/2\phi = 2$. An appropriate formula was derived for this case.

Krishnamurty *et al.* [448] investigated the Ⓣ thermal entrance problem for annular sector ducts. They employed the Lévêque method and derived the local dimensionless heat flux $\Phi_{x,T}$ as

$$\Phi_{x,1} = 1.067(x^*)^{-1/3}(\mu/\mu_w)^{0.14}F \tag{396}$$

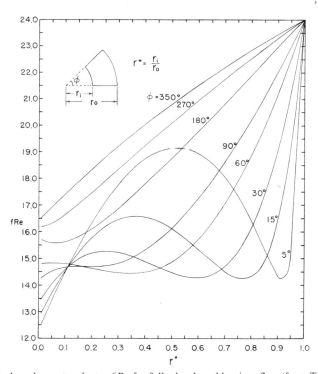

FIG. 75. Annular sector ducts: f Re for fully developed laminar flow (from Table 79).

where F is a function dependent on the local velocity gradient at the wall. Closed-form expressions are presented for F at each of the four walls of the annular sector duct. Extensive tabular values of F and the integrals of the necessary related functions are presented for a wide range of r^* and 2ϕ.

I. Moon-Shaped Ducts

A duct formed by two circular arcs, as shown by the solid lines in Fig. 76, is designated as a moon-shaped duct. Its geometrical properties are

$$P = 2(2a + b)\phi \tag{397}$$

$$A_c = (2a^2 - b^2)\phi + a^2 \sin 2\phi \tag{398}$$

$$D_h = 2a\left[\frac{(2 - \alpha^{*2})\phi + \sin 2\phi}{(2 + \alpha^*)\phi}\right] \tag{399}$$

$$\cos\phi = \frac{b}{2a}, \qquad \alpha^* = \frac{b}{a} \tag{400}$$

Based on the analogy with torsion theory [72], the fully developed laminar velocity profile for this duct is expressed by Shah and London [13] as

$$u = \frac{c_1}{4}(r^2 - b^2)\left(1 - \frac{2a\cos\theta}{r}\right) \tag{401}$$

and the mean velocity as

$$u_m = \frac{c_1 a^2\left(\frac{1}{2}\alpha^{*4} + 2\alpha^{*2} - 1\right)\phi - \frac{8}{3}\alpha^{*3}\sin\phi + (\alpha^{*2} - \frac{2}{3})\sin 2\phi - \frac{1}{12}\sin 4\phi}{(2 - \alpha^{*2})\phi + \sin 2\phi} \tag{402}$$

The friction factor is then determined from Eq. (87)

$$f\,\mathrm{Re} = -\frac{c_1 D_h^2}{2u_m}$$

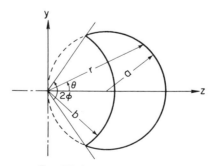

Fig. 76. A moon-shaped duct.

TABLE 80

Moon-Shaped Ducts: f Re for
Fully Developed Laminar Flow
(from Shah and London [13])

2ϕ	f Re	2ϕ	f Re
10	15.552	120	15.027
20	15.540	140	14.928
40	15.492	160	15.037
60	15.413	170	15.321
80	15.304	176	15.657
100	15.169	180	16.000

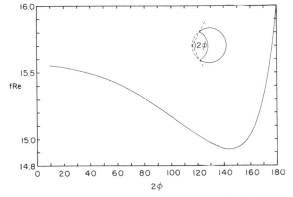

Fig. 77. Moon-shaped ducts: f Re for fully developed laminar flow (from Table 80).

where D_h and u_m are substituted from Eqs. (399) and (402). The fully developed laminar f Re factors for this duct are presented in Table 80 and Fig. 77 [13].

J. Circular Ducts with Diametrically Opposite Flat Sides

Cheng and Jamil [111] employed a point-matching method and analyzed fully developed laminar flow through a circular duct with diametrically opposite flat sides (refer to Fig. 78). The total number of matched points ranged from 36 for $2\phi = 10°$ to 73 for $2\phi = 170°$. Their f Re and Nu_{H1} are presented in Table 81 and Fig. 78 [449]. f Re and Nu_{H1} for $2\phi = 0°$ are determined by the method of Chapter XI.

K. Rectangular Ducts with Semicircular Short Sides

Zarling [134] employed a method that combined the Schwarz–Neumann successive approximations technique with a least squares point-matching method, and analyzed the (H1) fully developed laminar flow problem for

TABLE 81

Circular Ducts with Diametrically Opposite Flat Sides
and Rectangular Ducts with Semicircular Ends:
f Re and Nu_{H1} for Fully Developed Laminar Flow
(from Jamil [449] and Zarling [134], Respectively)

2ϕ	$\dfrac{2b}{2a}$	$f\text{Re}$	Nu_{H1}	$\dfrac{2b}{2a}$	$f\text{Re}$	Nu_{H1}
0	0	24.000	8.235	0	24.00	8.24
10	0.087	21.551	–	0.143	21.06	6.66
20	0.174	19.822	6.020	0.150	20.94	6.59
40	0.342	17.603	4.991	0.167	20.66	6.44
60	0.500	16.475	4.483	0.200	20.13	6.16
80	0.643	15.986	4.303	0.250	19.41	5.77
100	0.766	15.842	4.269	0.333	18.40	5.28
120	0.866	15.862	4.296	0.400	17.76	5.00
140	0.940	15.933	4.335	0.500	17.03	4.72
160	0.985	15.980	4.359	0.667	16.32	4.45
170	0.996	15.998	4.363	1	16.00	4.36
180	1	16.000	4.364			

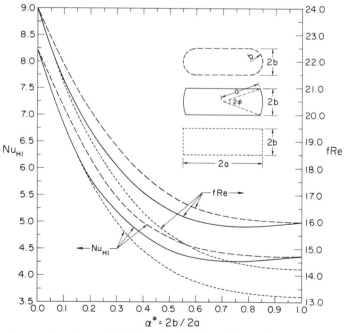

Fig. 78. Rectangular ducts with different end shapes: f Re and Nu_{H1} for fully developed
laminar flow (from Tables 42 and 81).

noncircular ducts. The Gram–Schmidt orthonormalization procedure was employed for the least squares approximation. As an example, $4f$ Re and Nu_{H1} were determined for a modified rectangular duct with short sides having a semicircular shape. These results are presented in Table 81 and Fig. 78.

Figure 78 compares f Re and Nu_{H1} for a sharp corner rectangular duct with two different rounded corner rectangular duct geometries. It is interesting to note that for $\alpha^* \to 0$, f Re and Nu_{H1} of rounded corner rectangular ducts approach those of the sharp corner rectangular duct. Also for $\alpha^* \to 1$, f Re and Nu_{H1} of rounded corner ducts approach those of the circular duct. For all intermediate values of α^*, the rounding of short sides increases f Re and Nu_{H1}. As an example, an increase of 4% for both f Re and Nu_{H1} obtains for the $\alpha^* = 0.125$ duct with semicircular ends.

L. Corrugated Ducts

Hu [129] analyzed fully developed laminar forced convection through corrugated ducts. This duct has n sinusoidal corrugations over the contour of a plain tube as shown in Fig. 79. The flow area for this duct geometry, with n corrugations and ϵ as an amplitude, is given by the closed-form formula

$$A_c = \pi a^2 (1 + \tfrac{1}{2}\epsilon^2) \tag{403}$$

However, the perimeter needs to be evaluated numerically. Subsequently, the hydraulic diameter is calculated from its definition and is presented in Table 82 for the different values of n and $e^*(=\epsilon/a)$.

Hu employed the method of conformal mapping and Green's function to determine f Re, Nu_{H1}, and Nu_{H2} as presented in Table 82. As discussed on

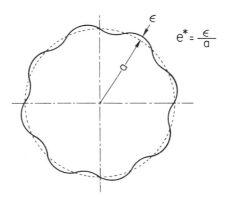

FIG. 79. A corrugated duct ($n = 8$).

TABLE 82

CORRUGATED DUCTS: $f \, Re$, Nu_{H1}, AND Nu_{H2} FOR FULLY
DEVELOPED LAMINAR FLOW (FROM HU [129,443])

n	e*	fRe	Nu_{H1}	Nu_{H2}	$D_h/2a$
8	0.02	15.990	4.356	4.357	0.9986
	0.04	15.962	4.334	4.335	0.9944
	0.06	15.915	4.297	4.299	0.9874
	0.08	15.850	4.244	4.246	0.9776
	0.10	15.765	4.176	4.177	0.9650
	0.12	15.678	4.090	4.089	0.9501
12	0.02	15.952	4.340	4.340	0.9966
	0.04	15.806	4.267	4.267	0.9863
	0.06	15.559	4.142	4.140	0.9689
	0.08	15.200	3.962	3.956	0.9439
	0.10	14.711	3.723	3.709	0.9107
16	0.02	15.887	4.316	4.316	0.9938
	0.04	15.542	4.168	4.167	0.9747
	0.06	14.943	3.912	3.906	0.9418
	0.08	14.051	3.540	3.527	0.8934
24	0.02	15.679	4.245	4.245	0.9856
	0.04	14.671	3.875	3.870	0.9402
	0.06	12.872	3.231	3.219	0.8583

p. 390, Nu_{H2} should be lower than or equal to Nu_{H1}. However, from a review of the Table 82 results, it is found that Nu_{H2} is virtually the same as Nu_{H1} for some corrugated ducts. This is not surprising in view of the close approach to the circular duct.

Hu also applied a perturbation method, an interior Galerkin variational method, and a method of boundary integral equations to analyze the same problem for corrugated ducts. He compared the results with those of Table 82 and found that the perturbation method would yield good results for small perturbations. When the perturbation is large, the results are useful only for the qualitative trends. The direct use of the interior Galerkin method in its standard way will not produce sufficiently accurate results no matter how many terms are used. A physical understanding of the problem in question is essential for the choice of the trial functions. The integral equation method is supposed to be an efficient numerical method for the solution of arbitrary domain problems, but when the velocity and temperature gradients along the duct wall are changing rapidly, the numerical schemes become less accurate or even no longer workable.

M. Cusped Ducts

A cusped duct, also referred to as a star-shaped duct, is made up of n concave circular arcs, as shown in Table 83. A bank of circular cylinders arranged in a square array and just touching each other forms an $n = 4$ cusped duct geometry. The $n = 3$ cusped duct geometry represents the case

TABLE 83

Cusped Ducts: f Re for
Fully Developed Laminar Flow
(from Shih [108])

n	fRe
3	6.503
4	6.606
5	6.634
6	6.639
8	6.629

n = 4

of touching cylindrical rods arranged in a triangular array, as for example, a fuel rod bundle used in a nuclear reactor. Leonard and Lemlich [450] analyzed fully developed laminar flow through a three-sided cusped duct geometry by employing a finite difference method, and determined f Re as 6.43. Shih [108] employed a 12-point matching method and evaluated the f Re factors and other flow parameters for the n-sided cusped duct geometry. Their f Re factors are presented in Table 83.

N. Cardioid Ducts

Tao [82] investigated the laminar forced convection through the cardioid duct of Fig. 80 for the $\widehat{H1}$ boundary condition. He included the effect of thermal energy sources within the fluid and employed the method of conformal mapping for the analysis. Tao extended his analysis for a Pascal's limaçon [83]. The circular tube and the cardioid duct are limiting cases of Pascal's limaçon. He presented closed-form formulas for the fluid velocity and temperature distributions, mean velocity, average and bulk mean temperatures, wall heat flux, and Nu_{H1}. Tyagi [64] extended Tao's work by including the effect of viscous dissipation, flow work, and uniform thermal energy sources within the fluid for the $\widehat{H1}$ boundary condition. For the special case of a cardioid duct (Fig. 80) with no viscous dissipation, no flow work, and no thermal energy sources within the fluid, f Re and Nu_{H1} are

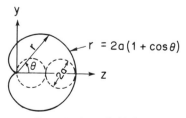

$r = 2a(1 + \cos\theta)$

FIG. 80. A cardioid duct.

given by

$$f\,\mathrm{Re} = \frac{27\pi^2}{17} = 15.675 \tag{404}$$

$$\mathrm{Nu_{H1}} = \frac{13005}{30503}\,\pi^2 = 4.208 \tag{405}$$

Note that $f\,\mathrm{Re}$ and $\mathrm{Nu_{H1}}$ are 2.0 and 5.4%, respectively, lower than the corresponding circular tube results.

Tao [84] also analyzed the cardioid duct for the (H2) boundary condition by using the conformal mapping method, where he included the effect of uniform thermal energy sources within the fluid. Tyagi [90,91] corrected the $\mathrm{Nu_{H2}}$ expression of Tao which was in error. Tyagi [90,91] extended the work of Tao [84] by including the effect of viscous dissipation within the fluid. Tyagi solved the problem by using a Laplacian operator instead of the biharmonic operator as did Tao. Tyagi presented formulas for the velocity and temperature distributions as well as the cross-sectional average and bulk mean fluid temperatures. The $\mathrm{Nu_{H2}}$ expression in [90,91] was based on the cross section average (instead of the bulk mean) fluid temperature.

Hu [129] analyzed the (H2) problem for the cardioid duct by employing the method of conformal mapping and Green's function. After correcting some algebraic errors in [129], Hu obtained the following expression for $\mathrm{Nu_{H2}}$ [443]:

$$\mathrm{Nu_{H2}} = \frac{3\pi^2}{8}\left[7 - 8\ln 2 - \frac{133847}{242760}\right]^{-1} = 4.097 \tag{406}$$

Note that $\mathrm{Nu_{H2}}$ is 6.1% lower than the value for the circular tube.

O. Miscellaneous Singly Connected Ducts

Gun and Darling [451] analyzed the singly connected ducts of Fig. 81 by employing a finite difference method. They presented graphically the velocity profiles for each of these duct geometries and obtained the $f\,\mathrm{Re}$ factors as 7.06, 6.50, and 6.50 for the corner, side, and center section ducts

FIG. 81. Singly connected duct geometries analyzed by Gunn and Darling [451].

of Fig. 81. A more accurate value of $f\,\mathrm{Re}$ for the center section is 6.606, from Table 83. Gun and Darling also experimentally determined the $f\,\mathrm{Re}$ factors to be within 5% of the theoretical values for these three geometries.

Sastry [85] employed the conformal mapping method and analyzed the laminar velocity and (H1) temperature problems for curvilinear polygonal ducts. He derived formulas for the velocity and temperature distribution for a general curvilinear polygonal duct. Specific expressions were derived for u and u_m for a Booth's lemniscate and a hexagonal duct.

Sastry [86] also employed a general power series mapping function to analyze the laminar velocity and (H1) temperature problems. Expressions were obtained for the local and mean velocity, local and mean temperature, and Nusselt number. Cardioid and ovaloid ducts were analyzed as specific examples.

Chapter XI

Small Aspect Ratio Ducts

Fully developed laminar flow through a duct with a small aspect ratio has been analyzed by employing an approximate method. Since the basic physical significance for this theory, as outlined in the literature [124–126], is not clear, some details of the approximate theory are presented as well as the final results.

Consider a duct (Fig. 82) with a small height-to-width ratio d/z_1 and with no abrupt variations in the height across the width. The height d is expressed as a function of z and is small everywhere compared to the width z_1. Thus the velocity and temperature gradients in the z direction are small compared to those in the y direction. Furthermore

$$\frac{\partial^2 u}{\partial z^2} \ll \frac{\partial^2 u}{\partial y^2}, \qquad \frac{\partial^2 t}{\partial z^2} \ll \frac{\partial^2 t}{\partial y^2} \tag{407}$$

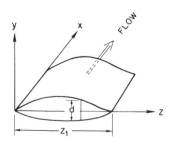

FIG. 82. A small aspect ratio duct.

280

Thus, these second derivatives may be neglected in the approximate analysis. Then resultant momentum and energy equations correspond to those for the parallel plates geometry. Consequently, the velocity $u(y, z)$ and temperature $t(y, z)$ at a section z are the same as those for parallel plates of height d. The mean velocity and bulk mean temperature at z (integrated only over the height d), designated as $u(z)$ and $t(z)$, are also those for the parallel plates duct of height d:

$$u(z) = -\frac{c_1}{3}\left(\frac{d}{2}\right)^2 \tag{408}$$

$$t_{H1}(z) = t_w - \frac{17}{35}\frac{u_m}{\alpha}\left(\frac{dt_m}{dx}\right)\left(\frac{d}{2}\right)^2 = t_w - \frac{17}{140}\left(\frac{q''D_h}{k}\right) \tag{409}$$

In addition, from the solution for the parallel plates duct, at any section z,

$$\int_0^d u^2\, dy = \frac{6}{5}\, u^2(z)d, \qquad \int_0^d u^3\, dy = \frac{54}{35}\, u^3(z)d \tag{410}$$

The mean velocity and bulk mean temperature of the fluid will be

$$u_m = \frac{1}{A_c}\int_{A_c} u(z)\, dA_c = -\frac{1}{12}\frac{c_1}{A_c}\int_0^{z_1} d^3\, dz \tag{411}$$

$$t_m = \frac{1}{A_c u_m}\int_{A_c} t(z)\, u(z)\, dA_c$$

$$= t_w - \left[\frac{17}{140}\frac{P^2}{4A_c}\frac{\left(\int_0^{z_1} d^7\, dz\right)}{\left(\int_0^{z_1} d^3\, dz\right)^2}\right]\frac{q''D_h}{k} \tag{412}$$

where

$$A_c = \int_0^{z_1} d\, dz \tag{413}$$

From the solution for the parallel plates duct, the maximum velocity is given by

$$u_{max} = -\frac{1}{2}c_1\left(\frac{d_{max}}{2}\right)^2 \tag{414}$$

so that

$$\frac{u_{max}}{u_m} = \frac{3}{2}d_{max}^2\frac{\int_0^{z_1} d\, dz}{\int_0^{z_1} d^3\, dz} \tag{415}$$

$K_d(\infty)$, $K_e(\infty)$, f Re, and Nu_{H1} for the small aspect ratio duct are evaluated from their definitions in Chapter III, the definition of $D_h = 4A_c/P$, and Eqs. (408), (410), (411), and (412). They are expressed as

$$K_d(\infty) = \frac{6}{5} \frac{\left(\int_0^{z_1} d\, dz \right) \left(\int_0^{z_1} d^5\, dz \right)}{\left(\int_0^{z_1} d^3\, dz \right)^2} \qquad (416)$$

$$K_e(\infty) = \frac{54}{35} \frac{\left(\int_0^{z_1} d\, dz \right)^2 \left(\int_0^{z_1} d^7\, dz \right)}{\left(\int_0^{z_1} d^3\, dz \right)^3} \qquad (417)$$

$$f\, Re = \frac{96}{P^2} \frac{\left(\int_0^{z_1} d\, dz \right)^3}{\left(\int_0^{z_1} d^3\, dz \right)} \qquad (418)$$

$$Nu_{H1} = \frac{140}{17} \frac{\left(\int_0^{z_1} d\, dz \right) \left(\int_0^{z_1} d^3\, dz \right)^2}{\frac{P^2}{4} \left(\int_0^{z_1} d^7\, dz \right)} \qquad (419)$$

$K(\infty)$ and L_{hy}^+ are subsequently determined from Eqs. (93) and (90), respectively. In these results, the perimeter P is $2z_1$. These formulas, Eqs. (416)–(419), are the same as those presented by Maclaine–Cross [125].

Based on the analysis of James [126], it can be shown that Nu_T for the duct of Fig. 82 is related to Nu_T for parallel plates by

$$Nu_T = Nu_{T,pp} \frac{D_h^2 u_m}{4 d_1^2 u_1} I \qquad (420)$$

where

$$I = \lim_{B \to \infty} \frac{\int_0^{z_1} \frac{d_1}{d} \exp[-B/(d^2 u/d_1^2 u_1)]\, dz}{\int_0^{z_1} (du/d_1 u_1) \exp[-B/(d^2 u/d_1^2 u_1)]\, dz} \qquad (421)$$

in which d_1 and u_1 are the height and fluid velocity at a specified reference section of the duct and B is a parameter related to the axial distance x [126]. If the y axis is assumed to pass through the maximum height of the duct so that $d_1 = d_{max}$, it can be shown that I approaches unity as B tends toward infinity for triangular and elliptical ducts. After substituting the values of

u_m and D_h in Eq. (420),

$$\mathrm{Nu_T} = 7.5407009 \, \frac{\int_0^{z_1} d\,dz \int_0^{z_1} d^3\,dz}{(P^2/4)\,d_{max}^4} \qquad (422)$$

Using the foregoing solutions, laminar fluid flow and heat transfer results were calculated by the present authors. These results are presented in Chapters VIII–X for the limiting geometries corresponding to small aspect ratio ducts.

Comparisons of limiting cases of triangular, sine, and elliptical ducts are of interest. These limiting cases are obtained when the aspect ratio $2b/2a$ approaches either 0 or ∞, with one dimension remaining finite. The corners of these limiting duct geometries are shown in Fig. 83.

f Re, $\mathrm{Nu_T}$, and $\mathrm{Nu_{H1}}$ for these duct geometries are summarized in Table 84. It is found that the corner shape has a strong influence on f Re, $\mathrm{Nu_T}$, and $\mathrm{Nu_{H1}}$. The more acute the passage is at the corner, the greater is the "corner effect." This is because the flow is more stagnant in the corner resulting in lower velocity and temperature gradients, and lower f Re, $\mathrm{Nu_T}$, and $\mathrm{Nu_{H1}}$ as a consequence.

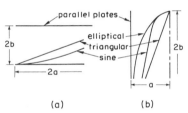

(a) (b)

FIG. 83. The corner shapes of limiting duct geometries with the aspect ratio (a) $2b/2a \to 0$, (b) $2b/2a \to \infty$.

TABLE 84

LIMITING TRIANGULAR, SINE, AND ELLIPTICAL DUCTS: f Re, $\mathrm{Nu_T}$, AND $\mathrm{Nu_{H1}}$ FOR FULLY DEVELOPED LAMINAR FLOW, DERIVED FROM EQS. (418), (422), AND (419)

Geometry	$2b/2a$	$D_h/2c^\dagger$	f Re	$\mathrm{Nu_T}$	$\mathrm{Nu_{H1}}$
Isosceles triangular ducts	∞	1	12.000	0.943	2.059
	0	1	12.000	0.943	2.059
Sine ducts[‡]	∞	1	15.303	0.739	2.521
	0	1	9.600	1.178	1.920
Elliptical ducts	$\infty, 0$	$\pi/2$	19.739	3.488	5.225
Parallel plates	$\infty, 0$	2	24.000	7.541	8.235

† $2c = $ a minimum of $2a$ or $2b$.

‡ $\mathrm{Nu_T}$ for the limiting sine duct geometry of $2b/2a \to \infty$ appears to be too low. The validity of Eq. (422) remains to be proved for this duct.

Chapter XII

Concentric Annular Ducts

The concentric circular annular duct[†] is an important geometry for many fluid flow and heat transfer devices. The simplest form of a two-fluid heat exchanger is a double pipe made up of two concentric circular tubes. One fluid flows through the inside tube, while the other fluid flows through the annular passage. The heat transfer and fluid friction problems for developing and developed profiles have been analyzed for a variety of boundary conditions for both the circular tube and concentric annular duct.

One limiting case of annular ducts, $r^* = 1$, is the parallel plates geometry. The other limiting case, $r^* = 0$, is a circular duct with an "infinitesimal thickness wire" at the center. Hence, the boundary conditions for the $r^* = 0$ annular duct are not identical to those for a circular duct. However, fortunately in most of the cases, the flow and heat transfer results for the $r^* = 0$ annular duct are identical to those for the circular duct. The circular duct and parallel plates are considered in detail in Chapters V and VI. Hence, the annular ducts of $0 < r^* < 1$ will be mainly considered in this chapter. A literature survey on the subject up to 1962 has been reported by Reynolds *et al.* [452].

A. Fully Developed Flow

1. FLUID FLOW

The solution to the hydrodynamic problem for fully developed laminar flow through an annular duct (Fig. 84) has long been known. It is found in

---

[†] For brevity, the concentric circular annular duct will be simply referred to as either the annular duct or the concentric annular duct.

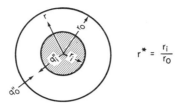

FIG. 84. A concentric circular annular duct.

the early textbooks such as those by Wien [453] and Lamb [454]. The velocity profile and corresponding friction factors are given by Lundberg et al. [41,42] as

$$u = -\frac{c_1 r_o^2}{4}\left[1 - \left(\frac{r}{r_o}\right)^2 + 2r_m^{*2}\ln\left(\frac{r}{r_o}\right)\right] \tag{423}$$

$$u_m = -\frac{c_1 r_o^2}{8}\left[1 + r^{*2} - 2r_m^{*2}\right] \tag{424}$$

$$\frac{u_{max}}{u_m} = \frac{2(1 - r_m^{*2} + 2r_m^{*2}\ln r_m^*)}{1 + r^{*2} - 2r_m^{*2}} \tag{425}$$

$$f_i\,\mathrm{Re} = -\frac{c_1 D_h}{u_m}\left(\frac{r_m^2 - r_i^2}{r_i}\right) \tag{426}$$

$$f_o\,\mathrm{Re} = -\frac{c_1 D_h}{u_m}\left(\frac{r_o^2 - r_m^2}{r_o}\right) \tag{427}$$

$$f\,\mathrm{Re} = \frac{16(1 - r^*)^2}{1 + r^{*2} - 2r_m^{*2}} \tag{428}$$

where

$$r_m^* = \frac{r_m}{r_o} = \left[\frac{1 - r^{*2}}{2\ln(1/r^*)}\right]^{1/2} \tag{429}$$

r_m designates the radius where the maximum velocity occurs ($\partial u/\partial r = 0$); f_i and f_o designate the Fanning friction factors at the inner and outer walls, respectively; f stands for the perimeter average Fanning friction factor, which is related to f_i and f_o as

$$f = \frac{f_i r_i + f_o r_o}{r_i + r_o} \tag{430}$$

All of the foregoing results were recalculated accurately by the present authors and are presented in Tables 85 and 86. The $f\,\mathrm{Re}$ factors are also presented in Fig. 85, and the $\mathrm{Nu_T}$ and $\mathrm{Nu_H}$ factors in Fig. 86. Lundgren

TABLE 85

Concentric Annular Ducts:
Flow Characteristics for Fully Developed
Laminar Flow (from Eq. (429)
and Lundgren et al. [51])

r^*	r_m^*	$\dfrac{u_{max}}{u_m}$	$K_d(\infty)$	$K_e(\infty)$
0	0	2.000	1.333	2.000
0.0001	0.232995	1.767	1.298	1.863
0.001	0.269040	1.725	1.287	1.823
0.01	0.329489	1.661	1.266	1.753
0.05	0.408028	1.598	1.243	1.675
0.10	0.463655	1.567	1.230	1.634
0.15	0.507570	1.549	1.222	1.610
0.20	0.546114	1.537	1.217	1.594
0.30	0.614748	1.522	1.210	1.573
0.40	0.677030	1.513	1.206	1.561
0.50	0.735534	1.508	–	–
0.60	0.791478	1.504	1.202	1.549
0.75	0.872002	1.501	–	–
0.80	0.898140	1.501	1.200	1.544
0.90	0.949561	1.500	1.200	1.543
1.00	1	1.500	1.200	1.543

TABLE 86

Concentric Annular Ducts: Flow and Heat Transfer
Characteristics for Fully Developed Laminar Flow
(from Eqs. (426)–(428), and Shah and London [13])

r^*	$f_i Re$	$f_o Re$	$f Re$	Nu_T	Nu_H
0	∞	16.000	16.000	3.657	4.364
0.0001	9742.834	16.973	17.945	–	5.185
0.001	1352.776	17.337	18.671	–	5.497
0.01	219.426	18.034	20.028	–	6.101
0.02	134.049	18.361	20.629	5.636	6.383
0.04	85.241	18.770	21.326	–	6.725
0.05	74.459	18.923	21.567	6.099	6.847
0.06	66.959	19.058	21.769	–	6.951
0.08	57.119	19.290	22.092	–	7.122
0.10	50.886	19.489	22.343	6.517	7.257
0.15	42.027	19.904	22.790	–	7.506
0.20	37.264	20.253	23.088	–	7.678
0.25	34.254	20.564	23.302	7.084	7.804
0.30	32.166	20.850	23.461	–	7.900
0.40	29.437	21.375	23.678	–	8.033
0.50	27.719	21.859	23.813	7.414	8.117
0.60	26.529	22.318	23.897	–	8.170
0.70	25.653	22.757	23.949	–	8.203
0.75	25.295	22.972	23.967	–	–
0.80	24.978	23.182	23.980	–	8.222
0.90	24.440	23.596	23.996	–	8.232
1.00	24.000	24.000	24.000	7.541	8.235

et al. [51] determined $K_d(\infty)$ and $K_e(\infty)$, which are presented in Table 85. It may be noted from Table 86 that the presence of a small circular core at the center ($r^* = 10^{-4}$) significantly increases the flow resistance f Re.

Natarajan and Lakshmanan [455] presented the following simple formula for f Re.

$$f\,Re = 24(r^*)^{0.035} \tag{431}$$

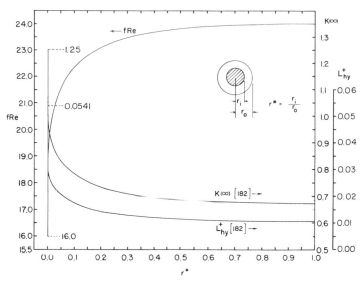

FIG. 85. Concentric annular ducts: f Re (from Table 86), and $K(\infty)$ and L_{hy}^{+} (from Tables 87 and 88), for fully developed laminar flow.

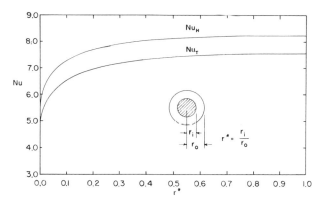

FIG. 86. Concentric annular ducts: Nu_T and Nu_H for fully developed laminar flow (from Table 86).

This agrees with the f Re factors of Table 86 within $\pm 2\%$ for $r^* \geq 0.005$.

The incremental pressure drop numbers for fully developed flow, $K(\infty)$, are compared in Table 87 for concentric annuli. $K(\infty)$ by Liu [182], Sparrow and Lin [163], Heaton et al. [44], and Sugino [157] are the asymptotic values of their hydrodynamic entry length solutions. $K(\infty)$ by Lundgren et al. [51] are obtained from an approximate analytical equation (93). $K(\infty)$ of Liu in Table 87 are believed to be the most accurate, because they are

TABLE 87

CONCENTRIC ANNULAR DUCTS:
$K(\infty)$ FOR FULLY DEVELOPED LAMINAR FLOW

r^*	$K(\infty)$				
	Liu [182]	Sparrow [163]	Heaton [44]	Sugino [157]	Lundgren [51]
0	1.25	–	–	–	1.333
0.0001	–	1.010	–	–	1.073
0.01	–	0.914	–	–	0.973
0.02	–	–	0.86	–	–
0.05	0.830	0.820	0.80	–	0.864
0.10	0.784	0.766	0.76	–	0.809
0.20	–	0.714	–	0.718	0.754
0.25	–	–	0.72	–	–
0.40	–	0.672	–	–	0.711
0.50	0.688	–	–	0.672	–
0.75	0.678	–	–	–	–
0.80	–	0.650	–	–	0.694
0.833	–	–	–	0.662	–
1.00	0.674	–	0.68	–	0.686

based on a rigorous numerical analysis. It may be noted that $K(\infty)$ for the circular tube is considerably higher than that for concentric annuli with $r^* > 0$.

The hydrodynamic entrance lengths L_{hy}^+ are compared in Table 88 for concentric annuli. Liu [182], Manohar [180], Roy [161], and Heaton et al. [44] defined L_{hy} as the distance from the entrance to the point where the maximum velocity reaches 99% of the fully developed value. Coney and El-Shaarawi [183] defined it as the distance required for the axial velocity profile to approach within $\pm 0.5\%$ of its fully developed value. Sparrow and Lin [163] defined L_{hy} as the duct length required to achieve $K(x)$ as 98% of

TABLE 88

CONCENTRIC ANNULAR DUCTS:
L_{hy}^+ FOR FULLY DEVELOPED LAMINAR FLOW

r^*	L_{hy}^+						
	Liu [182]	Manohar [180]	Roy [161]	Heaton [44]	Coney [183]	Sparrow [163]	McComas [48]
0	.0541	–	–	.0575	–	–	.0260
0.001	–	–	–	.0296	–	.0375	.0121
0.01	–	–	–	–	–	.0303	.00982
0.02	–	–	–	.0206	–	–	–
0.05	.0206	–	–	.0172	.0329	.0241	.00800
0.10	.0175	.0164	.0180	.0146	.0253	.0210	.00725
0.20	–	–	.0158	–	.0214	.0171	.00660
0.25	–	–	–	.0118	.0204	–	–
0.30	–	.0122	.0140	–	.0194	–	.00630
0.40	–	–	.0128	–	.0178	.0131	.00613
0.50	.0116	.0110	.0121	.0103	.0168	–	–
0.60	–	–	.0114	–	.0161	–	.00596
0.70	–	.0103	–	–	.0156	–	–
0.75	.0109	–	–	–	.0152	–	–
0.80	–	–	.0113	–	.0150	.0118	.00589
1.00	.0108	–	–	.0099	.0147	–	.00588

$K(\infty)$. The L_{hy}^{+} of Sparrow and Lin are tabulated by Tiedt [456]. McComas' L_{hy}^{+} are obtained from the approximate analytical equation (90) and are significantly lower than those of Liu (refer to p. 42). L_{hy}^{+} of McComas are included in Table 88 to emphasize that they are low by a factor of 20, and hence care must be exercised to use this equation for the duct geometries for which no rigorous solution is available. L_{hy}^{+} of Liu, based on the results of Table 93 and determined by the present authors, are believed to be the most accurate, because they are derived by a rigorous numerical analysis. Sugino [157] also determined L_{hy}^{+}, but his values are believed to be in error (see p. 297) and therefore are not reported in Table 88. It is observed that L_{hy}^{+} for a circular tube is considerably higher than that for concentric annuli with $r^{*} > 0$.

Refer to the section on hydrodynamically developing flow on p. 297 for details on the methods of analysis used by the aforementioned investigators.

2. HEAT TRANSFER

a. *Fundamental Solutions of Four Kinds*

Depending upon the temperature or heat flux specified at the inner and outer surfaces of the annulus, there are four possible fundamental boundary conditions as described by Fig. 8. For each kind of fundamental boundary condition, there are two solutions of the energy equation, one for each of the two surfaces heated, thus totaling eight heat transfer solutions. These solutions have been identified for ready reference in Table 4. Note that the fundamental boundary conditions of the fourth and fifth kinds are identical for concentric annular ducts heated symmetrically around each periphery. For the fundamental problem of the first kind, only one solution is required, since the other solution can readily be obtained from the relationships of Eqs. (432)–(435).

Jakob and Rees [457] obtained the temperature distributions for the fundamental solution of the second kind for hydrodynamically and thermally developed flow. Murakawa [458,459] presented an integral equation formulation for the solution of the first kind and a series solution approach to the same problem that included an arbitrary peripheral temperature variation. Murakawa did not present numerical results.

Lundberg *et al.* [41,42] systematically approached all four fundamental problems and their various combinations. They presented extensive results[†] for the dimensionless temperatures, heat fluxes, and Nusselt numbers as a

---

[†] Equations (II.D.23) and (II.D.24) of [41] were in error and were corrected to get the results reported in Table 89.

function of r^*. Based on their equations, the detailed results were recomputed by Shah and London [13]. Only partial results from [13] are presented in Tables 89 and 90, since the other results can be obtained from the following relationships for fully developed laminar flow:

$$\Phi_{oo}^{(1)} = -\Phi_{oi}^{(1)} = r^*\Phi_{ii}^{(1)} = -r^*\Phi_{io}^{(1)} \tag{432}$$

$$\theta_{mi}^{(1)} = 1 - \theta_{mo}^{(1)} \tag{433}$$

$$Nu_{io}^{(1)} = Nu_{ii}^{(1)} = \frac{\Phi_{ii}^{(1)}}{1 - \theta_{mi}^{(1)}} = \frac{\Phi_{ii}^{(1)}}{\theta_{mo}^{(1)}} \tag{434}$$

$$Nu_{oi}^{(1)} = Nu_{oo}^{(1)} = \frac{\Phi_{oo}^{(1)}}{1 - \theta_{mo}^{(1)}} = \frac{r^*\Phi_{ii}^{(1)}}{\theta_{mi}^{(1)}} \tag{435}$$

$$[\theta_{oi}^{(2)} - \theta_{mi}^{(2)}] = r^*[\theta_{io}^{(2)} - \theta_{mo}^{(2)}] \tag{436}$$

$$Nu_{ii}^{(2)} = \frac{1}{\theta_{ii}^{(2)} - \theta_{mi}^{(2)}} \tag{437}$$

$$Nu_{oo}^{(2)} = \frac{1}{\theta_{oo}^{(2)} - \theta_{mo}^{(2)}} \tag{438}$$

$$Nu_{oi}^{(2)} = Nu_{io}^{(2)} = 0 \tag{439}$$

$$\Phi_{ii}^{(3)} = \Phi_{oo}^{(3)} = 0 \tag{440}$$

$$\theta_{oi}^{(3)} = \theta_{io}^{(3)} = \theta_{mi}^{(3)} = \theta_{mo}^{(3)} = 1 \tag{441}$$

$$\Phi_{oi}^{(4)} = \frac{1}{\Phi_{io}^{(4)}} = -r^* \tag{442}$$

$$\theta_{ii}^{(4)} = r^*\theta_{oo}^{(4)} = \frac{1}{Nu_{ii}^{(1)}} + \frac{r^*}{Nu_{oo}^{(1)}} \tag{443}$$

$$Nu_{ii}^{(4)} = Nu_{io}^{(4)} = \frac{1}{\theta_{ii}^{(4)} - \theta_{mi}^{(4)}} = \frac{1}{r^*\theta_{mo}^{(4)}} = Nu_{ii}^{(1)} \tag{444}$$

$$Nu_{oo}^{(4)} = Nu_{oi}^{(4)} = \frac{1}{\theta_{oo}^{(4)} - \theta_{mo}^{(4)}} = \frac{r^*}{\theta_{mi}^{(4)}} = Nu_{oo}^{(1)} \tag{445}$$

The nomenclature for the foregoing dimensionless θ, Φ, and Nu is explained on p. 33.

Dwyer [460] also presented fully developed Nusselt numbers for the fundamental boundary condition of the second kind for laminar, slug, and turbulent flows. His results for laminar flow agree well with the results of Table 89.

The fully developed temperature profiles for each of the fundamental problems are delineated in Fig. 29 for parallel plates. Under fully developed

TABLE 89

CONCENTRIC ANNULAR DUCTS: FUNDAMENTAL SOLUTIONS OF THE
FIRST AND SECOND KINDS FOR FULLY DEVELOPED LAMINAR FLOW
(FROM SHAH AND LONDON [13])

r^*	$\phi_{ii}^{(1)}$	$Nu_{ii}^{(1)}$	$Nu_{oo}^{(1)}$	$\theta_{io}^{(2)}-\theta_{mo}^{(2)}$	$Nu_{ii}^{(2)}$	$Nu_{oo}^{(2)}$
0	∞	∞	2.66667	−.145833	∞	4.36364
0.001	289.24013	322.25436	2.82329	−.136168	337.04414	4.58657
0.01	42.99515	50.45396	2.90834	−.130725	54.01669	4.69234
0.02	25.05098	30.17942	2.94836	−.127945	32.70512	4.73424
0.04	14.91204	18.61387	2.99928	−.124122	20.50925	4.77803
0.05	12.68471	16.05843	3.01887	−.122568	17.81128	4.79198
0.06	11.13713	14.27970	3.03640	−.121144	15.93349	4.80323
0.08	9.10628	11.94251	3.06751	−.118559	13.46806	4.80270
0.10	7.81730	10.45870	3.09528	−.116214	11.90578	4.83421
0.15	5.97397	8.34163	3.15708	−.110999	9.68703	4.86026
0.20	4.97068	7.19736	3.21338	−.106390	8.49892	4.88259
0.25	4.32809	6.47139	3.26700	−.102207	7.75347	4.90475
0.30	3.87606	5.96628	3.31911	−.098363	7.24115	4.92801
0.40	3.27407	5.30511	3.42077	−.091495	6.58330	4.97917
0.50	2.88539	4.88896	3.52035	−.085513	6.18102	5.03653
0.60	2.61015	4.60183	3.61851	−.080249	5.91171	5.09922
0.70	2.40315	4.39137	3.71547	−.075579	5.72036	5.16618
0.80	2.24071	4.23035	3.81134	−.071409	5.57849	5.23654
0.90	2.10916	4.10311	3.90617	−.067665	5.46988	5.30955
1.00	2.00000	4.00000	4.00000	−.064286	5.38462	5.38462

TABLE 90

CONCENTRIC ANNULAR
DUCTS: FUNDAMENTAL
SOLUTIONS OF THE THIRD KIND
FOR FULLY DEVELOPED
LAMINAR FLOW (FROM SHAH
AND LONDON [13])

r^*	$Nu_{ii}^{(3)}$	$Nu_{oo}^{(3)}$
0	∞	3.6568
0.02	32.337	3.9934
0.05	17.460	4.0565
0.10	11.560	4.1135
0.25	7.3708	4.2321
0.50	5.7382	4.4293
1.00	4.8608	4.8608

conditions, for the first and fourth fundamental problems, heat enters at one wall and leaves at the other wall, and the temperature profile between the walls is linear. For the second fundamental problem, the heat entering at one wall is retained by the fluid since the other wall is adiabatic. As a result, the wall temperatures and fluid bulk mean temperature increase linearly in the flow direction. For the third fundamental problem, in the limit when L/D_h tends toward infinity, the fully developed temperature profile is uniform across the duct cross section and heat transfer approaches zero.

For the fundamental boundary condition of the second kind (the inner wall heated or cooled, the outer wall insulated), several early investigators, such as Davis [461] and Chen et al. [462], correlated heat transfer data empirically including variable fluid properties effects.

b. *Specified Constant Temperature or Heat Flux at Walls*

The direct practical applications of fundamental solutions described in the previous section are limited. However, these solutions can be combined to arrive at more practical solutions. Such solutions are described below for three problems: (i) constant temperatures specified on both walls, (ii) constant heat fluxes specified on both walls, and (iii) a constant temperature specified on one wall with a constant heat flux specified on the other wall. The effects of viscous dissipation, flow work, axial heat conduction, and thermal energy sources within the fluid are neglected in the above treatments. In the following formulas, θ and Φ are found either from Tables 89 and 90 or by the relationships of Eqs. (432)–(445). The superscripts 1a, 1b, 2a, 2b, 3a, and 3b are used in order to distinguish these special cases, as outlined in Table 5.

(i) *Constant Temperatures Specified on Both Walls.* For $x \geq 0$, the temperature at each wall is specified as

$$t = t_i \quad \text{on the inner wall}$$

$$t = t_o \quad \text{on the outer wall}$$

and $t = t_e$ for all r at $x < 0$. Fully developed heat transfer results depend upon whether t_i and t_o are different or equal.

If $t_i \neq t_o$, the fluid bulk mean temperature and heat fluxes at both walls, $t_m^{(1a)}$, $q_i''^{(1a)}$, and $q_o''^{(1a)}$, are presented by Eqs. (474)–(476), with all the thermal entrance variables replaced by the corresponding fully developed values (for example, $\theta_{x,mi}^{(1)}$ by $\theta_{mi}^{(1)}$). The Nusselt numbers at the inner and outer walls are obtained from Eqs. (477) and (478) and the relationships of Eqs. (432) and (433), as follows:

$$\mathrm{Nu}_i^{(1a)} = \Phi_{ii}^{(1)} / \theta_{mo}^{(1)} \tag{446}$$

$$\mathrm{Nu}_o^{(1a)} = \Phi_{oo}^{(1)} / \theta_{mi}^{(1)} \tag{447}$$

These Nusselt numbers are presented in Table 91 and Fig. 87.

When $t_i = t_o$, the nature of the heat transfer problem is changed, as was discussed for parallel plates on p. 155. For this case,[†] Nusselt numbers are obtained from Eqs. (477) and (478), with $t_o = t_e$, using $\Phi_{x,oo}^{(1)}$, $\theta_{x,mo}^{(1)}$, etc., from Lundberg *et al.* [41,42]. These Nusselt numbers are presented in Table 91 and Fig. 87 with 1b as a superscript. The perimeter average Nu_T is determined from the following expression incorporating the individual Nusselt numbers.

$$\mathrm{Nu}_T = \frac{\mathrm{Nu}_o^{(1b)} + r^* \mathrm{Nu}_i^{(1b)}}{1 + r^*} \tag{448}$$

The definition of h, Eq. (76), is employed to arrive at this equation. The Nusselt numbers calculated from Eq. (448) are presented in Table 86 and Fig. 86.

[†] Refer to the footnote on p. 35.

TABLE 91

CONCENTRIC ANNULAR DUCTS: NUSSELT NUMBERS
FOR CASES 1a AND 1b OF TABLE 5 FOR FULLY
DEVELOPED LAMINAR FLOW
(FROM SHAH AND LONDON [13])

r^*	$Nu_i^{(1a)}$	$Nu_o^{(1a)}$	$Nu_i^{(1b)}$	$Nu_o^{(1b)}$
0	∞	2.667	∞	3.657
0.0001	2355.258	2.779	–	–
0.001	322.254	2.823	–	–
0.01	50.454	2.908	–	–
0.02	30.179	2.948	58.084	4.587
0.04	18.614	2.999	–	–
0.05	16.058	3.019	31.234	4.843
0.06	14.280	3.036	–	–
0.08	11.943	3.068	–	–
0.10	10.459	3.095	20.430	5.125
0.15	8.342	3.157	–	–
0.20	7.197	3.213	–	–
0.25	6.471	3.267	12.633	5.696
0.30	5.966	3.319	–	–
0.40	5.305	3.421	–	–
0.50	4.889	3.520	9.441	6.401
0.60	4.602	3.619	–	–
0.70	4.391	3.715	–	–
0.80	4.230	3.811	–	–
0.90	4.103	3.906	–	–
1.00	4.000	4.000	7.541	7.541

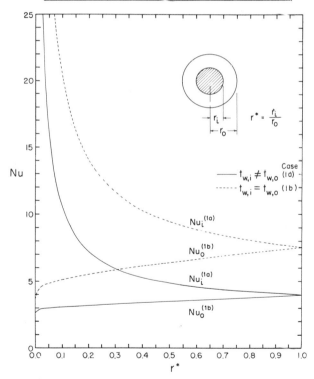

FIG. 87. Concentric annular ducts: Nusselt numbers for constant temperatures on both walls for fully developed laminar flow (from Table 91).

(ii) *Constant Heat Fluxes Specified on Both Walls.* For $x \geq 0$, the heat flux at each wall is specified as

$$q'' = q_i'' \quad \text{on the inner wall}$$

$$q = q_o'' \quad \text{on the outer wall}$$

and $t = t_e$ for all r at $x < 0$. For this case, the temperatures at the inner and outer walls, the bulk mean temperature, and the Nusselt numbers at the inner and outer walls are expressed by Eqs. (480)–(484), with all thermal entrance variables replaced by the corresponding fully developed values. These results are valid regardless of whether q_i'' and q_o'' are different or equal.

Two special cases of specified constant wall heat fluxes are (2a) constant and equal axial heat fluxes specified on both walls such that at any axial location the peripheral wall temperatures are constant but different at the inner and outer walls; (2b) constant but different wall heat fluxes specified on both walls such that at any axial location the peripheral wall temperatures at the inner and outer walls are constant and equal. The Nusselt numbers for both these cases are obtained by using Eqs. (480)–(484) for $x = \infty$ and the results of Table 89. These are presented in Table 92 and Fig. 88. Note that the heat flux is specified as positive if the heat transfer is from the wall to the fluid. Consequently, a negative Nusselt number means that heat transfer takes place from the fluid to the wall. An infinite Nusselt number at the inner wall means $t_i = t_m$ only and does not mean infinite heat flux. In case 2b, the ratio q_i''/q_o'' is unique for a given r^* and is also listed in Table 92.

For case 2b, the heat transfer coefficient can also be defined, similar to case 1b, as based on the total wall heat flux and the $(t_w - t_m)$ temperature difference, Eq. (75). The corresponding $\mathrm{Nu_H}$ were calculated from an equation similar to Eq. (448), with $\mathrm{Nu_i^{(2b)}}$ and $\mathrm{Nu_o^{(2b)}}$ of Table 92, and are presented in Table 86 and Fig. 86. As noted on p. 25, the (H1)–(H4) boundary conditions are the same for the concentric annular duct as they are for the circular tube and parallel plates, and hence they are designated as the (H) boundary condition. It may be noted that the presence of a small circular core at the center significantly increases $\mathrm{Nu_H}$.

Dwyer [463] analyzed the bilateral heat transfer in annuli for cases 2a and 2b. His tabulated values for Nusselt numbers for cases 2a and 2b and the (H) boundary condition agree well with the results of Table 92 and $\mathrm{Nu_H}$ of Table 86, respectively.

Urbanovich [464,465] included the effect of viscous dissipation and obtained a closed-form solution to the following two problems of fully developed laminar flow through the annulus: (a) Inside wall adiabatic, axial heat flux at outer wall constant. He presented formulas for the temperature distribution, heat flux, and heat transfer coefficient at the outer wall [464].

TABLE 92

CONCENTRIC ANNULAR DUCTS: NUSSELT NUMBERS FOR
CASES 2a AND 2b OF TABLE 5 FOR FULLY DEVELOPED
LAMINAR FLOW (FROM SHAH AND LONDON [13])

r^*	$Nu_i^{(2a)}$	$Nu_o^{(2a)}$	$Nu_i^{(2b)}$	$Nu_o^{(2b)}$	q_i''/q_o''
0	0	4.364	∞	4.364	∞
0.0001	−7.222	4.526	4051.526	4.781	847.448
0.001	−7.507	4.589	563.701	4.939	114.141
0.01	−8.912	4.721	91.097	5.251	17.348
0.02	−10.270	4.792	55.320	5.404	10.236
0.04	−13.269	4.894	34.772	5.603	6.206
0.05	−15.055	4.937	30.209	5.679	5.319
0.06	−17.128	4.977	27.026	5.747	4.703
0.08	−22.569	5.052	22.833	5.865	3.893
0.10	−31.036	5.122	20.162	5.967	3.379
0.15	−128.740	5.288	16.334	6.182	2.642
0.20	88.712	5.449	14.253	6.363	2.240
0.25	37.359	5.608	12.924	6.524	1.981
0.30	25.166	5.767	11.993	6.672	1.798
0.40	16.555	6.089	10.764	6.941	1.551
0.50	13.111	6.419	9.979	7.185	1.389
0.60	11.248	6.759	9.429	7.414	1.272
0.70	10.077	7.109	9.020	7.631	1.182
0.80	9.272	7.472	8.701	7.840	1.110
0.90	8.684	7.847	8.446	8.040	1.050
1.00	8.235	8.235	8.235	8.235	1.000

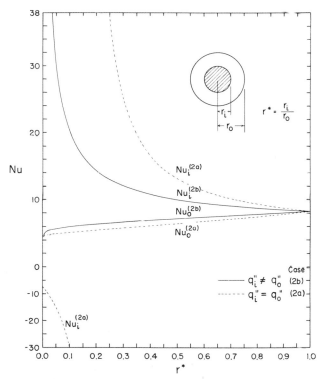

FIG. 88. Concentric annular ducts: Nusselt numbers for constant axial heat fluxes on both walls for fully developed laminar flow (from Table 92).

(b) Axial heat flux constant at the inside wall and variable at the outside wall:

$$q_o'' = 0, \qquad\qquad \delta < \phi < 2\pi - \delta \qquad\qquad (449)$$

$$q_o'' = q_c''(1 + m\cos\phi), \qquad -\delta \le \phi \le \delta \qquad\qquad (450)$$

where q_c'', m, and δ are constants. Closed-form expressions were presented for the fluid temperature distribution and the Nusselt number at the outer wall [465].

(iii) *Constant Temperature Specified on One Wall, and Constant Heat Flux Specified on the Other Wall.*[†]   For $x \ge 0$,

$$t = t_1 \qquad \text{on wall 1}$$

$$q'' = q_2'' \qquad \text{on wall 2}$$

and $t = t_e$ for all r at $x < 0$. In this case, the temperature at wall 2, the fluid bulk mean temperature, and the heat flux at wall 1 are given by Eqs. (485)–(487), with all the thermal entrance variables replaced by the corresponding fully developed values. The Nusselt numbers are obtained from Eqs. (488) and (489) and the relationships of Eqs. (440) and (441) as

$$\mathrm{Nu}_1 = -\Phi_{12}^{(4)}/\theta_{m2}^{(4)} \qquad\qquad (451)$$

$$\mathrm{Nu}_2 = 1/(\theta_{22}^{(4)} - \theta_{m2}^{(4)}) \qquad\qquad (452)$$

so that

$$\mathrm{Nu}_i^{(4a)} = \frac{1}{\theta_{ii}^{(4)} - \theta_{mi}^{(4)}}, \qquad \mathrm{Nu}_o^{(4a)} = -\Phi_{oi}^{(4)}/\theta_{mi}^{(4)} \qquad (453)$$

$$\mathrm{Nu}_i^{(4b)} = -\Phi_{io}^{(4)}/\theta_{mo}^{(4)}, \qquad \mathrm{Nu}_o^{(4b)} = \frac{1}{\theta_{oo}^{(4)} - \theta_{mo}^{(4)}} \qquad (454)$$

where the superscripts 4a and 4b are explained in Table 5. With the relationships of Eqs. (442)–(445), it can be shown that

$$\mathrm{Nu}_i^{(4a)} = \mathrm{Nu}_i^{(4b)} = \mathrm{Nu}_i^{(1a)} \qquad\qquad (455)$$

$$\mathrm{Nu}_o^{(4a)} = \mathrm{Nu}_o^{(4b)} = \mathrm{Nu}_o^{(1a)} \qquad\qquad (456)$$

These results have already been listed in Table 91.

c. *Exponential Wall Heat Flux,* (H5)

Gräber [244] analyzed the heat transfer problem for annular ducts with an axial exponential heat flux distribution (H5) on the inner or outer wall

---

[†] In this situation, no distinction is needed between inner and outer walls and so the subscripts 1 and 2 replace i and o.

only. He introduced the parameter F_o, defined on p. 84, which is related to the exponent m by Eq. (185). The ratio $Nu_{H5}/Nu_{jj}^{(2)}$ was presented graphically as functions of F_o (from -2 to 8) and r^* (from 0 to 1).

B. Hydrodynamically Developing Flow

One of the first solutions of a hydrodynamic entry length problem was conducted by Murakawa [374] for concentric annular ducts. He obtained an approximate entrance length solution by employing a series solution approach. The results, which involved an infinite series of Bessel functions, only partially satisfied the boundary conditions. More accurate solutions, based on linearization and finite difference techniques, are outlined below.

1. Solutions by Linearized Momentum Equations

Sugino [157] linearized the momentum equation, by employing the Langhaar method [153], and presented the hydrodynamic entry length solution in closed form in terms of the modified Bessel functions. He tabulated Δp^* as a function of $4x^+$, and $K(\infty)$, $f\,Re$, and $4L_{hy}^+$ for $r^* = 0.2, 0.5$, and 0.833. His Δp^* are in good agreement with those obtained from the $f_{app}\,Re$ factors of Table 93 for $x^+ < 0.0015$. However, for $x^+ > 0.0015$, his Δp^* values are higher than those obtained from Table 93. $K(\infty)$ of Sugino are presented in Table 87. His $f\,Re$ factors are identical to those of Table 86. Sugino found L_{hy}^+ to be 0.005 for the above r^* values. This is considerably lower than the Table 88 magnitude, and appears to be in error.

Heaton et al. [44] independently also employed the Langhaar-type linearization method, in a manner similar to the one described on p. 70, to obtain the hydrodynamic entry length solution. They determined the pressure drop from the mechanical energy integral equation. They tabulated values of $K(\infty)$ and entrance region Δp^* as a function of $x^* = 0.02, 0.05$, 0.10, 0.25, and 1.0. Their $K(\infty)$ are presented in Table 87. Shumway and McEligot [45] showed that the pressure drop of Heaton et al., which used the momentum integral equation, differed by 20% for $x^+ \lesssim 10^{-3}$, from the results based on the mechanical energy equation.[†] Thus, the velocity profiles and Δp^* values of Heaton et al. are unreliable in the early entry region. Shumway and McEligot obtained a finite difference solution for the hydrodynamic entry length problem for $r^* = 0.25$. They tabulated $f_{x,o}\,Re$, $f_{x,i}\,Re$, $f_x\,Re$, $f_m\,Re$, and $2x^+Prf_{app}\,Re$ (their $-P^+$) as functions of x^+/Pr for $Pr = 0.72$. Their $f_{app}\,Re$ has no counterpart in the present terminology. The $f_{app}\,Re$

[†] See footnote on p. 10 for the definition of the momentum and mechanical energy integral equations.

factors from their results for $r^* = 0.25$ fall between those for $r^* = 0.1$ and 0.5 in Fig. 89.

Chang and Atabek [160] and Roy [161] employed the linearization technique of Targ [159,159a] for the momentum equation, and obtained closed-form formulas for the axial velocity and pressure gradient in the entrance region. They also determined and presented L_{hy}^+ as a function of r^*. The results of Chang–Atabek and Roy are identical. Their L_{hy}^+ values are presented in Table 88.

Sparrow and Lin [163] linearized the momentum equation by introducing a stretched coordinate in the flow direction (refer to p. 71). They tabulated the first 30 eigenvalues for $r^* = 0.001, 0.01, 0.05, 0.1, 0.2, 0.4$, and 0.8 for the velocity profile expanded in terms of eigenfunctions. They also presented graphically the velocity profile u/u_m for $(r - r_i)/(r_o - r_i) = 0.1, 0.2, \ldots, 0.9$ as a function of $4x^+$, as well as $K(x)$ as a function of $4x^+$ for the foregoing values of r^*. The f_{app} Re factors calculated from $K(x)$ of Sparrow and Lin are slightly lower (within 2%) than those of Liu [182] in Table 93. Thus the stretched coordinate linearization method provides highly accurate results for the concentric annular duct hydrodynamic entry length problem. Quarmby [164] suggested a slightly different form of dimensionless axial and radial coordinates for the problem analyzed by Sparrow and Lin, which allows the case of the parallel plates duct, $r^* = 1$, to be recovered directly from the results of Sparrow and Lin.

2. SOLUTIONS BY FINITE DIFFERENCE METHODS

Manohar [180] solved the momentum equation (8) by a finite difference method for $r^* = 0.1, 0.3, 0.5$, and 0.7. The nonlinear momentum equation was solved iteratively. He presented graphically the velocity profiles, pressure distribution, and hydrodynamic entrance lengths. His tabulated L_{hy}^+, as defined on p. 41, are presented in Table 88. The f_{app} Re factors, based on Manohar's Δp^* values, are lower than those of Table 93. For $r^* = 0.1$, Manohar's f_{app} Re are 10 and 5% lower at $x^+ = 0.0025$ and 0.025, respectively. For $r^* = 0.5$, f_{app} Re factors are 19 and 9% lower for the same x^+ values.

Shah and Farnia [181] employed the finite difference method of Patankar and Spalding [248], as briefly described on p. 89, to analyze Eqs. (8) and (12) for concentric annuli. They presented graphically f_x Re and f_{app} Re as a function of $1/x^+$ for $r^* = 0, 0.005, 0.1$, and 1.0. The numerical results of Shah and Farnia are believed to be the most accurate. Their u_{max}/u_m and f_{app} Re, as tabulated by Liu [182], are presented in Table 93 and Fig. 89 for $r^* = 0.05$, $0.1, 0.5$, and 0.75. Similar results for $r^* = 0$ and 1 are presented in Tables 10 and 28, respectively.

TABLE 93

Concentric Annular Ducts: f_{app} Re and u_{max}/u_m for
Developing Laminar Flow (from Liu [182])

x^+	f_{app}Re	$\dfrac{u_{max}}{u_m}$	x^+	f_{app}Re	$\dfrac{u_{max}}{u_m}$
$r^*=0.05$			$r^*=0.10$		
0.0001992	241.2	1.092	0.0002103	233.3	1.094
0.0002989	197.7	1.112	0.0004207	166.6	1.132
0.0003986	171.7	1.129	0.0006309	137.1	1.161
0.0004980	154.1	1.144	0.0008410	119.5	1.185
0.0008945	116.0	1.190	0.001052	107.5	1.205
0.001293	97.54	1.227	0.001889	81.72	1.272
0.001691	86.10	1.259	0.002730	69.18	1.325
0.002090	78.13	1.286	0.003571	61.44	1.367
0.002489	72.19	1.311	0.004413	56.07	1.401
0.004075	57.98	1.389	0.005255	52.08	1.429
0.005669	50.38	1.443	0.008606	42.65	1.496
0.007262	45.52	1.482	0.01197	37.71	1.529
0.008857	42.08	1.510	0.01534	34.65	1.545
0.01045	39.50	1.530	0.01871	32.58	1.555
0.01680	33.46	1.572	0.02207	31.08	1.560
0.02319	30.38	1.587	0.03549	27.86	1.567
0.02956	28.54	1.593	0.04895	26.34	1.568
0.03593	27.32	1.596	0.06242	25.48	1.568
0.06770	24.65	1.598	0.07587	24.92	1.568
0.09320	23.80	1.598	0.08929	24.54	1.568
∞	21.57	1.598	∞	22.34	1.567
$r^*=0.50$			$r^*=0.75$		
0.0001893	248.3	1.090	0.0002839	202.5	1.109
0.0002366	222.4	1.100	0.0004259	166.0	1.133
0.0004250	166.2	1.132	0.0005679	144.3	1.152
0.0006143	138.9	1.158	0.0007097	129.5	1.169
0.0008039	122.0	1.180	0.001275	97.74	1.224
0.0009930	110.3	1.199	0.001843	82.35	1.267
0.001182	101.5	1.217	0.002411	72.86	1.304
0.001935	80.59	1.274	0.002979	66.28	1.335
0.002693	69.39	1.320	0.003547	61.38	1.362
0.003451	62.18	1.358	0.005807	49.76	1.431
0.004207	57.05	1.388	0.008084	43.58	1.466
0.004965	53.17	1.412	0.01036	39.71	1.484
0.007981	43.84	1.468	0.01263	37.07	1.493
0.01101	38.88	1.491	0.01490	35.15	1.497
0.01404	35.82	1.501	0.02395	31.01	1.502
0.01707	33.76	1.505	0.03304	29.08	1.502
0.02010	32.29	1.507	0.04212	27.98	1.502
0.03218	29.16	1.508	0.05120	27.27	1.502
0.04429	27.70	1.509	0.06031	26.78	1.502
0.08065	25.96	1.509	0.09653	25.77	1.502
∞	23.81	1.508	∞	23.97	1.501

Coney and El-Shaarawi [183] employed a modified implicit finite difference method of Bodoia and Osterle [172] to analyze the hydrodynamic entrance problem for concentric annuli. They obtained the solutions for fifteen different concentric annular ducts and presented graphically the dimensionless radial velocity profile $v/(v/r_o)$ and the ratio f_{app}/f_{fd} as functions of r^* and $4(1 - r^*)^2 x^+$; also provided were the hydrodynamic entry lengths of Table 88. The f_{app} Re factors calculated from their graphical results are 2–4% higher than those of Table 93. Their developing axial velocity profiles are in good agreement with those of Sparrow and Lin [163].

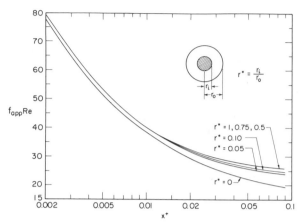

FIG. 89. Concentric annular ducts: f_{app} Re for developing laminar flow (from Table 93).

3. SOLUTIONS OF NAVIER–STOKES EQUATIONS

All the foregoing solutions were based on the boundary layer type idealizations. Fuller and Samuels [195] analyzed the complete set of Navier–Stokes equations. To model the flow, they employed a stream tube technique similar to the one used by Vrentas *et al.* [186] for a circular tube (see p. 92) and by Wang and Longwell [191] for parallel plates (see p. 166). They obtained the velocity profiles for various x/r_0 and Re for $r^* = 0.5$.

Fuller and Samuels showed that the effect of axial vorticity diffusion is to develop a nonuniform velocity profile at the entrance of the real tube ($x = 0$). This entry velocity profile is more developed for the decreasing values of the Reynolds number. It is about 50% fully developed for Re = 10. However, once in the tube, when axial diffusion is included, the velocity profile develops quite slowly and requires a greater entry length than when the axial diffusion is neglected, in spite of the head-start at $x = 0$.

4. CONCENTRIC ANNULAR DUCTS f_{app} Re FACTORS

The f_{app} Re factors for concentric annular ducts are presented in Table 93 and Fig. 89. The following observations may be made from these results. (1) For different concentric annuli, the f_{app} Re factors are in excellent agreement with each other for $r^* > 0$ and $x^+ \leq 0.01$. (2) The f_{app} Re factors for the circular tube ($r^* = 0$) are consistently low, even for low values of x^+ (about 2% low at $x^+ = 0.002$). This may be due to different boundary conditions employed at the centerline of a circular tube (maximum velocity) and of a concentric annular duct with $r^* \to 0$ (zero velocity). (3) The f_{app} Re factors for concentric annular ducts with $r^* \gtrsim 0.4$ are virtually the same as those for parallel plates even up to $x^+ = 0.1$.

For low values of x^+, the f_{app} Re factors of Table 93 are in excellent agreement with those of Eq. (188), varying from 1% higher at $x^+ = 0.0002$ to 1% lower at $x^+ = 0.001$. For a long duct with $x^+ \geq 0.02$, the f_{app} Re factors can be determined by Eq. (92) for the known values of f Re (Table 86) and $K(\infty)$ of Liu [182] (Table 87). Shah [259] combined the equations for f_{app} Re for low and high values of x^+ in a manner similar to that for the circular tube and parallel plates. He proposed Eq. (576) (see p. 399) for predicting f_{app} Re over the complete range of x^+ for concentric annuli. The values of $K(\infty)$, f Re, and C for concentric annuli are presented in Table 139. Equation (576) provides the f_{app} Re factors for concentric annuli that agree with the results of Table 93 within an rms error of 2%.

C. Thermally Developing and Hydrodynamically Developed Flow

Annular duct thermal entry length solutions for a fully developed laminar velocity profile are divided into six categories: (1) fundamental solutions of the first kind, (2) fundamental solutions of the second kind, (3) fundamental solutions of the third kind, (4) fundamental solutions of the fourth kind, (5) specified wall temperature and heat flux distribution, and (6) finite wall thermal resistance. All those solutions obtained up to 1960 are summarized by Lundberg et al. [41]. The definitions and descriptions of the four fundamental boundary conditions are presented on pp. 31–33. The six solutions are next discussed individually.

1. FUNDAMENTAL SOLUTIONS OF THE FIRST KIND

Fundamental boundary conditions of the first kind, outlined on p. 32, are (1) on one wall a constant temperature is specified as different from the entering fluid temperature; (2) the other wall is at the constant temperature of the entering fluid.

The solution to the energy equation with this boundary condition was first obtained by Murakawa [459,466] by employing an integral formulation. He also employed a series approach for the same problem including peripheral variations of temperature, but he did not present numerical results.

Lundberg et al. [41] analyzed the thermal entry length problem in detail for concentric annular ducts. In their analysis, axial heat conduction, viscous dissipation, flow work, and thermal energy sources within the fluid were neglected. The solutions were obtained by employing a modified Graetz method. The first four eigenvalues and constants, and asymptotic formulas for higher eigenvalues and constants were presented. Also, the fundamental solutions of the first kind were tabulated and graphed. Lundberg's solutions were recomputed, using their tabulated first four eigenvalues and constants and the remaining 120 from their asymptotic formulas based on the WKBJ and graphical methods. These results for $x^* \geq 10^{-2}$ are presented in Table 94. The other quantities of interest can be determined from the following equations:

$$\Phi_{x,oi}^{(1)} = -\theta_{x,mi}^{(1)} \, Nu_{x,oi}^{(1)}, \qquad \theta_{x,ii}^{(1)} = 1, \qquad \theta_{x,oi}^{(1)} = 0 \qquad (457)$$

$$\Phi_{x,io}^{(1)} = -\theta_{x,mo}^{(1)} \, Nu_{x,io}^{(1)}, \qquad \theta_{x,oo}^{(1)} = 1, \qquad \theta_{x,io}^{(1)} = 0 \qquad (458)$$

Viskanta [467,468] obtained two thermal entrance solutions of the first kind. He superimposed the solutions and presented $Nu_x^{(1)}$ and $\Phi_x^{(1)}$ at the inner and outer walls for different values of the inlet temperature ratio, $(t_o - t_e)/(t_i - t_e)$, and $r^* = 0.05, 0.2, 0.5,$ and 0.8.

Lundberg et al. [41] analyzed the Lévêque-type solutions for low values of x^* where the Graetz-type solutions converge very slowly and require a large number of terms in the series. They provided the first term of the Lévêque series for the solutions of the first kind. Worsøe-Schmidt [200] employed a perturbation method and extended the Lévêque-type solution by obtaining the first seven terms of the Lévêque series.

As described on p. 105, the boundary condition at the unheated wall is immaterial in the Lévêque method because the temperature signal affects only the fluid in the immediate vicinity of the heated wall. Hence the Lévêque-type results, for a step change in wall temperature, provide the small x^* approximations for both the first and third kinds of fundamental solutions. The fundamental solutions of the first kind for $x^* < 10^{-2}$ were recomputed using the extended Worsøe-Schmidt solutions.[†] Table 94 thus provides the augmented results of [41,200] with some typographical errors corrected.

It can be shown that

$$\theta_{x,mj}^{(3)} = \frac{r_j^*}{1 + r^*} (4x^* \Phi_{m,jj}^{(3)}) \qquad (458a)$$

where $\theta_{x,mj}^{(3)} = (t_m - t_e)/(t_w - t_e)$. Note the similarity between this equation and Eq. (124), where $(1 - \theta_m) = (t_m - t_e)/(t_w - t_e)$. For parallel plates, $r^* = r_j^* = 1$. Also for the Lévêque-type solution, $\Phi_{m,jj}^{(3)}$ is the same whether one or both walls of parallel plates are heated. Hence a comparison of Eqs. (458a) and (124) reveals that when only one wall is heated, $\theta_{x,mj}^{(3)}$ is reduced by a factor of $r_j^*/(1 + r^*) = 0.5$, as compared to when both walls are heated. As $Nu_{x,jj}^{(3)} = \Phi_{x,jj}^{(3)}/(1 - \theta_{x,mj}^{(3)})$, it is clear why $Nu_{x,jj}^{(3)}$ is lower than $Nu_{x,T}$ for parallel plates.

[†] As Worsøe-Schmidt [200] did not provide constants for Lévêque-type solutions for $r^* = 0.02$ and 0.05 (inner wall heated), the results presented in Tables 94 and 95 are based on the modified Graetz solution provided by Lundberg et al. [41].

TABLE 94

CONCENTRIC ANNULAR DUCTS: FUNDAMENTAL SOLUTIONS OF THE FIRST KIND FOR DEVELOPING TEMPERATURE AND DEVELOPED VELOCITY PROFILES (BASED ON THE RESULTS OF LUNDBERG et al. [41] AND WORSØE-SCHMIDT [200])

x^*	$\phi_{x,ii}^{(1)}$	$\theta_{x,mi}^{(1)}$	$Nu_{x,ii}^{(1)}$	$Nu_{x,oi}^{(1)}$	$\phi_{x,oo}^{(1)}$	$\theta_{x,mo}^{(1)}$	$Nu_{x,oo}^{(1)}$	$Nu_{x,io}^{(1)}$
				$r^* = 0.02$				
0.000010	–	–	–	–	51.081	0.00303	51.236	–
0.000015	–	–	–	–	44.465	0.00396	44.642	–
0.000025	–	–	–	–	37.306	0.00555	37.515	–
0.00005	–	–	–	–	29.350	0.00876	29.609	–
0.00010	78.5	0.0011	78.5	–	23.033	0.01380	23.355	–
0.00015	72.3	0.0014	72.4	–	19.959	0.01798	20.325	–
0.00025	65.4	0.0019	65.5	–	16.633	0.02509	17.061	–
0.0005	57.5	0.0031	57.7	–	12.934	0.03930	13.463	–
0.0010	50.87	0.00519	51.14	–	9.993	0.06134	10.646	–
0.0015	47.51	0.00711	47.85	–	8.560	0.07940	9.299	–
0.0025	43.73	0.01067	44.20	–	7.006	0.10958	7.869	–
0.005	39.28	0.01874	40.03	–	5.272	0.16848	6.340	–
0.010	35.475	0.03328	36.697	0.043	3.881	0.25664	5.220	0.567
0.015	33.495	0.04666	35.134	0.268	3.194	0.32527	4.734	2.149
0.025	31.207	0.07045	33.572	0.988	2.434	0.43240	4.289	8.199
0.05	28.381	0.11294	31.994	2.133	1.537	0.60540	3.896	19.962
0.10	26.124	0.15146	30.787	2.748	0.835	0.75734	3.440	27.500
0.15	25.399	0.16395	30.379	2.889	0.609	0.80651	3.148	29.364
0.25	25.087	0.16930	30.201	2.942	0.512	0.82760	2.972	30.096
0.5	25.051	0.16993	30.179	2.948	0.501	0.83006	2.948	30.179
∞	25.051	0.16993	30.179	2.948	0.501	0.83006	2.948	30.179
				$r^* = 0.05$				
0.000010	–	–	–	–	51.627	0.00297	51.781	–
0.000015	–	–	–	–	44.944	0.00389	45.120	–
0.000025	–	–	–	–	37.713	0.00545	37.920	–
0.00005	–	–	–	–	29.676	0.00860	29.934	–
0.00010	52.0	0.0014	52.1	–	23.296	0.01355	23.616	–
0.00015	47.0	0.0018	47.1	–	20.191	0.01766	20.554	–
0.00025	41.6	0.0027	41.7	–	16.831	0.02464	17.257	–
0.0005	35.4	0.0045	35.6	–	13.095	0.03862	13.621	–
0.0010	30.43	0.00759	30.67	–	10.125	0.06031	10.774	–
0.0015	27.95	0.01036	28.24	–	8.677	0.07809	9.412	–
0.0025	25.19	0.01539	25.59	–	7.108	0.10782	7.967	–
0.005	22.03	0.02652	22.63	–	5.356	0.16591	6.422	–
0.010	19.397	0.04606	20.334	0.054	3.951	0.25247	5.286	0.166
0.015	18.052	0.06366	19.279	0.311	3.258	0.32039	4.794	1.213
0.025	16.521	0.09426	18.241	1.080	2.491	0.42617	4.341	4.763
0.05	14.671	0.14681	17.195	2.241	1.590	0.59338	3.911	11.085
0.10	13.269	0.19140	16.409	2.841	0.915	0.73191	3.413	14.856
0.15	12.857	0.20457	16.164	2.970	0.717	0.77279	3.156	15.722
0.25	12.700	0.20961	16.068	3.015	0.641	0.78842	3.032	16.030
0.5	12.685	0.21009	16.058	3.019	0.634	0.78991	3.019	16.058
∞	12.685	0.21009	16.058	3.019	0.634	0.78991	3.019	16.058
				$r^* = 0.10$				
0.000010	80.290	0.00043	80.324	–	52.186	0.00287	52.336	–
0.000015	70.968	0.00056	71.008	–	45.438	0.00375	45.609	–
0.000025	60.873	0.00080	60.921	–	38.135	0.00525	38.337	–
0.00005	49.632	0.00129	49.696	–	30.019	0.00830	30.270	–
0.00010	40.682	0.00210	40.767	–	23.576	0.01308	23.888	–
0.00015	36.313	0.00280	36.415	–	20.441	0.01706	20.796	–
0.00025	31.567	0.00402	31.694	–	17.048	0.02380	17.464	–
0.0005	26.249	0.00662	26.424	–	13.275	0.03732	13.789	–
0.0010	21.949	0.01094	22.192	–	10.276	0.05832	10.912	–
0.0015	19.787	0.01472	20.083	–	8.814	0.07555	9.535	–
0.0025	17.290	0.02142	17.669	–	7.230	0.10440	8.073	–
0.005	13.833	0.03542	14.341	–	5.461	0.16087	6.509	–
0.010	12.918	0.06131	13.762	0.064	4.044	0.24530	5.359	0.155
0.015	11.851	0.08348	12.931	0.350	3.345	0.31166	4.860	0.933
0.025	10.651	0.12123	12.121	1.162	2.572	0.41480	4.395	3.394
0.05	9.227	0.18382	11.305	2.343	1.670	0.57485	3.927	7.491
0.10	8.199	0.23388	10.702	2.933	1.022	0.70058	3.413	9.792
0.15	7.921	0.24748	10.526	3.054	0.847	0.73472	3.193	10.286
0.25	7.825	0.25218	10.464	3.092	0.787	0.74651	3.103	10.446
0.5	7.817	0.25256	10.459	3.095	0.782	0.74744	3.095	10.459
∞	7.817	0.25256	10.459	3.095	0.782	0.74744	3.095	10.459

(continued)

TABLE 94 (*continued*)

x^*	$\phi^{(1)}_{x,ii}$	$\theta^{(1)}_{x,mi}$	$Nu^{(1)}_{x,ii}$	$Nu^{(1)}_{x,oi}$	$\phi^{(1)}_{x,oo}$	$\theta^{(1)}_{x,mo}$	$Nu^{(1)}_{x,oo}$	$Nu^{(1)}_{x,io}$
				$r^* = 0.25$				
0.000010	66.502	0.00079	66.555	–	53.276	0.00257	53.414	–
0.000015	58.354	0.00104	58.415	–	46.406	0.00337	46.563	–
0.000025	49.537	0.00147	49.609	–	38.972	0.00472	39.157	–
0.00005	39.733	0.00234	39.827	–	30.710	0.00746	30.940	–
0.00010	31.947	0.00375	32.067	–	24.150	0.01176	24.438	–
0.00015	28.156	0.00495	28.297	–	20.959	0.01535	21.286	–
0.00025	24.052	0.00702	24.222	–	17.506	0.02144	17.889	–
0.0005	19.482	0.01130	19.704	–	13.665	0.03368	14.141	–
0.0010	15.843	0.01826	16.138	–	10.613	0.05273	11.204	–
0.0015	14.067	0.02421	14.416	–	9.126	0.06841	9.796	–
0.0025	12.137	0.03460	12.572	–	7.515	0.09475	8.301	–
0.005	9.975	0.05639	10.571	–	5.717	0.14658	6.699	–
0.010	8.236	0.09229	9.073	0.083	4.277	0.22473	5.517	0.130
0.015	7.376	0.12296	8.410	0.410	3.568	0.28659	5.002	0.700
0.025	6.421	0.17352	7.769	1.292	2.786	0.38271	4.514	2.342
0.05	5.315	0.25334	7.118	2.526	1.885	0.52830	3.996	4.844
0.10	4.567	0.31231	6.641	3.120	1.276	0.63474	3.494	6.141
0.15	4.386	0.32662	6.513	3.233	1.129	0.66054	3.326	6.394
0.25	4.331	0.33093	6.474	3.265	1.085	0.66832	3.271	6.467
0.5	4.328	0.33120	6.471	3.267	1.082	0.66880	3.267	6.471
∞	4.328	0.33120	6.471	3.267	1.082	0.66880	3.267	6.471
				$r^* = 0.50$				
0.000010	60.470	0.00121	60.543	–	54.613	0.00220	54.733	–
0.000015	52.881	0.00158	52.964	–	47.601	0.00287	47.738	–
0.000025	44.669	0.00223	44.768	–	40.015	0.00403	40.177	–
0.00005	35.541	0.00354	35.668	–	31.583	0.00637	31.785	–
0.00010	28.295	0.00563	28.455	–	24.889	0.01007	25.142	–
0.00015	24.770	0.00739	24.954	–	21.633	0.01315	21.921	–
0.00025	20.954	0.01041	21.175	–	18.109	0.01840	18.448	–
0.0005	16.711	0.01658	16.993	–	14.190	0.02897	14.614	–
0.0010	13.339	0.02642	13.701	–	11.077	0.04549	11.605	–
0.0015	11.697	0.03472	12.118	–	9.561	0.05916	10.162	–
0.0025	9.916	0.04900	10.427	–	7.918	0.08222	8.627	–
0.005	7.930	0.07824	8.603	–	6.086	0.12797	6.979	–
0.010	6.341	0.12488	7.246	0.092	4.622	0.19773	5.761	0.116
0.015	5.559	0.16381	6.648	0.457	3.902	0.25362	5.227	0.591
0.025	4.698	0.22648	6.073	1.418	3.109	0.34086	4.717	1.884
0.05	3.713	0.32233	5.480	2.744	2.204	0.47154	4.171	3.751
0.10	3.073	0.38995	5.037	3.374	1.615	0.56325	3.698	4.671
0.15	2.928	0.40530	4.923	3.488	1.482	0.58407	3.563	4.841
0.25	2.888	0.40958	4.891	3.519	1.445	0.58987	3.523	4.887
0.5	2.885	0.40982	4.889	3.520	1.443	0.59018	3.520	4.889
∞	2.885	0.40982	4.889	3.520	1.443	0.59018	3.520	4.889
				$r^* = 1.0$				
0.000010	56.804	0.00171	56.901	–	56.804	0.00171	56.901	–
0.000015	49.570	0.00224	49.681	–	49.570	0.00224	49.681	–
0.000025	41.744	0.00314	41.876	–	41.744	0.00314	41.876	–
0.00005	33.046	0.00498	33.211	–	33.046	0.00498	33.211	–
0.00010	26.141	0.00788	26.349	–	26.141	0.00788	26.349	–
0.00015	22.782	0.01032	23.020	–	22.782	0.01032	23.020	–
0.00025	19.147	0.01447	19.428	–	19.147	0.01447	19.428	–
0.0005	15.106	0.02288	15.459	–	15.106	0.02288	15.459	–
0.0010	11.895	0.03613	12.341	–	11.895	0.03613	12.341	–
0.0015	10.332	0.04718	10.843	–	10.332	0.04718	10.843	–
0.0025	8.638	0.06595	9.248	–	8.638	0.06595	9.248	–
0.005	6.750	0.10371	7.531	–	6.750	0.10371	7.531	–
0.010	5.235	0.16249	6.251	0.064	5.235	0.16249	6.251	0.064
0.015	4.498	0.21032	5.696	0.500	4.498	0.21032	5.696	0.500
0.025	3.687	0.28586	5.162	1.597	3.687	0.28586	5.162	1.597
0.05	2.762	0.39926	4.597	3.112	2.762	0.39926	4.597	3.112
0.10	2.168	0.47770	4.151	3.835	2.168	0.47770	4.151	3.835
0.15	2.037	0.49507	4.035	3.965	2.037	0.49507	4.035	3.965
0.25	2.002	0.49976	4.002	3.998	2.002	0.49976	4.002	3.998
0.5	2.000	0.50000	4.000	4.000	2.000	0.50000	4.000	4.000
∞	2.000	0.50000	4.000	4.000	2.000	0.50000	4.000	4.000

The Nusselt number, based on the Lévêque-type solution of Lundberg *et al.* for $x^* < 10^{-2}$, is

$$\mathrm{Nu}_{x,jj}^{(1)} \simeq \mathrm{Nu}_{x,jj}^{(3)} = 0.427(4A)^{1/3}(x^*)^{-1/3}$$

$$= 0.427(f_j\,\mathrm{Re})^{1/3}(x^*)^{-1/3} \qquad (459)$$

where $f_j\,\mathrm{Re}$ are presented in Table 86, and A is given by Eq. (462).

Nunge *et al.* [201] investigated the limitations imposed by the idealizations of the Lévêque-type solutions (refer to p. 105). They concluded that the effect of curvature is important for intermediate values of x^*. However, the linear approximation to the velocity profile (instead of the fully developed profile) near the wall is excellent for concentric annular ducts. The latter conclusion is important and implies that the dependence on r^* of the thermal entrance solutions for annular ducts can be eliminated. They showed that, when the Nusselt number, defined in terms of D_j, is plotted against a newly defined axial distance σ, the r is effectively eliminated as a parameter in the entrance region. This newly defined Nusselt number is

$$\mathrm{Nu}_{x,jj}^{*(k)} = \frac{h_{x,jj}^{(k)}D_j}{k} = \frac{h_{x,jj}^{(k)}D_h}{k}\frac{r_j^*}{1-r^*} = \mathrm{Nu}_{x,jj}^{(k)}\frac{r_j^*}{1-r^*} \qquad (460)$$

where D_j is the diameter of the heated surface. Note that σ and x^* are related by

$$\sigma = \left(\frac{1-r^*}{r_j^*}\right)\left(\frac{36x^*}{A}\right)^{1/3} \qquad (461)$$

where A, the dimensionless velocity gradient evaluated at the heated wall, is

$$A = \frac{s(1-r^*)}{r_j^*}\left[\frac{\partial(u/u_\mathrm{m})}{\partial(r/r_j)}\right]_j = \frac{f_j\,\mathrm{Re}}{4} \qquad (462)$$

in which

$$s = \begin{cases} +1 & \text{for } j=i \\ -1 & \text{for } j=o \end{cases} \qquad (463)$$

The values of A could be obtained from $f_j\,\mathrm{Re}$ of Table 86. Nunge *et al.* showed that $\mathrm{Nu}_{x,jj}^{*(k)}$ is a function of σ only for all values of r^* in the thermal entrance region. It deviates from a single curve correlation as the asymptotic values of $\mathrm{Nu}_{x,jj}^{*(k)}$ are approached. This can be seen from Fig. 90 for $\mathrm{Nu}_{x,ii}^{*(1)}$. These Nusselt numbers were recomputed by the present authors.

2. FUNDAMENTAL SOLUTIONS OF THE SECOND KIND

Fundamental boundary conditions of the second kind, outlined on p. 32, are (1) on one wall, constant axial as well as peripheral wall heat flux is specified; (2) the other wall is insulated.

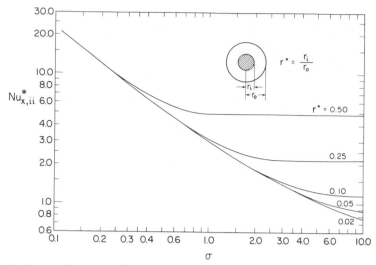

Fig. 90. Concentric annular ducts: Fundamental solutions of the first kind with inner wall heated as correlated by Nunge et al. [201].

Lundberg et al. [41,42] obtained fundamental solutions of the second kind with the same idealizations mentioned on p. 301 for the first fundamental problem. These solutions were recomputed using the first four eigenvalues and constants from [41] and the remaining 120 from the asymptotic formulas of their graphical method. The results for $x^* \geq 10^{-2}$ are presented in Table 95.[†] The other quantities of interest can be determined from the equations

$$\theta^{(2)}_{x,mi} = \left(\frac{4r^*}{1 + r^*}\right)x^*, \qquad \Phi^{(2)}_{x,ii} = 1, \qquad \Phi^{(2)}_{x,oi} = 0 \qquad (464)$$

$$\theta^{(2)}_{x,mo} = \left(\frac{4}{1 + r^*}\right)x^*, \qquad \Phi^{(2)}_{x,oo} = 1, \qquad \Phi^{(2)}_{x,io} = 0 \qquad (465)$$

Hatton and Quarmby [469] obtained thermal entry length solutions of the second and third kinds for the case of the heated inner wall. They presented eigenvalues and related constants in tabular form and $\mathrm{Nu}^{(2)}_{x,ii}$ as a function of $4x^*$ for $r^* = 1/6, 1/3, 1/2, 1/1.5$, and $20/21$. Their results are in excellent agreement with those of Lundberg et al.

Lundberg et al. also obtained the Lévêque-type solution applicable to small values of x^*. Worsøe-Schmidt [200] employed the perturbation method

[†] For $r^* = 0.10$ with inner wall heated, the results of Lundberg et al. are presented for $x^* \geq 10^{-3}$ for a proper match between the results of [41] and [200]. Also refer to the footnote on p. 302.

TABLE 95

Concentric Annular Ducts: Fundamental Solutions of the Second Kind for Developing Temperature and Developed Velocity Profiles (Based on the Results of Lundberg et al. [41] and Worsøe-Schmidt [200])

x^*	$\theta^{(2)}_{x,ii}$	$\theta^{(2)}_{x,oi}$	$Nu^{(2)}_{x,ii}$	$\theta^{(2)}_{x,oo}$	$\theta^{(2)}_{x.io}$	$Nu^{(2)}_{x,oo}$
			$r^* = 0.02$			
0.000025	–	–	–	0.021967	–	45.726
0.00005	–	–	–	0.027853	–	36.157
0.00010	0.0115	–	86.9	0.035376	–	28.584
0.00025	0.0142	–	70.4	0.048699	–	20.956
0.0005	0.0164	–	61.2	0.062242	–	16.589
0.0010	0.01859	–	54.01	0.079885	–	13.164
0.0025	0.02171	–	46.48	0.112095	–	9.776
0.005	0.02429	–	41.85	0.146217	–	7.898
0.010	0.027036	0.000002	38.093	0.193018	0.000097	6.502
0.015	0.028710	0.000026	36.319	0.228808	0.001278	5.883
0.025	0.030900	0.000234	34.556	0.286719	0.011714	5.300
0.05	0.034110	0.001567	33.126	0.401928	0.078339	4.858
0.10	0.038397	0.005296	32.729	0.603071	0.264808	4.741
0.15	0.042340	0.009206	32.706	0.799444	0.460325	4.735
0.25	0.050184	0.017049	32.705	1.191618	0.852448	4.734
0.5	0.069792	0.036657	32.705	2.172011	1.832839	4.734
1.0	0.109008	0.075872	32.705	4.132796	3.793624	4.734
∞	∞	∞	32.705	∞	∞	4.734
			$r^* = 0.05$			
0.000025	–	–	–	0.021735	–	46.210
0.00005	–	–	–	0.027554	–	36.544
0.00010	0.01725	–	58.0	0.034990	–	28.894
0.00025	0.02204	–	45.5	0.048151	–	21.187
0.0005	0.02602	–	38.6	0.061520	–	16.774
0.0010	0.03034	–	33.17	0.078920	–	13.314
0.0025	0.03674	–	27.58	0.110648	–	9.889
0.005	0.04227	–	24.20	0.144201	–	7.990
0.010	0.048370	0.000005	21.521	0.190125	0.000108	6.578
0.015	0.052171	0.000068	20.278	0.225170	0.001360	5.951
0.025	0.057227	0.000596	19.060	0.281745	0.011915	5.362
0.05	0.064801	0.003872	18.091	0.393912	0.077433	4.916
0.10	0.075142	0.012947	17.827	0.589334	0.258931	4.799
0.15	0.084713	0.022445	17.812	0.780094	0.448892	4.792
0.25	0.103763	0.041491	17.811	1.161063	0.829814	4.792
0.5	0.151382	0.089110	17.811	2.113444	1.792195	4.792
1.0	0.246620	0.184348	17.811	4.018208	3.686958	4.792
∞	∞	∞	17.811	∞	∞	4.792
			$r^* = 0.10$			
0.000025	0.013984	–	71.556	0.021503	–	46.703
0.00005	0.017254	–	58.019	0.027252	–	36.940
0.00010	0.021194	–	47.265	0.034596	–	29.212
0.00025	0.027597	–	36.356	0.047581	–	21.426
0.0005	0.033486	–	30.027	0.060756	–	16.967
0.0010	0.04043	–	24.96	0.077880	–	13.469
0.0025	0.05092	–	20.00	0.109038	–	10.005
0.005	0.06024	–	17.12	0.141892	–	8.083
0.010	0.070738	0.000012	14.903	0.186700	0.000116	6.652
0.015	0.077455	0.000139	13.889	0.220768	0.001387	6.016
0.025	0.086584	0.001172	12.904	0.275539	0.011723	5.416
0.05	0.100634	0.007462	12.128	0.383391	0.074621	4.961
0.10	0.120267	0.024794	11.918	0.570190	0.247944	4.841
0.15	0.138533	0.042927	11.907	0.752296	0.429271	4.835
0.25	0.174902	0.079288	11.906	1.115950	0.792877	4.834
0.5	0.265811	0.170197	11.906	2.025041	1.701967	4.834
1.0	0.447629	0.352015	11.906	3.843224	3.520150	4.834
∞	∞	∞	11.906	∞	∞	4.834

(continued)

TABLE 95 (continued)

x^*	$\theta^{(2)}_{x,ii}$	$\theta^{(2)}_{x,oi}$	$Nu^{(2)}_{x,ii}$	$\theta^{(2)}_{x,oo}$	$\theta^{(2)}_{x,io}$	$Nu^{(2)}_{x,oo}$
			$r^* = 0.25$			
0.000025	0.016881	–	59.307	0.021064	–	47.657
0.00005	0.021101	–	47.480	0.026676	–	37.713
0.00010	0.026327	–	38.099	0.033832	–	29.840
0.00025	0.035147	–	28.615	0.046453	–	21.904
0.0005	0.043591	–	23.153	0.059215	–	17.357
0.0010	0.053885	–	18.838	0.075739	–	13.786
0.0025	0.070894	–	14.515	0.105611	–	10.245
0.005	0.086819	–	12.075	0.136845	–	8.275
0.010	0.105856	0.000030	10.219	0.179005	0.000118	6.803
0.015	0.118598	0.000328	9.381	0.210720	0.001311	6.146
0.025	0.136648	0.002649	8.573	0.261101	0.010594	5.522
0.05	0.165971	0.016492	7.938	0.358324	0.065968	5.042
0.10	0.208792	0.054572	7.764	0.523546	0.218290	4.913
0.15	0.248963	0.094456	7.754	0.683863	0.377823	4.905
0.25	0.328974	0.174448	7.753	1.003882	0.697793	4.905
0.5	0.528974	0.374448	7.753	1.803883	1.497792	4.905
1.0	0.928974	0.774448	7.753	3.403883	3.097793	4.905
∞	∞	∞	7.753	∞	∞	4.905
			$r^* = 0.50$			
0.000025	0.018562	–	53.969	0.020549	–	48.822
0.00005	0.023344	–	42.960	0.025994	–	38.668
0.00010	0.029345	–	34.233	0.032919	–	30.626
0.00025	0.039675	–	25.418	0.045080	–	22.515
0.0005	0.049805	–	20.351	0.057314	–	17.864
0.0010	0.062474	–	16.356	0.073054	–	14.207
0.0025	0.084191	–	12.367	0.101237	–	10.574
0.005	0.105411	–	10.127	0.130327	–	8.548
0.010	0.131911	0.000052	8.433	0.168984	0.000105	7.027
0.015	0.150414	0.000563	7.668	0.197616	0.001127	6.345
0.025	0.177603	0.004452	6.931	0.242318	0.008904	5.693
0.05	0.224075	0.024263	6.353	0.326157	0.054896	5.186
0.10	0.294840	0.090599	6.192	0.464851	0.181605	5.046
0.15	0.361768	0.157245	6.182	0.598526	0.314515	5.037
0.25	0.495119	0.290577	6.181	0.865215	0.581152	5.037
0.5	0.828452	0.623910	6.181	1.531881	1.247819	5.037
1.0	1.495118	1.290576	6.181	2.865214	2.581152	5.037
∞	∞	∞	6.181	∞	∞	5.037
			$r^* = 1.0$			
0.000025	0.019758	–	50.740	0.019758	–	50.740
0.00005	0.024940	–	40.257	0.024940	–	40.257
0.00010	0.031498	–	31.950	0.031498	–	31.950
0.00025	0.042931	–	23.567	0.042931	–	23.567
0.0005	0.054322	–	18.754	0.054322	–	18.754
0.0010	0.068821	–	14.965	0.068821	–	14.965
0.0025	0.094345	–	11.193	0.094345	–	11.193
0.005	0.120121	–	9.081	0.120121	–	9.081
0.010	0.153517	0.000080	7.490	0.153517	0.000080	7.490
0.015	0.177641	0.000853	6.773	0.177641	0.000853	6.773
0.025	0.214324	0.006691	6.086	0.214324	0.006691	6.086
0.05	0.280307	0.041118	5.546	0.280307	0.041118	5.546
0.10	0.385362	0.136066	5.395	0.385362	0.136066	5.395
0.15	0.485691	0.235737	5.385	0.485691	0.235737	5.385
0.25	0.685714	0.435714	5.385	0.685714	0.435714	5.385
0.5	1.185714	0.935714	5.385	1.185714	0.935714	5.385
∞	∞	∞	5.385	∞	∞	5.385

and extended the Lévêque-type solution by obtaining the first seven terms ($N = 6$) of the series

$$Nu^{(2)}_{x,jj} = \frac{1}{\frac{1}{2}\sum_{n=0}^{N}(36x^*/A)^{(n+1)/3}\theta^{(2)}_{j,n}(0) - [4r_j^*/(1 + r^*)]x^*} \qquad (466)$$

where A is given by Eq. (462), and $\theta^{(2)}_{j,n}(0)$ are tabulated by Worsøe-Schmidt. The last term of the denominator of Eq. (466) corresponds to $\theta^{(2)}_{x,mj} = (t_m - t_e)/(q''D_h/k)$. Since $r^* = r_j^* = 1$ for parallel plates, this dimensionless bulk mean temperature equals $2x^*$ when one wall is heated. For the case of both walls heated, $\Theta_m = (t_m - t_e)/(q''D_h/k) = 4x^*$ from Eq. (311). Thus it should be clear why $Nu^{(2)}_{x,jj}$ is lower than $Nu_{x,H}$ for parallel plates.

For very small values of x^*, retaining only the first term of the series in Eq. (466) and neglecting the second term of the denominator, yields the following equation with $\theta^{(2)}_{j,o}(0) = 0.73849$:

$$Nu^{(2)}_{x,jj} = 0.517(4A)^{1/3}(x^*)^{-1/3} = 0.517(f_j Re)^{1/3}(x^*)^{-1/3} \qquad (467)$$

As discussed on p. 402, this equation is recommended to determine $Nu_{x,H1}$ for a singly connected noncircular duct.

Based on Worsøe-Schmidt's extension, fundamental solutions of the second kind were recomputed and are presented in Table 95 for $x^* < 10^{-2}$. Thus Table 95 provides the augmented results of [41,200]. As described on p. 305, Nunge et al. [201] found that the dependence of r^* on Nu can be eliminated (except for the asymptotic region), if $Nu^{*(2)}_{x,jj}$ defined by Eq. (460) is plotted against σ defined by Eq. (461).

In all of the foregoing solutions, the effect of fluid axial heat conduction was neglected. Hsu [470] included this effect [Eq. (31) with $v = 0$] and obtained fundamental solutions of the second kind for initial and boundary conditions corresponding to Fig. 23c (t_e = constant at $x = -\infty$). He solved the eigenvalue problem for each of the two semi-infinite regions and matched the temperature and axial temperature gradient at $x = 0$ properly. He presented local Nusselt numbers $Nu^{(2)}_{x,ii}$ and $Nu^{(2)}_{x,oo}$ graphically for $r^* = 0.1, 0.3, 0.5, 0.7,$ and 0.9, and $Pe = 1, 5, 10, 20, 30, 50,$ and ∞ as a function of $x^* (10^{-4} \leq x^* \leq 1)$. The temperature solutions corresponding to the limiting case of $Pe = \infty$ (no fluid axial heat conduction) are in excellent agreement with the results of Lundberg et al. [41]. The Nusselt number vs. x^* behavior for all r^* is similar to that observed for the circular tube (see Fig. 24). For $Pe < \infty$, the Nusselt number is finite at $x = 0$; it decreases with decreasing Pe near $x^* \simeq 0$. In contrast, for large x^*, $Nu^{(2)}_{x,ii}$ and $Nu^{(2)}_{x,oo}$ increase with decreasing Pe. The inflection point changes from $x^* \simeq 0.0075$ for $r^* = 0$ to $x^* \sim 0.004$ for $r^* = 0.9$. Additionally, Hsu found that $Nu^{(2)}_{x,ii}$ is larger than $Nu^{(2)}_{x,oo}$ for a specified r^*. This is the same trend found when the axial heat conduction is neglected (Table 95).

3. FUNDAMENTAL SOLUTIONS OF THE THIRD KIND

The fundamental boundary conditions of the third kind, as outlined on p. 32, are (1) on one wall a constant temperature, different from that of the entering fluid, is specified; (2) the other wall is insulated (adiabatic).

Lundberg *et al.* [41] obtained thermal entrance solutions for the fundamental boundary conditions of the third kind, employing the same idealizations as those mentioned on p. 301 for the fundamental solutions of the first kind. They tabulated the first four eigenvalues and presented asymptotic formulas for higher values for the Graetz-type solution. They also presented results graphically. Their fundamental solutions of the third kind were recomputed by the present authors employing the first 124 terms in the series, and these results are presented in Table 96.[†] The other quantities of interest are

$$\Phi_{x,oi}^{(3)} = 0, \qquad \theta_{x,ii}^{(3)} = 1, \qquad Nu_{x,oi}^{(3)} = 0 \qquad (468)$$

$$\Phi_{x,io}^{(3)} = 0, \qquad \theta_{x,oo}^{(3)} = 1, \qquad Nu_{x,io}^{(3)} = 0 \qquad (469)$$

Viskanta [467] also obtained the complete thermal entrance solutions of the third kind.

Hatton and Quarmby [469] investigated the fundamental solution of the third kind when the inner wall was heated. They presented the first ten eigenvalues and constants in tabular form and $Nu_{x,ii}^{(3)}$ as a function of $4x^*$ for $r^* = 1/6, 1/3, 1/2, 2/3$, and $20/21$. The first four eigenvalues of Hatton and Quarmby and those of Lundberg *et al.* [41,42] agree very well. Ziegenhagen [471] also obtained, by the WKBJ method,[‡] the first ten eigenvalues for the third fundamental problem with the inner wall heated. The first three eigenvalues differ but the rest agree very well with those of Hatton and Quarmby.

The Lévêque-type results do not depend upon the thermal boundary condition at the opposite wall. Hence the Lévêque-type solutions for the heated wall are identical for the first and third fundamental boundary conditions. The recomputed results based on the extended Lévêque-type solution by Worsøe-Schmidt are reported in Table 94 for $x^* < 10^{-2}$.

As described on p. 305, Nunge *et al.* [201] found that the dependence of r^* on Nu can be eliminated (except for the asymptotic region), if $Nu_{x,jj}^{*(2)}$ defined by Eq. (460) is plotted against σ defined by Eq. (461).

Employing the Lévêque-type approach, Venkata Rao *et al.* [472] derived a formula for the average Nusselt number for the boundary condition of the third kind. The effects of natural convection and temperature-dependent viscosity were included by introducing empirical factors.

[†] The present authors were unable to recalculate the results for $r^* = 0.02$ with inner wall heated using the eigenvalues and constants reported in [41] due to some believed typographical errors. Hence the results reported in Table 96 for $r^* = 0.02$ with inner wall heated are the same as those reported by Lundberg *et al.* [41].

[‡] See the footnote on p. 171.

TABLE 96

CONCENTRIC ANNULAR DUCTS: FUNDAMENTAL SOLUTIONS OF THE THIRD KIND FOR
DEVELOPING TEMPERATURE AND DEVELOPED VELOCITY PROFILES
(BASED ON THE RESULTS OF LUNDBERG et al.[41])

x	$\phi_{x,ii}^{(3)}$	$\theta_{x,oi}^{(3)}$	$\theta_{x,mi}^{(3)}$	$Nu_{x,ii}^{(3)}$	$\phi_{x,oo}^{(3)}$	$\theta_{x,io}^{(3)}$	$\theta_{x,mo}^{(3)}$	$Nu_{x,oo}^{(3)}$
				$r^* = 0.02$				
0.010	35.394	.00012	.0331	36.775	3.8810	0.00118	0.25616	5.217
0.015	33.412	.00129	.0470	35.209	3.1941	0.01264	0.32493	4.732
0.025	31.122	.01019	.0739	33.690	2.4341	0.08227	0.43369	4.298
0.05	28.207	.05729	.13407	32.574	1.5241	0.34276	0.62191	4.031
0.10	24.666	.16699	.23740	32.345	0.68894	0.69722	0.82751	3.994
0.15	21.723	.26613	.32824	32.338	0.31481	0.86161	0.92117	3.993
0.25	16.857	.43053	.47872	32.337	0.06575	0.97109	0.98353	3.993
0.50	8.9413	.69793	.72350	32.337	0.00131	0.99942	0.99967	3.993
1.0	2.5157	.91501	.92220	32.337	0.00000	1.00000	1.00000	3.993
∞	0	1	1	32.337	0	1	1	3.993
				$r^* = 0.05$				
0.010	19.405	.00020	.04562	20.332	3.9517	0.00140	0.25254	5.287
0.015	18.060	.00204	.06341	19.282	3.2583	0.01355	0.32063	4.796
0.025	16.528	.01488	.09620	18.288	2.4911	0.08436	0.42858	4.359
0.05	14.605	.07801	.16980	17.592	1.5705	0.34310	0.61631	4.093
0.10	12.273	.21677	.29722	17.464	0.71766	0.69444	0.82311	4.057
0.15	10.391	.33670	.40489	17.461	0.33135	0.85888	0.91832	4.057
0.25	7.4510	.52437	.57327	17.460	0.07066	0.96991	0.98258	4.057
0.5	3.2443	.79290	.81419	17.460	0.00148	0.99937	0.99963	4.057
1.0	0.6151	.96074	.96477	17.460	0.00000	1.00000	1.00000	4.057
∞	0	1	1	17.460	0	1	1	4.057
				$r^* = 0.10$				
0.010	12.920	.00032	.061221	13.762	4.0442	0.00153	0.24535	5.359
0.015	11.853	.00301	.083659	12.935	3.3450	0.01394	0.31196	4.862
0.025	10.652	.02049	.124349	12.165	2.5716	0.08400	0.41803	4.419
0.05	9.1601	.10117	.21362	11.648	1.6427	0.33688	0.60416	4.150
0.10	7.3655	.26918	.36295	11.562	0.76967	0.68400	0.81292	4.114
0.15	5.9683	.40768	.48371	11.560	0.36427	0.85041	0.91145	4.114
0.25	3.9200	.61095	.66090	11.560	0.08162	0.96648	0.98016	4.114
0.5	1.3705	.86398	.88144	11.560	0.00194	0.99920	0.99953	4.114
1.0	0.1675	.98337	.98551	11.560	0.00000	1.00000	1.00000	4.114
∞	0	1	1	11.560	0	1	1	4.114
				$r^* = 0.25$				
0.010	8.2382	.00058	.09229	9.076	4.2773	0.00159	0.22480	5.518
0.015	7.3767	.00507	.12336	8.415	3.5684	0.01349	0.28706	5.005
0.025	6.4209	.03177	.17815	7.813	2.7859	0.07846	0.38738	4.548
0.05	5.2488	.14493	.29346	7.429	1.8451	0.31296	0.56785	4.269
0.10	3.8763	.36106	.47417	7.372	0.92804	0.64888	0.78074	4.233
0.15	2.8861	.52418	.60844	7.371	0.47144	0.82159	0.88860	4.232
0.25	1.6004	.73615	.78287	7.371	0.12169	0.95395	0.97125	4.232
0.5	0.3664	.93959	.95028	7.371	0.00412	0.99844	0.99903	4.232
1.0	0.0192	.99683	.99739	7.371	0.00000	1.00000	1.00000	4.232
∞	0	1	1	7.371	0	1	1	4.232
				$r^* = 0.50$				
0.010	6.3404	.00087	.1250	7.246	4.6214	0.00147	0.19789	5.762
0.015	5.5585	.00729	.1644	6.652	3.9013	0.01207	0.25428	5.232
0.025	4.6968	.04351	.2322	6.117	3.1083	0.06933	0.3466	4.757
0.05	3.6492	.18808	.3692	5.785	2.1535	0.27967	0.51803	4.468
0.10	2.4675	.44405	.5700	5.739	1.1814	0.59887	0.73330	4.430
0.15	1.6829	.62075	.7067	5.738	0.65443	0.77776	0.85225	4.429
0.25	0.7831	.82354	.8635	5.738	0.20086	0.93179	0.95465	4.429
0.5	0.1156	.97394	.9798	5.738	0.01048	0.99644	0.99763	4.429
1.0	0.0025	.99943	.9996	5.738	0.00003	0.99999	0.99999	4.429
∞	0	1	1	5.738	0	1	1	4.429
				$r^* = 1.0$				
0.010	5.2421	.00119	.16254	6.260	5.2421	0.00119	0.16254	6.260
0.015	4.5014	.00982	.21094	5.705	4.5014	0.00982	0.21094	5.705
0.025	3.6868	.05685	.29193	5.207	3.6868	0.05685	0.29193	5.207
0.05	2.7028	.23561	.44867	4.902	2.7028	0.23561	0.44867	.902
0.10	1.6468	.52780	.66124	4.861	1.6468	0.52780	0.66124	861
0.15	1.0127	.70956	.79166	4.861	1.0127	0.70956	0.79166	4.861
0.25	0.3831	.89014	.92119	4.861	0.3831	0.89014	0.92119	4.861
0.5	0.0337	.99033	.99306	4.861	0.0337	0.99033	0.99306	4.861
∞	0	1	1	4.861	0	1	1	4.861

4. FUNDAMENTAL SOLUTIONS OF THE FOURTH KIND

The fundamental boundary conditions of the fourth kind, as outlined on p. 32, are (1) on one wall constant heat flux is specified; (2) the other wall is at the constant temperature of the entering fluid.

Lundberg et al. [41] obtained the Graetz-type and Lévêque-type solutions for this problem. They reported results in graphical and tabular forms. Their recomputed results for the Graetz-type solution (based on 124 terms) are listed in Table 97 for $x^* \geq 10^{-2}$. The other quantities of interest are

$$\Phi_{x,oi}^{(4)} = -\theta_{x,mi}^{(4)} \, \mathrm{Nu}_{x,oi}^{(4)}, \qquad \Phi_{x,ii}^{(4)} = 1, \qquad \theta_{x,oi}^{(4)} = 0 \qquad (470)$$

$$\Phi_{x,io}^{(4)} = -\theta_{x,mo}^{(4)} \, \mathrm{Nu}_{x,io}^{(4)}, \qquad \Phi_{x,oo}^{(4)} = 1, \qquad \theta_{x,io}^{(4)} = 0 \qquad (471)$$

As mentioned earlier, the Lévêque-type solution does not depend upon the thermal boundary condition at the opposite wall. Hence the Lévêque-type solutions for the heated wall are identical for the second and fourth fundamental boundary conditions. The recomputed results based on the extended Lévêque-type solutions by Worsøe-Schmidt [200] are reported in Table 95 for $x^* < 10^{-2}$.

5. SPECIFIED WALL TEMPERATURE AND HEAT FLUX DISTRIBUTION

When viscous dissipation, thermal energy sources, flow work, and fluid axial heat conduction terms are eliminated from Eq. (24), the result is a linear and homogeneous equation. Consequently, a solution can be established for boundary conditions, synthesized from the four fundamental problems, by the superposition of the eight solutions as outlined in Table 4. The following three problems with axisymmetric boundary conditions, synthesized from the fundamental problems of the first, second, and fourth kinds, are of engineering interest: (a) constant temperatures specified on both walls, (b) constant heat fluxes specified on both walls, and (c) constant temperature specified on one wall and constant heat flux specified on the other wall. The temperature, wall heat flux, and Nusselt number relationships for these three problems are presented below. The Nusselt numbers for the outer and inner walls are defined as

$$\mathrm{Nu}_o = \frac{h_o D_h}{k}, \qquad \text{where} \quad h_o = \frac{q_o''}{t_o - t_m} \qquad (472)$$

$$\mathrm{Nu}_i = \frac{h_i D_h}{k}, \qquad \text{where} \quad h_i = \frac{q_i''}{t_i - t_m} \qquad (473)$$

In the following formulas, θ and Φ are found from Table 94, 95, 96, or 97. For the extended problem of an arbitrarily prescribed axisymmetric heat

TABLE 97

CONCENTRIC ANNULAR DUCTS: FUNDAMENTAL SOLUTIONS OF THE
FOURTH KIND FOR DEVELOPING TEMPERATURE AND DEVELOPED
VELOCITY PROFILES (BASED ON THE RESULTS OF LUNDBERG et al. [41])

x^*	$\theta^{(4)}_{x,mi}$	$Nu^{(4)}_{x,ii}$	$Nu^{(4)}_{x,oi}$	$\theta^{(4)}_{x,mo}$	$Nu^{(4)}_{x,oo}$	$Nu^{(4)}_{x,io}$
			$r^* = 0.02$			
0.010	0.0007837	38.093	0.030	0.039215	6.809	–
0.015	0.0011737	36.315	0.216	0.058823	6.130	3.28
0.025	0.0019243	34.512	0.855	0.098039	5.490	6.34
0.05	0.0034689	32.689	1.976	0.196078	4.938	15.128
0.10	0.0052665	31.260	2.648	0.37264	4.529	22.761
0.15	0.0060899	30.663	2.830	0.52557	4.256	25.540
0.25	0.0066382	30.279	2.926	0.77896	3.871	27.751
0.5	0.0067801	30.181	2.948	1.1911	3.375	29.338
1.0	0.0067830	30.179	2.948	1.5257	3.057	29.995
∞	0.0067834	30.179	2.948	1.6568	2.948	30.179
			$r^* = 0.05$			
0.010	0.0019050	21.522	0.037	0.037508	6.552	–
0.015	0.0028515	20.276	0.237	0.056550	5.930	0.513
0.025	0.0046715	19.028	0.903	0.094383	5.337	3.030
0.05	0.0084154	17.776	2.038	0.18544	4.807	8.355
0.10	0.012796	16.798	2.713	0.34771	4.379	12.442
0.15	0.014824	16.392	2.897	0.48524	4.096	13.862
0.25	0.016192	16.128	2.995	0.70032	3.720	14.967
0.5	0.016555	16.060	3.018	1.0081	3.289	15.728
1.0	0.016563	16.059	3.019	1.2005	3.067	16.006
∞	0.016563	16.058	3.019	1.2455	3.019	16.058
			$r^* = 0.10$			
0.010	0.0036361	14.902	0.042	0.036254	6.646	0.034
0.015	0.0054426	13.886	0.256	0.054414	6.011	0.517
0.025	0.0089163	12.875	0.942	0.090404	5.401	2.244
0.05	0.016093	11.864	2.093	0.17599	4.836	5.739
0.10	0.024625	11.072	2.778	0.32377	4.350	8.309
0.15	0.028670	10.740	2.966	0.44364	4.039	9.187
0.25	0.031492	10.521	3.069	0.61953	3.656	9.860
0.5	0.032288	10.460	3.095	0.83845	3.271	10.304
1.0	0.032307	10.459	3.095	0.94176	3.116	10.442
∞	0.032307	10.459	3.095	0.95614	3.095	10.459
			$r^* = 0.25$			
0.010	0.0079989	10.225	0.050	0.031985	6.802	0.066
0.015	0.011974	9.382	0.282	0.047949	6.143	0.419
0.025	0.019636	8.548	0.999	0.079018	5.504	1.598
0.05	0.035729	7.710	2.190	0.15253	4.880	3.800
0.10	0.055780	7.040	2.908	0.27133	4.321	5.322
0.15	0.065984	6.748	3.113	0.35987	3.992	5.826
0.25	0.073803	6.541	3.231	0.47491	3.633	6.200
0.5	0.076431	6.474	3.266	0.58532	3.344	6.421
1.0	0.076523	6.471	3.267	0.61639	3.271	6.469
∞	0.076523	6.471	3.267	0.61810	3.267	6.471
			$r^* = 0.50$			
0.010	0.013332	8.433	0.055	0.026664	7.026	0.064
0.015	0.019960	7.665	0.302	0.039962	6.343	0.364
0.025	0.032796	6.906	1.057	0.066002	5.672	1.318
0.05	0.060343	6.137	2.317	0.12508	4.998	3.007
0.10	0.096747	5.503	3.095	0.21530	4.400	4.125
0.15	0.11694	5.211	3.325	0.27690	4.076	4.484
0.25	0.13433	4.984	3.468	0.34758	3.759	4.739
0.5	0.14163	4.894	3.518	0.40000	3.554	4.870
1.0	0.14203	4.889	3.520	0.40889	3.521	4.889
∞	0.14203	4.889	3.520	0.40908	3.520	4.889
			$r^* = 1.0$			
0.010	0.02009	7.495	0.254	0.02009	7.495	0.254
0.015	0.03001	6.774	0.408	0.03001	6.774	0.408
0.025	0.04936	6.062	1.171	0.04936	6.062	1.171
0.05	0.09211	5.341	2.559	0.09211	5.341	2.559
0.10	0.15285	4.723	3.453	0.15285	4.723	3.453
0.15	0.19025	4.416	3.730	0.19025	4.416	3.730
0.25	0.22740	4.148	3.914	0.22740	4.148	3.914
0.5	0.24801	4.013	3.993	0.24801	4.013	3.993
1.0	0.24998	4.000	4.000	0.24998	4.000	4.000
∞	0.25000	4.000	4.000	0.25000	4.000	4.000

flux or temperature on either wall, the solutions can be determined by the method of superposition, as shown in detail by Lundberg et al. [41,42].

a. *Constant Temperatures Specified on Both Walls*

For $x \geq x_e$,

$$t = t_i \qquad \text{on the inner wall}$$

$$t = t_o \qquad \text{on the outer wall}$$

$$t = t_e \qquad \text{for all} \quad r \text{ at } x < 0$$

In this case,

$$t_{x,m}^{(1a)} = (t_i - t_e)\theta_{x,mi}^{(1)} + (t_o - t_e)\theta_{x,mo}^{(1)} + t_e \tag{474}$$

$$q_{x,i}''^{(1a)} = \frac{k}{D_h}[(t_i - t_e)\Phi_{x,ii}^{(1)} + (t_o - t_e)\Phi_{x,io}^{(1)}] \tag{475}$$

$$q_{x,o}''^{(1a)} = \frac{k}{D_h}[(t_i - t_e)\Phi_{x,oi}^{(1)} + (t_o - t_e)\Phi_{x,oo}^{(1)}] \tag{476}$$

$$\mathrm{Nu}_{x,i}^{(1a)} = \frac{(t_i - t_e)\Phi_{x,ii}^{(1)} + (t_o - t_e)\Phi_{x,io}^{(1)}}{(t_i - t_e)(1 - \theta_{x,mi}^{(1)}) - (t_o - t_e)\theta_{x,mo}^{(1)}} \tag{477}$$

$$\mathrm{Nu}_{x,o}^{(1a)} = \frac{(t_o - t_e)\Phi_{x,oo}^{(1)} + (t_i - t_e)\Phi_{x,oi}^{(1)}}{(t_o - t_e)(1 - \theta_{x,mo}^{(1)}) - (t_i - t_e)\theta_{x,mi}^{(1)}} \tag{478}$$

The thermal entrance solution for the technically important case of equal wall temperatures can be obtained from the above relationships for $t_o = t_i$ and the results of Table 94. To distinguish this special case, the resulting variables are designated with the superscript (1b), as outlined in Table 5. The perimeter average heat transfer coefficient h for this case is defined by Eq. (76). Subsequently $\mathrm{Nu}_{x,T}$ can be determined from the expression

$$\mathrm{Nu}_{x,T} = \frac{\mathrm{Nu}_{x,o}^{(1b)} + r^*\mathrm{Nu}_{x,i}^{(1b)}}{1 + r^*} \tag{479}$$

Fuller and Samuels [195] analyzed the Ⓣ thermal entrance problem with fully developed laminar flow for the concentric annular duct of $r^* = 0.5$. They included the effect of fluid axial heat conduction and obtained a solution for the initial and boundary conditions of Fig. 17b on each wall. Their $\mathrm{Nu}_{x,T}$ for $\mathrm{Pe} = \infty$ are in excellent agreement with those of Lundberg et al. [41,42].

Fuller and Samuels concluded that the effect of fluid axial heat conduction on the thermal entrance solution is as follows: (1) As Pe decreases, the effect of axial heat conduction propagates further upstream; the temperature profile at the entrance of a real tube, $x = 0$, becomes more nonuniform; and the assumption of a uniform inlet temperature is not valid even for Pe $= 200$. However, for Pe > 50, only small errors will occur in the downstream temperature profiles if a flat profile is assumed at the entrance. (2) For $x > 0$, the axial heat conduction has the effect of raising the temperature in the center of the flow passage while decreasing it near the wall. (3) Axial heat conduction has a more significant effect on $Nu_{x,T}$ than it does on the temperature profiles. The lower Pe, the higher $Nu_{x,T}$ becomes for a specified x^*. A higher thermal entrance length $L^*_{th,T}$ also results.

b. *Constant Heat Fluxes Specified on Both Walls*

$$q'' = q_i'' \qquad \text{on the inner wall}$$

$$q'' = q_o'' \qquad \text{on the outer wall}$$

$$t = t_e \qquad \text{for all} \quad r \text{ at } x < 0$$

The thermal entrance solution for this case is obtained by superimposing the results of Table 95:

$$t_{x,i}^{(2a)} = \frac{D_h}{k} \left[q_i'' \theta_{x,ii}^{(2)} + q_o'' \theta_{x,io}^{(2)} \right] + t_e \tag{480}$$

$$t_{x,o}^{(2a)} = \frac{D_h}{k} \left[q_i'' \theta_{x,oi}^{(2)} + q_o'' \theta_{x,oo}^{(2)} \right] + t_e \tag{481}$$

$$t_{x,m}^{(2a)} = \frac{D_h}{k} \left[q_i'' \theta_{x,mi}^{(2)} + q_o'' \theta_{x,mo}^{(2)} \right] + t_e \tag{482}$$

$$Nu_{x,i}^{(2a)} = \frac{q_i''}{q_i'' \left[\theta_{x,ii}^{(2)} - \theta_{x,mi}^{(2)} \right] - q_o'' \left[\theta_{x,mo}^{(2)} - \theta_{x,io}^{(2)} \right]} \tag{483}$$

$$Nu_{x,o}^{(2a)} = \frac{q_o''}{q_o'' \left[\theta_{x,oo}^{(2)} - \theta_{x,mo}^{(2)} \right] - q_i'' \left[\theta_{x,mi}^{(2)} - \theta_{x,oi}^{(2)} \right]} \tag{484}$$

For the case of thermally and hydrodynamically developed flow, the $\circledH$ boundary condition has been designated as constant but different heat fluxes specified on both walls, such that the peripheral wall temperatures at the inner and outer walls are constant and the same at any axial location. For this case, the unique q_i''/q_o'' ratio for a given r^* is listed in Table 92. For thermally developing flow, however, the q_i''/q_o'' ratio is also a function of x^* at a specified r^*. Hence the foregoing $\circledH$ boundary condition does not obtain. Moreover, since it would not be realized in a practical application, no results are presented for such boundary conditions.

c. *Constant Temperature Specified on One Wall, and Constant Heat Flux Specified on the Other Wall*

For $x > x_e$[†]

$$t = t_1 \qquad \text{on wall 1}$$

$$q'' = q_2'' \qquad \text{on wall 2}$$

$$t = t_e \qquad \text{for all} \quad r \text{ at } x < x_e$$

The thermal entrance solution for this case is obtained by superimposing the results of Tables 96 and 97:

$$t_{x,2} = (t_1 - t_e)\theta_{x,21}^{(3)} + \frac{D_h}{k} q_2'' \theta_{x,22}^{(4)} + t_e \tag{485}$$

$$t_{x,m} = (t_1 - t_e)\theta_{x,m1}^{(3)} + \frac{D_h}{k} q_2'' \theta_{x,m2}^{(4)} + t_e \tag{486}$$

$$q_{x,1}'' = \frac{k}{D_h}(t_1 - t_e)\Phi_{x,11}^{(3)} + q_2'' \Phi_{x,12}^{(4)} \tag{487}$$

$$\mathrm{Nu}_{x,1} = \frac{(t_1 - t_e)\Phi_{x,11}^{(3)} + (q_2'' D_h/k)\Phi_{x,12}^{(4)}}{(t_1 - t_e)(1 - \theta_{x,m1}^{(3)}) - (q_2'' D_h/k)\theta_{x,m2}^{(4)}} \tag{488}$$

$$\mathrm{Nu}_{x,2} = \frac{1}{(\theta_{x,22}^{(4)} - \theta_{x,m2}^{(4)}) + [(t_1 - t_e)k/q_2'' D_h](\theta_{x,21}^{(3)} - \theta_{x,m1}^{(3)})} \tag{489}$$

d. *Peripherally Constant But Axially Arbitrary Distribution*

As discussed earlier, once the eight solutions to the four fundamental problems (Table 4) are available, the solution for any axially arbitrary, but peripherally constant, boundary condition can be obtained by Duhamel's superposition technique. Lundberg *et al.* [42] solved the following two problems by the superposition technique: (1) heat flux specified on one wall, temperature specified on the other, and (2) heat flux specified on both walls. Hatton and Quarmby [469] analyzed the following three problems by the superposition technique: (1) axially linear increase in temperature on the inner wall, (2) axially linear increase in heat flux on the inner wall, and (3) half-sine wave heat input variation superimposed on a uniform heat input on the inner wall. In all three cases, the outer wall was insulated.

[†] No distinction is needed between inner and outer walls, and so the subscripts 1 and 2 replace i and o.

e. *Both Peripherally and Axially Arbitrary Wall Temperature
   or Heat Flux Distribution*

In all of the preceding thermal entry length problems, the peripheral wall
temperature and heat flux were considered constant (axisymmetric). Bhattacharyya *et al.* [293,473] analyzed the thermal entry problems of the first
and second kinds with arbitrary peripheral wall temperature and heat flux,
respectively. The method of separation of variables was employed to obtain
the solution, and necessary eigenvalues and constants were tabulated for
$r^* = 0.2$, 0.5, and 0.8. They extended the analysis further using Duhamel's
superposition theorem for the case of arbitrary temperature or heat flux
both peripherally and axially along the duct wall.

Bhattacharyya *et al.* obtained numerical results for their thermal entrance
solutions for the following four peripheral wall temperature distributions
(for $r^* = 0.2$, 0.5, and 0.8):

$$\text{(a)} \quad t_i = t_{i,m}(1 + 0.1 \cos\theta), \qquad t_o = 0 \tag{490}$$

$$\text{(b)} \quad t_i = t_{i,m}(1 + 0.1 \cos 2\theta), \qquad t_o = 0 \tag{491}$$

with

$$(t_{i,m} - t_e)/t_e = 0.5 \tag{491a}$$

and

$$\text{(c)} \quad t_o = t_{o,m}(1 + 0.1 \cos\theta), \qquad t_i = 0 \tag{492}$$

$$\text{(d)} \quad t_o = t_{o,m}(1 + 0.1 \cos 2\theta), \qquad t_i = 0 \tag{493}$$

with

$$(t_{o,m} - t_e)/t_e = 0.5 \tag{494}$$

Here $t_{i,m}$ and $t_{o,m}$ are mean temperatures of inner and outer walls, respectively,
and they are treated as axially constant. These authors presented graphically
$Nu_{x,ii}^{(1)}$ and $Nu_{x,oo}^{(1)}$ as a function of x^* and arrived at the following conclusions:
(1) the effect of peripherally variable temperature on the Nusselt number is
greater for an inner wall variation than for an outer wall variation. (2) the
effect of peripherally variable wall temperature on the local Nusselt number
increase as r^* decreases. These two conclusions are valid for the thermal
entrance region $x^* \lesssim 0.01$, while the opposite effects are observed for
$x^* \gtrsim 0.05$. (3) The effect of the second harmonic term of the wall temperature
on the peripheral local Nusselt number is of nearly the same order as the
first harmonic term for $r^* \gtrsim 0.5$. Also, this effect is more pronounced for the
thermal entrance region, $x^* \lesssim 0.01$, than that for the developed region,

$x^* \gtrsim 0.05$. (4) The peripherally variable wall temperature has very little effect on the thermal entry length. The latter remains approximately the same as that for the case of peripherally constant wall temperature.

Bhattacharyya [293] also obtained the fundamental solutions of the second kind for arbitrary variable peripheral wall heat flux. Numerical thermal entrance solutions were obtained for four specific peripheral wall heat flux distributions:

$$\text{(a)} \quad q_i'' = q_{i,m}''(1 + 0.1 \cos \theta), \qquad q_o'' = 0 \qquad (495)$$

$$\text{(b)} \quad q_i'' = q_{i,m}''(1 + 0.1 \cos 2\theta), \qquad q_o'' = 0 \qquad (496)$$

$$\text{(c)} \quad q_o'' = q_{o,m}''(1 + 0.1 \cos \theta), \qquad q_i'' = 0 \qquad (497)$$

$$\text{(d)} \quad q_o'' = q_{o,m}''(1 + 0.1 \cos 2\theta), \qquad q_i'' = 0 \qquad (498)$$

The wall temperatures and Nusselt numbers were presented graphically for the foregoing examples as a function of x^* for $r^* = 0.2, 0.5$, and 0.8. Bhattacharyya arrived at the following conclusions: (1) The effect of the first harmonic heat flux on the axial temperature at both walls is more pronounced than that of the second harmonic heat flux. However, for $r^* \gtrsim 0.5$, the first and second harmonic heat fluxes have almost identical effects, provided the amplitudes are the same. (2) The effect of the first harmonic heat flux on the Nusselt number is more pronounced than that of the second harmonic. (3) The peripherally variable wall heat flux has very little effect on the thermal entry length compared to the case of peripherally constant wall heat flux for small r^*. However, the thermal entry length for the variable peripheral wall heat flux is large for large r^* ($\gtrsim 0.5$).

6. THERMAL ENTRANCE LENGTH $L_{th,j}^{*(k)}$

Fundamental solutions of four kinds for developing temperature and developed velocity profiles are presented in preceding sections. The thermal entrance lengths, based on these solutions, are presented in Table 98. The thermal entrance length $L_{th,j}^{*(k)}$ for kth fundamental solution and jth wall heated is the value of x^* at which $\text{Nu}_{x,jj}^{(k)} = 1.05 \, \text{Nu}_{jj}^{(k)}$. As mentioned in the footnote on p. 310, the present authors were unable to recalculate $\text{Nu}_{x,ii}^{(3)}$ for $r^* = 0.02$; $\text{Nu}_{ii}^{(3)}$ for $r^* = 0.02$ was calculated as 29.829. Based on this solution, $L_{th,i}^{*(3)}$ for $r^* = 0.02$ has been determined as 0.02252. In Table 98, it is not clear why $L_{th,o}^{*(2)}$ for $r^* = 0.05$ is slightly lower than that for $r^* = 0.02$.

7. FINITE WALL THERMAL RESISTANCE, (T3)

Hsu and Huang [474] considered the finite thermal resistance, the (T3) boundary condition, at the inner and outer walls and obtained the thermal entry length solution for fully developed laminar flow through an annulus.

TABLE 98

CONCENTRIC ANNULAR DUCTS: $L_{th,j}^{*(k)}$ FOR DEVELOPING
TEMPERATURE AND DEVELOPED VELOCITY PROFILES
(DETERMINED BY THE PRESENT AUTHORS)

r^*	$L_{th,i}^{*(1)}$	$L_{th,o}^{*(1)}$	$L_{th,i}^{*(2)}$	$L_{th,o}^{*(2)}$
0.02	0.05840	0.1650	0.02699	0.03901
0.05	0.06488	0.1458	0.03043	0.03886
0.10	0.06953	0.1311	0.03334	0.03911
0.25	0.07621	0.1126	0.03726	0.04006
0.50	0.08237	0.1003	0.03975	0.04090
1.00	0.09023	0.09023	0.04101	0.04101

r^*	$L_{th,i}^{*(3)}$	$L_{th,o}^{*(3)}$	$L_{th,i}^{*(4)}$	$L_{th,o}^{*(4)}$
0.02	0.02252	0.03001	0.07962	0.04241
0.05	0.02429	0.02970	0.09493	0.6638
0.10	0.02558	0.02960	0.1101	0.5284
0.25	0.02720	0.02964	0.1309	0.3770
0.50	0.02829	0.02956	0.1721	0.2875
1.00	0.02913	0.02913	0.2201	0.2201

The first ten eigenvalues and constants were presented for $r^* = 0.5$ and the following combinations of $R_{w,i}$ and $R_{w,o}$: $(0.5, \infty)$, $(0.1, \infty)$, $(0.05, \infty)$, $(0.5, 1)$, $(0.5, 0.5)$, $(0.5, 0.25)$, $(0.1, 0.2)$, $(0.1, 0.1)$, $(0.05, 0.1)$, $(0.1, 0.05)$, $(0.05, 0.05)$, and $(0.05, 0.025)$. They presented graphically local Nusselt numbers as a function of $4(1 - r^*)^2 x^*/r^{*2}$ for some of the foregoing combinations of $R_{w,i}$ and $R_{w,o}$.

D. Simultaneously Developing Flow

Murakawa [466] presented an analysis for the combined hydrodynamic and thermal entry length problem when the inner surface of the annular duct was heated. The analysis is complicated, and the numerical results are quoted for one radius ratio only.

Heaton et al. [44] obtained the fundamental solution of the second kind for the combined entry length problem. The hydrodynamic entry length solution was obtained by the Langhaar-type linearization method. As described on p. 297, even though their velocity profiles were inaccurate, Heaton et al. used physical arguments to readjust their theoretical temperature profiles. As a result, their theoretical Nusselt numbers agreed with their experimental measurements. Shumway and McEligot [45] showed that the heat transfer results of Heaton et al. [44] are within 1% of their predictions. Heaton et al. tabulated Nusselt numbers, wall temperatures, and fluid bulk mean temperatures as functions of x^* for $r^* = 0$, 0.02, 0.05, 0.10, 0.25, 0.50, and 1.00 and Pr = 0.01, 0.70, 10.0, and ∞. Only the fundamental solutions of the second kind for $r^* = 0.25$ and Pr = 0.7 are presented as an example in Table 99. This provides a ready comparison with the results of Table 95 for the hydrodynamically developed flow case.

Shumway and McEligot [45] employed a finite difference method to solve the combined entry problem for constant and variable physical

TABLE 99

CONCENTRIC ANNULAR DUCTS: FUNDAMENTAL SOLUTIONS OF
THE SECOND KIND FOR SIMULTANEOUSLY DEVELOPING FLOW
FOR $Pr = 0.7$ $(r^* = 0.25)$ (FROM HEATON et al. [44])

x^*	$\theta^{(2)}_{x,ii}$	$\theta^{(2)}_{x,oi}$	$Nu^{(2)}_{x,ii}$	$\theta^{(2)}_{x,oo}$	$\theta^{(2)}_{x,io}$	$Nu^{(2)}_{x,oo}$
0.00010	0.01818	–	55.4	0.01842	–	52.4
0.00015	0.02172	–	46.3	0.02358	–	43.2
0.00025	0.0272	–	37.1	0.0302	–	34.0
0.00050	0.0364	–	27.8	0.0420	–	24.7
0.0010	0.0480	–	21.2	0.0582	–	18.2
0.0015	0.0563	–	18.16	0.0703	–	15.27
0.0025	0.0678	–	15.20	0.0889	–	12.36
0.0050	0.0860	–	12.20	0.1226	–	9.38
0.010	0.1053	–	10.28	0.1667	–	7.43
0.015	0.1184	–	9.40	0.1990	–	6.62
0.025	0.1367	0.0027	8.57	0.2516	0.0106	5.83
0.050	0.1659	0.0165	7.94	0.3538	0.0660	5.16
0.10	0.2087	0.0546	7.77	0.5230	0.2180	4.93
0.15	0.2490	0.0945	7.75	0.6840	0.3780	4.91
0.25	0.3290	0.1745	7.75	1.004	0.6980	4.91
0.50	0.5290	0.3745	7.75	1.804	1.498	4.91
1.0	0.9290	0.7745	7.75	3.404	3.098	4.91
∞	∞	∞	7.75	∞	∞	4.91

properties. The fundamental solutions of the first, third, and fourth kinds were obtained for the combined entry problem, for $r^* = 0.25$ and $Pr = 0.72$. Their results are presented in Table 99a. The results for the other r^* can be estimated from the results of Heaton et al. [44] and Table 99a.

Coney and El-Shaarawi [46] analyzed numerically the fundamental problem of the third kind for simultaneously developing flow. The solutions were obtained for concentric annular ducts of $r^* = 0.1$, 0.25, 0.5, and 0.9 with $Pr = 0.7$ by employing an implicit finite difference technique. The local Nusselt numbers and the fluid bulk mean average temperature were presented graphically for $r^* = 0.1$ and 0.5. Their results for $r^* = 0.25$ are in reasonable agreement with those of Shumway and McEligot [45].

We should recall at this point that even though the fundamental solutions of all four kinds are available for the combined entry length problem, superposition techniques cannot be employed for any specified variations in axial wall temperature or heat flux. The reasoning leading to this caution was presented previously (p. 34). The combined entry length problem for the concentric annular duct for any specified variations in wall temperature or heat flux can be directly solved numerically by the method and the computer program outlined by Shumway [475].

Fuller and Samuels [195] analyzed the Ⓣ combined entry length problem for a concentric annular duct of $r^* = 0.5$. They showed that the heat transfer rate and $Nu_{x,T}$ in the thermal entrance region increase for the developing velocity field in comparison to the developed velocity field. They also showed that the effect of the radial velocity component is small for low and high Péclet numbers. The maximum effect of radial convection for $Re = 5$ was found to occur between $Pe = 50$ and 500.

TABLE 99a

Concentric Annular Ducts: Fundamental Solutions of the First, Second, and Fourth Kinds for Simultaneously Developing Flow for Pr = 0.72 ($r^* = 0.25$)

(from Shumway and McEligot [45])

First Kind, Inner Wall Heated

x^*	$\phi_{x,ii}^{(1)}$	$-\phi_{x,oi}^{(1)}$	$\theta_{x,mi}^{(1)}$	$Nu_{x,ii}^{(1)}$	$Nu_{x,oi}^{(1)}$
0.0001	41.40	—	0.00615	41.65	—
0.00025	28.19	—	0.01087	28.48	—
0.0005	21.39	—	0.01482	21.71	—
0.001	16.48	—	0.02215	16.85	—
0.0025	12.02	—	0.03857	12.50	0.0001
0.005	9.730	0.0001	0.05987	10.35	0.00083
0.01	8.050	0.00088	0.09475	8.892	0.0936
0.025	6.383	0.2294	0.1744	7.731	1.315
0.05	5.310	0.6410	0.2533	7.112	2.530
0.1	4.567	0.9735	0.3120	6.639	3.120

First Kind, Outer Wall Heated

x^*	$\phi_{x,oo}^{(1)}$	$-\phi_{x,io}^{(1)}$	$\theta_{x,mo}^{(1)}$	$Nu_{x,io}^{(1)}$	$Nu_{x,oo}^{(1)}$
0.0001	36.86	—	0.02310	—	37.73
0.00025	23.61	—	0.03660	—	24.51
0.0005	16.91	—	0.05205	—	17.84
0.001	12.11	—	0.07430	—	13.08
0.0025	7.839	0.0001	0.11195	0.0001	8.893
0.005	5.663	0.03041	0.1718	0.0024	6.838
0.01	4.119	0.0508	0.2476	0.2050	5.474
0.025	2.681	1.029	0.3980	2.585	4.454
0.05	1.841	2.650	0.5350	4.947	3.965
0.1	1.268	2.916	0.6362	6.154	3.485
0.25	1.085	4.321	0.6684	6.465	3.271

Third Kind, Inner Wall Heated

x^*	$\phi_{x,ii}^{(3)}$	$\theta_{x,oi}^{(3)}$	$\theta_{x,mi}^{(3)}$	$Nu_{x,oi}^{(3)}$	$Nu_{x,ii}^{(3)}$
0.0001	41.40	—	0.00615	—	41.65
0.00025	28.19	—	0.01482	—	28.48
0.0005	21.39	—	0.02215	—	21.71
0.001	16.48	—	0.03857	—	16.85
0.0025	12.02	0.0001	0.05987	0.0001	12.50
0.005	9.730	0.00067	0.09480		10.35
0.01	8.050	0.03287	0.1793		8.893
0.025	6.382	0.1461	0.2941		7.777
0.05	5.240	0.3615	0.4744		7.424
0.1	3.872	0.7355	0.7823		7.367
0.25	1.60				7.366

Third Kind, Outer Wall Heated

x^*	$\phi_{x,oo}^{(3)}$	$\theta_{x,io}^{(3)}$	$\theta_{x,mo}^{(3)}$	$Nu_{x,io}^{(3)}$	$Nu_{x,oo}^{(3)}$
0.0001	36.86	—	0.02310	—	37.73
0.00025	23.61	—	0.03660	—	24.51
0.0005	16.91	—	0.05205	—	17.84
0.001	12.11	—	0.07430	—	13.08
0.0025	7.830	0.0001	0.11195	0.0001	8.893
0.005	5.663	0.00025	0.1718		6.838
0.01	4.119	0.00285	0.2477		5.474
0.025	2.680	0.09247	0.4038		4.496
0.05	1.795	0.3294	0.5785		4.259
0.1	0.9069	0.6570	0.7857		4.231
0.25	0.1205	0.9544	0.9715		4.231

Fourth Kind, Inner Wall Heated

x^*	$\theta_{x,ii}^{(4)}$	$-\phi_{x,oi}^{(4)}$	$\theta_{x,mi}^{(4)}$	$Nu_{x,ii}^{(4)}$	$Nu_{x,oi}^{(4)}$
0.0001	0.01808	—	0.00008	55.55	—
0.00025	0.02717	—	0.00020	37.08	—
0.0005	0.03642	—	0.00040	27.76	—
0.001	0.04815	—	0.00080	21.12	—
0.0025	0.06787	—	0.00200	15.18	—
0.005	0.08578	0.0001	0.00399	12.23	0.001
0.01	0.1061	0.00043	0.00799	10.19	0.0537
0.025	0.1369	0.07809	0.01961	8.523	0.9997
0.05	0.1655	0.07809	0.03567	7.704	2.189
0.1	0.1978	0.1620	0.05570	7.038	2.908
0.25	0.2267	0.2383	0.07374	6.539	3.231

Fourth Kind, Outer Wall Heated

x^*	$\theta_{x,oo}^{(4)}$	$-\phi_{x,io}^{(4)}$	$\theta_{x,mo}^{(4)}$	$Nu_{x,io}^{(4)}$	$Nu_{x,oo}^{(4)}$
0.0001	0.01959	—	0.00032	—	51.89
0.00025	0.03066	—	0.00080	—	33.48
0.0005	0.04284	—	0.00160	—	24.25
0.001	0.05972	—	0.00320	—	17.69
0.0025	0.0921	—	0.00799	—	11.90
0.005	0.1266	0.0001	0.01598	0.001	9.041
0.01	0.1731	0.00302	0.03196	0.0946	7.088
0.025	0.2690	0.1317	0.07944	1.657	5.539
0.05	0.3575	0.5812	0.1525	3.811	4.879
0.1	0.5027	1.443	0.2712	5.322	4.320
0.25	0.7497	2.941	0.4745	6.208	3.622

Chapter XIII
Eccentric Annular Ducts

An eccentric instead of concentric annular duct is sometimes used as a fluid-flow and heat-transfer device. The eccentricity may stem from design constraints, but more frequently it is a result of manufacturing tolerances or deformation in service for the nominally concentric annular duct configuration. As will be shown later, measurement of the fully developed f Re factors provides a sensitive test for the magnitude of the eccentricity, particularly for large r^* annular ducts.

The fully developed laminar fluid flow and heat transfer problems have been analyzed in detail for eccentric annular ducts. Hydrodynamic entry length solutions are available for five eccentric annuli. Thermal entry length problems with developed and developing velocity profiles are also available for some eccentric annuli.

A thorough literature survey on the subject up to 1967 has been reported by Tiedt [456].

A. Fully Developed Flow

1. FLUID FLOW

Macdonald [476] derived an expression for the torsion moment of an eccentric annulus by using the bipolar and bicircular isometric coordinates, which contained both circles as parameter curves. Caldwell [73] and Piercy et al. [74] independently showed that Macdonald's equation for the torsion moment was comparable to the equation for the volumetric flow rate through

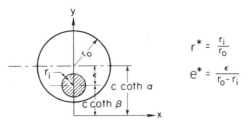

FIG. 91. An eccentric annular duct.

the eccentric annulus for the fully developed laminar flow case, except for the supplementary term Z of Eq. (510). The results for the velocity profile by Piercy *et al.* are in simpler form and are given below. Consider the eccentric annular duct of Fig. 91. Note that x in this section represents the transverse direction at a duct cross section. By applying the transformation

$$x + iy = c \tan \tfrac{1}{2}(\xi + i\eta) \tag{499}$$

the momentum equation (4) is transformed into bipolar coordinates. Subsequently, based on the corresponding solution for the torsion problem, Piercy *et al.* arrived at the following velocity distribution:

$$u = -c_1 r_0^2 S^2 \left[C + A\eta + B - \frac{\cosh \eta - \cos \xi}{4(\cosh \eta + \cos \xi)} \right], \tag{500}$$

where

$$C = \sum_{n=1}^{\infty} \frac{(-1)^n \cos n\xi}{\sinh[n(\beta - \alpha)]} \{ e^{-n\beta} \coth \beta \sinh[n(\eta - \alpha)]$$
$$- e^{-n\alpha} \coth \alpha \sinh[n(\eta - \beta)] \} \tag{501}$$

$$A = \frac{\coth \alpha - \coth \beta}{2(\alpha - \beta)} \tag{502}$$

$$B = \frac{\beta(1 - 2\coth \alpha) - \alpha(1 - 2\coth \beta)}{4(\alpha - \beta)} \tag{503}$$

$$S = \left(\frac{1 - r^*}{2e^*} \right)(1 - e^{*2})^{1/2} \left[\left(\frac{1 + r^*}{1 - r^*} \right)^2 - e^{*2} \right]^{1/2} \tag{504}$$

$$\alpha = \sinh^{-1} S \tag{505}$$

$$\beta = \sinh^{-1}(S/r^*) \tag{506}$$

$$c = r_0 S \tag{507}$$

The bipolar coordinates ξ and η can be presented as functions of x and y from Eq. (499) by equating the real and imaginary parts:

$$e^{-2\eta} = \frac{x^2 + (y - c)^2}{x^2 + (y + c)^2} \tag{508}$$

$$\tan \xi = \frac{2cx}{x^2 + y^2 - c^2} \tag{509}$$

The velocity distribution expressed by Eq. (500) is quite nonuniform over the cross section. For example, Piercy et al. [74] plotted constant-velocity contours for $r^* = 0.25$ and $e^* = 0.33$ and found the ratio of the maximum velocity in the wide part of the channel to the maximum velocity in the narrow part to be 3. This ratio is 1.4 for $r^* = 0.4$, $e^* = 0.1$, and 8 for $r^* = 0.4$, $e^* = 0.5$. This nonuniform velocity profile causes considerable wall temperature nonuniformity in the angular direction.

Both the mean velocity u_m [obtained by carrying out the necessary integration of Eq. (500)] and the flow rate Q of Caldwell and Piercy et al. agree with each other identically. Based on these results, f Re factors for the eccentric annular ducts $(0 < e^* < 1, 0 \le r^* < 1)$ are given by

$$f \, \mathrm{Re} = 16(1 - r^{*2})(1 - r^*)^2$$
$$\times \left[1 - r^{*4} + Z - 8e^{*2}(1 - r^*)^2 S^2 \sum_{n=1}^{\infty} \frac{ne^{-n(\alpha + \beta)}}{\sinh[n(\beta - \alpha)]} \right]^{-1} \tag{510}$$

where S, α, and β are given by Eqs. (504)–(506), and

$$Z = \frac{4e^{*2}(1 - r^*)^2}{\alpha - \beta} S^2 \tag{511}$$

Two limiting cases that are excluded in the above formulation are: (1) $e^* = 1$, $0 \le r^* < 1$, and (2) $0 \le e^* \le 1$, $r^* = 1$; the associated formulas for these cases are presented below. The third limiting case of $e^* = 0$, $0 \le r^* \le 1$, corresponds to the concentric annular duct geometry, for which the f Re factors are presented by Eq. (428).

Based on the exact solution of torsion in a hollow shaft of unit eccentricity by Stevenson [477], the f Re factors for $e^* = 1$, $0 \le r^* < 1$ are given by Tiedt [456] as

$$f \, \mathrm{Re} = \frac{16(1 - r^{*2})(1 - r^*)^2}{1 - r^{*4} - 4r^{*2}\psi'[1/(1 - r^*)]} \tag{512}$$

where ψ' is the so-called Trigamma function with argument $[1/(1 - r^*)]$:

$$\psi'\left(\frac{1}{1-r^*}\right) = \sum_{n=0}^{\infty} \left\{\frac{1}{n + [1/(1-r^*)]}\right\}^2 \qquad (513)$$

Becker [478] derived the expression for the volumetric flow rate through an eccentric annulus for $0 \le e^* \le 1$, $r^* \to 1$. Based on these results, Tiedt [456] showed that

$$f\,\mathrm{Re} = \frac{24}{1 + 1.5e^{*2}} \qquad (514)$$

From Eqs. (510), (512), and (514), Tiedt [456] accurately determined $4f\,\mathrm{Re}$ factors in detail for eccentric annular ducts. His partial results, in terms of $f\,\mathrm{Re}$, are presented in Table 100 and Fig. 92.

The other investigations into the velocity problem of eccentric annuli are summarized now. Heyda [479] determined Green's function in bipolar coordinates for the potential equation and obtained the velocity distribution in the form of an infinite series. His main interest was establishing the locus of maximum velocity for fully developed laminar flow in an eccentric annulus. Redberger and Charles [480] solved the momentum equation by a finite difference method after transforming it into bipolar coordinates. They presented graphically the ratio of the mass flow rate of eccentric to concentric annuli of the same diameter ratio. Their results are in satisfactory agreement with those of Tiedt [456].

Snyder and Goldstein [481] arrived at a closed-form solution, in a form different from Eq. (500), for the velocity distribution in bipolar coordinates. They determined local wall shear stresses, $f_i\,\mathrm{Re}$, $f_o\,\mathrm{Re}$ and $f\,\mathrm{Re}$ for $r^* = 1/2$, 5/6, and $0.1 \le e^* \le 0.9$. Their $f\,\mathrm{Re}$ factors are in good agreement with those of Table 100. Jonsson and Sparrow [482] also independently analyzed laminar flow through eccentric annular ducts. Numerical results were presented graphically for wall shear stresses and $f\,\mathrm{Re}$ for $r^* = 0.05$, 0.10, 0.25, 0.75, and 0.95, and $0.01 \le e^* \le 0.99$. The $f\,\mathrm{Re}$ factors of Jonsson and Sparrow are in excellent agreement with those of Table 100.

Cheng and Hwang [110], while studying the (H1) temperature problem by a point-matching method, also obtained the velocity distribution for fully developed laminar flow through eccentric ducts. Their results for various flow characteristics are in excellent agreement with those of Table 100. Similarly, Trombetta [117], while analyzing the three fundamental problems for eccentric annular ducts, also obtained the velocity distribution by a discrete least squares method. His $f\,\mathrm{Re}$ factors are in excellent agreement with those of Table 100.

TABLE 100

ECCENTRIC ANNULAR DUCTS: f Re FOR FULLY DEVELOPED
LAMINAR FLOW (FROM TIEDT [456])

r^*	fRe					
	$e^* = .05$	$e^* = 0.1$	$e^* = 0.2$	$e^* = 0.3$	$e^* = 0.4$	$e^* = 0.5$
0	16.000	16.000	16.000	16.000	16.000	16.000
0.005	19.504	19.444	19.210	18.844	18.380	17.857
0.01	20.004	19.932	19.654	19.221	18.674	18.061
0.02	20.600	20.512	20.174	19.650	18.993	18.261
0.03	20.993	20.893	20.509	19.918	19.179	18.360
0.04	21.289	21.180	20.758	20.110	19.304	18.414
0.05	21.528	21.409	20.955	20.258	19.393	18.442
0.06	21.726	21.600	21.116	20.376	19.459	18.454
0.08	22.044	21.905	21.369	20.553	19.548	18.450
0.10	22.292	22.140	21.561	20.680	19.599	18.423
0.15	22.731	22.555	21.887	20.877	19.647	18.317
0.20	23.023	22.829	22.093	20.985	19.641	18.197
0.25	23.231	23.024	22.234	21.048	19.615	18.081
0.30	23.387	23.168	22.335	21.086	19.582	17.975
0.40	23.598	23.362	22.465	21.125	19.515	17.800
0.50	23.729	23.481	22.541	21.139	19.458	17.671
0.60	23.811	23.555	22.587	21.144	19.415	17.579
0.70	23.861	23.601	22.615	21.145	19.386	17.518
0.80	23.891	23.627	22.631	21.145	19.367	17.480
0.90	23.908	23.642	22.639	21.145	19.358	17.460
1.00	23.910	23.645	22.642	21.145	19.355	17.455
	$e^* = 0.6$	$e^* = 0.7$	$e^* = 0.8$	$e^* = 0.9$	$e^* = .95$	$e^* = 1$
0	16.000	16.000	16.000	16.000	16.000	16.000
0.005	17.317	16.802	16.352	16.012	15.900	15.842
0.01	17.432	16.833	16.309	15.907	15.769	15.690
0.02	17.514	16.805	16.182	15.693	15.517	15.399
0.03	17.527	16.737	16.041	15.486	15.277	15.126
0.04	17.511	16.656	15.900	15.288	15.051	14.869
0.05	17.479	16.569	15.761	15.099	14.836	14.627
0.06	17.439	16.479	15.625	14.919	14.633	14.399
0.08	17.345	16.301	15.369	14.584	14.257	13.978
0.10	17.243	16.129	15.131	14.280	13.917	13.600
0.15	16.990	15.739	14.610	13.629	13.198	12.808
0.20	16.760	15.406	14.181	13.105	12.625	12.184
0.25	16.558	15.125	13.825	12.677	12.161	11.683
0.30	16.384	14.887	13.528	12.325	11.781	11.276
0.40	16.107	14.517	13.073	11.791	11.210	10.669
0.50	15.909	14.256	12.755	11.422	10.818	10.254
0.60	15.770	14.075	12.537	11.170	10.551	9.973
0.70	15.678	13.955	12.392	11.004	10.375	9.788
0.80	15.622	13.882	12.304	10.903	10.268	9.675
0.90	15.593	13.843	12.258	10.850	10.213	9.617
1.00	15.584	13.833	12.245	10.835	10.196	9.600

Bourne et al. [483] experimentally derived the f Re factors for the annuli of unit eccentricity ($e^* = 1$). They found the f Re factors decreasing with increasing r^* up to $r^* = 0.75$ (f Re $\simeq 12.0$ at $r^* = 0.75$) and then increasing with increasing r^*, approaching 24 at $r^* = 1$, in contrast to 9.6 from Fig. 92. They made a comparison with what they called Caldwell's theoretical solution, which seemed to confirm their experimental findings. However, Caldwell's results [73], though valid for a doubly connected duct ($e^* < 1$), are not valid for the singly connected duct ($e^* = 1$). As presented in Fig. 92, the f Re vs. r^* curve for $e^* = 1$ does not show a minimum, in contradiction to Bourne's results. Tiedt [484] discusses this contradiction in detail. Accord-

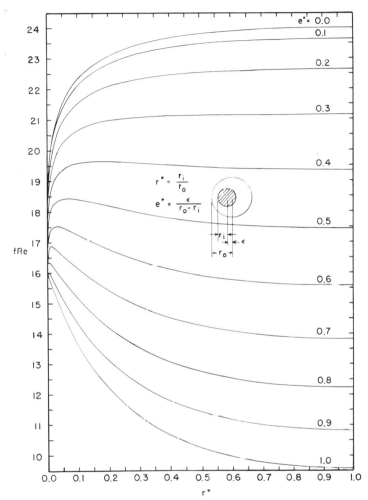

FIG. 92.   Eccentric annular ducts: f Re for fully developed laminar flow (from Table 100).

ing to his experiments, f Re factors are very sensitive to the experimental setup for large r^*, and the experimental results of Bourne *et al.* for $r^* > 0.75$ are believed to be in error.

An inspection of Fig. 92 reveals that for small eccentricities ($e^* < 0.1$) the f Re characteristics of eccentric annuli are not significantly different from those of concentric annuli. However, the f Re factor is very sensitive to e^* for values greater than 0.2. The large e^* is achieved either by large eccentricities ϵ/r_o at small r^* or by small eccentricities ϵ/r_o at large r^* ($r^* \rightarrow 1$), as is

$$e^* = \frac{\epsilon}{r_o - r_i} = \frac{\epsilon/r_o}{1 - r^*} \qquad (515)$$

Thus, for large r^*, a more precise measurement of the eccentricity ϵ is required to fix e^* accurately. Initially, Tiedt [456] was not able to correlate his experimental $f\,Re$ factors for fully developed laminar flow through an eccentric annulus of large r^*, because his initial measurements of ϵ were not sufficiently precise. This dramatic influence of eccentricity on the flow resistance (or flow rate) through an annular slit has been long realized in a number of engineering applications, e.g., a journal bearing.

In the case of concentric annuli, it was shown that there is a considerable increase in the flow resistance to viscous flow by introducing along the axis of a straight pipe a small r_i core ($r^* \geq 0.001$). This behavior is repeated in Fig. 92 as the $e^* = 0$ characteristic. This flow resistance, if desired, can be significantly reduced below the concentric case by locating the core with an eccentricity $e^* > 0.5$.

2. HEAT TRANSFER

One of the first analytical studies of laminar forced convection heat transfer through eccentric annuli was conducted by Sastry [485], who employed the conformal mapping method to solve the axially constant wall heat flux problem. However, he specified the boundary condition for the velocity problem as $u = 0$ on the outer circle, $u = u_o = $ constant on the inner circle. Hence, his results are not relevant for ducts with stationary walls.

Cheng and Hwang [110] analyzed the (H1) temperature problem for eccentric annuli. As defined in Table 5, the (H1) boundary condition for a doubly connected duct means axially constant but different heat fluxes for each wall, such that uniform and equal peripheral wall temperatures result at any axial location. For fully developed flow conditions, this wall temperature will vary linearly. Cheng and Hwang employed a 20-point matching method and determined Nu_{H1}, based on h defined by Eq. (76). Their results are reported in Table 101 and Fig. 93. Cheng and Hwang also presented Nu_{H1} graphically as a function of thermal energy sources with e^* as a parameter.

Trombetta [117] conducted a thorough study of the fundamental problems of the first, second, and fourth kinds (Table 4) for eccentric annuli. He employed a discrete least squares method and obtained detailed solutions for the velocity and temperature problems. His results for the velocity problem are in excellent agreement with those of Tiedt [456]. His results for the temperature problems are accurate to within 1% for most of the range of r^* and e^*. Trombetta presented graphically $Nu_{ii}^{(1)}$, $Nu_{oo}^{(1)}$, $Nu_{ii}^{(2)}$,

TABLE 101

Eccentric Annular Ducts: Nu_{H1} for Fully
Developed Laminar Flow (from Cheng and
Hwang [110])

e^*	Nu_{H1}			
	$r^*=0.25$	$r^*=0.50$	$r^*=0.75$	$r^*=0.90$
0.00	7.804	8.117	8.214	8.232
0.01	7.800	8.111	8.208	8.226
0.10	7.419	7.608	7.659	7.667
0.20	6.524	6.473	6.432	6.422
0.40	4.761	4.393	4.227	4.192
0.60	3.735	3.247	3.024	2.975
0.80	3.203	2.644	2.384	2.324
0.90	3.038	2.446	2.171	2.106
0.99	2.925	2.305	2.016	1.947

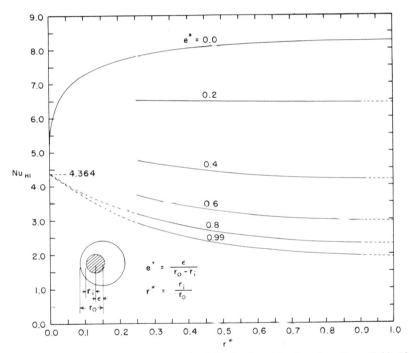

FIG. 93. Eccentric annular ducts: Nu_{H1} for fully developed laminar flow (from Table 101).

$Nu_{oo}^{(2)}$, $Nu_{ii}^{(4)}$, $Nu_{oi}^{(4)}$, $Nu_{oo}^{(4)}$, and $Nu_{io}^{(4)}$ as functions of r^* and e^*. Based on detailed computer solutions [486], Tables 102–104 summarize Trombetta's results.

The concentric annular duct relationships for the fundamental solutions of the first and second kinds [Eqs. (432)–(439)] are also valid for eccentric annular ducts, with a possible exception of Eq. (436). If r^* is replaced by

TABLE 102

Eccentric Annular Ducts: Fundamental Solutions of
the First Kind for Fully Developed Velocity Profile
(from Trombetta [486])

r^*	$\phi_{ii}^{(1)}$	$Nu_{ii}^{(1)}$	$Nu_{oo}^{(1)}$	$\phi_{ii}^{(1)}$	$Nu_{ii}^{(1)}$	$Nu_{oo}^{(1)}$
	$e^* = 0.20$			$e^* = 0.40$		
0.02	25.30	30.20	3.120	26.17	30.46	3.715
0.05	12.84	16.08	3.188	13.38	16.28	3.764
0.10	7.932	10.48	3.257	8.325	10.66	3.797
0.20	5.056	7.232	3.360	5.347	7.414	3.837
0.30	3.948	6.008	3.453	4.193	6.202	3.882
0.40	3.337	5.354	3.543	3.554	5.562	3.936
0.50	2.943	4.945	3.633	3.138	5.166	3.998
0.60	2.663	4.664	3.724	2.843	4.899	4.065
0.70	2.452	4.459	3.814	2.620	4.707	4.136
0.80	2.287	4.303	3.904	2.444	4.563	4.210
0.90	2.153	4.181	3.993	2.301	4.452	4.287
0.95	2.094	4.129	4.038	2.239	4.406	4.325
0.99	2.051	4.091	4.074	2.193	4.372	4.356
	$e^* = 0.60$			$e^* = 0.80$		
0.02	28.11	31.57	5.129	33.10	35.58	9.493
0.05	14.61	17.07	5.076	17.89	19.92	8.747
0.10	9.222	11.32	4.971	11.63	13.61	8.004
0.20	6.012	8.004	4.830	7.787	9.921	7.240
0.30	4.750	6.777	4.764	6.229	8.555	6.875
0.40	4.044	6.136	4.744	5.340	7.846	6.688
0.50	3.582	5.746	4.754	4.749	7.420	6.596
0.60	3.250	5.487	4.783	4.320	7.144	6.559
0.70	2.998	5.306	4.826	3.992	6.956	6.557
0.80	2.799	5.174	4.878	3.730	6.825	6.579
0.90	2.636	5.075	4.936	3.514	6.732	6.617
0.95	2.565	5.035	4.968	3.420	6.696	6.640
0.99	2.513	5.007	4.993	3.350	6.672	6.661
	$e^* = 0.90$			$e^* = 0.95$		
0.02	39.60	41.39	18.24	45.70	46.95	34.26
0.05	20.72	24.62	14.69	28.24	30.11	22.71
0.10	15.34	17.48	12.55	20.25	22.74	18.52
0.20	10.52	13.09	10.69	14.42	17.73	15.41
0.30	8.486	11.43	9.887	11.78	15.70	14.13
0.40	7.308	10.56	9.481	10.17	14.59	13.44
0.50	6.515	10.05	9.266	9.083	13.92	13.07
0.60	5.937	9.716	9.157	8.284	13.49	12.88
0.70	5.490	9.495	9.112	7.664	13.20	12.79
0.80	5.132	9.345	9.109	7.167	13.01	12.76
0.90	4.837	9.243	9.134	6.755	12.88	12.79
0.95	4.708	9.205	9.155	6.575	12.83	12.81
0.99	4.612	9.181	9.175	6.440	12.80	12.83

the ratio of the inner to outer perimeter of the annulus, Eq. (436) is valid
for concentric annular and confocal elliptical ducts. Hence, by inference,
it may also be valid for eccentric annular ducts.

The relationships of Eqs. (443)–(445) for the concentric annular duct
fundamental solutions of the fourth kind are *not* valid for eccentric annular
ducts. Instead, the following relationships apply:

$$Nu_{ii}^{(4)} = \frac{1}{\theta_{ii}^{(4)} - \theta_{mi}^{(4)}} \quad , \qquad Nu_{oo}^{(4)} = \frac{1}{\theta_{oo}^{(4)} - \theta_{mo}^{(4)}} \tag{516}$$

$$Nu_{oi}^{(4)} = \frac{r^*}{\theta_{mi}^{(4)}} \quad , \qquad\qquad Nu_{io}^{(4)} = \frac{1}{r^*\theta_{mo}^{(4)}} \tag{517}$$

TABLE 103

ECCENTRIC ANNULAR DUCTS: FUNDAMENTAL SOLUTIONS OF THE SECOND KIND FOR
A FULLY DEVELOPED VELOCITY PROFILE (FROM TROMBETTA [486])

r^*	$\theta_{oi}^{(2)}-\theta_{mi}^{(2)}$	$\theta_{io}^{(2)}-\theta_{mo}^{(2)}$	$Nu_{ii}^{(2)}$	$Nu_{oo}^{(2)}$	$\theta_{oi}^{(2)}-\theta_{mi}^{(2)}$	$\theta_{io}^{(2)}-\theta_{mo}^{(2)}$	$Nu_{ii}^{(2)}$	$Nu_{oo}^{(2)}$
	$e^*=0.20$				$e^*=0.40$			
0.02	−0.00194	−0.09706	31.16	4.602	−0.00040	−0.02026	27.61	4.358
0.05	−0.00434	−0.08686	16.61	4.584	−0.00007	−0.00160	14.09	4.223
0.10	−0.00727	−0.07269	10.75	4.497	−0.00272	0.02710	8.627	3.962
0.20	−0.00852	−0.04260	7.136	4.221	0.01949	0.09741	5.140	3.347
0.30	−0.00100	−0.00333	5.487	3.825	0.06030	0.2010	3.504	2.662
0.40	0.02157	0.05392	4.290	3.299	0.1455	0.3638	2.391	1.977
0.50	3.07323	0.1465	3.218	2.645	0.3196	0.6392	1.552	1.353
0.60	0.1895	0.3159	2.200	1.903	0.6942	1.157	0.9223	0.835
0.70	0.4795	0.6851	1.284	1.158	1.611	2.301	0.4766	0.445
0.80	1.407	1.759	0.5670	0.531	4.522	5.653	0.1927	0.184
0.90	6.933	7.703	0.1344	0.130	21.81	24.24	0.0435	0.043
0.95	30.28	31.87	0.0322	0.032	94.85	99.04	0.0103	0.010
0.99	−	−	0.0012	−	−	−	0.0004	−
	$e^*=0.60$				$e^*=0.80$			
0.02	0.00145	0.07048	23.63	4.207	0.00332	0.1526	19.78	4.195
0.05	0.00469	0.09227	11.63	4.017	0.00918	0.1707	9.447	4.000
0.10	0.01300	0.1298	6.840	3.677	0.02176	0.2100	5.427	3.639
0.20	0.04654	0.2322	3.838	2.932	0.06738	0.3329	2.971	2.831
0.30	0.1178	0.3924	2.479	2.182	0.1611	0.5384	1.889	2.036
0.40	0.2605	0.6513	1.612	1.513	0.3475	0.8685	1.214	1.356
0.50	0.5479	1.096	1.005	0.971	0.7224	1.445	0.7516	0.837
0.60	1.163	1.938	0.5790	0.567	1.525	2.542	0.4320	0.472
0.70	2.665	3.808	0.2932	0.289	3.487	4.982	0.2191	0.234
0.80	7.432	9.291	0.1173	0.116	9.715	12.14	0.0881	0.092
0.90	35.74	39.71	0.0264	0.026	46.70	51.89	0.0200	0.020
0.95	155.3	163.5	0.0063	0.006	202.9	213.6	0.0048	0.005
0.99	−	−	0.0002	−	−	−	0.0002	−
	$e^*=0.90$				$e^*=0.95$			
0.02	0.00448	0.1900	18.05	4.229	0.00541	0.2131	17.32	4.249
0.05	0.01176	0.2063	8.458	4.015	0.01350	0.2284	8.035	4.029
0.10	0.02657	0.2462	4.766	3.653	0.02955	0.2634	4.511	3.677
0.20	0.07747	0.3797	2.578	2.829	0.08324	0.4077	2.382	2.828
0.30	0.1829	0.6052	1.633	2.006	0.1974	0.6527	1.492	1.993
0.40	0.3917	0.9785	1.050	1.312	0.4220	1.055	0.9584	1.281
0.50	0.8126	1.625	0.6514	0.794	0.8750	1.752	0.5950	0.765
0.60	1.714	2.858	0.3756	0.440	1.846	3.078	0.3439	0.419
0.70	3.919	5.599	0.1914	0.215	4.220	6.029	0.1757	0.203
0.80	10.92	13.65	0.0774	0.083	11.75	14.70	0.0713	0.078
0.90	52.49	58.32	0.0177	0.018	5.651	62.79	0.0164	0.017
0.95	228.1	240.1	0.0042	0.004	245.5	258.4	0.0039	0.004
0.99	−	−	0.0002	−	−	−	0.0002	−

The relationship of Eq. (442) for concentric annuli is also valid for eccentric annuli.

As mentioned on p. 292 for concentric annular ducts, additional solutions of practical interest can be obtained by the superposition of the fundamental solutions. The three examples considered there are (1) constant temperatures specified on both walls, (2) constant heat fluxes specified on both walls, and (3) constant temperature specified on one wall, constant heat flux specified on the other. For eccentric annuli, some of these solutions can be obtained in the same manner, but the constant and equal wall temperatures solution cannot be obtained by the superposition of the results of Table 102. This solution may be derived from the asymptotic thermal entrance solution (refer to the footnote on p. 35).

As defined in Table 5, the (H2) boundary condition for a doubly connected duct means both axially and peripherally constant but different heat fluxes

TABLE 104

Eccentric Annular Ducts: Fundamental Solutions of the Fourth Kind for a Fully Developed Velocity Profile (from Trombetta [486])

r^*	$\theta_{mi}^{(4)}$	$\theta_{mo}^{(4)}$	$Nu_{ii}^{(4)}$	$Nu_{oo}^{(4)}$	$\theta_{mi}^{(4)}$	$\theta_{mo}^{(4)}$	$Nu_{ii}^{(4)}$	$Nu_{oo}^{(4)}$
	$e^* = 0.20$				$e^* = 0.40$			
0.02	0.006411	1.702	30.20	3.181	0.005387	1.831	30.46	4.005
0.05	0.01569	1.291	16.08	3.276	0.01333	1.419	16.28	4.179
0.10	0.03078	1.001	10.49	3.377	0.02665	1.126	10.68	4.362
0.20	0.06009	0.7377	7.247	3.524	0.05429	0.8530	7.463	4.600
0.30	0.08855	0.5984	6.040	3.643	0.08347	0.7026	6.304	4.754
0.40	0.1163	0.5070	5.405	3.746	0.1140	0.5997	5.728	4.855
0.50	0.1432	0.4407	5.017	3.840	0.1455	0.5221	5.407	4.919
0.60	0.1693	0.3898	4.756	3.927	0.1774	0.4604	5.221	4.958
0.70	0.1946	0.3491	4.571	4.009	0.2093	0.4098	5.114	4.979
0.80	0.2188	0.3157	4.434	4.088	0.2406	0.3673	5.055	4.991
0.90	0.2420	0.2879	4.327	4.166	0.2711	0.3312	5.024	4.999
0.95	0.2533	0.2756	4.283	4.204	0.2859	0.3152	5.014	5.003
0.999	0.2621	0.2645	4.250	4.242	–	0.3006	–	5.008
	$e^* = 0.60$				$e^* = 0.80$			
0.02	0.003907	2.038	31.57	6.059	0.002119	2.162	35.58	15.19
0.05	0.009952	1.623	17.07	6.357	0.005951	1.905	19.85	14.85
0.10	0.02085	1.316	11.33	6.658	0.01406	1.588	13.43	13.04
0.20	0.04614	1.018	8.056	7.016	0.03663	1.226	9.612	12.62
0.30	0.07599	0.8448	6.902	7.200	0.06675	1.012	8.226	12.52
0.40	0.1097	0.7227	6.369	7.268	0.1032	0.8605	7.607	12.52
0.50	0.1466	0.6285	6.122	7.246	0.1450	0.7447	7.379	12.40
0.60	0.1856	0.5521	6.041	7.154	0.1910	0.6510	7.412	12.05
0.70	0.2259	0.4881	6.068	7.010	0.2402	0.5720	7.652	11.47
0.80	0.2666	0.4335	6.169	6.834	0.2911	0.5038	8.069	10.74
0.90	0.3068	0.3863	6.313	6.648	0.3423	0.4442	8.626	9.970
0.95	0.3264	0.3652	6.393	6.559	0.3674	0.4172	8.939	9.604
0.999	–	0.3459	–	6.476	–	0.3925	–	9.269
	$e^* = 0.90$				$e^* = 0.95$			
0.02	0.001085	–	41.63	–	0.000484	–	48.73	–
0.05	0.003773	–	24.24	–	0.002629	–	29.49	–
0.10	0.01045	1.769	16.39	26.78	0.006641	1.850	19.48	–
0.20	0.03158	1.354	11.36	18.90	0.02900	1.424	12.80	25.08
0.30	0.06159	1.105	9.500	17.65	0.05889	1.155	10.44	21.58
0.40	0.09912	0.9342	8.678	17.64	0.09689	0.9724	9.421	21.52
0.50	0.1431	0.8053	8.391	17.68	0.1419	0.8362	9.063	21.89
0.60	0.1924	0.7017	8.472	17.24	0.1929	0.7274	9.160	21.55
0.70	0.2460	0.6147	8.869	16.19	0.2486	0.6362	9.656	20.13
0.80	0.3022	0.5394	9.569	14.72	0.3075	0.5572	10.56	17.94
0.90	0.3593	0.4731	10.56	13.15	0.3675	0.4876	11.88	15.59
0.95	0.3874	0.4430	11.14	12.42	0.3972	0.4559	12.68	14.52
0.999	–	0.4154	–	11.76	–	0.4268	–	13.57

specified on each wall. The magnitude of peripheral wall heat fluxes is such that it results in equal *average* temperatures $t_{w,m}$ for each wall at every cross section, with $t_{w,m}$ a linear function of the flow length x. To obtain Nu_{H2} for eccentric annuli, first the ratio q_i''/q_o'' is obtained by equating Eqs. (480) and (481), with $x = \infty$:

$$\frac{q_i''}{q_o''} = \frac{\theta_{oo}^{(2)} - \theta_{io}^{(2)}}{\theta_{ii}^{(2)} - \theta_{oi}^{(2)}} = \frac{(\theta_{oo}^{(2)} - \theta_{mo}^{(2)}) - (\theta_{io}^{(2)} - \theta_{mo}^{(2)})}{(\theta_{ii}^{(2)} - \theta_{mi}^{(2)}) - (\theta_{oi}^{(2)} - \theta_{mi}^{(2)})} \quad (518)$$

Note that the temperatures on the left-hand sides of Eqs. (480) and (481) represent the peripherally average wall temperatures at section x.

Using q_i''/q_o'' of Eq. (518), the Nusselt numbers at each wall for this case are next determined from the following equations, which are dervied from

Eqs. (483) and (484), for $x = \infty$:

$$\text{Nu}_i^{(2b)} = \frac{1}{(\theta_{ii}^{(2)} - \theta_{mi}^{(2)}) + (q_o''/q_i'')(\theta_{io}^{(2)} - \theta_{mo}^{(2)})} \tag{519}$$

$$\text{Nu}_o^{(2b)} = \frac{1}{(\theta_{oo}^{(2)} - \theta_{mo}^{(2)}) + (q_i''/q_o'')(\theta_{oi}^{(2)} - \theta_{mi}^{(2)})} \tag{520}$$

Finally, Nu_{H2} is computed:

$$\text{Nu}_{H2} = \frac{\text{Nu}_o^{(2b)} + r^*\text{Nu}_i^{(2b)}}{1 + r^*} \tag{521}$$

Nu_{H2} for $e^* = 0.2$ to 0.95 were determined using the foregoing relationships and the results of Table 103. The Nu_{H2} for $e^* = 0.05$ and 0.10 were determined from the results provided by Trombetta [486]. These are presented in Table 105 and Fig. 94.

Nu_{H1} cannot be determined from Trombetta's results by employing superposition because the nonexisting fundamental solutions of the fifth kind are needed, as mentioned in Table 5. However, Cheng and Hwang [110] have obtained Nu_{H1} directly, as reported in Table 101 and Fig. 93.

B. Hydrodynamically Developing Flow

Feldman [184] analyzed laminar developing flow through eccentric annuli by first transforming the pertinent equations into a bipolar coordinate system and then employing a finite difference method.

Since the eccentric annular duct is not axisymmetric, the three velocity components u, v, and w exist in the entrance region. For concentric annular ducts, only u and v exist in the entrance region due to the axisymmetrical geometry. As mentioned on p. 10, to obtain a rigorous solution to the entry length problem, a third equation is necessary. Feldman obtained the third equation, relating v and w velocity components, based on the following hypothesis of the transverse flow. He idealized the transverse flow as emanating radially from the surface, bending away from the radial line due to the presence of the other surface, and consequently transferring fluid from the narrow to the wider part of the annulus as shown in Fig. 95. The surface along which the two transverse flows merged was approximated by a circular cylinder, which approximately coincided with the ridges of the maximum axial velocity for fully developed flow.

Feldman employed a finite difference iterative method with variable grid and step sizes, and obtained the entry length solutions for five eccentric annular ducts using this model. The f_{app} Re factors are presented in Table 106 and Fig. 96. In order to investigate the sensitivity of the choice of trans-

TABLE 105

ECCENTRIC ANNULAR DUCTS: Nu_{H2} FOR FULLY DEVELOPED LAMINAR FLOW
(BASED ON EQ. (521) AND TABLE 103)

r^*	Nu_{H2}							
	$e^*=.05$	$e^*=.1$	$e^*=.2$	$e^*=.4$	$e^*=.6$	$e^*=.8$	$e^*=.9$	$e^*=.95$
0.02	6.347	6.224	5.826	4.913	4.537	4.157	4.149	4.160
0.05	6.777	6.582	5.956	4.701	4.289	3.849	3.827	3.833
0.10	7.139	6.815	5.854	4.228	3.518	3.334	3.322	3.341
0.20	7.430	6.793	5.191	3.191	2.517	2.360	2.361	2.370
0.30	7.452	6.395	4.241	2.244	1.694	1.571	1.571	1.575
0.40	7.267	5.688	3.198	1.484	1.080	0.990	0.988	0.989
0.50	6.839	4.691	2.206	0.918	0.650	0.590	0.586	0.586
0.60	6.064	3.463	1.364	0.522	0.362	0.326	0.323	0.322
0.70	4.776	2.154	0.725	0.262	0.179	0.160	0.158	0.157
0.80	3.273	1.002	0.299	0.104	0.070	0.062	0.061	0.061
0.90	0.887	0.246	0.068	0.023	0.015	0.013	0.013	0.013
0.95	0.229	0.059	0.016	0.005	0.003	0.003	0.003	0.003

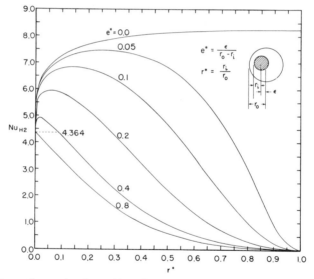

FIG. 94. Eccentric annular ducts: Nu_{H2} for fully developed laminar flow (from Table 105).

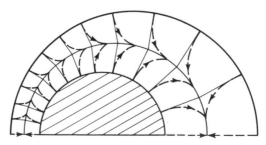

FIG. 95. The transverse flow model for an eccentric annulus.

TABLE 106

ECCENTRIC ANNULAR DUCTS: f_{app} Re FOR DEVELOPING LAMINAR
FLOW, AND $K(\infty)$, L_{hy}^+, AND u_{max}/u_m FOR DEVELOPED LAMINAR
FLOW (FROM FELDMAN [487])

x^+	$e^*=.5$ $r^*=.5$	$e^*=.5$ $r^*=.1$	$e^*=.7$ $r^*=.3$	$e^*=.9$ $r^*=.1$	$e^*=.9$ $r^*=.5$
	f_{app}Re				
0.001	113.65	109.58	114.75	116.10	120.16
0.002	82.03	79.10	83.05	81.72	83.47
0.004	–	58.33	–	58.18	–
0.005	54.81	–	–	–	52.17
0.006	–	49.51	52.16	–	–
0.008	–	44.43	–	–	–
0.010	42.35	41.05	42.39	38.51	37.46
0.015	37.13	–	–	–	–
0.020	–	–	32.18	29.23	27.70
0.025	–	30.71	–	–	–
0.030	30.21	–	–	25.29	23.58
0.040	27.90	–	–	23.03	21.21
0.050	26.32	25.52	23.45	21.54	19.66
0.060	25.18	–	–	20.49	18.55
0.070	24.30	–	–	19.69	17.71
0.075	–	23.35	–	–	–
0.080	23.61	–	–	19.08	17.06
0.090	23.05	–	–	–	16.53
0.100	–	22.16	19.57	–	16.09
0.104	22.42	–	–	18.04	–
0.150	–	20.91	–	–	14.68
0.184	20.51	–	–	16.46	–
0.200	–	–	17.29	–	13.92
0.300	–	–	–	–	13.12
0.304	19.42	–	–	15.63	–
0.400	–	–	–	–	12.71
0.424	18.93	–	–	15.26	–
0.500	–	–	15.83	–	12.45
1.000	–	–	15.35	–	–
1.544	–	–	15.17	–	–
∞	17.67	18.35	14.86	14.33	11.42
$K(\infty)$	2.143	1.535	1.959	1.571	2.060
L_{hy}^+	0.254	0.0897	0.156	0.106	0.313
u_{max}/u_m	2.373	2.149	2.277	2.163	2.324

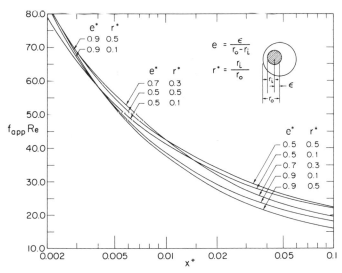

FIG. 96. Eccentric annular ducts: f_{app} Re for developing laminar flow (from Table 106).

verse flow model on the solutions, he also analyzed two less rigorous approximate models for the transverse flow. He demonstrated that the results are not sensitive to the choice of transverse flow model. For $e^* = 0.001$, $r^* = 0.4$, Feldman's results are in good agreement with those of Sparrow and Lin [163] for $e^* = 0$, $r^* = 0.4$. His results for $e^* = 0.9$, $r^* = 0.1$ are approximately the same as for the circular tube by Hornbeck [173]. The lack of closer agreement is due to the dissimilarity between the two geometries. Feldman's fully developed f Re factors agree with those of Table 100 within 0.4%.

Feldman also obtained $K(\infty)$, L_{hy}^+, and u_{max}/u_m for five eccentric annuli as presented in Table 106. A comparison with concentric annuli of the same r^* (Tables 87, 88, and 85) reveals that $K(\infty)$, L_{hy}^+, and u_{max}/u_m for eccentric annuli are greater by a factor of about 2.5, 10, and 1.5, respectively. $K(\infty)$ for concentric annuli decreases monotonically with increasing r^* as found from Table 87. However, for eccentric annuli with $e^* = 0.5$ and 0.9, $K(\infty)$ increases with increasing r^*. For a given r^*, $K(\infty)$ increases about two- to threefold between $e^* = 0$ and 0.5. One of the striking findings of the analytical solutions is that the hydrodynamic entry length is significantly increased for eccentric annular ducts. Thus it may be grossly in error to employ the entry lengths of concentric annuli for eccentric annuli.

C. Thermally Developing and Hydrodynamically Developed Flow

Vilenskii et al. [488] obtained the thermal entry length solutions for the fundamental problem of the second kind for eccentric annuli. They employed a finite difference method with a variable size grid both along the cross section and the flow length. They presented graphically the difference between wall and fluid bulk mean temperature, related to $\theta_{x,lj}^{(2)} - \theta_{x,mj}^{(2)}$, along the peripheral direction for various values of $4x^*$ for three eccentric annular ducts: (1) $e^* = 0.5$, $r^* = 0.4$, (2) $e^* = 0.1$, $r^* = 0.8$, and (3) $e^* = 0.5$, $r^* = 0.8$. They observed that even for small eccentricities in relatively wide ducts (low r^*), the wall temperature has a considerable angular nonuniformity, which increases for increasing x^*. Increasing e^* and r^* leads to an increase in the angular nonuniformity of the temperature for a given x^*. This behavior is associated with the considerable nonuniformity of the velocity profile.

Vilenskii et al. [488] presented $Nu_{x,ii}^{(2)}$ graphically as a function of $4x^*$ for five eccentric annuli. For eccentric annuli with small e^* and r^*, the Nusselt numbers are approximately the same as those for the corresponding concentric annuli for small x^*. With increasing e^* and r^*, the deviation between Nusselt numbers of eccentric and concentric annuli starts at smaller and smaller values of x^*.

Vilenskii *et al.* [488] also presented fully developed $Nu_{ii}^{(2)}$ and $Nu_{oo}^{(2)}$ for eccentric annuli. Their graphical results are in fair agreement with those in Table 103. Comparing fully developed Nusselt numbers, they concluded that increasing e^* and r^* increases the thermal entry length significantly.

Feldman [184] obtained the four fundamental solutions for an eccentric annular duct of $e^* = 0.5$, $r^* = 0.5$. He employed an implicit alternating direction method to solve the finite difference form of the energy equation at each axial step. A partial set of results for the four fundamental problems, provided to the present authors, is presented in Tables 107–109 [487]. In addition, Feldman analyzed the fundamental problem of the first kind with the inner wall heated for five eccentric annuli. The results for the local fluid bulk mean temperature are presented in Table 110.

The thermal entry length solution depends upon the velocity profile. For hydrodynamically developed flow, the maximum velocity in the narrower part is lower than the maximum velocity in the wider part of an eccentric annular duct. Thus the velocity profile could be substantially nonuniform across the cross section, and this nonuniformity increases with increasing e^* at a given r^*. This nonuniformity results in a minimum for $Nu_{x,lj}^{(1)}$ (for l, j = i or o) as explained below. For the fundamental solution of the first kind, high heat fluxes near the duct inlet diminish rapidly with axial distance. However, eccentricity causes much of the initial fluid temperature rise to occur in the narrower part of the duct, where the axial velocities are the lowest. Hence the fluid bulk mean temperature is increased slowly near the duct inlet. This results in a sharp drop in $Nu_{x,jj}^{(1)}$ near the duct inlet. Further away from the duct inlet, however, the wall heat flux levels off, while the fluid bulk mean temperature rises, thereby causing the Nusselt number to increase from its minimum value. Feldman [184] defined the thermal entry length as the value of x^* at which the difference between the maximum temperature $\theta_{max,j}^{(k)}$ and the bulk mean temperature $\theta_{x,mj}^{(k)}$ is 99% of the corresponding fully developed temperature profile value. This value of x^* is designated as $X_{th,j}^*$ in contrast to $L_{th,j}^*$ as defined on p. 50. Here the suffix j stands for the heated wall. Feldman obtained $X_{th,j}^*$ for several eccentric annuli, and they are presented in Táble 111 for $e^* = 0.5$, $r^* = 0.5$. $X_{th,o}^*$ are presented in Table 110 for five eccentric annuli. It is found that increasing e^* at a given r^* increases $X_{th,o}^{*(1)}$, a conclusion that is in agreement with that of Vilenskii *et al.* [488] for $L_{th}^{*(2)}$.

D. Simultaneously Developing Flow

Feldman [184] obtained a combined entry length solution by a finite difference method. He analyzed the fundamental problem of the first kind

TABLE 107

Eccentric Annular Duct ($e^* = 0.5$, $r^* = 0.5$): Fundamental Solutions of the First Kind for a Fully Developed Velocity Profile (from Feldman [487])

x^*	$\theta_{x,mi}^{(1)}$	$Nu_{x,ii}^{(1)}$	$Nu_{x,oi}^{(1)}$	$\theta_{x,mo}^{(1)}$	$Nu_{x,oo}^{(1)}$	$Nu_{x,io}^{(1)}$
0.00001	0.001049	53.22	0.21(-23)	0.00196	49.57	0.23(-23)
0.00002	0.001659	42.46	0.95(-17)	0.00311	38.88	0.10(-16)
0.00005	0.003070	31.48	0.13(-10)	0.00569	28.53	0.14(-10)
0.0001	0.004908	25.00	0.191(-6)	0.00895	22.57	0.209(-6)
0.0002	0.007814	19.92	0.0004848	0.01410	17.85	0.0005373
0.0005	0.01450	14.85	0.2502	0.02576	13.03	0.2359
0.001	0.02317	11.90	2.105	0.04037	10.27	2.416
0.002	0.03659	9.584	5.989	0.06283	8.111	6.976
0.005	0.06454	7.329	8.874	0.1096	5.971	10.453
0.01	0.09604	6.206	8.367	0.1618	4.884	9.933
0.02	0.1399	5.514	7.068	0.2333	4.195	8.478
0.05	0.2213	5.122	5.546	0.3621	3.797	6.779
0.10	0.2946	5.142	4.844	0.4754	3.856	6.002
0.15	0.3333	5.225	4.588	–	–	–
0.20	0.3549	5.287	4.464	0.5687	4.108	5.572
0.25	0.3673	5.327	4.398	–	–	–
0.30	0.3745	5.352	4.360	0.5992	4.233	5.451
0.35	0.3788	5.367	4.339	–	–	–
0.40	0.3812	5.377	4.326	0.6096	4.281	5.411
0.50	–	–	–	0.6132	4.299	5.397
1.00	–	–	–	0.6152	4.308	5.390
1.56	0.3848	5.390	4.308	0.6152	4.308	5.390

TABLE 108

Eccentric Annular Duct ($e^* = 0.5$, $r^* = 0.5$): Fundamental Solutions of the Second, and Third Kinds for a Fully Developed Velocity Profile (from Feldman [487])

x^*	$\theta_{max,o}^{(2)}$	$Nu_{x,oo}^{(2)}$	$\theta_{x,mi}^{(3)}$	$Nu_{x,ii}^{(3)}$	$\theta_{x,mo}^{(3)}$	$Nu_{x,oo}^{(3)}$
0.00001	0.02066	62.47	0.001049	53.22	0.001959	49.57
0.00002	0.02618	48.61	0.001659	42.46	0.003114	38.88
0.00003	–	–	0.002175	37.21	–	–
0.00005	0.03605	35.23	0.003070	31.48	0.005690	28.53
0.0001	0.04606	27.75	0.004908	25.00	0.008952	22.57
0.0002	0.05894	21.90	0.007814	19.92	0.01410	17.85
0.0003	–	–	0.01026	17.49	–	–
0.0005	0.08210	16.05	0.01450	14.85	0.02576	13.03
0.001	0.1063	12.69	0.02319	11.90	0.04040	10.27
0.002	0.1395	10.07	–	–	0.06321	8.111
0.003	–	–	0.04848	8.454	–	–
0.005	0.2114	7.450	0.06801	7.175	0.1130	5.879
0.01	0.3149	5.951	0.1064	5.619	0.1717	4.535
0.02	0.4978	4.807	0.1628	4.332	0.2544	3.500
0.03	–	–	0.2065	3.752	0.3165	3.048
0.05	0.9331	3.845	0.2760	3.196	0.4115	2.620
0.1	1.464	3.687	0.4031	2.685	0.5726	2.249
0.2	2.203	–	0.5729	2.394	0.7588	2.083
0.3	2.746	–	0.6875	2.299	0.8609	2.047
0.5	3.576	–	0.8292	2.240	0.9532	2.032
0.7	–	–	0.9058	2.227	0.9842	2.030
1.0	5.134	–	0.9613	2.227	–	–
1.51	–	–	0.9915	2.227	–	–

TABLE 109

ECCENTRIC ANNULAR DUCT ($e^* = 0.5$, $r^* = 0.5$): FUNDAMENTAL SOLUTIONS OF THE FOURTH KIND FOR A FULLY DEVELOPED VELOCITY PROFILE (FROM FELDMAN [487])

x^*	$\theta^{(4)}_{x,mi}$	$Nu^{(4)}_{x,ii}$	$Nu^{(4)}_{x,oi}$	$\theta^{(4)}_{x,mo}$	$Nu^{(4)}_{x,oo}$	$Nu^{(4)}_{x,io}$
0.00001	0.0000130	64.99	0.11(-23)	0.0000250	62.47	0.11(-23)
0.00002	0.0000260	51.49	0.54(-17)	0.0000510	48.61	0.58(-17)
0.00005	0.0000652	38.01	0.70(-11)	0.0001297	35.23	0.74(-11)
0.0001	0.0001313	30.22	0.109(-6)	0.0002606	27.75	0.117(-6)
0.0002	0.0002637	24.04	0.000316	0.0005228	21.90	0.000342
0.0005	0.0006625	17.86	0.1571	0.001314	16.05	0.1730
0.001	0.001331	14.30	1.836	0.002636	12.69	2.071
0.002	0.002649	11.49	5.992	0.005264	10.07	6.942
0.005	0.006335	8.698	10.31	0.01288	7.456	12.36
0.01	0.01173	7.179	10.45	0.02462	6.017	12.87
0.02	0.02092	6.134	8.769	0.04574	4.993	11.01
0.05	0.04270	5.418	6.149	0.09975	4.250	
0.1	0.06934	5.358	4.797	0.1731	-	5.868
0.2	0.1026	5.760	3.987	0.2822	-	4.686
0.5	0.1378	6.785	3.503	0.4558	-	3.811
1.0	0.1458	7.152	3.423	0.5469	-	3.549
1.5	0.1463	7.179	3.418	0.5675	-	3.501

TABLE 110

ECCENTRIC ANNULAR DUCTS: $\theta^{(1)}_{x,mi}$ FOR A FULLY DEVELOPED VELOCITY PROFILE (FROM FELDMAN [487])

x^*	$\theta^{(1)}_{x,mi}$				
	$e^* = 0.5$ $r^* = 0.1$	$e^* = 0.9$ $r^* = 0.5$	$e^* = 0.5$ $r^* = 0.5$	$e^* = 0.1$ $r^* = 0.5$	$e^* = 0.5$ $r^* = 0.9$
0.00001	0.0003758	0.000801	0.001049	0.001179	0.001430
0.00002	0.0006057	0.001243	0.001659	0.001866	0.002259
0.00005	0.001154	0.002226	0.003070	0.003481	0.004135
0.0001	0.001897	0.003477	0.004908	0.005512	0.006558
0.0002	0.003112	0.005418	0.007814	0.008786	0.01042
0.0005	0.005999	0.009695	0.01450	0.01650	0.01911
0.001	0.009939	0.01508	0.02317	0.02637	0.03022
0.002	0.01652	0.02340	0.03659	0.04203	0.04739
0.005	0.03164	0.04118	0.06454	0.07790	0.08263
0.01	0.04984	0.06272	0.09604	0.1242	0.1219
0.02	0.07566	0.09472	0.1399	0.1951	0.1761
0.05	0.1237	0.1582	0.2213	0.3153	0.2750
0.1	0.1653	0.2242	0.2946	0.3831	0.3638
0.2	0.1944	0.2938	0.3549	0.4063	0.4391
0.4	0.2028	0.3389	0.3812	0.4085	0.4754
0.5	0.2032	0.3456	0.3836	0.4086	0.4793
1.0	0.2033	0.3516	0.3848	0.4086	0.4819
1.56	0.2033	0.3517	0.3848	0.4086	0.4819
$x^{*(1)}_{th,o}$	0.302	0.575	0.393	0.175	0.433

TABLE 111

ECCENTRIC ANNULAR DUCT ($e^* = 0.5$, $r^* = 0.5$): $X^{*(k)}_{th,j}$ FOR DEVELOPING TEMPERATURE AND DEVELOPED VELOCITY PROFILES (FROM FELDMAN [184])

	$X^{*(1)}_{th,j}$	$X^{*(2)}_{th,j}$	$X^{*(3)}_{th,j}$	$X^{*(4)}_{th,j}$
$j = i$	0.390	1.10	0.785	1.31
$j = o$	0.393	1.15	1.46	0.821

with the inner wall heated for an eccentric annular duct of $e^* = 0.5$, $r^* = 0.5$. He employed the velocity distribution in the entrance region as discussed on p. 333. To approximately cover the Prandtl number range, he obtained thermal entry length solutions for $Pr = 0.05$ (liquid metals), 1.0 (gases), and ∞ (highly viscous liquids). His numerical results for $\theta_{x,mi}^{(1)}$ and $Nu_{x,ii}^{(1)}$ for $Pr = 0.05$ and 1.0 are presented in Table 112. The results for $Pr = \infty$ have been presented in Table 107.

TABLE 112

ECCENTRIC ANNULAR DUCT ($e^* = 0.5$, $r^* = 0.5$): FUNDAMENTAL SOLUTIONS
OF THE FIRST KIND FOR SIMULTANEOUSLY DEVELOPING FLOW
(FROM FELDMAN [487])

x^*	$Pr = 0.05$			$Pr = 1.0$		
	$\theta_{x,mi}^{(1)}$	$Nu_{x,ii}^{(1)}$	$Nu_{x,oi}^{(1)}$	$\theta_{x,mi}^{(1)}$	$Nu_{x,ii}^{(1)}$	$Nu_{x,oi}^{(1)}$
0.00001	0.003774	165.88	–	0.003605	90.89	–
0.00002	0.005487	109.80	–	0.004911	66.21	–
0.00004	0.007828	75.21	–	0.006793	49.24	–
0.0001	0.01245	47.33	–	0.01074	33.76	–
0.0002	0.01767	34.15	–	0.01522	25.14	–
0.0004	0.02529	25.08	–	0.02186	18.89	–
0.001	0.04126	16.73	0.00014	0.03618	13.08	0.00042
0.002	0.05994	12.58	0.01846	0.05313	10.33	0.04291
0.004	0.08740	9.682	0.4251	0.07890	8.537	0.7373
0.01	0.1432	7.144	2.470	0.1313	6.858	3.460
0.02	0.2003	5.980	3.806	0.1822	5.708	4.774
0.04	0.2650	5.413	4.301	0.2400	5.226	4.867
0.1	–	–	–	0.3203	5.184	4.601

Chapter XIV

Other Doubly Connected Ducts

In this chapter, we describe the remaining solutions for doubly connected duct geometries, other than those discussed in Chapters XII and XIII. Only the fully developed laminar fluid flow and heat transfer problems have been investigated for the five geometries that follow.

A. Confocal Elliptical Ducts

1. FLUID FLOW

Fully developed laminar flow through confocal ellipses, Fig. 97, (also referred to as concentric ellipses) has been analyzed by Piercy et al. [74], Sastry [80,89], Shivakumar [489], and Topakoglu and Arnas [490]. The conformal mapping method was used by the first three investigators to arrive at the flow characteristics. Closed-form solutions to the momentum and energy equations were obtained by Topakoglu and Arnas by the use of elliptical coordinates.

While Piercy et al. employed $z' = c \cosh \zeta$ as the mapping function, Sastry [89] employed $z' = c(\zeta + 1/\zeta)$. The equations for the velocity profile

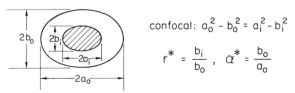

confocal: $a_o^2 - b_o^2 = a_i^2 - b_i^2$

$$r^* = \frac{b_i}{b_o}, \quad \alpha^* = \frac{b_o}{a_o}$$

FIG. 97. A confocal elliptical duct.

341

and volumetric flow rate of these investigators are so highly complex that numerical values were not computed. Shivakumar [489] employed $z' = c(\zeta + \lambda/\zeta)$ as a mapping function, and obtained numerical results for seven different confocal elliptical geometries. The $f\,Re$ factors, calculated from Shivakumar's volumetric flow rate $-8\pi Q/c_1$, are in excellent agreement with those of Table 113.

Topakoglu and Arnas [490] obtained closed-form solutions to the momentum and energy equations written in an elliptical coordinate system. They presented equations for the velocity profile, coordinates for the points of maximum velocity, and the mass flow rate through the cross section. Note that in the present terminology the definition of Reynolds number employed by Topakoglu and Arnas is $-\rho c_1(a_0 + b_0)^3/(32\mu)$. The following expression for $f\,Re$ is derived from the equations of Topakoglu and Arnas:

$$f\,Re = \frac{256 A_c{}^3}{\pi I_{oo} P^2 (a_o + b_o)^4} \tag{522}$$

where

$$I_{oo} = \frac{1}{4}(1 - \omega^4)\left(1 + \frac{m^8}{\omega^4}\right) - 2m^4 \frac{(1 - \omega^2)}{(1 + \omega^2)}$$

$$+ \frac{1}{4\ln\omega}(1 - \omega^2)^2\left(1 - \frac{m^4}{\omega^2}\right)^2 \tag{523}$$

$$\frac{A_c}{(a_o + b_o)^2} = \frac{\pi}{4}(1 - \omega^2)\left(1 + \frac{m^4}{\omega^2}\right) \tag{524}$$

$$\frac{P}{(a_o + b_o)} = 2\left[(1 + m^2)E_1 + \left(1 + \frac{m^2}{\omega^2}\right)\omega E_\omega\right] \tag{525}$$

$$m = \left(\frac{1 - \alpha^*}{1 + \alpha^*}\right)^{1/2} \tag{526}$$

$$\omega = \frac{a_i + b_i}{a_o + b_o}$$

$$= \frac{\alpha^* r^* + [1 - \alpha^{*2}(1 - r^*)^2]^{1/2}}{1 + \alpha^*} \tag{527}$$

E_1 and E_ω are the complete integrals of the second kind, which are evaluated for the arguments $(1 - b_o{}^2/a_o{}^2)$ and $(1 - b_i{}^2/a_i{}^2)$, respectively, where

$$\frac{b_i}{a_i} = \left(1 - \frac{m^2}{\omega^2}\right)\bigg/\left(1 + \frac{m^2}{\omega^2}\right) \tag{528}$$

The $f\,Re$ factors, calculated by the present authors, are presented in Table 113 and Fig. 98.

TABLE 113

CONFOCAL ELLIPTICAL DUCTS: $f\,\mathrm{Re}$ FOR FULLY DEVELOPED LAMINAR FLOW
[FROM EQ. (522)]

r^*	fRe					
	$\alpha^*=0.2$	0.40	0.60	0.80	0.90	0.95
0.02	19.419	19.468	20.291	21.766	22.436	22.620
0.05	19.433	19.534	20.454	22.071	22.825	23.049
0.10	19.452	19.622	20.662	22.388	23.151	23.366
0.20	19.478	19.759	20.965	22.750	23.454	23.643
0.30	19.495	19.871	21.201	22.974	23.610	23.777
0.40	19.507	19.973	21.404	23.135	23.708	23.855
0.50	19.516	20.072	21.585	23.257	23.773	23.903
0.60	19.525	20.171	21.749	23.353	23.819	23.934
0.70	19.534	20.268	21.896	23.429	23.851	23.953
0.80	19.544	20.365	22.029	23.490	23.874	23.966
0.90	19.555	20.460	22.148	23.539	23.890	23.973
0.95	19.561	20.506	22.203	23.560	23.896	23.975
0.98	19.565	20.534	22.234	23.572	23.900	23.976

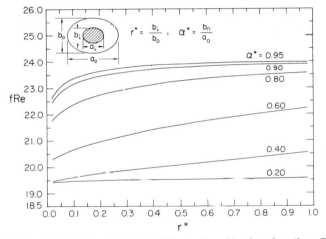

FIG. 98. Confocal elliptical ducts: $f\,\mathrm{Re}$ for fully developed laminar flow (from Table 113).

2. HEAT TRANSFER

Sastry [80] analyzed the (H1) temperature problem for a region bounded by confocal ellipses by employing the method of complex variables. He used the boundary condition for the velocity problem as $u = 0$ on the outer ellipse, $u = u_0 \neq 0$ on the inner ellipse. Hence, his results for the velocity and temperature problems are not relevant to stationary ducts.

Topakoglu and Arnas [490] analyzed the fundamental problem of the fifth kind[†] for confocal ellipses. The solution to the energy equation was

---

[†] One wall at axially constant q', the other wall insulated; the temperatures along the peripheries uniform at a section but different (see Table 4).

obtained in closed form by the use of elliptical coordinates. Based on their equations, the numerical results for the fundamental solutions of the fifth kind were determined by the present authors and are presented in Table 114. The other quantities of interest are

$$[\theta_{mi}^{(5)} - \theta_{oi}^{(5)}] = P^*[\theta_{mo}^{(5)} - \theta_{io}^{(5)}] \tag{529}$$

$$Nu_{ii}^{(5)} = \frac{1}{\theta_{ii}^{(5)} - \theta_{mi}^{(5)}} \tag{530}$$

$$Nu_{oo}^{(5)} = \frac{1}{\theta_{oo}^{(5)} - \theta_{mo}^{(5)}} \tag{531}$$

$$Nu_{oi}^{(5)} = Nu_{io}^{(5)} = 0 \tag{532}$$

where P^* is the ratio of inner to outer ellipse perimeters.

TABLE 114

CONFOCAL ELLIPTICAL DUCTS: FUNDAMENTAL SOLUTIONS OF THE FIFTH KIND FOR
FULLY DEVELOPED LAMINAR FLOW (CALCULATED BY THE PRESENT AUTHORS
BASED ON THE WORK OF TOPAKOGLU AND ARNAS [490])[†]

r^*	$\theta_{mo}^{(5)} - \theta_{io}^{(5)}$	$Nu_{ii}^{(5)}$	$Nu_{oo}^{(5)}$	P_i/P_o	$\theta_{mo}^{(5)} - \theta_{io}^{(5)}$	$Nu_{ii}^{(5)}$	$Nu_{oo}^{(5)}$	P_i/P_o
	$\alpha^*=0.20$				$\alpha^*=0.40$			
0.02	.0084733	5.0556	4.7885	.93277	.018465	5.1361	4.3924	.79690
0.05	.0084179	5.0571	4.7914	.93313	.018190	5.1449	4.4062	.79852
0.10	.0083411	5.0566	4.7957	.93421	.017833	5.1466	4.4271	.80321
0.20	.0082406	5.0481	4.8031	.93782	.017434	5.1210	4.4649	.81765
0.30	.0082038	5.0327	4.8097	.94292	.017370	5.0760	4.5010	.83636
0.40	.0082233	5.0126	4.8160	.94915	.017561	5.0240	4.5370	.85764
0.50	.0082925	4.9895	4.8224	.95628	.017946	4.9711	4.5736	.88046
0.60	.0084058	4.9645	4.8289	.96411	.018473	4.9206	4.6108	.90413
0.70	.0085579	4.9383	4.8358	.97251	.019101	4.8740	4.6485	.92820
0.80	.0087441	4.9116	4.8430	.98135	.019796	4.8319	4.6864	.95236
0.90	.0089584	4.8849	4.8505	.99054	.020531	4.7946	4.7242	.97635
0.95	.0089174	4.8679	4.8507	.99524	.020920	4.7779	4.7433	.98823
	$\alpha^*=0.60$				$\alpha^*=0.80$			
0.02	.040344	5.4154	4.1682	.62794	.078352	6.0394	4.3262	.42663
0.05	.039484	5.4462	4.2063	.63250	.075656	6.1164	4.4038	.43907
0.10	.038432	5.4553	4.2620	.64482	.072468	6.1019	4.5023	.46833
0.20	.037328	5.4003	4.3593	.67905	.068620	5.9296	4.6477	.53686
0.30	.037073	5.3156	4.4480	.71917	.066252	5.7568	4.7622	.60658
0.40	.037311	5.2323	4.5322	.76138	.064555	5.6206	4.8599	.67360
0.50	.037818	5.1597	4.6128	.80391	.063199	5.5169	4.9459	.73703
0.60	.038453	5.0994	4.6897	.84585	.062035	5.4378	5.0224	.79675
0.70	.039129	5.0506	4.7627	.88668	.060992	5.3766	5.0911	.85282
0.80	.039790	5.0116	4.8315	.92610	.060036	5.3285	5.1530	.90534
0.90	.040407	4.9806	4.8960	.96392	.059152	5.2901	5.2088	.95439
0.95	.040696	4.9677	4.9267	.98218	.058731	5.2737	5.2345	.97762
	$\alpha^*=0.90$				$\alpha^*=0.95$			
0.02	.099392	6.7910	4.5920	.29979	.107724	7.7718	4.7438	.21908
0.05	.094830	6.8271	4.6769	.32361	.102188	7.5342	4.8133	.25646
0.10	.089767	6.6459	4.7654	.37128	.096232	7.0391	4.8760	.31993
0.20	.083280	6.2544	4.8791	.46681	.088414	6.4203	4.9588	.43272
0.30	.078792	5.9841	4.9648	.55461	.082898	6.0796	5.0268	.53017
0.40	.075289	5.7999	5.0380	.63476	.078598	5.8656	5.0884	.61683
0.50	.072399	5.6687	5.1033	.70833	.075087	5.7187	5.1454	.69524
0.60	.069942	5.5714	5.1626	.77619	.072141	5.6119	5.1986	.76690
0.70	.067816	5.4970	5.2168	.83894	.069629	5.5311	5.2483	.83273
0.80	.065955	5.4385	5.2665	.89701	.067462	5.4680	5.2946	.89331
0.90	.064318	5.3918	5.3121	.95064	.065582	5.4178	5.3375	.94900
0.95	.063569	5.3717	5.3333	.97585	.064735	5.3964	5.3578	.97508

[†] K_2 of Eq. (6.2) of [490] should have $(1 - \omega^4)$ instead of $(1 - \omega^2)$.

As mentioned on p. 292 for concentric annular ducts, additional solutions of practical interest can be obtained by the superposition of the fundamental solutions. The solution for the case of heat flux specified constant on both walls is obtained in the same manner as that for concentric annular ducts (described on p. 294), except that the superscript designating the kind of fundamental solution is changed from (2) to (5). When constant axial heat fluxes $q_o''(=q_o'/P_o)$ and $q_i''(=q_i'/P_i)$ are specified on each wall, the corresponding Nusselt numbers are obtained from Eqs. (519) and (520) with the superscript 2 replaced by 5. The Nusselt numbers for the cases of $q_o'/q_i' = -0.5$ and -2.0 analyzed by Topakoglu and Arnas can be obtained by the use of Eqs. (519) and (520) and the results of Table 114. Similarly, the Nusselt numbers $Nu_i^{(5b)}$ and $Nu_o^{(5b)}$ for the equal wall temperature case analyzed by Topakoglu and Arnas can also be determined from Eqs. (519) and (520)

TABLE 115

Confocal Elliptical Ducts: Nu_{H1} for Fully Developed
Laminar Flow [from Eq. (533)]

r^*	Nu_{H1}					
	$\alpha^*=0.2$	0.40	0.60	0.80	0.90	0.95
0.02	5.1237	5.1231	5.4782	6.5083	7.1933	7.4100
0.05	5.1248	5.1311	5.5121	6.6178	7.3679	7.6216
0.10	5.1252	5.1395	5.5534	6.7384	7.5273	7.7940
0.20	5.1230	5.1479	5.6162	6.8973	7.6945	7.9574
0.30	5.1185	5.1541	5.6770	7.0218	7.7961	8.0427
0.40	5.1130	5.1626	5.7441	7.1325	7.8696	8.0955
0.50	5.1072	5.1751	5.8179	7.2325	7.9259	8.1306
0.60	5.1016	5.1922	5.8966	7.3224	7.9699	8.1546
0.70	5.0965	5.2137	5.9779	7.4023	8.0046	8.1711
0.80	5.0921	5.2390	6.0597	7.4724	8.0320	8.1825
0.90	5.0885	5.2676	6.1404	7.5336	8.0536	8.1901
0.95	5.0788	5.2836	6.1801	7.5606	8.0621	8.1928

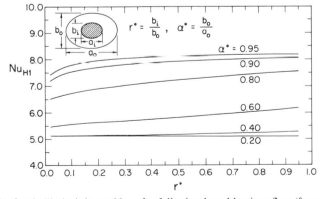

Fig. 99. Confocal elliptical ducts: Nu_{H1} for fully developed laminar flow (from Table 115).

(with superscript 5) with q_i''/q_o'' obtained from Eq. (518) (with superscript 5). Nu_{H1} is then computed from the following equation, where h is defined by Eq. (76):

$$Nu_{H1} = \frac{Nu_o^{(5b)} + P^*Nu_i^{(5b)}}{1 + P^*} \tag{533}$$

Nu_{H1} for confocal elliptical ducts were determined from Eq. (533) and the results of Table 114 by the described procedure, and are presented in Table 115 and Fig. 99.

B. Regular Polygonal Ducts with Central Circular Cores

Gaydon and Nuttall [491] proposed a method for estimating the volumetric flow rate for fully developed laminar flow through a cylindrical multiply connected duct. The method yields the upper and lower bounds of the flow rate derived from the Schwartz inequality of the variational method. They obtained such bounds for a square and a hexagonal duct with central circular cores. These upper bound f Re factors are in excellent agreement with the more accurate f Re factors of Table 116.

Cheng and Jamil [109] analyzed fully developed laminar flow through regular polygonal ducts having a central circular core (see Fig. 100). They employed a point-matching method for the analysis. Ten points on one-half side of the regular polygon were used. For the heat transfer problem, they considered the (H1) thermal boundary condition, as defined in Table 5 and also on p. 328 for doubly connected ducts, and tabulated f Re, Nu_{H1}, and the flow rate for $0 < a/\xi_1 \leq 0.5$, $3 \leq n \leq 20$. Here a is the radius of the circular core and ξ_1 is the radius of an inscribed circle in an n-sided regular polygon. Cheng and Jamil also presented graphically shear stress distribution, fluid temperature gradients at the wall, and fluid velocity and temperature profiles for typical duct geometries. Their f Re factors for the limited range of a/ξ_1 are in excellent agreement with those of Table 116. Nu_{H1} of Cheng and Jamil are presented in Table 117 and Fig. 101.

Ratkowsky and Epstein [112] independently investigated the flow characteristics of the same problem by a discrete least squares method. Their f Re factors [492] are presented in Table 116 and Fig. 100. Two limiting cases of this geometry (see Fig. 100), $a/\xi_1 = 0$, and 1, are of interest. When a/ξ_1 equals zero, the corresponding geometry is the n-sided regular polygonal duct, which has been discussed in Chapter X. The f Re factors for this case are in excellent agreement with those of Table 75. Ratkowsky and Epstein have analyzed the other limiting case ($a/\xi_1 = 1$) in detail. The f Re factors for $n = 3$ through 18 are presented in Table 116. They showed

TABLE 116

Regular Polygonal Ducts with Central Circular Cores: f Re for Fully Developed Laminar Flow (from Ratkowsky [492])

$\dfrac{a}{\xi_1}$	fRe				
	n=3	n=4	n=6	n=8	n=18
1.000	7.80	7.10	6.62	6.48	6.48
0.991	−	−	−	−	17.82
0.987	−	−	−	11.04	−
0.981	−	−	−	−	21.59
0.979	−	−	10.89	−	−
0.969	9.18	9.86	−	−	−
0.950	−	−	15.35	18.79	23.53
0.900	−	−	19.68	22.04	23.89
0.875	12.93	16.21	−	−	−
0.800	15.32	19.08	22.48	23.41	23.92
0.750	16.58	20.25	22.92	23.54	−
0.700	−	−	23.14	23.59	−
0.600	18.98	21.82	23.23	23.54	−
0.511	−	−	−	−	23.70
0.500	19.69	22.02	23.12	23.39	−
0.400	19.90	21.89	22.88	23.18	−
0.318	−	−	−	−	23.36
0.300	19.73	−	22.53	22.88	−
0.200	19.22	20.92	22.01	22.42	−
0.100	−	−	21.13	21.59	22.16
0.050	−	−	20.31	20.79	−
0.025	16.89	18.40	19.59	20.07	−
0	13.33	14.23	15.05	15.41	15.86

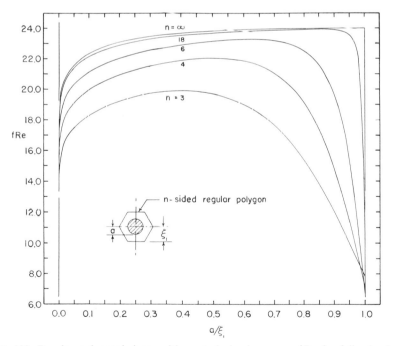

FIG. 100. Regular polygonal ducts with central circular cores: f Re for fully developed laminar flow (from Table 116).

TABLE 117

Regular Polygonal Ducts with Central Circular Cores: Nu_{H1} for
Fully Developed Laminar Flow (from Cheng and Jamil [109])

a/ξ_1	Nu_{H1}								
	n=3	n=4	n=5	n=6	n=7	n=8	n=9	n=10	n=20
0.00	3.111	3.608	3.859	4.002	4.102	4.153	4.196	4.227	4.329
0.05	4.938	5.723	6.094	6.303	6.431	6.525	6.588	6.635	6.778
0.1	5.296	6.104	6.467	6.670	6.794	6.885	6.946	6.992	7.142
1/9	5.354	6.163	6.523	6.722	6.845	6.935	6.995	7.040	7.189
0.125	5.417	6.228	6.582	6.778	6.898	6.986	7.045	7.090	7.236
1/7	5.486	6.297	6.644	6.834	6.951	7.037	7.094	7.138	7.280
1/6	5.560	6.370	6.705	6.887	6.998	7.080	7.136	7.177	7.314
0.2	5.626	6.438	6.755	6.924	7.027	7.102	7.154	7.192	7.321
0.25	5.655	6.474	6.760	6.906	6.995	7.060	7.104	7.138	7.251
1/3	5.511	6.351	6.593	6.697	6.757	6.801	6.831	6.854	6.933
0.5	4.466	5.317	5.488	5.498	5.480	5.465	5.453	5.444	5.419

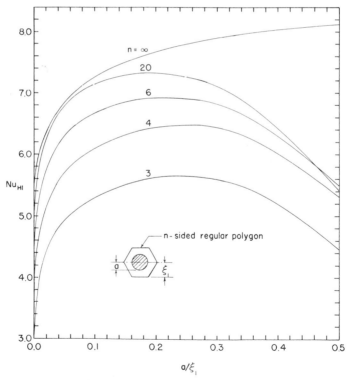

Fig. 101. Regular polygonal ducts with central circular cores: Nu_{H1} for fully developed laminar flow (from Table 117).

that when $n \to \infty$, $f\,\mathrm{Re} \to 56/9 = 6.222$ for this case. However, the number of sides n of a regular polygon cannot be smaller than 3; the corresponding case is discussed in the following section.

It is interesting to note that the presence of a small circular core ($a/\xi_1 \simeq 0$), centrally located in a regular polygonal duct, dramatically increases $f\,\mathrm{Re}$ and $\mathrm{Nu_{H1}}$. This behavior is similar to that observed for concentric annular ducts.

C. Isosceles Triangular Ducts with Inscribed Circular Cores

As mentioned in the preceding section, Ratkowsky and Epstein [112] also analyzed laminar flow through a section bounded by a regular polygon and its inscribed circle ($a/\xi_1 = 1$), and reported $f\,\mathrm{Re}$ factors for $n \ge 3$. They also considered the isosceles triangular duct with the side angle $\phi < 60^\circ$ (see figure with Table 118). They reported the study of Bowen [493] for this geometry. Bowen employed a finite difference method to obtain the $f\,\mathrm{Re}$ factors presented in Table 118.

TABLE 118

Isosceles Triangular Ducts with Inscribed Circular Cores: $f\,\mathrm{Re}$ for
Fully Developed Laminar Flow (from Bowen [493])

ϕ	fRe	ϕ	fRe	
0	12.0	20	9.79	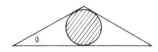
2	11.4	30	9.14	
5	11.3	40	8.69	
10	10.65	50	8.26	
15	10.16	60	7.80	

D. Elliptical Ducts with Central Circular Cores

Sastry [88,89] and Shivakumar [489] analyzed fully developed laminar flow through an elliptical duct with a central circular core by the conformal mapping method. Sastry employed the Schwarz–Neumann alternating method, an approximate method of conformal mapping, and determined the velocity profile with the first four approximations. Shivakumar also employed the conformal mapping method, although a different one, to analyze the same problem. He presented an equation for the volumetric flow rate through this cross section and evaluated numerical results for some geometries. Based on his tabulated results, $f\,\mathrm{Re}$ factors were determined and are presented in Table 119 [494]. Table II of Shivakumar [489] has some errors; the values for the area of the cross section should be multiplied by π, and the flow rate values in the last three columns should be divided by π [494].

TABLE 119

Elliptical Ducts with Central Circular Cores: f Re for Fully Developed
Laminar Flow (from Shivakumar [489,494])

α^*	r^*	fRe
0.9	0.5	23.519
0.9	0.6	23.435
0.9	0.7	23.159
0.9	0.95	16.816
0.7	0.5	21.694
0.7	0.7	19.402
0.5	0.5	19.321

$$\alpha^* = \frac{2b_o}{2a_o}$$

$$r^* = \frac{a}{b_o}$$

E. Circular Ducts with Central Regular Polygonal Cores

Cheng and Jamil [109,111] employed a point-matching method to analyze fully developed laminar flow through an annulus having a circle as the outside boundary and a concentric regular polygon as the inside boundary. Both surfaces of the annulus were heated to obtain the (H1) boundary condition defined in Table 5 and also on p. 328. They presented f Re and Nu_{H1} graphically. The f Re factors were obtained for the limited range $0 < \xi_1/a \le 0.5$ and are in excellent agreement with those of Table 120. The tabulation of f Re factors of Cheng and Jamil is available in [13,449]. Their Nu_{H1} are presented in Table 121 and Fig. 103.

Hagen and Ratkowsky [113] studied the flow characteristics of the same problem by a discrete least squares method. Their f Re factors are reported in Table 120 and Fig. 102. Note that while Cheng and Jamil used ξ_1/a as an independent variable, Hagen and Ratkowsky employed ξ_2/a. Here, ξ_1 and ξ_2 are the radii of inscribed and circumscribed circles, respectively.

Cheng and Jamil [111] encountered some difficulties with their point-matching method as the number of sides of the regular polygonal core decreased from 20 to 3. They found that the shear stress distributions and normal temperature gradients along the inner regular polygonal boundary exhibited a wavy character, with a very small region having a negative shear stress distribution. In spite of the difficulty with local values, Cheng and Jamil stated that the integrated overall quantities, such as f Re and Nu_{H1} were sufficiently accurate for practical purposes. Hagen and Ratkowsky [113] did not encounter the negative shear stress distribution and its wavy character; however, the least squares matching procedure became more difficult as the number of sides of the polygon became smaller.

It may be noted that the presence of a small regular polygonal core at the center of the circular duct significantly increases the flow resistance (f Re) and heat transfer (Nu_{H1}).

TABLE 120

CIRCULAR DUCT WITH CENTRAL REGULAR POLYGONAL
CORES: f Re FOR FULLY DEVELOPED LAMINAR FLOW
(FROM RATKOWSKY [492])

$\dfrac{\xi_2}{a}$	fRe				
	n=3	n=4	n=6	n=8	n=18
1.000	15.75	15.67	15.60	15.57	15.52
0.975	-	16.89	-	-	22.93
0.950	17.15	17.96	19.74	21.21	23.66
0.925	-	18.90	-	-	-
0.900	18.42	19.71	21.71	23.47	23.93
0.85	19.50	20.94	22.71	-	-
0.80	20.36	21.80	23.21	23.68	23.94
0.75	21.05	22.34	23.44	23.76	-
0.70	21.60	22.72	23.54	23.78	23.90
0.65	21.92	22.93	23.58	23.74	-
0.60	22.20	23.10	23.57	23.71	23.84
0.55	22.34	23.09	23.52	23.70	-
0.50	22.36	23.10	23.49	23.60	23.75
0.45	22.38	23.04	23.41	23.54	-
0.40	22.35	22.98	23.34	23.46	23.62
0.35	22.37	22.91	23.25	23.37	-
0.30	22.09	22.83	23.12	23.26	23.41
0.25	22.11	22.68	22.98	23.11	-
0.20	21.98	22.55	22.80	22.92	23.05
0.15	21.90	22.31	22.55	22.65	-
0.100	21.64	21.95	22.17	22.24	22.32
0.075	-	21.74	-	-	-
0.050	21.08	-	21.54	21.54	21.56
0.000	16.00	16.00	16.00	16.00	15.00

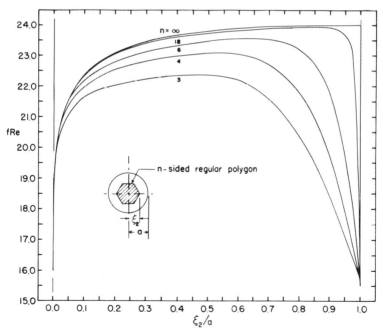

FIG. 102. Circular duct with central regular polygonal cores: f Re for fully developed laminar flow (from Table 120).

TABLE 121

CIRCULAR DUCT WITH CENTRAL REGULAR POLYGONAL CORES:
Nu_{H1} FOR FULLY DEVELOPED LAMINAR FLOW
(FROM JAMIL [449])

ξ_1/a	Nu_{H1}					
	$n=4$	$n=5$	$n=6$	$n=7$	$n=8$	$n=9$
0.0	4.364	4.364	4.364	4.364	4.364	4.364
1/9	7.337	7.321	7.318	7.320	–	7.325
1/8	7.398	7.385	7.385	7.388	7.390	7.395
1/7	7.467	7.457	7.459	7.464	7.467	7.057
1/6	7.538	7.537	7.543	7.550	7.555	7.558
1/5	7.616	7.626	7.638	7.648	7.654	7.660
1/4	7.693	7.723	7.744	7.759	7.769	7.776
1/3	7.738	7.814	7.858	7.884	7.898	7.909
1/2	–	7.718	7.910	7.989	8.027	8.049

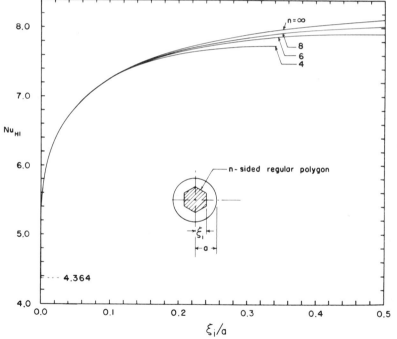

FIG. 103. Circular duct with central regular polygonal cores: Nu_{H1} for fully developed laminar flow (from Table 121).

F. Miscellaneous Doubly Connected Ducts

Sastry [87] employed the Schwarz–Neumann alternating method, an approximate method of conformal mapping, to analyze the fully developed laminar velocity and (H1) temperature problems for a circular tube with central elliptical cores. He employed boundary conditions for the velocity problem as $u = 0$ on the circle and $u = u_o \neq 0$ on the ellipse. Hence his results for the velocity and temperature problems are not relevant for stationary ducts.

Sastry [88] investigated fully developed laminar flow for a circular tube having a central square core with rounded corners, and considered the no-slip boundary condition at both inner and outer boundaries. He analyzed the fundamental boundary conditions of the fifth kind (see Table 4) for the temperature problem. He employed the Schwarz–Neumann alternating method and presented the approximate solutions for the velocity and temperature distributions, mean velocity and cross-sectional average fluid temperature. He did not present explicit formulas for $f\,\mathrm{Re}$ and $\mathrm{Nu}_{ii}^{(5)}$.

Chapter XV

Longitudinal Flow
over Circular Cylinders

One of the common types of fuel arrangement employed in a nuclear power reactor is a bundle of circular section fuel rods located inside a round or a hexagonal tube. The coolant or heat transfer fluid flows longitudinally over the rods. The rods are generally arranged in regular triangular or square arrays or are concentrically arranged in a ring around a central rod. Other applications of this flow geometry are also encountered in some shell-and-tube heat exchangers.

Longitudinal laminar flow over such a fuel rod bundle has been analyzed in detail for the fully developed case. No hydrodynamic entrance solutions are available in the literature. Only the Lévêque-type thermal entry length solutions are available for the (T) and (H1) boundary conditions.

When the analysis is carried out for the coolant (fluid) medium with the appropriate boundary conditions at the rod walls, it is referred to as the *single region* analysis. In contrast, for a *multiregion* analysis, instead of imposing the temperature boundary condition at the tube wall, the mathematical modeling is performed for three zones simultaneously—the nuclear fuel region (within the rod), the rod wall region, and the coolant region—treating the temperature and heat flux as continuous functions across the interfaces. The multiregion problem has been previously referred to as the conjugated problem (p. 11). The single region analysis is adequate for the fluid flow problem; however, both the single region and the more complex multiregion analyses are desirable for the temperature problem. A summary of fluid flow and heat transfer results for the rod bundle geometry up to 1968 is provided by Axford [495].

A. Fully Developed Flow

1. FLUID FLOW

a. *Triangular and Square Array Rod Bundles*

Longitudinal fully developed laminar flow over the triangular and square array rod bundles has been analyzed by Sparrow and Loeffler [103], Sholokhov *et al.* [496], Schmid [497], Rehme [498,499], and Meyder [102]. The triangular array rod bundle has also been analyzed by Axford [500,501] and Ullrich [118]. A hexagonal rod bundle usually has rods arranged in a triangular array and surrounded by a hexagonal duct.

Sparrow and Loeffler [103] employed algebraic–trigonometric poly-nomials to solve the Laplace equation. They obtained numerical solutions by a six-point matching method for the infinite triangular and square array rod bundles, as shown in Fig. 104.

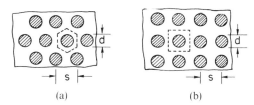

(a)   (b)

FIG. 104. Triangular and square array arrangements for an infinite rod bundle.

They presented a pressure-drop–flow parameter $(-dp/dx)(d^4/\mu)/16Q$ and also $(f\,\mathrm{Re})(d/D_h)$ graphically as a function of porosity o (range 0.1–1.0). The porosity σ is defined as the ratio of free flow area to frontal area. The frontal area equals the free flow area plus the cross-sectional area of rods. The porosity σ and pitch-to-diameter ratio s/d are related to each other as

$$\sigma = \begin{cases} 1 - \dfrac{\sqrt{3}\pi}{6(s/d)^2}, & \text{for triangular array} \qquad (534) \\[3mm] 1 - \dfrac{\pi}{4(s/d)^2}, & \text{for square array} \qquad (535) \end{cases}$$

They developed the following approximate expression for friction factors for a limited range of σ ($\sigma > 0.8$, $s/d > 2.1$, for the triangular array; $\sigma > 0.9$, $s/d > 2.8$, for the square array):

$$(f\,\mathrm{Re})\left(\frac{d}{D_h}\right) = \frac{8\sigma^2}{2(1-\sigma) - \ln(1-\sigma) - 0.5(1-\sigma)^2 - 1.5} \qquad (536)$$

where

$$\frac{D_h}{d} = \begin{cases} \frac{2\sqrt{3}}{\pi}\left(\frac{s}{d}\right)^2 - 1, & \text{for triangular array} \qquad (537) \\[3ex] \frac{4}{\pi}\left(\frac{s}{d}\right)^2 - 1, & \text{for square array} \qquad (538) \end{cases}$$

While analyzing the multiregion temperature problem, Axford [500,501] also solved the velocity problem for longitudinal laminar flow over an infinite triangular bundle. He employed a finite Fourier cosine transform and a point-matching method for the solution. He used 6 to 15 boundary points to obtain the unknown coefficients of the series. Axford's equations for the velocity profile and the coefficients agree with the results of Sparrow and Loeffler [103]. Axford tabulated various parameters of interest. The following equation for the friction factors for the infinite triangular rod bundle is derived from his results:

$$f\,\text{Re} = \frac{\pi}{6M(s/d)^4}\left[\frac{2\sqrt{3}}{\pi}\left(\frac{s}{d}\right)^2 - 1\right]^3 \qquad (539)$$

The values of the parameter M, a function of s/d, are tabulated by Axford to five significant figures. Using Eq. (539), $f\,\text{Re}$ factors for the triangular array are presented in Table 122 and Fig. 105. For $s/d > 2.1$, $f\,\text{Re}$ factors can also be evaluated accurately from Eq. (536).

Sholokhov et al. [496] obtained an electric analog solution and a finite difference solution for the same problem. Sholokhov et al. presented graphically the pressure-drop–flow parameter and velocity distribution for a wide range of pitch-to-diameter ratios. The results compare favorably with those of Sparrow and Loeffler.

The limiting case of touching rods corresponds to the cusped passages for which $f\,\text{Re}$ factors are provided in Chapter X.

The foregoing results for the $f\,\text{Re}$ factors were for fully developed laminar flow over an infinite rod bundle arranged in a triangular array. If the bundle is divided into hexagonal "cells" (see Fig. 104a) and the cell is idealized as isolated, the flow rate through the cell is not influenced by the neighboring cells. For a finite bundle, however, there is a wall effect. To investigate the influence of the wall, Schmid [497] considered a semi-infinite rod bundle arranged in a square array and limited by a fixed wall on one side. He found that the flow rate through only the first and second row of cells is influenced by the wall. The influence of wall on the flow rate through the third and higher row cells is negligible. The influence in the first and second row of cells exceeds a value of only 1% for $s/d > 2.5$, $\xi > 1.3$, and for $s/d > 2.5$, $\xi < 0.6$. Here ξ is the ratio of the hydraulic diameter of the first or second

TABLE 122

LONGITUDINAL FLOW OVER AN INFINITE TRIANGULAR
ARRAY OF CIRCULAR CYLINDERS:
f Re, Nu_{H1}, AND Nu_{H2} FOR FULLY DEVELOPED LAMINAR FLOW

$\frac{s}{d}$	fRe [13]	Nu_{H1} [96]	Nu_{H2} [96]	$\frac{s}{d}$	fRe [13]	Nu_{H1} [96]	Nu_{H2} [96]
1.000	6.503	–	–	1.20	24.950	7.48	6.90
1.001	–	1.26	0.149	1.25	26.301	–	–
1.004	7.35	–	–	1.30	27.417	9.19	9.03
1.01	8.634	1.52	0.263	1.40	–	10.34	10.28
1.02	10.629	1.82	0.404	1.50	31.035	11.26	11.22
1.03	12.441	2.14	0.580	1.60	–	12.08	12.05
1.04	14.055	2.48	0.795	1.75	–	13.28	13.26
1.05	15.478	2.82	1.06	1.80	–	13.68	13.66
1.06	–	3.18	1.36	1.9	0 –	14.47	14.46
1.07	–	3.54	1.70	2.00	39.384	15.27	15.26
1.10	20.377	4.62	2.94	3.011	58.46	–	–
1.15	23.141	–	–	4.00	79.803	–	–

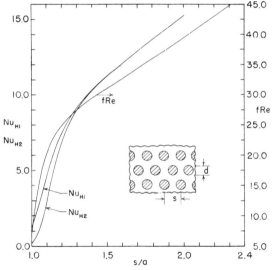

FIG. 105. Longitudinal flow over an infinite triangular array of circular cylinders: f Re, Nu_{H1}, and Nu_{H2} for fully developed laminar flow (from Table 122).

row cell to the hydraulic diameter of a cell far from the wall. Schmid's volumetric flow rate for a cell far from the wall is in good agreement with that of Sparrow and Loeffler [103].

A number of different fuel rod arrangements may be employed for hexagonal, round, or square tubes in a nuclear reactor. The influence of the wall on fluid friction and heat transfer characteristics could be different depending upon the arrangement. Gunn and Darling [451] investigated the flow characteristics of a four-rod bundle in a square array ($s/d = 1.31$,

$w/d = 1.16$) and determined $f\,\mathrm{Re}$ as 14.50. Rehme [498,499] took the same approach for the solution of a seven-rod bundle in a triangular array. He found that a finite difference solution for a bundle with more than seven rods requires excessively high computer time and a large storage. Hence, he proposed an analysis based on the superposition of the local subcell solutions as described below.

Rehme divided the regions of a rod bundle into central, wall, and corner channels of hexagonal and square tubes, as shown in Fig. 106. The velocity gradient at the lines of the subdivision was considered zero. A finite difference solution to the flow problem was obtained using the so-called method of cyclic reduced block overrelaxation. The $f\,\mathrm{Re}$ factors for these channels are functions of the relevant parameters: pitch-to-diameter ratio s/d and wall distance-to-diameter ratio w/d. The $f\,\mathrm{Re}$ factors are presented in Tables 123 and 124 [502]. The $f\,\mathrm{Re}$ factors for the central channel of the hexagonal tube agree within 2% with those of the infinite triangular array of Table 122.

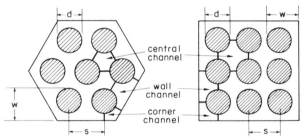

central channel

wall channel

corner channel

FIG. 106. Central, corner, and wall channels of a rod bundle in hexagonal and square tubes.

The $f\,\mathrm{Re}$ factors for a rod bundle made up of m subchannels can be obtained by the following equation, which contains the results of individual channels [499]:

$$\frac{1}{f\,\mathrm{Re}} = \sum_{i=1}^{m} \frac{1}{(f\,\mathrm{Re})_i} \left(\frac{P}{P_i}\right)^2 \left(\frac{A_{c,i}}{A_c}\right)^3 \tag{540}$$

By employing the definition of the hydraulic diameter, Eq. (540) can be alternatively expressed as

$$\frac{1}{f\,\mathrm{Re}} = \sum_{i=1}^{m} \frac{1}{(f\,\mathrm{Re})_i} \left(\frac{D_{h,i}}{D_h}\right)^2 \left(\frac{A_{c,i}}{A_c}\right) \tag{541}$$

Here P and A_c are the total wetted perimeter and total free flow area of the rod bundle in question, and those quantities with the suffix i are for the ith subchannel. The wetted perimeter and flow areas associated with different channels are provided in Table 125. Equation (540) was derived considering (1) a constant pressure drop in all subchannels, and (2) a total flow rate

TABLE 123

CENTRAL AND CORNER CHANNELS: f Re FOR FULLY DEVELOPED
LAMINAR FLOW (FROM REHME [499,502])

central channel				corner channel			
triangular array		square array		triangular array		square array	
s/d	fRe	s/d	fRe	s/d	fRe	s/d	fRe
1.0	6.50	1.00	6.60	1.00	6.68	1.00	7.14
1.02	10.66	1.02	8.40	1.025	15.58	1.025	11.23
1.05	15.58	1.05	10.97	1.05	20.79	1.05	14.32
1.1	20.81	1.12	16.04	1.10	22.56	1.10	18.04
1.15	23.22	1.22	21.19	1.15	23.09	1.15	19.89
1.2	25.20	1.32	25.18	1.25	23.30	1.25	21.43
1.4	29.66	1.42	27.93	1.50	23.20	1.50	22.09
1.6	32.29	1.52	30.36	2.00	22.74	1.75	21.97
1.8	35.97	1.75	35.46			2.30	21.42
2.0	38.55	2.00	40.38				
2.5	47.01						

TABLE 124

WALL CHANNELS: f Re FOR FULLY DEVELOPED LAMINAR FLOW (FROM REHME [499,502])

$\frac{s}{d}$	fRe								
	w/d								
	1.00	1.02	1.05	1.1	1.2	1.5	1.75	2.0	3.0
1.00	6.56	8.46	10.98	13.99	16.68	17.27	–	16.46	15.46
1.02	7.22	9.15	11.70	14.75	–	17.91	–	17.20	14.42
1.05	7.98	9.93	12.52	15.70	–	19.16	–	18.00	16.98
1.10	8.74	10.68	13.31	16.69	–	–	19.94	19.26	17.98
1.20	8.88	10.67	13.24	16.89	21.39	23.63	22.98	22.18	20.24
1.50	7.35	8.54	10.35	13.35	18.63	26.56	27.72	27.68	26.21
1.75	6.93	–	9.17	11.49	15.94	25.10	28.35	29.55	–
2.00	6.99	7.70	8.85	10.69	–	23.24	27.43	29.94	31.94
3.00	8.66	9.14	9.85	–	13.15	18.88	22.80	25.96	33.11
5.00	12.12	–	12.93	–	14.98	–	–	–	28.76
10.00	16.67	16.88	17.16	17.60	–	19.94	–	21.79	24.77

TABLE 125

THE WETTED PERIMETER AND FLOW AREAS FOR CENTRAL, CORNER, AND
WALL CHANNELS OF TRIANGULAR AND SQUARE ARRAYS

		Central channel	Wall channel	Corner channel
Triangular array	perimeter	$\frac{1}{2}\pi d$	$\frac{1}{2}\pi d + s$	$\frac{1}{6}\pi d + \frac{2}{\sqrt{3}}\left(w - \frac{d}{2}\right)$
	flow area	$\frac{\sqrt{3}}{4}s^2 - \frac{\pi}{8}d^2$	$\left(w - \frac{d}{2}\right)s - \frac{\pi}{8}d^2$	$\frac{1}{\sqrt{3}}\left(w - \frac{d}{2}\right)^2 - \frac{\pi}{24}d^2$
Square array	perimeter	πd	$\frac{1}{2}\pi d + s$	$\frac{\pi}{4}d + 2\left(w - \frac{d}{2}\right)$
	flow area	$s^2 - \frac{\pi}{4}d^2$	$\left(w - \frac{d}{2}\right)s - \frac{\pi}{8}d^2$	$\left(w - \frac{d}{2}\right)^2 - \frac{\pi}{16}d^2$

equal to the sum of subchannel flow rates. It may be noted that Eqs. (541) and (592) are identical, if one considers the relationship of N_i of Eq. (592) to m by Eq. (584). Rehme's f Re factor for the seven-rod bundle from the direct computation and from the subchannel analysis agreed within $\pm 2\%$ for most cases, supporting the accuracy of the numerical analysis and the validity of subchannel analysis. Rehme's subchannel analysis yields f Re of 14.65 relative to 14.50 from Gunn and Darling's analysis of a four-rod bundle in a square tube [451]. Rehme applied his subchannel analysis to the 16- and 19-rod bundles. According to Ullrich [118], Rehme's predicted f Re factors agree with the experimental values of Galloway and Epstein [503] within 4% for the 19-rod bundles in a triangular array. Rehme's f Re factors based on his subchannel analysis for the 19- and 37-rod bundles are also in excellent agreement (within $\pm 1.4\%$) with the results of Ullrich [118] obtained by a highly sophisticated method. Hence, it may be concluded that Rehme's subchannel analysis is a very appropriate tool to predict the f Re factors for longitudinal laminar fully developed flow through finite rod bundles. It may be mentioned that Mottaghian and Wolf [504] stated that the f Re factors based on Rehme's subchannel analysis for the 16- and 19-rod bundles differed from the experimental results of Galloway and Epstein [503] by at least 10%. This statement is invalid because it is based on Table 2 of Rehme [499], which later turned out to be wrong with respect to the w/d values reported. Ullrich's foregoing comparison is based on the corrected values provided by Rehme [502].

Meyder [102] numerically analyzed fully developed laminar flow over infinite rod bundles of triangular and square arrays. In his finite difference solution, the flow area was represented by a curvilinear orthogonal mesh-grid, as briefly described on p. 65. Meyder presented constant velocity contours for the "central channel" of the triangular and square array rod bundles for several pitch-to-diameter ratios. His calculated f Re factors for the central channel are in excellent agreement with those of Rehme in Table 123.

Ullrich [118] analyzed fully developed laminar flow in hexagonal tube bundles of arbitrary size. He employed a continuous least squares method for the analysis, and obtained u, τ, and τ_m at the rod and cassette surfaces, and u_m, Δp, and $4f$ Re for several hexagonal bundles with rods in regular triangular arrangement. The influence of rod pitch, wall distance, porosity, and number of rods was investigated in detail. The f Re factors predicted by Rehme's subchannel analysis agreed very closely with his analytical values as mentioned above.

b. Concentric Rod Bundles

Fully developed laminar flow over an n-rod bundle, in which $(n - 1)$ rods are concentrically arranged around a central rod (Fig. 107), has been in-

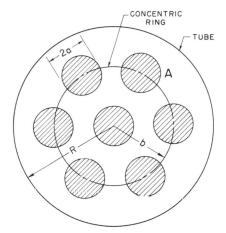

Fig. 107. A one-ring concentric seven-rod bundle in a tube.

vestigated by Axford [505], Min *et al.* [506], Min [507], Chen [508], Zarling [509], and Mottaghian and Wolf [504,510].

Axford [505] employed the periodic harmonic Howland functions. He derived the equations for u and u_m and numerically determined the detailed velocity profile u for $n = 3, 4, 5, 6$, and 7 tubes for two arrangements: (1) $a/R = 0.242$, $b/R = 0.606$, and (2) $a/R = 0.263$, $b/R = 0.632$. However, no results were reported for u_m or f Re.

Min *et al.* [506,507] employed the method of conjugate functions, which involved the use of a mapping function and transformation of the complex potential. The velocity profile was specified by an infinite series. The coefficients of the series were determined by a discrete least squares method. Numerical results were obtained for the velocity and shear stress distribution, pressure drop, and friction factors for 13-rod clusters. Their f Re factors are within 1% of those in Table 126.

Chen [508] used the same method as Axford [505] and extended Axford's analysis to (1) one rod in the concentric ring, such as A in Fig. 107, displaced from the symmetric position, and (2) all $(n - 1)$ rods around the central rod displaced such that the circular ring connecting the center of these rods becomes eccentrically oriented in the tube with respect to the central rod. While Axford solved only the flow problem, Chen also analyzed the heat transfer problem for the above cases, in addition to the one-ring symmetrical problem. His sample numerical calculations for the flow problem are in excellent agreement with those for Min's geometry ($a/R = 0.25$, $b/R = 0.675$) and Axford's geometry ($a/R = 0.26$, $b/R = 0.63$). Chen then obtained the velocity distribution, friction factors, temperature and wall heat flux distributions, and local Nusselt numbers for a seven-rod cluster with $a/R = 0.2$, $b/R = 0.6$. He did not present the peripheral average Nusselt numbers.

TABLE 126

Concentric n-Rod Bundle ($a/R = 0.2$):
f Re for Fully Developed Laminar Flow
(from Zarling [511])

$\dfrac{b}{R}$	fRe			
	n=5	n=7	n=9	n=11
0.44	18.65	14.12	–	–
0.48	20.38	17.04	–	–
0.52	21.51	20.55	–	–
0.56	21.81	23.66	18.05	–
0.60	21.22	25.27	21.51	–
0.64	19.84	24.31	22.61	–
0.65	–	–	–	15.45
0.68	17.96	21.29	20.27	14.22
0.72	15.89	17.43	16.12	11.12
0.76	13.90	13.80	11.95	8.03
0.77	–	–	–	7.42

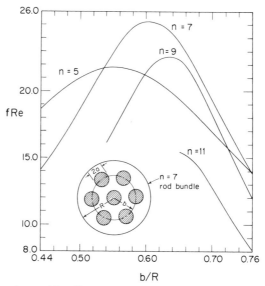

Fig. 108. Concentric n-rod bundle (a/R) = 0.2: f Re for fully developed laminar flow (from Table 126).

Chen investigated in detail the effect of displacement of one rod on the shear stress distribution and friction factors at the peripheral rods, at the central rod, and at the tube wall. He found that the greatest influence existed only for the neighboring two rods. The overall channel pressure drop and f Re factors decreased monotonically with increasing displacement of the rod from the symmetrical position. Chen studied the flow problem with a cluster eccentricity $\epsilon/R = 0.4$. Compared to the concentric case, he found considerable changes in the velocity distribution.

While analyzing the heat transfer problem, Zarling [509] also obtained numerical results for the velocity problem for longitudinal laminar flow over a concentric rod bundle. Zarling employed the same method as that of Axford [505] and Chen [508]. His $f\,\mathrm{Re}$ factors are presented in Table 126 [511] and Fig. 108.

The preceding investigations [505–509] were limited to the one-ring problem. Mottaghian and Wolf [504,510] considered a tube bundle consisting of an arbitrary number of rods, with different radii, placed on concentric rings around a central rod. The bundle was idealized as having a characteristic symmetry with respect to the angular direction. They devised a seminumerical method, which consisted of superposing the solutions, then transforming them into each individual rod polar coordinate system, and then matching the no-slip boundary condition at a finite number of points. Their results for a one-ring seven-rod problem are in close agreement with those of Min [506,507] and Chen [508]. Mottaghian and Wolf presented graphically the $f\,\mathrm{Re}$ factors for a two-ring 19-rod problem and showed the influence of pitch-to-diameter ratio, tube wall spacing, radial displacement of a rod ring, and the angle of the characteristic symmetry segment. Further results and details of the analytical method employed are described by Mottaghian [512].

2. HEAT TRANSFER

Sparrow *et al.* [104] analyzed the (H1) problem for longitudinal laminar flow over rods arranged in an equilateral triangular array. They presented $\mathrm{Nu_{H1}}$ graphically for a pitch-to-diameter ratio s/d of 1.1 to 4.0. For $s/d > 2$, they showed that $\mathrm{Nu_{H1}}$ can be accurately calculated from an equivalent annulus model as described below.

The hexagonal flow area associated with each rod, as shown in Fig. 104a, is approximated by a circle of equal area. The flow and heat transfer behavior of the system is considered to be the same as that in the area between the inner radius and the radius of maximum velocity of a circular annulus. This model implicitly assumes that the transverse flow of heat is entirely in the radial direction, which means that there is no circumferential variation of either the surface temperature of the rod or its surface heat flux.

In tube bundles, the inside or outside tube wall surface temperature is generally not peripherally uniform. A similar statement also applies to the heat fluxes on the inside and outside tube wall, particularly for small s/d [500,501]. Since the nuclear reactor design criteria favor smaller s/d (approaching unity) for the fuel subassemblies, a knowledge of the temperature fields in all regions of the subassembly is required in order to determine thermal stresses, hot spots, corrosion rates, and possible eutectic formation at the fuel–cladding interface. To predict quantitatively the magnitude of the circumferential variations of the inside and outside tube wall temperatures and heat fluxes, Axford [500,501] carried out a multiregion analysis as described on p. 354. He employed the method of finite Fourier cosine transforms to obtain an infinite series solution. The coefficients of this series (truncated at N terms) were obtained by a point-matching method. He found that the peripheral variations of the outside tube wall temperature and heat fluxes became quite sensitive to changes in the pitch-to-diameter ratio s/d of the tube bundles as this ratio approached unity. Hence, he concluded that the single region analysis with the (H1) or (H2) boundary conditions (for example, as done by Sparrow et al. [104]) contained an inherent source of error for s/d approaching unity. Unfortunately, Axford did not present Nusselt numbers.

Dwyer and Berry [96], by a finite difference method, investigated the (H2) temperature problem for longitudinal laminar flow over an infinite triangular rod bundle. They tabulated both Nu_{H1} and Nu_{H2} as a function of pitch-to-diameter ratio s/d from 1.001 to 2. These results are presented in Table 122 and Fig. 105. For $s/d > 2$, Nu_{H1} and Nu_{H2} are almost identical, and can be evaluated from the equivalent annulus model analysis of Sparrow et al. [104].

Hsu [513] extended the analysis of Dwyer and Berry by investigating the influence of rod displacement on Nu_{H1} and Nu_{H2} of the displaced rod. The Nusselt numbers are presented graphically for $s/d \leq 2$ and a wide range of dimensionless displacements. It is shown that the rod displacement reduces the Nusselt number more for the (H2) boundary condition than for the (H1) boundary condition.

Gräber [244] analyzed the (H5) heat transfer problem for an infinite triangular array rod bundle. He employed the equivalent annulus model together with his results for concentric annular ducts. He presented graphically the ratio Nu_{H5}/Nu_{H1} as functions of F_0 (defined on p. 84) and s/d.

Chen [508] analyzed the fully developed laminar forced convection problem for a concentric seven-rod cluster in a tube. He considered uniform thermal energy generation in all seven rods, and the containing tube as an adiabatic wall. The conjugated problem was solved for the fluid and rods by considering the temperature and the wall heat flux as continuous across the solid wall–fluid interface. Numerical results were obtained for a seven-rod cluster with $a/R = 0.2$, $b/R = 0.5$, and $k_w/k_f = 0.5$. Chen presented graphically the temperature, heat flux, and local Nusselt number distribution around the central and peripheral rods.

Zarling [509] studied fully developed laminar forced convection for a concentric n-rod cluster in a tube. He considered two thermal boundary conditions: (1) the rods were maintained at one constant temperature while the shell was maintained at another constant temperature, and (2) the rods were held at a uniform heat flux while the shell was held at a constant temperature. In both boundary condition sets, he assumed that whatever heat was transferred from the rods to the fluid was also transferred out through the shell. As a result, the term $\rho c_p u(\partial t/\partial x)$ in Eq. (24) was treated as zero. In the absence of thermal energy generation, viscous dissipation, and flow work within the fluid, the resulting energy equation is the Laplace equation. Zarling obtained the solution by employing a method similar to the one used by Axford and Chen to analyze their velocity problems. He presented the average Nusselt numbers at the shell, peripheral rods, and central rod for the foregoing two boundary conditions.

B. Thermally Developing and Hydrodynamically Developed Flow

Zarling and Min [509,514] analyzed the T thermal entry length problem for longitudinal laminar flow over five, seven, and nine-rod bundles. They obtained the first three terms of the Lévêque-type similarity solution for the central and peripheral rods of three different arrangements, $b/R = 0.6, 0.667$, and 0.72. They tabulated and graphically presented $Nu_{x,T}$ and $Nu_{m,T}$ as a function of x^* ($0.0001 \leq x^* \leq 0.0020$) for the foregoing nine geometries. Zarling and Min [509,515] also studied the H1 thermal entry length problem for the aforementioned nine bundles using the same Lévêque method.

Chapter XVI

Longitudinal Fins
and Twisted Tapes within Ducts

Longitudinal fins within ducts are widely used in compact heat exchanger applications [6]. Such ducts are also referred to as internally finned tubes. Laminar fully developed convection in tubes with longitudinal rectangular and triangular fins has been studied analytically, employing some simplified idealizations. In this chapter, the "fin efficiency" is considered as 100% for all fin geometries, except for the twisted tapes. For the flow area evaluation, the fin thickness is considered as zero for all longitudinal thin fins. These idealizations are needed to make the problem amenable to mathematical analysis. Additionally, the fluid properties are treated as constant. Fully developed and developing laminar flow forced convention problems have been analyzed for the circular tube having a twisted tape. Only fully developed laminar fluid flow and heat transfer problems have been investigated for all other geometries considered.

The friction factor–Reynolds number product and Nusselt numbers for internally finned tubes, designated as $(f\,\mathrm{Re})_\mathrm{d}$ and $\mathrm{Nu}_{\mathrm{bc,d}}$, are defined using a free flow area and a hydraulic diameter identical to those of the corresponding *finless* tube. For example, f_d and Re_d of $(f\,\mathrm{Re})_\mathrm{d}$ and $\mathrm{Nu}_{\mathrm{bc,d}}$ are computed from

$$f_\mathrm{d} = \Delta p^* \left(\frac{D_{\mathrm{h,finless}}}{4L} \right) = \left(\frac{g_\mathrm{c}\rho\,\Delta p}{2LW^2} \right)(A_\mathrm{c}{}^2 D_\mathrm{h})_{\mathrm{finless}} \tag{542}$$

$$\mathrm{Re}_\mathrm{d} = \frac{G D_{\mathrm{h,finless}}}{\mu} = \left(\frac{W}{\mu} \right)\left(\frac{D_\mathrm{h}}{A_\mathrm{c}} \right)_{\mathrm{finless}} \tag{543}$$

$$\mathrm{Nu}_{\mathrm{bc,d}} = \frac{q'' D_{\mathrm{h,finless}}}{k(t_\mathrm{w} - t_\mathrm{m})} = \frac{q'}{4k(t_\mathrm{w} - t_\mathrm{m})}\left(\frac{D_\mathrm{h}{}^2}{A_\mathrm{c}} \right)_{\mathrm{finless}} \tag{544}$$

where the formula for $Nu_{bc,d}$ is intended for the axially constant heat flux boundary conditions.

The reason for this choice is that finned and finless tubes would occupy the same space in a heat exchanger; hence one can compare how much better the finned tube performs relative to the finless tube.

For longitudinal fins within a tube, $(f\,Re)_d$ and $Nu_{bc,d}$ based on the finless tube geometrical dimensions, and $f\,Re$ and Nu_{bc} based on the actual geometrical dimensions, are related to each other as follows:

$$(f\,Re)_d = (f\,Re)\left(\frac{D_{h,finless}}{D_{h,finned}}\right)^2 \left(\frac{A_{c,finless}}{A_{c,finned}}\right) \tag{545}$$

$$Nu_{bc,d} = Nu_{bc}\left(\frac{D_{h,finless}}{D_{h,finned}}\right)^2 \left(\frac{A_{c,finned}}{A_{c,finless}}\right) \tag{546}$$

Note that the flow area ratios are generally different from unity; only for the zero thickness will the flow area ratios be unity.

A. Longitudinal Thin Fins within a Circular Duct

Hu and Chang [129,442] analyzed fully developed laminar flow through a circular tube having n longitudinal rectangular fins equally spaced along the wall, as shown in the insert in Fig. 109. For the velocity problem, they considered the usual no-slip boundary condition all along the tube wall and the fin surface. For the temperature problem, they employed the (H2) boundary condition, namely, constant heat flux axially as well as peripherally along the tube wall and fins. The fin efficiency was treated as 100%, and the fin thickness was treated as zero for the flow area calculations and velocity profile determination. Hu and Chang obtained the solutions to the differential momentum and energy equations in the forms of integral representations by means of Green's function. The integral representation for the velocity involves a boundary velocity from the tube center to the fin tip, which is defined by an integral equation. The integral equation for the boundary velocity was solved numerically.

The $(f\,Re)_d$ factors and $Nu_{H2,d}$ of Hu and Chang [129,442] are presented in Tables 127 and 128 and Figs. 109 and 110. Note that $(f\,Re)_d/16$ and $Nu_{H2,d}/4.364$ are used as the ordinates of Figs. 109 and 110 in order to provide a ready indication of the increase in $(f\,Re)_d$ and $Nu_{H2,d}$ for the finned geometries relative to the unfinned duct. The actual hydraulic diameter D_h of the finned tube is related to the inside diameter of the tube and the number of fins of the height l as follows (for the fin thickness treated as zero):

$$D_{h,finned} = \frac{\pi d^2}{\pi d + 2nl} \tag{547}$$

TABLE 127

Longitudinal Thin Fins within a Circular Duct: $(f\,Re)_d$ for
Fully Developed Laminar Flow

n	$(fRe)_d$ from Hu [443]							
	ℓ* =.2	0.4	0.6	0.7	0.79	0.795	0.8	0.9
2	17.28	20.83	27.42	31.89	35.68	35.98	36.64	40.54
8	21.22	42.87	101.10	139.55	161.03	162.03	164.84	172.70
12	–	–	–	–	–	286.66	–	–
16	25.99	69.57	219.54	348.86	434.40	439.37	448.43	481.12
20	–	–	–	–	607.72	616.52	632.11	–
22	–	–	–	–	701.75	712.76	732.60	–
24	–	–	–	–	–	813.67	838.23	–
28	–	–	–	–	–	1025.6	1062.7	–
32	30.43	91.65	372.37	773.69	1221.0	1251.6	1298.7	1546.8

n	$(fRe)_d$ from Masliyah [516]						
	ℓ* = .2	0.4	0.6	0.7	0.8	0.9	1.0
4	19.13	29.04	50.31	59.31	73.43	–	77.26
8	22.39	47.02	110.81	137.35	170.47	–	175.96
12	25.55	64.22	179.77	233.09	303.16	–	314.99
16	28.20	77.86	246.08	336.15	469.92	491.04	495.24
20	30.33	88.04	305.47	439.13	668.23	708.29	716.79
24	32.01	95.57	356.04	537.09	894.15	965.09	980.77

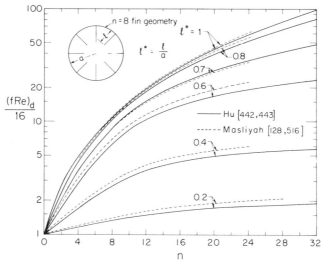

FIG. 109. Longitudinal thin fins within a circular duct: $(f\,Re)_d/16$ for fully developed laminar flow (based on the results of Table 127).

Hu and Chang found the optimum value of $Nu_{H2,d}$ to be 86.82 for $l/a = 0.795$ and $n = 22$. For this fin geometry, $(f\,Re)_d$ is 712.76. Thus the ratio of $Nu_{H2,d}$ to $(f\,Re)_d$ is 0.116 for this so-called optimum geometry. This ratio for the finless circular tube is 0.273 (4.364/16). Clearly, only when the pressure drop

TABLE 128

Longitudinal Thin Fins within a Circular Duct: $Nu_{H1,d}$ and $Nu_{H2,d}$ for Fully Developed Laminar Flow

n	$Nu_{H2,d}$ from Hu [443]							
	$\ell* = .2$	0.4	0.6	0.7	0.79	0.795	0.8	0.9
2	4.25	4.32	4.88	5.38	6.11	6.16	6.23	6.93
8	4.27	4.67	8.66	16.79	29.49	30.10	30.65	27.26
12	–	–	–	–	–	53.65	–	–
16	4.12	4.04	7.29	21.65	72.66	73.48	71.06	31.85
20	–	–	–	–	81.89	83.60	80.41	–
22	–	–	–	–	84.11	86.82	84.02	–
24	–	–	–	–	–	85.00	83.70	–
28	–	–	–	–	–	75.32	78.06	–
32	3.84	3.39	4.10	8.62	55.76	62.43	67.05	25.15

n	$Nu_{H1,d}$ from Masliyah [516]						
	$\ell* = .2$	0.4	0.6	0.7	0.8	0.9	1.0
4	4.58	6.05	11.82	15.34	19.30	–	19.08
8	4.74	6.98	21.10	34.27	42.58	–	40.68
12	4.77	6.65	20.52	40.92	72.27	–	68.80
16	4.74	6.09	16.22	34.45	106.50	105.00	103.40
20	4.68	5.64	12.73	26.07	138.35	147.20	144.60
24	4.62	5.32	10.41	19.83	156.92	195.40	192.40

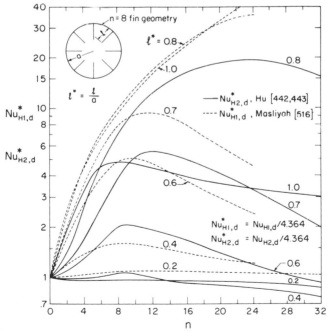

Fig. 110. Longitudinal thin fins within a circular duct: $Nu^*_{H1,d}$ and $Nu^*_{H2,d}$ for fully developed laminar flow (based on the results of Table 128).

is of no concern does the foregoing finned tube geometry offer optimum heat transfer (in the case of fin efficiency approaching 100%).

Hu and Chang also examined the effects of viscous dissipation and thermal energy generation within the fluid on the basis of the same total heat transfer rate without these effects. They found that the effect of viscous dissipation is insignificant in all cases. However, a superimposed rate of thermal energy generation appreciably decreases the Nusselt number. For $Sa/q_w'' > 2.4$, their optimum number of fins is reduced from 22 to 16.

As described on p. 375, Masliyah and Nandakumar [128] analyzed the (H1) boundary condition for the longitudinal triangular fins. When the angle 2ϕ approaches zero (Fig. 117), the longitudinal triangular fin geometry reduces to the longitudinal rectangular fin geometry analyzed by Hu and Chang. Hence $(f\,Re)_d$ and $Nu_{H1,d}$ for this geometry, obtained from Masliyah [516], are presented in Tables 127 and 128 for the case of $2\phi = 0°$. $(f\,Re)_d/16$ and $Nu_{H1,d}/4.364$ are presented in Figs. 109 and 110. Some interesting contrasts with the (H2) solutions are evident. The comparison in Fig. 109 shows that $(f\,Re)_d$ of Hu and Masliyah agree within about 5% for $l^* \geq 0.7$ and within about 11% for $l^* < 0.7$. The differences are due to different numerical methods employed.

B. Longitudinal Thin Fins within Square and Hexagonal Ducts

Chen [133] extended the study of Hu [129] by considering longitudinal fins within a duct of arbitrary cross section. By the use of exact or approximate conformal mapping functions, a circle can be mapped onto the desired arbitrary cross section. Knowing the solutions for the circular tube, the solutions for the arbitrary duct geometries can be obtained with appropriate transformations. Chen employed the method of conformal mapping and Green's functions to obtain numerical solutions for square and hexagonal ducts with longitudinal fins. These ducts have rounded corners and non-straight sides as shown in Fig. 111. If the square duct of Fig. 111a is approximated as having straight sides with rounded corners (radius = 0.40a), the

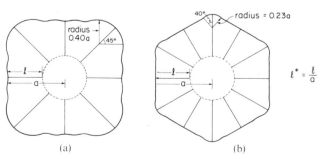

(a)   (b)

Fig. 111. Longitudinal thin fins within square and hexagonal ducts.

hydraulic diameter of this duct with 8 and 16 fins is given by

$$\frac{D_h}{2a} = \begin{cases} \dfrac{7.72531}{9.30150 + 16l^*}, & \text{for} \quad n = 8 \qquad (548) \\[2ex] \dfrac{7.72531}{10.6198 + 32l^*}, & \text{for} \quad n = 16 \qquad (549) \end{cases}$$

The effect on the hydraulic diameter of rounding of the corners of the hexagonal duct is negligible. Thus, using sharp corners and straight sides approximations, the hydraulic diameter of the hexagonal duct (Fig. 111b) with 12 fins is given by

$$\frac{D_h}{2a} = \frac{6.92820}{8.78461 + 24l^*}, \qquad \text{for} \quad n = 12 \qquad (550)$$

Chen [133] presented f Re and $Nu_{H2,d}$ for these duct geometries. By employing the relationship of Eq. (545), the $(f \, Re)_d$ factors were calculated by the present authors as presented in Table 129 along with $Nu_{H2,d}$.

TABLE 129

Longitudinal Thin Fins within Square and Hexagonal Ducts:
$(f \, Re)_d$ and $Nu_{H2,d}$ for Fully Developed Laminar Flow
(from Chen [133])

l^*	square duct n = 8		square duct n = 18		hexagonal duct n = 12	
	$(fRe)_d$	$Nu_{H2,d}$	$(fRe)_d$	$Nu_{H2,d}$	$(fRe)_d$	$Nu_{H2,d}$
0.2	25.59	4.01	43.75	3.74	30.14	3.73
0.4	48.00	4.32	70.52	4.57	67.18	3.96
0.6	105.4	8.34	140.2	13.51	175.1	8.17
0.7	138.6	17.21	175.5	51.83	247.2	23.28
0.8	157.5	28.30	188.4	76.0	291.7	49.53
0.9	161.4	28.04	184.6	34.78	302.7	35.90

Gangal and Aggarwala [131] analyzed fully developed laminar combined free- and forced-convection flow through a square duct having four equal internal fins as shown in Fig. 112. They considered constant heat flux axially with peripheral constant wall and fin temperatures. The fin efficiency was considered as 100%. The fin thickness was idealized as zero in the calculations of the flow area, and the velocity and temperature distributions. They

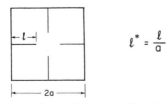

$$\ell^* = \frac{\ell}{a}$$

Fig. 112. A square duct with four equal longitudinal thin fins.

TABLE 130

Longitudinal Four Thin Fins
within a Square Duct:
$(f\,Re)_d$ and $Nu_{H1,d}$ for Fully
Developed Laminar Flow (from
Gangal and Aggarwala [130])

$\dfrac{\ell}{a}$	$(fRe)_d$	$Nu_{H1,d}$
0	14.261	3.609
0.125	15.285	3.721
0.250	18.281	4.160
0.375	23.630	5.172
0.500	31.877	7.309
0.625	42.527	11.096
0.750	52.341	14.025
1	56.919	14.431

employed their method in [130] to analyze the square duct problem and obtained numerical values for $2(f\,Re)_d$ and $Nu_{H1,d}$ for l^* varying from 0 to 1. Their $(f\,Re)_d$ and $Nu_{H1,d}$ are represented in Table 130.

C. Longitudinal Thin Fins from Opposite Walls within Rectangular Ducts

Aggarwala and Gangal [130] analyzed fully developed laminar combined free- and forced-convection flow through a rectangular duct having internal fins as shown in Fig. 113. They considered constant heat flux axially with peripheral constant wall and fin temperatures. The fin efficiency was considered as 100%; the fin thickness was idealized as zero in the calculations of the flow area, and the velocity and temperature distributions. The coupled momentum and energy equations were combined in a nonhomogeneous Helmholtz wave equation in the complex domain by introducing a complex function. This complex function is directly related to the velocity and temperature distributions. The solution of the equation gives rise to dual series equations, which are reduced to Fredholm integral equations of the second kind in the complex domain. These latter equations were solved numerically.

Aggarwala and Gangal presented $2(f\,Re)_d$ and $Nu_{H1,d}$ for (1) a square duct with two longitudinal fins, (2) a square duct with four longitudinal

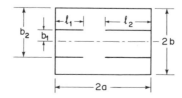

Fig. 113. A rectangular duct with four longitudinal thin fins from opposite walls.

fins, and (3) finless rectangular ducts with $0.1 \le \alpha^* \le 1$ [426]. Their f Re and Nu_{H1} for finless rectangular ducts for a Raleigh number of 1 are within 0.24 and 0.03% of the exact values of Table 42. $(f\,Re)_d$ and $Nu_{H1,d}$ for the square duct with two and four internal fins for a Rayleigh number of 1 are presented in Table 131. Partial results from this table for fins with equal heights $(l_1 = l_2)$ are shown in Fig. 114. Gangal [426] obtained $(f\,Re)_d$ and $Nu_{H1,d}$ for two fins with $l_1 = l_2 = a$ as 34.997 and 9.278, in comparison to the

TABLE 131

LONGITUDINAL THIN FINS FROM OPPOSITE WALLS
WITHIN A SQUARE DUCT: $(f\,Re)_d$ AND $Nu_{H1,d}$
FOR FULLY DEVELOPED LAMINAR FLOW
(FROM AGGARWALA AND GANGAL [131,426])

$\dfrac{\ell_1}{a}$	$\dfrac{\ell_2}{a}$	Two fins		Four fins	
		$(fRe)_d$	$Nu_{H1,d}$	$(fRe)_d$	$Nu_{H1,d}$
0.0	0.0	14.261	3.609	14.261	3.609
0.0	0.2	14.853	3.683	15.265	3.684
0.0	0.4	16.329	3.932	17.737	3.934
0.0	0.6	18.506	4.395	21.497	4.398
0.0	0.8	21.325	5.151	26.797	5.232
0.2	0.2	15.483	3.766	16.388	3.777
0.2	0.4	17.055	4.044	19.166	4.081
0.2	0.6	19.368	4.557	23.408	4.646
0.2	0.8	22.340	5.385	29.385	5.674
0.4	0.4	18.860	4.407	22.719	4.549
0.4	0.6	21.472	5.058	28.132	5.430
0.4	0.8	24.741	6.067	35.652	7.069
0.6	0.6	24.466	5.926	35.313	6.970
0.6	0.8	27.917	7.108	44.579	9.752
0.8	0.8	31.366	8.288	55.398	14.136
1.0	1.0	34.983	9.277	68.359	19.179

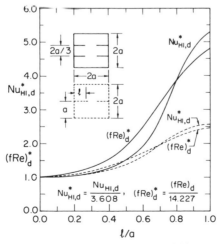

FIG. 114. Longitudinal thin fins from opposite walls within a square duct: $(f\,Re)_d^*$ and $Nu_{H1,d}^*$ for fully developed laminar flow (based on the results of Table 131).

more accurate values of 34.983 and 9.277 based on Table 42. Similar results for four fins with $l_1 = l_2 = a$ by Gangal are 68.365 and 19.179, in comparison to the values of 68.359 and 19.179 based on Table 42.

D. Longitudinal Thin V-Shaped Fins within a Circular Duct

Kun [132] analyzed fully developed laminar flow through a circular duct having longitudinal V-shaped thin fins, as shown in the insert in Figs. 115

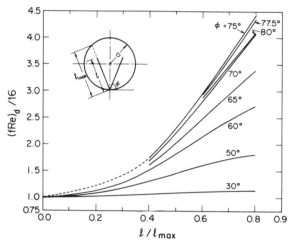

FIG. 115. Longitudinal thin V-shaped fins within a circular duct: $(f\,Re)_d/16$ for fully developed laminar flow (based on the results of Table 132).

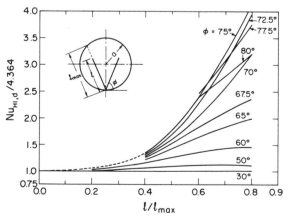

FIG. 116. Longitudinal thin V-shaped fins within a circular duct: $Nu_{H1,d}/4.364$ for fully developed laminar flow (based on the results of Table 132).

TABLE 132

LONGITUDINAL THIN V-SHAPED FINS WITHIN A CIRCULAR DUCT:
$(f\,Re)_d$ AND $Nu_{H1,d}$ FOR FULLY DEVELOPED LAMINAR FLOW
(FROM KUN [517])

ϕ	$\dfrac{\ell}{\ell_{max}}$	$(f Re)_d$	$Nu_{H1,d}$	ϕ	$\dfrac{\ell}{\ell_{max}}$	$(f Re)_d$	$Nu_{H1,d}$
30	0.20	16.353	4.359	67.5	0.40	26.724	5.501
	0.40	17.059	4.386		0.60	42.620	7.990
	0.60	17.776	4.400		0.80	60.129	10.885
	0.80	18.203	4.403	70	0.40	27.328	5.647
40	0.20	16.799	4.385		0.60	45.009	8.856
	0.40	18.567	4.472		0.80	65.356	13.981
	0.60	20.497	4.535	72.5	0.40	27.810	5.768
	0.80	21.731	4.521		0.60	46.607	9.771
50	0.20	17.444	4.426		0.80	69.095	17.218
	0.40	21.034	4.681	75	0.40	28.072	5.872
	0.60	25.735	4.895		0.60	47.342	10.469
	0.80	29.046	4.887		0.80	70.942	18.010
60	0.20	18.154	4.496	77.5	0.60	47.078	10.787
	0.40	24.351	5.071		0.80	69.702	16.353
	0.60	34.527	5.994	80	0.60	46.070	10.477
	0.80	43.552	6.376		0.80	65.779	13.967
65	0.40	26.022	5.341				
	0.60	39.903	7.209				
	0.80	54.368	8.703				

and 116. He employed the (H1) boundary condition for the heat transfer problem. He idealized the fins as 100% efficient, and considered the fin thickness as zero for the calculations of the flow area and the velocity and temperature distributions.

Employing Green's function, Kun first presented the velocity and temperature distributions in terms of integral equations, and subsequently obtained the solutions numerically. He presented graphically Nu_{H1} as a function of the dimensionless fin height l/a with the fin angle ϕ as a parameter. His $(f\,Re)_d$ and $Nu_{H1,d}$ are presented in Table 132. The ratios $(f\,Re)_d/16$ and $Nu_{H1,d}/4.364$ are presented in Figs. 115 and 116, where

$$l_{max} = 2a \sin \phi \tag{551}$$

$$D_h = \frac{2a}{1 + (2l/\pi a)} \tag{552}$$

E. Longitudinal Triangular Fins within a Circular Duct

Masliyah and Nandakumar [127,128] analyzed fully developed laminar flow through a circular tube having n triangular fins equally spaced along the wall, as shown in Fig. 117. The flow area and wetted perimeter for this

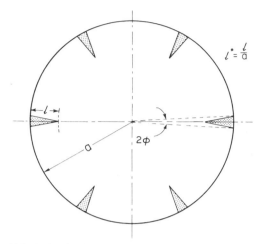

FIG. 117. A circular duct with $n = 6$ triangular longitudinal fins.

finned geometry are given by

$$A_{\text{c,finned}} = \pi a^2 - n[a^2\phi - a(a - l)\sin\phi] \tag{553}$$

$$P_{\text{finned}} = 2\pi a + 2nl' - 2n\phi a \tag{554}$$

where

$$l' = [a^2 + (a - l)^2 - 2a(a - l)\cos\phi]^{1/2} \tag{555}$$

Then the hydraulic diameter can be calculated from its definition, $D_{\text{h,finned}} = 4(A_c/P)_{\text{finned}}$.

For the velocity problem, they considered the usual no-slip boundary condition all along the tube wall and the fin surface. For the temperature problem, they considered the (H1) boundary condition, namely, constant heat flux axially and uniform temperature along the tube wall and fins. The fin efficiency was considered as 100%. Note that the fin thickness was considered finite, as shown in Fig. 117. They divided the smallest symmetrical region of the tube into triangular elements and employed a finite element method to obtain solutions for the velocity and temperature problems.

Nandakumar and Masliyah [127] presented a solution to the velocity problem for the limited range of l^*, ϕ, and n. Their $(f\,\text{Re})_d$ factors are presented in Table 133 and Fig. 118.

Masliyah and Nandakumar [128] further extended their work in order to cover a wider range of parameters for $f\,\text{Re}$ and also to obtain the (H1) heat

TABLE 133

LONGITUDINAL TRIANGULAR FINS WITHIN A CIRCULAR DUCT: $(f\,Re)_d$ FOR
FULLY DEVELOPED LAMINAR FLOW (FROM MASLIYAH [516])

n	$(fRe)_d$ for $2\phi = 3°$						$(fRe)_d$ $2\phi=12°$
	$\ell^* = .1$	$\ell^*=.2$	$\ell^*=.4$	$\ell^*=.6$	$\ell^*=.7$	$\ell^*=.8$	$\ell^*=.1$
4	–	19.20	29.35	51.73	69.56	77.24	–
8	18.27	22.56	48.16	118.4	174.2	194.2	18.27
12	19.36	25.78	66.13	198.3	326.3	376.7	19.32
16	20.36	28.50	80.36	277.4	523.5	638.1	20.27
20	21.25	30.61	90.66	348.0	759.4	990.2	–
24	22.02	32.26	98.15	407.3	1021.8	1440.5	–
	$(fRe)_d$ for $2\phi = 6°$						$\ell^*=.2$
4	–	19.29	29.74	53.25	71.91	81.41	–
8	10.11	22.77	49.38	126.9	193.0	223.4	23.34
12	19.11	26.09	68.25	219.2	388.5	479.4	27.06
16	20.03	28.86	82.79	311.8	665.5	903.8	29.82
20	20.85	31.02	93.25	392.7	1013.3	1560.9	31.97
24	21.55	32.65	100.6	456.8	1402.2	2502.2	33.53

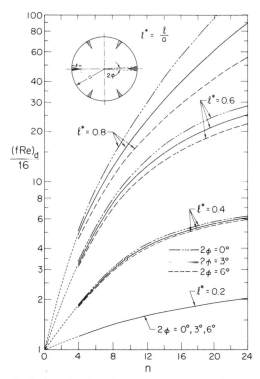

FIG. 118. Longitudinal triangular fins within a circular duct: $(f\,Re)_d/16$ for fully developed
laminar flow (based on the results of Tables 127 and 133).

transfer results. Their $(f\,Re)_d$ and $Nu_{H1,d}$ are presented in Tables 133 and 134 and Figs. 118 and 119 [128] The case of $2\phi = 0°$ represents the longitudinal fins of zero thickness and has already been discussed on p. 370.

TABLE 134

Longitudinal Triangular Fins within a
Circular Duct: $Nu_{H1,d}$ for Fully Developed Laminar Flow
(from Masliyah [516])

n	$\ell*$				
	0.2	0.4	0.6	0.7	0.8
	$Nu_{H1,d}$ for $2\phi = 3°$				
4	4.58	6.03	11.72	18.02	19.29
8	4.73	6.85	20.01	43.26	43.60
12	4.75	6.42	17.82	64.58	76.17
16	4.71	5.82	13.15	63.21	112.11
20	4.64	5.37	9.95	46.17	131.90
24	4.58	5.07	8.02	30.86	117.50
	$Nu_{H1,d}$ for $2\phi = 6°$				
4	4.58	6.00	11.63	17.82	19.29
8	4.71	6.71	18.87	43.31	44.88
12	4.73	6.19	15.30	58.88	80.81
16	4.67	5.57	10.64	45.28	112.45
20	4.60	5.14	7.91	26.75	98.20
24	4.54	4.87	6.42	16.01	56.04

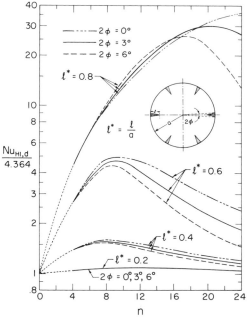

Fig. 119. Longitudinal triangular fins within a circular duct: $Nu_{H1,d}/4.364$ for fully developed laminar flow (based on the results of Tables 128 and 134).

F. Circular Duct with a Twisted Tape[†]

A twisted tape is sometimes inserted in a circular tube to establish swirl flow and thereby to increase the heat transfer coefficient on the inside tube surface. This tape, generally a thin metal strip, is twisted about its longitudinal axis as indicated in Fig. 120. In the analytical results presented below, the width of the tape is considered as equal to the internal diameter of the tube. When the tape twist ratio $X_L (= H/d)$ approaches infinity, the duct geometry becomes two semicircular straight ducts in parallel.

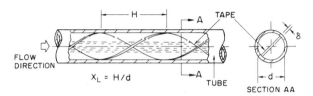

FIG. 120. A circular tube with a twisted tape.

1. FULLY DEVELOPED FLOW

a. *Fluid Flow*

Date and Singham [99] and Date [100,101] employed a finite difference method and used the so-called upwind difference scheme for the grid points to analyze constant property fully developed laminar flow through a circular tube containing a twisted tape. They found that the flow with a twisted tape is quite different from the flow through a straight duct because of secondary or swirl flow v and w velocity components. The secondary flow velocity is largest in regions close to the twisted tape surface and decreases in the direction of tube surface. The magnitude of the secondary flow velocities increases with an increase in Re_d and a decrease in X_L. Date [100] found it to be between 10 and 35% of the mean axial velocity. The secondary flow increases with the axial distance, reaching its maximum value when the flow is fully developed.

Date and Singham [99] showed typical velocity profiles in twisted tape ducts with $X_L = 5.24$, 3.14, and 2.25 for $Re_d = 1200$. The velocity profiles across the duct cross section are asymmetric. As the tape twist increases, they become flatter and flatter. At very high twists (low X_L, e.g., 2.25), they disintegrate into a pattern that demonstrates two peaks of axial velocity. A similar phenomenon is observed at a constant tape twist ratio X_L when Re_d is increased.

[†] The authors are grateful to Professors A. E. Bergles and A. W. Date for reviewing this section.

Date and Singham [99] showed that the friction factor depends upon the Reynolds number and the tape twist ratio X_L. They were the first investigators to recognize the parameter Re_d/X_L, which accounts for the centrifugal force effect. When small, the parameter Re_d/X_L represents an effect similar to that of the Dean number in curved pipe flow. However, for large values of Re_d/X_L, $(f\,\mathrm{Re})_d$ depends upon X_L in addition to Re_d/X_L, as shown later. When $X_L \to \infty$, the limiting geometry consists of two semicircular ducts with $f\,\mathrm{Re} = 15.523$. This is only 1.5% lower than the exact value of 15.767 (Table 77). For $X_L \to \infty$, Date and Singham correlated their numerical results within $\pm 5\%$ by Eqs. (557) and (558). Equation (556) is proposed by the present authors, based on the numerical results of Date and Singham and the theory for the straight semicircular duct:

$$(f\,\mathrm{Re})_d = \begin{cases} 42.23, & \text{for} \quad (\mathrm{Re}_d/X_L) < 6.7 & (556) \\ 38.4(\mathrm{Re}_d/X_L)^{0.05}, & \text{for} \quad 6.7 \le (\mathrm{Re}_d/X_L) \le 100 & (557) \\ C(\mathrm{Re}_d/X_L)^{0.3}, & \text{for} \quad (\mathrm{Re}_d/X_L) > 100 & (558) \end{cases}$$

where

$$C = 8.8201X_L - 2.1193X_L{}^2 + 0.2108X_L{}^3 - 0.0069X_L{}^4 \qquad (559)$$

Equation (557) is a corrected version of the equation originally presented by Date and Singham [99].

In the foregoing correlations, it is idealized that the thickness of the twisted tape is zero. However, it may not be negligible if the tube diameter is small. The thickness of the tape reduces the flow area and increases the pressure drop. The foregoing correlations are modified by the present authors, as follows, in order to account for the finite thickness δ of the twisted tape:

$$(f\,\mathrm{Re})_d = \begin{cases} 42.23\zeta, & \text{for} \quad \mathrm{Re}_d/X_L < 6.7 & (560) \\ 38.4(\mathrm{Re}_d/X_L)^{0.05}\zeta, & \text{for} \quad 6.7 \le \mathrm{Re}_d/X_L \le 100 & (561) \\ C(\mathrm{Re}_d/X_L)^{0.3}\zeta, & \text{for} \quad \mathrm{Re}_d/X_L > 100 & (562) \end{cases}$$

where C is given by Eq. (559) and ζ is given by

$$\zeta = \frac{(D_h{}^2 A_c)_{\delta=0}}{(D_h{}^2 A_c)_{\delta \ne 0}} = \left(\frac{\pi}{\pi+2}\right)^2 \left(\frac{\pi+2-2\delta/d}{\pi-4\delta/d}\right)^2 \left(\frac{\pi}{\pi-4\delta/d}\right) \qquad (563)$$

The factor ζ was obtained from the ratio of $(f\,\mathrm{Re})_{d,\delta \ne 0}$ to $(f\,\mathrm{Re})_{d,\delta=0}$. These quantities were determined from Eq. (545) with $f\,\mathrm{Re}$ considered as approximately constant.[†]

[†] Note that the finite thickness of the tape would change the shape of a semicircular duct to a circular segment duct with the segment angle $2\phi < 180°$. However, a review of Table 77 reveals that the $f\,\mathrm{Re}$ factors are constant within 1% for the segment angle 2ϕ varying from 180 to 0°.

Hong and Bergles [446,518] experimentally determined the friction factors for fully developed twisted-tape flow with $X_L = 3.125$ and ∞ by using ethylene glycol and water as the test fluids. They employed δ/d of 0.045, so that $\zeta = 1.152$. Their results[†] are shown in Fig. 121. For $X_L = \infty$, these results are in excellent agreement with the theoretical predictions. For $X_L = 3.125$, f_d is lower than the theoretical value (about 17 and 30% at $Re_d = 1000$ for ethylene glycol and water, respectively). It is interesting to note that the theoretical increase in the friction factor above the semicircular duct value is only 40% at $Re_d = 1000$. As will be discussed later, the increase in the Nusselt number for $X_L < \infty$ can be very large, depending upon the fluid Prandtl number.

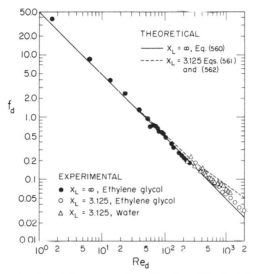

FIG. 121. Circular tube with a twisted tape: f_d as a function of Re_d for fully developed laminar flow (from Hong and Bergles [518]).

b. *Heat Transfer*

Date and Singham [99] and Date [100,101] considered a circular tube subjected to axially constant wall heat flux with a peripherally constant wall temperature at any cross section, namely the (H1) thermal boundary condition. However, the tape was considered as a fin attached to and in perfect contact with the circular tube. Thus, in the heat transfer analysis,

---

[†] Note that $4(f\,Re)_d$ is 194.6 based on Eq. (560) and $\zeta = 1.152$. Hong and Bergles [446, 518] considered the same flow area A_c with and without the twisted tape, so that $\zeta = 1.086$ after neglecting the last term on the right-hand side of Eq. (563). Hence they arrived at $4(f\,Re)_d$ as 183.6.

the tape was treated as a fin with a peripherally variable heat flux and temperature. The resulting fin efficiency was found to be less than 100%. Because of this, the temperature of the tape differed from the temperature of the circular tube wall at a specified cross section. This boundary condition does not match any described in Chapter II; thus, in order to avoid confusion, no boundary condition subscript is used for the corresponding Nusselt numbers.

Date and Singham showed that the Nusselt number for the circular duct with twisted tapes is dependent upon four groups: Re_d/X_L, C_{fin}, Pr, and X_L, where $C_{fin} = (k_m\delta/k_f d)$ and k_m is the thermal conductivity of the tape material. $C_{fin} = 0$ signifies that the tape is a heat nonconductor, and $C_{fin} = \infty$ signifies that the tape is a perfect heat conductor. A higher Re_d/X_L, C_{fin}, or Pr results in higher Nusselt numbers, while X_L has a weak influence on Nusselt numbers. All heat transfer results of Date and Singham [99] and Date [100,101] are in error, because their q' relates to the semicircular duct and *not* the circular tube with a twisted tape. Hence, all of the Nusselt numbers should be multiplied by a factor of 2, irrespective of the fin parameter used [519].

Date and Singham presented temperature profiles at $Re_d = 378$, 809, and 1600 for $X_L = 5.24$, Pr = 1.0, $C_{fin} = 1.85$. The temperature profiles across the duct cross section are asymmetric. As the Reynolds number increases, the temperature profiles become flatter, confirming an increase in the Nusselt number. The effects of increasing Pr and decreasing X_L at a constant Re_d are similar.

Table 135 and Fig. 122 are based on the graphical results of Date [100] for $X_L = 2.25$ and 5.24. The reason for the choice of the coordinate for the abscissa of Fig. 122 will be clear after the discussion of Eq. (564). For $Re_d/X_L = 0$ and $C_{fin} = \infty$, Date determined Nu_d as 10.8 and Hong and Bergles [446,518] determined it as 11.00, in contrast to a more precise value of 10.95 based on Nu_{H1} of Table 77 for $2\phi = 180°$. For $Re_d/X_L = 0$ and $C_{fin} = 0$, Date determined Nu_d as 5.188, while Hong and Bergles obtained it as 5.172.

A review of Fig. 122 reveals that the influence of Pr and C_{fin} is quite significant on Nusselt numbers. The influence of Pr on Nusselt numbers of laminar fully developed twisted-tape flow is very similar to that for the curved pipe flow [13], i.e., at a given Re_d/X_L (Dean number in the case of curved pipe flow), the higher fluid Prandtl numbers result in higher Nusselt numbers. Date's analysis shows that C_{fin} also has a strong effect on Nusselt number, as shown in Fig. 122.

Hong and Bergles [446,518] experimentally evaluated the heat transfer performance for a laminar twisted-tape flow. The tube was electrically heated but the twisted tape was electrically isolated from the wall. Hence

TABLE 135

Circular Tube with a Twisted Tape: Nu_d as Functions of Re_d/X_L, C_{fin}, and Pr for
Fully Developed Laminar Flow (from the Graphical Results of Date [100])

$\dfrac{Re_d}{X_L}$	$C_{fin} = 0$		$C_{fin} = 1.85$				$C_{fin} = \infty$			
	Pr=1	Pr=10	Pr=.1	Pr=1	Pr=10	Pr=100	Pr=.1	Pr=1	Pr=10	Pr=100
0	5.19	5.19	7.60	7.60	7.60	7.60	10.8	10.8	10.8	10.8
2	5.50	5.58	7.70	7.80	8.0	10.8	–	11.0	–	–
5	5.58	7.04	7.74	8.04	9.3	17.2	–	11.1	–	–
10	5.72	8.28	7.78	8.25	10.7	25.0	11.0	11.3	17.0	48.8
20	5.96	9.72	7.82	8.55	12.6	37.0	11.5	11.9	20.2	79.0
30	6.18	10.7	7.86	8.76	14.0	46.6	11.8	12.4	23.0	105.
40	6.35	11.6	7.90	8.90	15.2	55.5	12.0	13.0	25.4	130.
50	6.50	12.3	7.95	9.06	16.2	64.0	12.2	13.5	28.0	153.
60	6.66	13.0	7.98	9.18	17.0	72.0	12.4	14.0	30.4	176.
80	6.92	14.4	8.05	9.40	18.6	86.0	12.7	14.9	35.6	220.
100	7.15	15.5	8.12	9.65	20.0	100.	12.9	15.8	40.2	262.
150	7.68	17.9	8.40	10.2	23.0	131.	13.3	18.2	52.6	365.
200	8.22	–	8.66	10.8	25.7	161.	13.7	20.8	65.0	470.
300	9.56	–	9.34	12.2	30.8	220.	14.2	25.8	90.0	690.
400	11.0	–	10.1	13.8	36.4	280.	–	30.0	–	–
500	12.5	–	11.1	15.7	42.8	350.	–	33.8	–	–
600	13.9	–	12.3	17.8	50.3	420.	–	37.4	–	–

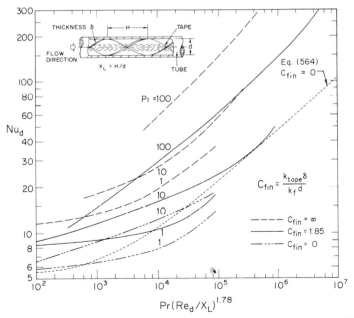

FIG. 122. Circular tube with a twisted tape: Nu_d for fully developed laminar flow [based on
the results of Table 135 and the experimental correlation of Eq. (564)].

their results are for the case of $C_{fin} \simeq 0$ ($C_{fin} = 0.124$ for ethylene glycol). They obtained test results for $X_L = 2.45$ and 5.08 by using distilled water and ethylene glycol as the test fluids (Pr $\simeq 3$ to 191). They correlated the test results by the following expression:

$$\mathrm{Nu_d} = 5.172\{1 + 0.005484[\mathrm{Pr}(\mathrm{Re_d}/X_L)^{1.78}]^{0.7}\}^{0.5} \tag{564}$$

In Fig. 122, this correlation is compared with the results of Date and Singham. Using an abscissa of $\mathrm{Pr}(\mathrm{Re_d}/X_L)^{1.78}$, Hong and Bergles were able to correlate all of their test data for Pr ranging from 3 to 191 within $\pm 20\%$, while Date's analytical results show a separate influence of Pr on $\mathrm{Nu_d}$. The analytical results of Date and the correlation of Hong and Bergles differ substantially. The strong Prandtl number effect for $C_{fin} > 0$, predicted by Date, needs experimental verification.

An examination of $(f\,\mathrm{Re})_d$ and $\mathrm{Nu_d}$ for the laminar twisted-tape flow reveals that with increasing $\mathrm{Re_d}/X_L$ there is a dramatic increase in heat transfer with only a slight increase in pressure drop. Both analysis and experiments show this increase in heat transfer. More analyses and experiments are required to determine more precisely the augmentation characteristics for laminar twisted-tape flow in a circular tube.

2. Developing Flows

Date [100,520] investigated simultaneously developing flow through a circular tube containing a twisted tape. For the heat transfer problem, he considered the tube at constant temperature axially and peripherally. He showed that Δp^* and θ_m are functions of the following parameters:

$$\Delta p^* = \Delta p^*(\mathrm{Re_d}, X_L, x/a) \tag{565}$$

$$\theta_m = \theta_m(\mathrm{Re_d}, \mathrm{Pr}, X_L, x/a) \tag{566}$$

The results have been presented graphically by Date.

Chapter XVII

Discussion—An Overview for the Designer and the Applied Mathematician

Chapters V–XVI provide an up-to-date compilation, in a common format, of available analytical solutions for laminar fluid flow and forced convection heat transfer through straight, constant cross-sectional ducts. Table 136 provides a ready reference to locate these solutions in tabular/graphical forms.

These solutions are valid for steady flow of a constant property Newtonian fluid. They include hydrodynamically developed and developing flows, thermally developed and developing flows, and hydrodynamically and thermally developing flows. Wall heat transfer effects due to axial heat conduction, viscous dissipation, flow work, and thermal energy sources within the fluid have been included for some of the duct geometries, whenever available. However, the influences of natural convection, change of phase, mass transfer, and chemical reactions have not been considered. All forms of body forces (except the centrifugal force for twisted-tape flow), magnetohydrodynamic flows, electrically conducting flows, and heat radiating flows are also omitted. As long as the surface roughness of the inside wall does not significantly affect the cross section of the duct geometry, it does not influence the friction factor and Nusselt number in laminar flow. Hence the solutions presented are valid for both smooth and somewhat rough walls.

The purpose of this chapter is to provide an overview of the subject for the designer. First, the results presented in the preceding chapters are compared so that the designer can understand some of the idealizations involved

TABLE 136

Summary Index of Developed and Developing Laminar Flow Solutions
Reported in the Text

Geometry	$f\,Re$	Nu_{fd}	Hydrodynamic entry length solutions	Thermal entry length solutions	
				Developed velocity profile	Developing velocity profile
⬭ circle	16	Tables 6–9 Fig. 12 Eqs. (179, 184)	Tables 10, 11 Fig. 13	Tables 13, 15, 16, 18, 19 Figs. 15, 19, 20, 24	Tables 20–23, 24a Figs. 25–28
parallel plates	24	Tables 25, 26 Eqs. (273)– (275), (278)	Tables 27–29	Tables 31, 34, 35 Figs. 31, 33	Tables 36–38 Figs. 34, 35
rectangle	Table 42 Fig. 37	Tables 42–44, 46 Figs. 38–40	Table 47 Fig. 41	Tables 48, 49, 49a, 51 Fig. 42	Tables 52–54 Fig. 43
equilateral triangle	Table 57 Fig. 47	Tables 57–59 Figs. 48–50	Table 62 Fig. 55	Table 63 Fig. 56	Table 63 Fig. 56
right triangle	Table 60 Fig. 52	Table 60 Fig. 52	Table 62 Fig. 55	Table 64 Fig. 57	Table 64 Fig. 57
ellipse	Table 65 Fig. 59	Tables 65, 66 Fig. 60	None	[434,435,35]	None
annulus (large)	Table 86 Fig. 85	Tables 86, 89–92 Figs. 86–88	Table 93 Fig. 89	Tables 94–97 Fig. 90	Table 99
annulus (small)	Table 100 Fig. 92	Tables 101– 105 Figs. 93, 94	Table 106 Fig. 96	Tables 107–110	Table 112
semicircle	15.767	$Nu_{H1} = 4.089$ $Nu_{H2} = 2.923$	None	Table 78	None

(continued)

TABLE 136 (*continued*)

SUMMARY INDEX OF FULLY DEVELOPED LAMINAR FLOW SOLUTIONS FOR
SINGLY CONNECTED DUCTS REPORTED IN THE TEXT

Geometry	f Re	Nu_{fd}	Geometry	f Re	Nu_{fd}
	Table 55	Table 55		Table 77 Fig. 73	Table 77 Fig. 73
	Table 61 Fig. 53	Table 61 Fig. 54		Table 79 Fig. 75	None
	Table 68 Fig. 62	Table 68 Fig. 63		Table 80 Fig. 77	None
	Table 71 Fig. 65	Table 71 Fig. 66		Table 81 Fig. 78	Table 81 Fig. 78
	Table 73 Fig. 67	Table 73 Fig. 68		Table 81 Fig. 78	Table 81 Fig. 78
	Table 74	Table 74		Table 82	Table 82
	Table 75 Fig. 70	Table 75 Fig. 70		Table 83	None
	Table 76 Fig. 72	Table 76 Fig. 72		15.675	$Nu_{H1} = 4.208$ $Nu_{H2} = 4.097$

(*continued*)

TABLE 136 (*continued*)

SUMMARY INDEX OF FULLY DEVELOPED LAMINAR FLOW SOLUTIONS FOR DOUBLY CONNECTED DUCTS AND INTERNALLY FINNED TUBES REPORTED IN THE TEXT

Geometry	f Re	Nu_{fd}	Geometry	f Re	Nu_f
	Table 113 Fig. 98	Table 114 Table 115 Fig. 99		Table 127 Fig. 109	Table Fig. 11
	Table 116 Fig. 100	Table 117 Fig. 101		Table 129 Table 130	Table Table
	Table 118	None		Table 129	Table
	Table 119	None		Table 131 Fig. 114	Table Fig. 11
	Table 120 Fig. 102	Table 121 Fig. 103		Table 132 Fig. 115	Table 1 Fig. 11
	Table 122 Table 123 Fig. 105	Table 122 Fig. 105			
	Table 123	None		Table 133 Fig. 118	Table 1 Fig. 11
	Table 126 Fig. 108	None		Twisted-tape Eqs. (560)–(562)	Table 1 Fig. 12

and determine which passage geometries best suit his needs. Second since a heat exchanger consists of many ducts in parallel, but the solutions are derived for a single duct, application of the theory needs to be qualified. If all of the passages do not have identical shapes, the effective friction factor and Nusselt number could be substantially different from those for single-duct behavior. Thus a procedure is presented for determining the effective friction factor and Nusselt number for the multipassage heat exchanger. Next, a guideline is presented for the designer regarding when the free convection effects in low Reynolds number laminar flow could be important. Since the transport properties of most fluids vary with temperature, the results of the preceding chapters need to be modified for heat transfer applications; thus the influence of temperature-dependent properties is summarized next. Comments are then made to clarify the format of the published literature on laminar flow forced convection in ducts. As a final consideration, areas of future research are suggested.

A. Comparisons of Solutions

1. COMPARISONS OF THERMAL BOUNDARY CONDITIONS

a. *A Comparison of Axially Constant Wall Heat Flux Boundary Conditions* (H1), (H2), *and* (H4)

The thermal boundary condition of approximately constant axial heat rate per unit duct length ($q' \simeq$ constant) is realized in many cases, such as electric resistance heating, nuclear heating, and counterflow heat exchangers with equal thermal capacity rates (Wc_p). There are three idealized constant q' boundary conditions, namely, (H1), (H2), and (H4), that have been analyzed to varying degrees. These three boundary conditions become identical for straight ducts having constant peripheral curvature and no "corners effects," e.g., the circular duct, parallel plates, and the concentric annular duct. These boundary conditions for ducts having corners and/or different curvature around the periphery (e.g., rectangular, triangular, sine, and elliptical ducts) are different and yield different Nusselt numbers. Implicit idealizations made for wall thermal conductivity for these boundary conditions are summarized in Table 137. Since only the wall temperature or the fluid temperature gradient at the wall–fluid interface is needed as a boundary condition, the radial thermal conductivity of the wall is not involved in the analysis. From Table 137, it can be seen that the (H1) and (H2) boundary conditions are the special cases of the (H4) boundary condition. As indicated above, they become identical for ducts such as the straight circular duct, because then there is no

TABLE 137

Implied Idealizations of Wall
Thermal Conductivity for Some
Thermal Boundary Conditions
(Refer to Table 3 for
Further Details)

Boundary condition	Axial k_w	Peripheral k_w
(H1)	zero	infinite
(H2)	zero	zero
(H4)	zero	finite
(T)	infinite	infinite

peripheral variation of wall temperature and the magnitude of k_w is of no significance.

For a heat exchanger with highly conductive materials, e.g., copper or aluminum, the (H1) boundary condition may be a good approximation when q' is constant. However, for a heat exchanger with low thermal conductivity materials, e.g., glass ceramic or teflon, the (H2) boundary condition may be more realistic if the wall thickness all around the periphery is uniform. For a more general problem with constant q', the (H4) boundary condition is more appropriate.

The Nusselt number for the (H1) boundary condition is higher than that for the (H2) boundary condition with (H4) in between. For the (H1) boundary condition, based on the definition of Eq. (104),

$$h_{H1} = \left(\frac{1}{t_w - t_m}\right)\frac{q'}{P}, \qquad \begin{cases} q' = \text{constant with } x \\ t_w = \text{constant with } s \end{cases} \tag{567}$$

Similarly, the heat transfer coefficient for the (H2) boundary condition is found as

$$h_{H2} = \left(\frac{1}{(1/P)\int_\Gamma t_w ds - t_m}\right)\frac{q'}{P}, \qquad \begin{cases} q' = \text{constant with } x, \\ q'' = q'/P, \quad \text{constant with } s. \end{cases} \tag{568}$$

For the (H2) boundary condition, $q'' = -k(\partial t/\partial n)$ is constant around the periphery and is transferred to the fluid away from the walls. The fluid mean effective conduction path length δ_f for this thermal energy transfer is highest from the corners relative to the wall surface away from the corners:

$$q'' = k\frac{\Delta t}{\Delta n} = k\frac{(t_w - t_m)}{\delta_f} = \text{constant} \tag{569}$$

Consequently, the corner temperature difference $(t_w - t_m)$ is higher, and this increases the peripheral average t_w in Eq. (568) with a reduction of h_{H2} relative to its counterpart h_{H1}.

Nu_{H2} can be significantly lower than Nu_{H1} for an acute cornered duct geometry. Consequently, judgment must be exercised before applying the theoretical results for the (H1) boundary condition. This also suggests that more solutions are needed for both the (H2) and (H4) noncircular duct heat transfer problems. Additionally, more experimental measurements of peripheral wall temperatures would be useful.

b. *A Comparison of* (H1) *and* (T) *Boundary Conditions*

In the previous section, examples were mentioned where the (H1) boundary condition may be realized in a practical situation.

The (T) boundary condition may be approximated for heat transfer in a condenser, evaporator, or a liquid-to-gas heat exchanger with high hA on the liquid side. In these cases, the temperature of the fluid on one side is fairly uniform and constant, and the thermal resistances of the wall and constant fluid temperature side are relatively small. For this situation, it is implicitly assumed that k_w in the axial direction is arbitrary, but that k_w in the radial direction is infinite. The other possible idealized situation for the (T) boundary condition is infinite k_w axially and peripherally, as described in Table 137. The radial thermal conductivity or the radial wall temperature profile for this case is not involved in the analysis.

Nu_{H1} is higher than Nu_T for all duct geometries. The physical reasoning for this behavior can be obtained from a consideration of the fluid temperature profiles, for example, as shown in Fig. 123, which pictures a circular tube with heat transfer from the wall to the fluid.

It can be seen that there is an inflection point, as obtained from a rigorous analytical solution, in the dimensionless fluid temperature profile for the (T) boundary condition. This is because t_w is constant and the fluid bulk mean temperature t_m "catches up" with it. For the constant q' boundary condition, t_w is continuously "running away" from t_m so that an inflection in the profile does not develop. From Eq. (110), the heat transfer coefficient is

$$h = k\left(\frac{\partial t}{\partial r}\right)_{r=a} \bigg/ (t_w - t_m) \tag{570}$$

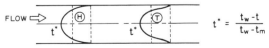

FIG. 123. Fluid temperature profiles for the (H) and (T) boundary conditions.

for the circular tube. From Fig. 123, for the same $(t_w - t_m)$, the fluid temperature gradient at the wall is smaller for the ⓣ boundary condition because of the inflection. Thus, from Eq. (570), h and consequently Nu are lower for the ⓣ boundary condition. This result may be generalized to apply to noncircular ducts.

2. Comparisons of f Re and Nu for Fully Developed Flows

a. *Ranges of f Re and Nu*

Fully developed flow friction–Reynolds number products and Nusselt numbers for various boundary conditions are presented in Chapters V–XVI for 30 singly connected ducts, seven doubly connected ducts, and three multiply connected duct geometries, as summarized in Table 136.

A review of these results shows that f Re factors for singly connected ducts vary from 6.503 for a three-sided cusped duct to 24 for parallel plates. For doubly connected ducts, f Re factors vary from 6.222 for an n-sided regular polygonal duct with an inscribed circle and n approaching infinity to 24 for a concentric annular duct approaching parallel plate geometry $(r^* \rightarrow 1)$. For multiply connected ducts, such as a bundle of circular cylinders, the lower limit of f Re factors may correspond to the limiting case of a three- or four-sided cusped duct (see Table 122, Fig. 105). There is no upper limit to the magnitude of the f Re factor because D_h can go to infinity and hence f Re (which is proportional to $D_h{}^2$) can go to infinity.

The Nusselt numbers for the ⓗ① boundary condition for singly connected ducts vary fourfold, from 1.920 for a sine duct with an aspect ratio $2b/2a$ approaching zero to 8.235 for parallel plates. Nu_{H1} for a three-sided cusped duct (not presently available) would be even smaller than that for the sine duct. For doubly connected ducts, Nu_{H1} also varies fourfold from 1.947 for an eccentric annular duct with $e^* = 0.99$ and $r^* = 0.90$ to 8.235 for the concentric annular duct approaching parallel plates $(r^* \rightarrow 1)$. Here again, the n-sided regular polygonal duct with an inscribed circle, as mentioned above, may be considered as having a lower Nu_{H1} magnitude. For multiply connected ducts, the lower limit of Nu_{H1}, similar to f Re factors, corresponds to the limiting case of a singly connected duct (see Fig. 105). There is no upper limit to Nu_{H1}.

Nu_{H2} approaches zero for the limiting geometries of ducts having sharp corners, such as triangular, sine, rhombic, circular sector, and circular segment ducts. The highest magnitude of Nu_{H2} for singly and doubly connected ducts corresponds to 8.235 for the parallel plates geometry (Table 138). Note that Nu_{H2} for a rectangular duct with an aspect ratio approaching zero $(2b/2a \rightarrow 0)$ *does not* approach the value for parallel plates (see Table 42, Fig. 38). This is because the imposed constant heat flux on the short sides of the rectangular duct affects Nu_{H2} even when $2b/2a$ approaches zero. No such boundary condition is imposed for the parallel plates problem.

Fully developed Nusselt numbers for the Ⓣ boundary condition are evaluated for a limited number of duct geometries. For singly connected ducts, the lowest value of Nu_T is reported as 0.739 for a sine duct having $2b/2a \rightarrow \infty$, and is the highest for parallel plates at 7.541. For doubly connected ducts, Nu_T is evaluated only for the concentric annular ducts ranging from 3.657 for $r^* = 0$ to 7.541 for $r^* = 1$. No Nu_T is reported for multiply connected ducts.

Flow friction factors and Nusselt numbers for several singly connected duct geometries of technical interest are summarized in Table 138. As expected from the discussion of the previous section, both Nu_{H2} and Nu_T are lower than Nu_{H1}. For the duct geometries having strong corner effects, Nu_{H2} is even lower than Nu_T, but otherwise Nu_{H2} is higher. The ratio of maximum to minimum Nu_{H1}, Nu_{H2}, and Nu_T is found as 2.7, 5.6, and 3.2, respectively, for the geometries considered in Table 138. The ratio Nu_{H1}/Nu_T varies from 1.09 to 1.26. The ratio of maximum to minimum $f\,Re$ is 1.9.

b. *Area and Volume "Goodness" Factor Comparisons*

The performance of heat exchange surfaces made up of different idealized passage geometries may be contrasted in the following two complementary different ways: a comparison of the flow area "goodness" factors j/f, and a comparison of the core volume "goodness" factors h_{std} versus E_{std}.

There are many different criteria for selecting and optimizing heat exchanger passage geometries, as outlined by Bergles et al. [521]. The two goodness factors relate only to the core flow area and core volume. These factors are easy to understand and apply, and may serve a function of screening the selection of surfaces before other design criteria are applied. These two goodness factors are described in more detail below.

(i) *Area Goodness Factor Comparison.* This factor is

$$\frac{j}{f} = \frac{Nu\,Pr^{-1/3}}{f\,Re} = \frac{1}{A_c^2}\left[\frac{Pr^{2/3}}{2g_c\rho}\frac{N_{tu}W^2}{\Delta p}\right] \tag{571}$$

From the first equality in the above equation, j/f is constant for the fully developed laminar flow of a specified fluid (Table 138). The second equality provides the significance of j/f as being inversely proportional to A_c^2 ($A_c =$ core free flow area) with the bracketed quantities being constant. The dimensionless j and f are independent of the scale of the geometry (D_h). Hence, when compared for different surfaces, the factor j/f reveals the influence of the duct cross-sectional shape regardless of the magnitude of the hydraulic diameter.

TABLE 138

Solutions for Heat Transfer and Friction for Fully Developed Flow Through Specified Ducts

Geometry $(L/D_{\rm h} > 100)$	Nu_{H1}	Nu_{H2}	Nu_T	$f\,Re$	$\dfrac{j_{H1}{}^{\ddagger}}{f}$	$\dfrac{Nu_{H1}}{Nu_T}$
$2b$ — isosceles triangle, $\dfrac{2b}{2a} = \dfrac{\sqrt{3}}{2}$, base $2a$	3.014	1.474	$2.39^{\dagger}$	12.630	0.269	1.26
$2b$ — equilateral triangle $60°$, $\dfrac{2b}{2a} = \dfrac{\sqrt{3}}{2}$, base $2a$	3.111	1.892	2.47	13.333	0.263	1.26
$2b$ — square, $\dfrac{2b}{2a} = 1$, $2a$	3.608	3.091	2.976	14.227	0.286	1.21
hexagon	4.002	3.862	$3.34^{\dagger}$	15.054	0.299	1.20
$2b$ — rectangle, $\dfrac{2b}{2a} = \dfrac{1}{2}$, $2a$	4.123	3.017	3.391	15.548	0.299	1.22
circle	4.364	4.364	3.657	16.000	0.307	1.19
$2b$ — rectangle, $\dfrac{2b}{2a} = \dfrac{1}{4}$, $2a$	5.331	2.930	4.439	18.233	0.329	1.20
$2b$ — rectangle, $\dfrac{2b}{2a} = \dfrac{1}{8}$, $2a$	6.490	2.904	5.597	20.585	0.355	1.16
parallel plates, $\dfrac{2b}{2a} = 0$	8.235	8.235	7.541	24.000	0.386	1.09
$\dfrac{b}{a} = 0$, insulated	5.385		4.861	24.000	0.253	1.11

† Interpolated values.
‡ This heading is the same as $Nu_{H1}\,Pr^{-1/3}/f\,Re$ with $Pr = 0.7$.

j_{H1}/f is presented in Fig. 124 for most of the duct geometries of Table 138. The duct geometry with a higher j/f is "good," because it will require a lower flow area and hence a lower frontal area for the exchanger. For the geometries of Table 138, this factor ranges from 0.265 (equilateral triangular duct) to 0.390 (parallel plates). Thus parallel plates relative to the triangular duct has 47% (0.390/0.265 − 1) higher j/f, and consequently, an 18% (1 − $1/\sqrt{1.47}$) smaller free flow area requirement. The porosity of the exchanger must be considered in order to translate this free flow area advantage into a frontal area improvement. Note that in the flow area goodness factor comparison, no estimate of total heat transfer area or volume can be inferred. Such estimates may be derived from the volume goodness factor described below.

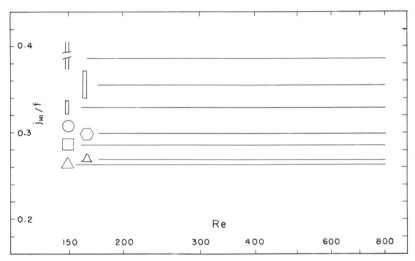

FIG. 124. Flow area goodness factor for some duct geometries of Table 138.

(ii) *Volume Goodness Factor Comparison.*   The excellence of a particular surface geometry in terms of core volume is characterized by a high position on the plot of h_{std} vs. E_{std}, where

$$h_{std} = \frac{k}{D_h} \text{Nu} = \frac{c_p\mu}{\text{Pr}^{2/3}} \frac{1}{D_h} j \text{Re} \tag{572}$$

$$E_{std} = \frac{W\Delta p}{\rho A} = \frac{\mu^3}{2g_c\rho^2} \frac{1}{D_h{}^3} f \text{Re}^3 = u_m r_h \frac{\Delta p}{L} \tag{573}$$

where the suffix std stands for the standard set of fluid property conditions (e.g., dry air at one atmospheric pressure and 500°F or 260°C temperature). From the first equality in Eq. (572), h_{std} is constant for fully developed laminar flow through a constant cross-sectional duct. The expression for E_{std} in Eq. (573) applies rigorously to constant fluid density fully developed laminar flow. For such a flow, Δp is proportional to W; hence E_{std} is proportional to W^2.

The dimensionless heat transfer in a heat exchanger is measured by the exchanger effectiveness, which in turn depends upon the number of transfer units, N_{tu}, for fixed flow rates. In a "balanced" heat exchanger, the thermal resistances of both sides of a heat exchanger are of the same order of magnitude. Hence N_{tu} is proportional to hA or $h_{std}A$ of the one side in question. Thus the higher h_{std} for a specified E_{std}, the lower the heat transfer area A will be for the specified exchanger effectiveness; and since $A = \alpha V = (4\sigma/D_h)V$, the smaller will be the heat exchanger volume for a given porosity σ and hydraulic diameter D_h.

The h_{std} vs. E_{std} plot for most of the geometries of Table 138 is presented in Fig. 125. For easy reference, Reynolds numbers of 100, 200, and 500 are marked on each curve. Unlike the nondimensional j/f factor, the dimensional h_{std} and E_{std} are strongly dependent upon the scale of the surface geometry (D_h), as evident from Eqs. (572) and (573). Consequently, a common hydraulic diameter $D_h = 0.002$ ft (0.61 mm) is used in Fig. 125 to provide a comparison of the different surface geometries having the same scale. For a screening process for surface selection, however, the designer may specify the scale for each surface considered since existing tooling and other manufacturing limitations may impose such constraints.

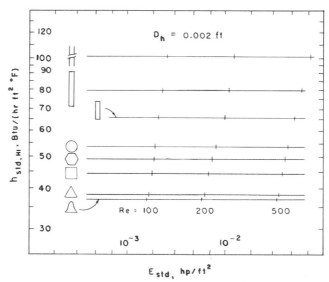

FIG. 125. Volume goodness factors for some duct geometries of Table 138 [dry air at 1 atm and 500°F properties: Pr = 0.680, μ = 0.06748 lbm/hr ft (2.789 × 10^{-5} Pa s), c_p = 0.2476 Btu/lbm °F (1.037 KJ/kg K), k = 0.02458 Btu/(hr ft °F)(0.04254 W/m K), ρ = 0.04132 lbm/ft^3 (0.6619 kg/m^3)].

From this figure, it can be seen that h_{std} varies from 37.0 to 101.2 Btu/(hr ft^2 °F) (a factor of 2.7), with $h_{std} = 38.2$ Btu/(hr ft^2 °F) for the equilateral triangular heat exchanger.[†] Thus the parallel plates heat exchanger would require 62% $(1 - 38.2/101.2)$ less heat transfer area compared to the equilateral triangular heat exchanger. Additionally, if both exchangers are designed for the same E_{std} and W, the parallel plates heat exchanger will also have 62% less pressure drop. However, it should be pointed out that these separate effects of the flow area and volume goodness factors are not cumulative. If the above advantages in A, V, and Δp are to be realized simultaneously, the required free flow area for parallel plates must be larger. This may be seen from Eq. (571):

$$A_c \propto (f/j\,\Delta p)^{1/2} \tag{574}$$

Thus the parallel plates heat exchanger requires 34% $\{[1/(1.47 \times 0.377)]^{1/2} - 1\}$ larger free flow area, and for the same porosities, 34% more frontal area than the equilateral triangular matrix. However, for a common pressure drop (not the same E_{std}), the previous gain (described under the area goodness factor) of the 18% reduction in frontal area will still apply as will the 62% reduction of volume, with the same W, N_{tu}, and σ.

The parallel plates heat exchanger may prove impractical. But it is clear from Figs. 124 and 125 that there are several other configurations that possess significant advantages over the triangular and sine duct geometries. These latter geometries are more readily fabricated than the parallel plates exchanger and have been used in early gas turbine regenerators. The rectangular duct family with $\alpha^* \leq 0.25$ appears to be a promising configuration, as seen from Figs. 124 and 125. However, care must be exercised in the fabrication of the passages, because a departure from idealized rectangular to say trapezoidal passages seriously deteriorates the performance of the heat exchanger, as discussed by Shah [116].

3. COMPARISONS OF f_{app} Re AND $Nu_{x,bc}$ FOR DEVELOPING FLOWS

a. A Comparison of f_{app} Re

Hydrodynamic entry length solutions are available for a circular tube, parallel plates, rectangular ducts, isosceles triangular ducts, and concentric and eccentric annular ducts. f_{app} Re for the parallel plates, circular tube, square duct, and equilateral triangular duct are compared in Fig. 126 for the case of a uniform velocity profile at the entrance of the duct $(x = 0)$. If the velocity profile is linear at the entrance, f_{app} Re would be lower than that for the uniform velocity profile at the entrance, as discussed on p. 164.

[†] 1 Btu/(hr ft^2 °F) = 5.678 W/(m^{2}K), 1 hp/ft^2 = 8026.6 W/m^2.

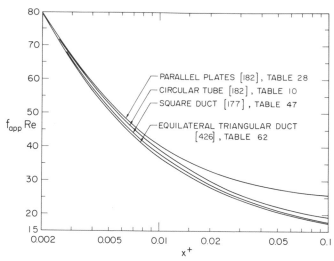

FIG. 126. A comparison of f_{app} Re for parallel plates, and circular, square, and equilateral triangular ducts for developing laminar flow.

At $x^+ = 0.1$, f_{app} Re = 25.8, 19.2, 17.8, and 17.47, respectively, for the parallel plates, circular tube, square duct, and equilateral triangular duct. In contrast, the fully developed f Re factors for these geometries are 24, 16, 14.23, and 13.33, respectively. Thus, even though the flow is fully developed over part of the flow length, as $L_{hy}^+ < 0.1$ for the foregoing geometries, the effect of the entrance region is significant, 7.5, 20, 25, and 31%, respectively, over f Re. In a typical compact heat exchanger of $L_{hy}^+ = 0.1$ ($L/D_h = 100$, Re = 1000), the theoretical f_{app} Re thus could be significantly higher than the fully developed value of f Re for the case of idealized uniform flow distribution through the exchanger.

Shah [259] proposed Eq. (576) to determine the entrance region f_{app} Re for constant cross-sectional ducts having no sharp corners. The basis of this equation is discussed below.

At the entrance of the duct, the velocity boundary layer starts developing at each wall under the imposed flow acceleration. As long as the thickness of the boundary layer is small compared to the duct dimensions, the boundary layers from different walls do not affect each other appreciably. Based on experiments with circular tubes, Shapiro et al. [14] state that the boundary layer behavior in the tube entry ($x^+ \lesssim 0.001$) is substantially identical with that on a flat plate (external flow), and the tube curvature has negligible effect. The resultant f_{app} Re in the entrance region, as presented on p. 86, is given by

$$f_{app} Re = 3.44(x^+)^{-0.5} \qquad \text{for} \quad x^+ \lesssim 0.001 \qquad (575)$$

f_{app} Re for parallel plates, based on the numerical results of Bodoia and Osterle [172], are also in excellent agreement with the above equation for $x^+ \leq 0.001$. The above equation is also recommended for rectangular and concentric annular ducts as discussed on p. 211 and 301, respectively. For a "long" duct, f_{app} Re can be accurately calculated from Eq. (92).

$$f_{app} \text{ Re} = K(\infty)/4x^+ + f \text{ Re} \tag{575a}$$

Extending Bender's idea [258], the following simplified equation may be used for singly and doubly connected ducts when the entrance velocity profile is uniform.

$$f_{app} \text{ Re} = 3.44(x^+)^{-0.5} + \frac{K(\infty)/(4x^+) + f \text{ Re} - 3.44(x^+)^{-0.5}}{1 + C(x^+)^{-2}} \tag{576}$$

Here, $K(\infty)$ and f.Re are for hydrodynamically fully developed flow; C is a constant dependent upon the duct geometry and is generally obtained by comparing the above approximate f_{app} Re with that of the analytical entry length solution. The values of $K(\infty)$, f Re, and C are summarized for quick reference in Table 139. In comparison to the more accurate solutions described in preceding chapters, the error of the f_{app} Re prediction resulting from the application of Eq. (576), is also reported in this table.

TABLE 139

Some Singly and Doubly Connected Duct
Geometries $K(\infty)$, f Re, and C to Be Used
in Eq. (576) (from Shah [259])

	K(∞)	fRe	C	% error
α*	Rectangular ducts			
1.00	1.43	14.227	0.00029	±2.3
0.50	1.28	15.548	0.00021	±1.9
0.20	0.931	19.071	0.000076	±1.7
0.00	0.674	24.000	0.000029	±2.4
2φ	Equilateral triangular duct			
60°	1.69	13.333	0.00053	±2.4
r*	Concentric annular ducts			
0	1.25	16.000	0.000212	±1.9
0.05	0.830	21.567	0.000050	±2.0
0.10	0.784	22.343	0.000043	±1.9
0.50	0.688	23.813	0.000032	±2.2
0.75	0.678	22.967	0.000030	±2.1
1.00	0.674	24.000	0.000029	±2.4

b. *Comparisons of Nusselt Numbers*

Thermal entry length solutions are reported in the text for a limited number of duct geometries for developed velocity profiles and developing velocity profiles. Comparisons of these two cases are made separately below.

(i) *Thermally Developing and Hydrodynamically Developed Flow.* A number of solutions are reported in the text for the (H1) and (T) boundary conditions for some singly and doubly connected ducts. These solutions for the circular tube, parallel plates, rectangular, and isosceles triangular ducts are compared in Figs. 127 and 128. It is evident that the order of duct geometries for decreasing values of $Nu_{x,bc}$ at a fixed x^* is the same as that for the case of fully developed solutions summarized in Table 138.

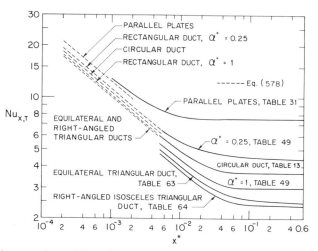

FIG. 127. A comparison of $Nu_{x,T}$ for parallel plates, and circular, rectangular, and isosceles triangular ducts for developed velocity and developing temperature profiles.

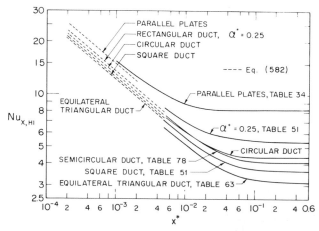

FIG. 128. A comparison of $Nu_{x,H1}$ for parallel plates, and circular, rectangular, equilateral triangular, and semicircular ducts for developed velocity and developing temperature profiles.

The thermal entrance Nusselt numbers for noncircular ducts for $x^* \leq$ 0.001 may be determined, based on the Lévêque-type approximation, by Eqs. (578) and (582), as discussed below.

The principal idealization in the Lévêque-type approximation is that the thermal boundary layer is very thin compared to the momentum boundary layer in the thermal entrance region near $x^* = 0$. Hence the axial velocity component u in the differential energy equation is replaced by a linear velocity distribution that employs the velocity gradient at the wall, as discussed on p. 105. The local Nusselt number near the $x^* = 0$ region depends upon the velocity gradient at the wall. For the circular tube and parallel plates, $Nu_{x,T}$ by Lévêque [199] is given as

$$Nu_{x,T} = 0.53837 B^{1/3}(x^*)^{-1/3} \qquad (577)$$

where $B = [\partial(u/u_m)/\partial(n/D_h)]_w$ is the velocity gradient at the wall in nondimensional form. For the circular tube, $B = 8$; and for the parallel plates, $B = 12$. Since the $f\,Re$ factor is proportional to the velocity gradient at the wall, as shown in Eq. (87), the foregoing equation could be represented as

$$Nu_{x,T} = 0.427(f\,Re)^{1/3}(x^*)^{-1/3} \qquad (578)$$

This equation is also applicable to the family of concentric annular ducts [see Eq. (459)]:

$$Nu_{x,jj}^{(3)} = 0.427(f_j\,Re)^{1/3}(x^*)^{-1/3} \qquad (579)$$

where the superscript 3 on the Nusselt number denotes the fundamental solution of the third kind (see p. 32 for definition), and the suffix j stands for the heated wall.

Equation (578) agrees with $Nu_{x,T}$ of Table 13 for the circular tube within 6% for $x^* \leq 0.02$. It agrees with $Nu_{x,T}$ of Table 31 for parallel plates within 4% for $x^* \leq 0.001$. From these limiting geometries for the concentric annular duct family, it is apparent that the range of x^* for the applicability of Eq. (578) decreases with increasing r^*, where r^* is the ratio of the inner to outer annulus radius.

When no thermal entrance solution is available for a singly connected noncircular duct, Eq. (578) may be used as an approximation for $Nu_{x,T}$ in the entrance region ($x^* \leq 0.001$) using the peripheral average fully developed value of f. A sharp cornered duct geometry will increase the degree of approximation. Equation (578) is plotted in Fig. 127 for the circular duct, parallel plates, rectangular ducts of $\alpha^* = 0.25$ and 1, and the equilateral triangular duct. It provides a good approximation for $Nu_{x,T}$ at $x^* \leq 0.005$ for the rectangular duct of $\alpha^* = 0.25$. However, $Nu_{x,T}$ of Eq. (578) for the square duct at $x^* = 0.005$ is 12% higher than the more accurate value by Chandrupatla

[407], Table 49a. It is important to emphasize that $Nu_{x,T}$ of Eq. (578) may be considerably higher than the true value for a noncircular duct having sharp corner angles $<60°$. For example, $Nu_{x,T}$ of Eq. (578) for an equilateral triangular duct is 17.7% higher than the Wibulswas results [220], Table 63. Since the f Re factors for the equilateral and right-angled isosceles triangular ducts are the same, within 1.4%, $Nu_{x,T}$ from Eq. (578) for these geometries are plotted as a single line in Fig. 127.

The mean Nusselt number from the integration of Eq. (578) is

$$Nu_{m,T} = 0.641(f\,Re)^{1/3}(x^*)^{-1/3} \tag{580}$$

A comparison with $Nu_{m,T}$ of Tables 13 and 31 shows that this equation provides $Nu_{m,T}$ for the concentric annular duct family within 1% accuracy for $x^* \leq 0.001$.

For the fundamental boundary condition of the second kind (refer to p. 32 for the definition), the Lévêque solution for the concentric annular duct family is given by [see Eq. (467)]

$$Nu_{x,jj}^{(2)} = 0.517(f_j\,Re)^{1/3}(x^*)^{-1/3} \tag{581}$$

which reduces to the following form for the circular tube and parallel plates for the (H1) boundary condition:

$$Nu_{x,H1} = 0.517(f\,Re)^{1/3}(x^*)^{-1/3} \tag{582}$$

This equation agrees with $Nu_{x,H}$ of Table 18 for the circular tube within 4% for $x^* \leq 0.01$. It agrees with $Nu_{x,H}$ of Table 34 for parallel plates within 3% for $x^* \leq 0.001$. Thus, within about 4% error, the range of x^* for the applicability of Eq. (582) decreases with increasing r^* for the concentric annular duct family.

When no thermal entrance solution is available for a singly connected noncircular duct, Eq. (582) may be used as an approximation to calculate $Nu_{x,H1}$ in the entrance region ($x^* \leq 0.001$) using the peripheral average fully developed value of f. A sharp cornered duct geometry will increase the degree of approximation. Equation (582) is plotted in Fig. 128 for the circular duct, parallel plates, rectangular ducts of $\alpha^* = 0.25$ and 1, and the equilateral triangular duct. It is found that this equation provides a good approximation for $Nu_{x,H1}$ for rectangular ducts at $x^* \lesssim 0.005$. As for $Nu_{x,T}$ of Eq. (578), the approximate $Nu_{x,H1}$ of Eq. (582) are higher than the true values for a sharp cornered geometry. $Nu_{x,H1}$ for the equilateral triangular duct is 11.2% higher at $x^* = 0.005$ than the more accurate value provided by Wilbulswas [220], Table 63.

The mean Nusselt number from the integration of Eq. (582) is

$$Nu_{m,H1} = 0.775(f\,Re)^{1/3}(x^*)^{-1/3} \tag{583}$$

A comparison with $Nu_{m,H1}$ of Tables 18 and 34 shows that this equation provides $Nu_{m,H1}$ for the concentric annular duct family within 1% accuracy for $x^* \leq 0.001$.

In summary, Eqs. (578), (580), (582), and (583) may be used to approximate $Nu_{x,T}$, $Nu_{m,T}$, $Nu_{x,H1}$, and $Nu_{m,H1}$ for $x^* \leq 0.001$ for a singly connected noncircular duct (having no sharp corners). As a test, it has been shown that for the circular tube, parallel plates, and rectangular ducts, these equations provide local and mean Nusselt numbers within 10 and 5%, respectively. It is emphasized that the velocity profile is considered as fully developed to provide f Re in the above formulations.

(ii) *Simultaneously Developing Flows.* In simultaneously developing flow, both the velocity and temperature profiles develop in the entrance region. If they are uniform at the duct inlet, the fluid velocity, velocity gradients, and temperature gradients near the wall in the entrance region will be higher compared to the case of already developed velocity profiles. The higher velocities near the wall convect more thermal energy in the flow direction, and heat transfer in the thermal entrance region is higher for the case of developing velocity profiles. As an example, Fig. 25 compares the circular duct $Nu_{x,T}$ for the developed and developing velocity profiles. The curve for $Pr = \infty$ in this figure represents the solution for any fluid ($Pr \leq \infty$) having its velocity profile fully developed before the temperature profile starts developing. It is clear from Fig. 25 that the Nusselt number is higher for the simultaneously developing flow situation ($Pr < \infty$).

Additionally, in simultaneously developing flow, the rate of development of the temperature boundary layer relative to the velocity boundary layer does depend upon the fluid Prandtl number. If the velocity and temperature profiles are uniform at the duct entrance, the lower the fluid Prandtl number, the faster the development of the temperature boundary layer will be in comparison to the velocity boundary layer in the entrance region of the duct. This would result in lower temperature gradients at the wall and in turn decrease the Nusselt number and heat transfer at a given $x^+ = x/D_h$ Re. Thus, the lower the Prandtl number, the lower the Nusselt number will be at a given x^+ for a specified duct geometry. However, if the axial coordinate is stretched or compressed by considering $x^* = x^+/Pr$, it is found that the lower the Prandtl number, the higher the Nusselt number will be at a given x^*. The effect of the fluid Prandtl number on the entrance region Nusselt numbers is shown in Figs. 25–27 for the circular tube, in Figs. 34 and 35 for parallel plates, in Fig. 43 for an $\alpha^* = 0.5$ rectangular duct, and in Figs. 56 and 57 for isosceles triangular ducts.

$Nu_{x,H1}$ for simultaneously developing flow for the circular duct, parallel plates, rectangular ducts, and equilateral triangular ducts are compared in Fig. 129 for $Pr = 0.7$. For the circular duct and parallel plates, this Nusselt

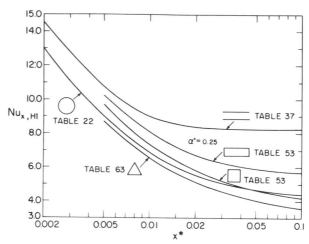

Fig. 129. A comparison of $Nu_{x,H1}$ for parallel plates, and circular, rectangular, and equilateral triangular ducts for simultaneously developing flow, $Pr = 0.7$.

number is derived by a rigorous analysis, which included the transverse velocity component v in the solution. However, $Nu_{x,H1}$ for rectangular and equilateral triangular ducts in Fig. 129 are based on the less rigorous analyses of Montgomery and Wibulswas [219] in which the transverse velocity components v and w are neglected. As discussed on pp. 139 and 146, the neglect of transverse velocity components for the circular tube results in $Nu_{x,H1}$ (for $x^* < 0.04$) that are higher than those obtained by a rigorous analysis. Similar behavior is also anticipated for noncircular ducts. Hence, $Nu_{x,H1}$ for the rectangular and equilateral triangular ducts in Fig. 129 are believed to be higher than actual for $x^* < 0.04$. This may explain why $Nu_{x,H1}$ for the square duct is higher than that for the circular duct in Fig. 129. Although not shown in this figure, the curve for the right-angled isosceles triangular duct (based on Table 64) follows closely the curve for the square duct, for $x^* < 0.05$. One would have anticipated the right-angled isosceles triangular duct curve to be below the curve for the equilateral triangular duct. Thus $Nu_{x,H1}$ for $Pr = 0.72$ in Table 64 appears to be of questionable accuracy.

As discussed on pp. 143 and 148, Churchill and Ozoe [273,306] showed that the flat plate solution for simultaneously developing flow can be applied to the circular duct entrance region where the boundary layers are thin compared to the duct radius. However, as mentioned above, the growth rate of velocity and temperature boundary layers depends upon the fluid Prandtl number, and hence either one or both boundary layers could be thin. Additionally, only a limited number of solutions for simultaneously developing flow are available, and they do not cover the low range of x^* to reveal how the solutions behave in relation to the flat plate solution. Hence, similar to Eq. (576) for f_{app} Re, a simplified equation for Nusselt numbers in the entrance region of the different duct geometries is not feasible at present. Such simplified equations may have a functional dependency the same as that for the circular duct as presented in Eqs. (254) or (255) and (257) or (258).

B. Heat Exchangers with Multi-Geometry Passages in Parallel

The theoretical results outlined in Chapters V–XVI are valid for a single duct and have been verified experimentally for a variety of noncircular ducts. A heat exchanger generally consists of many such passages in parallel. If all of these passages are uniform and identical, the performance of such an exchanger will meet the theoretical predictions; however, manufacturing and fabrication of such an exchanger results in passages that are never quite identical. An approximate method is outlined below to calculate the effective f Re and Nu for a heat exchanger consisting of n different geometry passages in parallel. Fully developed and developing flows are considered separately below.

1. FULLY DEVELOPED FLOWS

When the flow is fully developed over most of the flow length, the idealization of flow being fully developed throughout the exchanger is a good approximation. Such is the case for vehicular gas turbine regenerators that employ highly compact surfaces having continuous cylindrical passages of triangular, sine, rectangular, trapezoidal, etc., shapes generally with rounded corners. The effective f Re and Nu_{bc} designated as $(f Re)_e$ and $Nu_{bc,e}$ for the multi-geometry passages may be obtained by the following approximate procedure of Shah and London [522], which is based on the method of London [523].

Consider fully developed laminar flow through an m-passage heat exchanger core having n different shaped passages in parallel. The total number of the ith shaped passages is N_i, so that

$$N_1 + N_2 + \cdots + N_i + \cdots + N_n = m \qquad (584)$$

If $\chi_i = N_i/m$, it denotes the fractional distribution, so that

$$\sum_{i=1}^{n} \chi_i = 1 \qquad (585)$$

Idealize the fluid properties ρ, μ, k, and c_p as constant and the same for all passages; all passages have the same flow length L; the entrance and exit pressure losses in the heat exchanger are so small that the flow distribution within the heat exchanger core is solely dependent upon the nonuniform passage distribution. The procedures to determine $(f\,\text{Re})_e$ and $\text{Nu}_{bc,e}$ are described below.

a. *Flow Distribution, Pressure Drop, and* $(f\,\text{Re})_e$

The pressure drop for fully developed laminar flow through any one of the ith shaped passages is given by

$$\Delta p = f_i \frac{4L}{D_{h,i}} \frac{\rho u_{m,i}^2}{2g_c} = \left(\frac{2\mu L}{g_c \rho}\right)\left(\frac{(f\,\text{Re})_i}{A_{c,i} D_{h,i}^2}\right) W_i \tag{586}$$

If $(f\,\text{Re})_e$, P, A_c, and W represent the effective f Re factor, total wetted perimeter, total flow cross section area, and total fluid flow rate, respectively, for the m-passage heat exchanger, then

$$\Delta p = \left(\frac{2\mu L}{g_c \rho}\right)\left(\frac{(f\,\text{Re})_e}{A_c D_h^2}\right) W \tag{587}$$

where

$$A_c = \sum_{i=1}^{n} N_i A_{c,i} = \sum_{i=1}^{n} m\chi_i A_{c,i} \tag{588}$$

$$P = \sum_{i=1}^{n} N_i P_i = \sum_{i=1}^{n} m\chi_i P_i \tag{589}$$

$$D_h = \frac{4A_c}{P} \tag{590}$$

From the continuity equation,

$$W = \sum_{i=1}^{n} N_i W_i \tag{591}$$

Because the pressure drop is the same for each individual passage, substituting Eqs. (586) and (587) into Eq. (591) and rearranging it yields

$$\frac{1}{(f\,\text{Re})_e} = \sum_{i=1}^{n} \frac{1}{(f\,\text{Re})_i}\left(\frac{D_{h,i}}{D_h}\right)^2\left(\frac{N_i A_{c,i}}{A_c}\right) \tag{592}$$

Once $(f\,\text{Re})_e$ is determined from the above equation, the flow distribution can be obtained by equating Eqs. (586) and (587):

$$\frac{W_i}{W} = \left[\frac{(f\,\text{Re})_e}{(f\,\text{Re})_i}\right]\left[\frac{D_{h,i}}{D_h}\right]^2\left[\frac{A_{c,i}}{A_c}\right] \tag{593}$$

This equation provides the ratio of the flow rate through one passage to the total flow rate through the heat exchanger. The flow rate through N_i passages, all of the ith shape, is then given by

$$\text{fractional flow rate through the } i\text{th shaped passages} = N_i W_i / W \quad (594)$$

Thus, if the distribution $m\chi_i$ or N_i of the flow passages is given, and if $(f\,\text{Re})_i$ of these passages are known, one can determine the flow distribution from Eq. (594), the effective $f\,\text{Re}$ factor from Eq. (592), and the pressure drop through the heat exchanger from Eq. (586) or (587).

b. Heat Transfer and $\text{Nu}_{bc,e}$

Heat transfer through differently shaped passages would be different, which in turn would produce different temperature differences from the hot to the cold fluid. Subsequently, one cannot consider different passages in parallel to arrive at an effective h as a weighting of conductances h_1, h_2, etc. To properly arrive at the effective h (and hence $\text{Nu}_{bc,e}$) for the multipassage geometry heat exchanger, the passage geometrical properties, fluid physical properties, exchanger flow arrangement, and ε-N_{tu} relationship must be considered. The procedure is outlined below for two specific heat exchanger systems: (i) a counterflow heat exchanger with $C_{min}/C_{max} = 1$, as is closely realized in a vehicular gas turbine regenerator, and (ii) a heat exchanger of any flow arrangement with $C_{min}/C_{max} = 0$. This case is closely realized in condensers, evaporators, and in some liquid-to-gas heat exchangers having a high capacity rate and high heat transfer coefficient on the phase change or liquid side. These two extreme cases should provide the range of $\text{Nu}_{bc,e}$ for the multipassage geometry heat exchanger. It will be explained later why the following procedure cannot be generalized to the other flow arrangements. N_{tu} is related to Nu_{bc}, D_h, P, and W of the ith passage geometry for any heat exchanger, as follows:

$$N_{tu,l} = \left(\frac{hA}{Wc_p}\right)_i = \left(\frac{kL}{c_p}\right)\left(\frac{\text{Nu}_{bc,i}}{D_{h,i}}\frac{P_i}{W_i}\right) \quad (595)^\dagger$$

(i) *A Counterflow Heat Exchanger with* $C_{min}/C_{max} = 1$. In both direct transfer and periodic flow (rotary) gas turbine regenerators of the counterflow type with $C_{min}/C_{max} \simeq 1$, the wall temperature variation parallels the fluid temperature as shown in Fig. 130a. In this case, the exchanger effectiveness ε is related to N_{tu} as [6]

$$\varepsilon = \frac{N_{tu}}{1 + N_{tu}} \quad (596)$$

† To evaluate $N_{tu,i}$, the ith passage is compared with the nominal reference passage of known Nu_{bc}, P, D_h, W, and $N_{tu,ref}$.

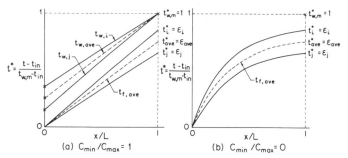

FIG. 130. Idealized wall and fluid temperature variations for the $N_{tu,e}$ evaluation; t_{in} is the inlet fluid temperature at $x = 0$; $t_{w,m}$ is the wall temperature at $x = L$; subscripts i and j stand for the ith and jth shaped passages.

In a two-fluid heat exchanger, the wall temperature at any given point depends upon the hot and cold fluid temperatures, and the thermal resistances of the hot and cold fluid sides. In the case of a multipassage geometry core, if the thermal resistance on the other side is either very low or of the same order of magnitude as that on the first side, the wall temperature of all differently shaped passages would be approximately the same at $x = L$, as shown in Fig. 130a. Hence, the effectiveness of the ith passage would be

$$\varepsilon_i = \frac{t_{out,i} - t_{in}}{t_{w,m} - t_{in}} = \frac{N_{tu,i}}{1 + N_{tu,i}} \tag{597}$$

The effective outlet temperature of the fluid (in Fig. 130a) then depends upon the flow rates through each passage and is given by the energy balance as

$$Wc_p t_{out,ave} = c_p \sum_{i=1}^{n} N_i W_i t_{out,i}$$

$$= c_p \sum_{i=1}^{n} N_i W_i [\varepsilon_i (t_{w,m} - t_{in}) + t_{in}] \tag{598}$$

or, alternatively,

$$\varepsilon_{ave} = \sum_{i=1}^{n} \frac{N_i W_i}{W} \varepsilon_i \tag{599}$$

Subsequently, the effective N_{tu} is determined, using Eq. (596), as

$$N_{tu,e} = \frac{\varepsilon_{ave}}{1 - \varepsilon_{ave}} \tag{600}$$

and the effective Nusselt number is then obtained from

$$\mathrm{Nu}_{bc,e} = \left(\frac{c_p}{kL}\right)\left(\frac{WD_h N_{tu,e}}{P}\right) \tag{601}$$

where P and D_h for the m-passage exchanger are expressed by Eqs. (589) and (590).

(ii) *A Heat Exchanger with* $C_{min}/C_{max} = 0$. In this case, the relationship of effectiveness and N_{tu} for a heat exchanger of any flow arrangement reduces to the following expression [6]:

$$\varepsilon = 1 - \exp(-N_{tu}) \tag{602}$$

Hence, from Fig. 130b, the effectiveness of the ith passage will be

$$\varepsilon_i = 1 - \exp(-N_{tu,i}) \tag{603}$$

Subsequently, using this ε_i, the average effectiveness and the effective N_{tu} are given by

$$\varepsilon_{ave} = \sum_{i=1}^{n} \frac{N_i W_i}{W} \varepsilon_i \tag{604}$$

$$N_{tu,e} = \ln\left(\frac{1}{1 - \varepsilon_{ave}}\right) \tag{605}$$

The effective Nusselt number is then determined from Eq. (601).

In the foregoing two special cases, it was idealized that $t_{w,m}$ for all differently shaped passages at $x = L$ in Fig. 130 was the same. As a result, all the ε_i magnitudes were based on the same initial temperature difference $(t_{w,m} - t_{in})$. This in turn provided a way to calculate ε_{ave} from Eq. (599) or (604), and subsequently determined $N_{tu,e}$ and $\mathrm{Nu}_{bc,e}$. However, for any other flow arrangements, the $(t_{w,m} - t_{in})$ difference for each of the multi-geometry passages will not be the same. As a consequence, one cannot arrive at ε_{ave} and hence $N_{tu,e}$ and $\mathrm{Nu}_{bc,e}$.

The thermal boundary conditions for the foregoing two special cases are the axially constant heat flux (H1) and constant wall temperature (T), respectively. Thus, one can establish the range of $\mathrm{Nu}_{bc,e}$ and make a judicious choice of a value for the application at hand, which may have a different flow arrangement or a different value of C_{min}/C_{max}.

2. Developing Flows†

Equations (592) and (601) were derived for fully developed laminar flow $(f\,\mathrm{Re})_e$ and $\mathrm{Nu}_{bc,e}$ in an exchanger having n different-shaped passages. In that case, both $(f\,\mathrm{Re})_i$ and $\mathrm{Nu}_{bc,i}$ were constant, independent of the flow

† Flow uniformity tends to be assured for highly interrupted heat exchanger surfaces so that the passage-to-passage nonuniformity is not of great importance.

rate or the exchanger length. If the velocity and/or temperature profiles are developing, as is the case with interrupted heat exchanger surfaces, the aforementioned theory is valid if $(f \, \text{Re})_i$ and $\text{Nu}_{\text{bc},i}$ are replaced by $(f_{\text{app}} \, \text{Re})_i$ and $(\text{Nu}_{\text{m,bc}})_i$. However, these quantities are not constant; they are dependent upon $L_i^{+}(=L/D_h \, \text{Re})$ and $L_i^*(=L/D_h \, \text{Pe})$, which are dependent upon the ith passage L/D_h and Re for a given Pr. Once L_i^{+} and L_i^* are known, $(f_{\text{app}} \, \text{Re})_i$ and $(\text{Nu}_{\text{m,bc}})_i$ are determined from the respective entry length solutions. However, the flow distribution must be known to determine L_i^{+} and L_i^*. Hence the calculation procedure becomes an iterative one as follows.

Start with a uniform flow distribution. Determine Re_i, L_i^{+}, and $(f_{\text{app}} \, \text{Re})_i$ for the ith shaped passages. Subsequently calculate $(f_{\text{app}} \, \text{Re})_e$ using Eq. (592) and the flow distribution from Eq. (593). With this flow distribution, calculate new values of Re_i, L_i^{+}, and $(f_{\text{app}} \, \text{Re})_i$. Repeat the above procedure until the solution converges to the desired degree. Employing the converged flow distribution solution and the known thermal entry length solution will provide $(\text{Nu}_{\text{m,bc}})_e$, in a straightforward manner (no iterations), for the two special cases of heat exchangers with $C_{\text{min}}/C_{\text{max}} = 0$ and 1.

C. Influence of Superimposed Free Convection

All of the results presented and the literature reviewed in the preceding chapters are derived for pure forced-convective flow. As will be demonstrated later, free-convection effects are generally quite negligible for compact heat exchanger applications. However, in those applications for which the flow velocity (Reynolds number) is sufficiently low, or if high temperature differences are employed, or if the passage geometry has a large hydraulic diameter, the free-convection effects may be important. The effect of the superimposed free convection could be important primarily in the laminar flow region of a duct.

The effect of free convection is correlated by combinations of the following dimensionless numbers: Grashof number Gr, Rayleigh number Ra, Prandtl number Pr and L/D along with the Reynolds number Re. The Grashof and Rayleigh numbers are defined as

$$\text{Gr} = \frac{g\rho^2 D_h^{3}\beta(t_w - t_m)}{\mu^2} \tag{606}$$

$$\text{Ra} = \text{Gr} \, \text{Pr} = \frac{g\rho^2 D_h^{3}\beta c_p(t_w - t_m)}{\mu k} \tag{607}$$

where β is the coefficient of thermal expansion and g is the local gravitational acceleration. The free-convection effects are generally more for increasing values of Gr or Ra. Theoretically, Gr or Ra of zero represents the pure forced-convection flow.

The subject of combined free and forced convection is complex, and a comprehensive review is outside the scope of this monograph. A summary up to 1968 on the subject is provided by Kays and Perkins [12]. Only a limited consideration of free convection superimposed on forced convection is presented here, as a guide to the designer, for horizontal and vertical circular tubes.

1. Effects of Superimposed Free Convection in Horizontal Circular Tubes

For horizontal tubes, free convection sets up secondary flows at a cross section that aids the convection process. Hence the heat transfer coefficient and Nusselt number for the combined convection are higher than those for the pure forced convection. The maximum heat transfer occurs at the bottom of the tube. When the free-convection effects are significant in laminar flow, large temperature gradients exist near the wall, and the temperature variation in the vertical direction is significant. The measured velocity distribution in horizontal and vertical direction is also markedly different from the velocity distribution for Poiseuille flow. Metais and Eckert [524] have classified free-, mixed-, and forced-convection regimes, as shown in Fig. 131, for horizontal pipes with the axially constant wall temperature boundary condition. The limits of the forced- and mixed-convection regimes are defined in such a manner that free-convection effects

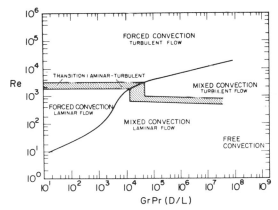

Fig. 131. Horizontal circular tube: Free-, forced-, and mixed-convection flow regimes for $10^2 < Pr(D/L) < 1$.

contribute only about 10% to the heat flux. Figure 131 may therefore be used as a guide to determine whether or not free convection is important.

Brown and Thomas [525] experimentally determined the laminar combined free and forced convection to water flowing through a horizontal tube with constant wall temperature. They could not reconcile the correlations of McAdams [56], Eubank and Proctor [526], and Oliver [527] with their results. Based on the dimensional analysis and their experimental results, they derived a Nusselt number relationship as

$$\text{Nu}_\text{T} = 1.75[\text{Gz} + 0.012(\text{Gz}\,\text{Gr}^{1/3})^{4/3}]^{1/3}(\mu_\text{w}/\mu_\text{m})^{-0.14} \qquad (608)$$

where $\text{Gz} = \pi/(4x^*)$. All of the fluid properties in Nu_T, Gz, Gr, and μ_m are evaluated at the fluid bulk mean temperature. This equation correlates the experimental data of Brown and Thomas for water within $\pm 8\%$. It agrees within $\pm 50\%$ with the majority of the published experimental data for water, ethyl alcohol, and highly viscous oils.

The wall-to-fluid bulk mean temperature difference is the largest at the onset of heating or cooling for the axially constant wall temperature boundary condition. However, it takes some length before the secondary flows set up at a cross section. The secondary flow enhances the convection process so that the temperature of the fluid more rapidly approaches the tube wall temperature. Hence, after the entrance length for the combined convection, the effect of superimposed free convection decreases with increasing tube length for the constant wall temperature case. In contrast, the influence of superimposed free convection remains constant for the axially constant heat flux boundary condition, since $(t_\text{w,m} - t_\text{m})$ remains constant axially for fully developed flow.

As noted on p. 25, (H1)–(H4) are different boundary conditions for the case of axially constant wall heat transfer rate. These four boundary conditions become identical for the circular tube for the pure forced-convection case, because the temperature profile across a cross section is symmetrical, independent of the peripheral direction. However, when free convection is superimposed, secondary flows set up that are symmetrical to the vertical diameter. As the local fluid temperature is different near the tube wall (which sets up the secondary flows), it causes the temperature gradients in the tube wall in the peripheral direction. Hence, for the circular tube, the (H1)–(H4) boundary conditions are now different for the combined convection problem. The most realistic boundary condition is (H4).

Morcos and Bergles [528] presented the following correlation for combined convection in a horizontal circular tube for the (H4) boundary condition:

$$\text{Nu}_\text{H4} = \{(4.36)^2 + [0.145(\text{Gr}^*\,\text{Pr}^{1.35}K_\text{p}^{0.25})^{0.265}]^2\}^{1/2} \qquad (609)$$

where

$$Gr^* = Gr\,Nu = g\beta D_h^4 q''/v^2 k \qquad (610)$$

and $K_p = k_w a'/kD_h$ is the peripheral heat conduction parameter. This correlation was derived for various fluids and tube wall materials, and was based on their own experimental data as well as others in the literature. All of the fluid properties used in Nu_{H4}, Gr^*, Pr, and K_p are evaluated at the *film* temperature.

The thermal entrance length is significantly reduced in the presence of the free-convection effects. Thus, for most cases when free convection is superimposed, the combined convection flow is fully developed. Hence Eq. (609) is sufficient to determine the Nusselt number for combined convection. The transition criterion for the onset of significant free-convection effects for a horizontal circular pipe with the (H4) boundary condition, as suggested by Bergles and Simonds [529], is

$$Ra_{cr} = 1.8 \times 10^4 + 55(x^*)^{-1.7} \qquad \text{for} \quad x^* > 10^{-4} \qquad (611)$$

for water. However, for air, the onset of significant free convection effects occur at [530]

$$Ra\,Re = 1000 \qquad (612)$$

The friction factors are also higher for the combined convection case. Based on the ethylene glycol test data, Morcos and Bergles [528] presented the following empirical correlation:

$$f/f_{iso} = [1 + (0.195\,Ra^{0.15})^{15}]^{1/15} \qquad (613)$$

where $f_{iso} = 16/Re$ is the isothermal friction factor. Both f and Ra are based on the fluid properties evaluated at the *film* temperature.

2. EFFECTS OF SUPERIMPOSED FREE CONVECTION IN VERTICAL CIRCULAR TUBES

Unlike horizontal tubes, the effect of superimposed free convection for vertical tubes is dependent upon the flow direction and whether or not the fluid is heated or cooled. For fluid heating with upward flow or fluid cooling with downward flow, free convection aids forced convection, and the resultant heat transfer coefficient is higher than the pure forced convection coefficient. However, for fluid cooling with upward flow or fluid heating with downward flow, free convection counters forced convection and a lower heat transfer coefficient results. The flow regime chart of Metais and Eckert [524] for vertical tubes, as shown in Fig. 132, provides guidelines to determine the significance of the superimposed free convection. The results of

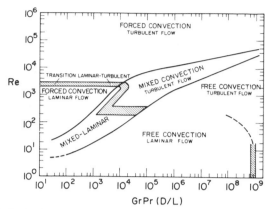

FIG. 132. Vertical circular tube: Free-, forced-, and mixed-convection flow regimes for $10^2 < \mathrm{Pr}(D/L) < 1$.

Fig. 132 are applicable for both constant heat flux and constant wall temperature boundary conditions.

There are numerous correlations and analytical results for vertical tubes, and since authors are unable to recommend any particular set, the reader is referred to the pertinent literature. For example, Kays and Perkins [12], Martinelli and Boelter [531], and Zeldin and Schmidt [327] discuss combined convection in vertical tubes for the constant wall temperature boundary condition. Kays and Perkins [12], Hallman [532], Kemeny and Somers [533], Scheele and Hanratty [534], Worsøe-Schmidt and Leppert [535], and Lawrence and Chato [536] have given results of combined convection in vertical tubes with the axially constant wall heat flux boundary condition.

3. APPLICATION TO COMPACT HEAT EXCHANGERS

A typical value for the Grashof number in a compact heat exchanger, such as the air flow side of an automobile radiator, is 60–100. With $\mathrm{Pr}/(L/D)$ of the order of 10^{-2}, the abscissa magnitude in Fig. 131 is of the order of one, and is off the scale; Fig. 131 is based on the available experimental data. However, since Re is much greater than 10^2, it is clear that free-convection effects will be negligible. However, for gas flows note that the $\rho^2 \beta$ term in Gr varies as P^2/T^3. Thus, because of the high pressures and low absolute temperatures (T) encountered in cryogenic heat exchangers, there well may be a significant free-convection effect.

D. Influence of Temperature-Dependent Fluid Properties

One of the basic idealizations made in all the analytical results presented in the preceding chapters is that the fluid properties are constant. Since the transport properties of most fluids vary with temperature, the constant property solution provides the best approximation for heat transfer problems involving small temperature differences. For those problems involving large temperature differences, constant property analytical solutions deviate substantially from experimental results.

The fluid properties vary differently with temperature for different fluids. Additionally, the fluid properties distort the velocity profile, which in turn affects the temperature profile. Hence, the temperature-dependent property problem is a complex one. A limited number of solutions for such a problem have appeared in the literature. As a thorough review of this problem is outside the scope of this monograph, only a brief summary is provided, based on the review of Kays and Perkins [12], Kays [7], and Kays and London [6].

For engineering applications, it is convenient to employ constant property analytical solutions, or experimental data obtained with small temperature differences, and then to apply some kind of correction to account for property variations. Two schemes for such correction are commonly used: the property ratio method and the reference temperature method. In the property ratio method, all properties are evaluated at the bulk mean temperature, and then all of the variable properties effects are lumped into a function of a ratio of some pertinent property evaluated at the surface temperature to that property evaluated at the bulk temperature. In the reference temperature method, a temperature is specified at which the properties appearing in the dimensionless groups may be evaluated such that the constant property results may be used to evaluate variable property behavior. Typically, this may be the film average temperature or the surface temperature.

As discussed by Kays [7] and Kays and London [6], for design purposes, the property ratio method leads to less ambiguity and awkwardness for the internal duct flow problem. Based on this method, temperature-dependent results are summarized now separately for gases and liquids because they involve different property ratio corrections.

For gases, the viscosity, thermal conductivity, and density are functions of the absolute temperature. This absolute temperature dependence is similar for different gases except for the temperature extremes. The temperature-dependent property effects for gases are adequately correlated by the following equations:

$$\frac{\text{Nu}}{\text{Nu}_{\text{cp}}} = \frac{\text{St}}{\text{St}_{\text{cp}}} = \left(\frac{T_{\text{w}}}{T_{\text{m}}}\right)^n \tag{614}$$

$$\frac{f}{f_{\text{cp}}} = \left(\frac{T_{\text{w}}}{T_{\text{m}}}\right)^m \tag{615}$$

where the suffix cp refers to the constant property variable. All of the properties in dimensionless groups of Eqs. (614) and (615) are evaluated at the *bulk mean* temperature. The values of the exponents n and m for laminar flow through a circular tube are recommended as [12]

$$n = 0.0, \quad m = 1.00 \qquad \text{for} \qquad 1 < T_w/T_m < 3 \quad \text{(heating)} \qquad (616)$$

$$n = 0.0, \quad m = 0.81 \qquad \text{for} \qquad 0.5 < T_w/T_m < 1 \quad \text{(cooling)} \qquad (617)$$

These exponents are valid for developed and developing flows for the Ⓣ and Ⓗ boundary conditions.

For liquids, viscosity is the only property of importance that varies greatly with temperature. Thus it is found that temperature-dependent effects for liquids are adequately correlated by

$$\frac{\text{Nu}}{\text{Nu}_{cp}} = \frac{\text{St}}{\text{St}_{cp}} = \left(\frac{\mu_w}{\mu_m}\right)^n \qquad (618)$$

$$\frac{f}{f_{cp}} = \left(\frac{\mu_w}{\mu_m}\right)^m \qquad (619)$$

The exponents n and m for laminar flow through a circular tube are recommended as [12]

$$n = -0.14, \quad m = +0.58 \qquad \text{for} \quad \mu_w/\mu_m < 1 \quad \text{(heating)} \qquad (620)$$

$$n = -0.14, \quad m = +0.50 \qquad \text{for} \quad \mu_w/\mu_m > 1 \quad \text{(cooling)} \qquad (621)$$

These exponents are valid for developed and developing flows for the Ⓣ and Ⓗ boundary conditions.

Since the information on flow through noncircular ducts with variable properties is meager, it is recommended that the above information may also be used for noncircular ducts.

E. Comments on the Format of Published Papers

Usually, applied mathematicians *obtain* solutions to the internal flow forced convection problem, but engineers *use* them. Some recommendations follow on the presentation of solutions so that they may be more useful. These relate to the problem statement, solution techniques, and presentation of theoretical solutions [17].

1. PROBLEM STATEMENT

The authors had an opportunity to review from worldwide sources over 450 papers on the laminar duct flow forced convection heat transfer problem. A number of these papers are unclear and confusing regarding the nature of the problem and what is achieved by the analysis presented.

Most papers presented a solution to one of the many variations of the conventional convection problem. However, the thermal boundary conditions used in the solution are given in equation form only and are not described physically. Sometimes it is also unclear which method of analysis is used, what its significance, usefulness, and advantages are, and what is the contribution of a particular paper to the literature. For the readers' benefit, all of the aforementioned points should be highlighted. Presumably this will provide the engineer with some degree of confidence in using the results.

2. Solution Techniques

A variety of methods have been devised, as found in the literature and briefly summarized in Chapter IV, for the solution of the conventional convection problem. Some of these methods are very powerful, are computationally very fast, and provide highly accurate results. Usually, however, their application is shown only for those ducts and thermal boundary conditions for which solutions already exist, such as the circular tube and parallel plates for the $\textcircled{H}$ and $\textcircled{T}$ thermal boundary conditions. It is strongly recommended that after qualifying a new method with known results, new solutions should be obtained for other duct geometries in order to demonstrate the applicability and usefulness of the method and to extend the body of knowledge usable by the engineer. It is suggested that the highlights of the new method and its comparison with the existing solution techniques be presented in the text of the paper, with all mathematical details summarized in the appendix.

3. Presentation of Theoretical Solutions

To an engineer, the value of the solutions lies in their applicability in the design of heat transfer equipment. The engineer may not have enough mathematical background or he may not be interested in the details of the theoretical analysis. What is of major interest is the physical interpretation of the idealizations of the actual problem implicitly built into the mathematical model used in the analysis. This understanding is a requirement for proper application. The final results should be presented in a conventional or standardized manner if at all feasible. Three points are discussed in more detail, to emphasize the importance of the format of the solutions: (a) incomplete solutions, (b) results in terms of confusing dimensionless groups, and (c) qualitative instead of quantitative results.

a. *Incomplete Solutions*

Many papers published in the literature either do not go so far as to present the final results in terms of local and mean Nusselt numbers and/or the wall heat flux, or such final results are presented only in terms of complex equations that do not readily provide the general trends of behavior.

One of the powerful techniques used to solve the thermal entry length problem is the separation of variables method. A number of papers in the literature simply presented eigenvalues, eigenfunctions, and constants of an infinite series solution. However, a considerable amount of time, effort, and skill are needed to convert this information into the local and mean Nusselt numbers and/or the wall heat flux. Since the engineer may not have the time or facility to convert such information, the valuable results presented sometimes receive only limited use.

b. *Results in Terms of Confusing Dimensionless Groups*

A natural tendency exists to present the results in terms of a particular set of dimensionless groups based on the analyst's particular normalization of the differential equations and boundary conditions. Many different sets of dimensionless groups can be formulated. And such results may not be understood by an engineer because he cannot readily interpret them in terms of a set that is more conventional and hence familiar to him. For example, the Nusselt number, defined in the literature as follows, has many different interpretations:

$$Nu = \frac{hD_h}{k_f} = \frac{q''D_h}{k_f(\Delta t)} \tag{622}$$

It may be for developed or developing flows, and may represent a peripheral local or average value. Additionally, (Δt) has been defined variously as follows: $(t_w - t_m)$, $(t_{w,m} - t_m)$, $(t_w - t_e)$, $(t_{wo} - t_e)$, $(t_w - t_{a,m})$, $(\Delta t)_{lm}$, etc. Consequently, the reader has to study the paper exhaustively before the specific definition of Nu is properly clarified.

An effort to standardize the definitions of dimensionless groups for laminar flow through a duct has been made by Shah and London [54]. These groups are summarized in Chapter III. It is recommended that these or a similar set arrived at by general agreement be used in future work.

c. *Qualitative Instead of Quantitative Results*

Many of the papers published in the literature present analytical results in terms of small-scale graphs. Such presentations are useful for showing qualitative behavior. In many cases, however, the solution is obtained by employing sophisticated mathematical and computational methodologies, and it may be a refinement of existing solutions for which qualitative behavior is already known. If such a solution is presented only in graphical form, its usefulness is diminished. It is strongly recommended that the theoretical results be presented in tabular form, or at least that the tabular results be forwarded to an agency[†] from which such information is readily obtainable.

F. The Complete Solution

Generally, an analyst determines the velocity distribution first and then the temperature distribution for the laminar flow forced convection heat transfer problem. However, the complete solution to this problem, in terms of engineering usage, also requires the determination of one or more of the following: (1) local and mean wall heat fluxes, (2) local and mean Nusselt numbers, (3) the thermal entrance length, (4) incremental heat transfer rate in the entrance region over and above the fully developed values, and (5) the peripheral maximum and minimum wall temperatures for hot- and cold-spot information as well as for a thermal stress analysis.

In addition, from the solution to the velocity problem, the determination of one or more of the following items would be useful for design purpose: (1) the apparent friction factors, the pressure drop, or the incremental pressure drop number $K(x)$, (2) the momentum flux correction factor K_d, (3) the kinetic energy correction factor K_e, and (4) the ratio of maximum to mean velocity. The foregoing information is useful for both the developed and developing velocity profiles. Additionally, the hydrodynamic entry length, as defined on p. 41, should be determined.

G. Areas of Future Research

Table 136 provides a summary of laminar flow solutions presented in Chapters V–XVI. It also suggests where further solutions are needed. Some of these areas of future research are indicated in the following discussion.

1. Conjugated Problem

While all practical forced convection heat transfer problems are in reality conjugated problems, they have been idealized to conventional convection problems in order to arrive at mathematical solutions. While several solutions of very simplified conjugated problems have been obtained, as outlined in the circular tube and parallel plates sections, more are needed for a variety of conjugated problems. Hopefully, these will serve to bring out the

[†] One such agency in the U.S. is the National Auxiliary Publications Service, c/o Microfiche Publications, 440 Park Avenue South, New York 10016.

significance of some of the neglected effects in the conventional treatments as this is one of the more important fields of future research.

2. Thermal Entry Length Solutions

Even for constant-property flows, only a limited number of thermal entrance solutions exists, as seen from the summary of Table 136. A larger number of such solutions would be worthwhile for developed and developing velocity profiles in different duct geometries with different thermal boundary conditions. To the authors' knowledge, for example, no solution is available for the thermal boundary condition of Eq. (68).

3. Thermal Entrance Lengths

Fully developed flows prevail in a number of heat exchangers over most of the flow length. In such cases, only a knowledge of the thermal entry length magnitude is required. Sandall and Hanna [58] were the first investigators to devise an approximate method to determine the (H1) thermal entrance length without solving the complete entry length problem. Improvements in this method for circular and noncircular ducts, as well as new methods for other thermal boundary conditions are needed.

4. Incremental Heat Transfer Rate

The effect of the thermal entrance region is to increase the heat transfer rate and/or Nusselt number compared to that for thermally developed flow. This increment in the heat transfer rate over and above the fully developed value is designated as the incremental heat transfer number $N_{bc}(x)$, defined by Eq. (130). No method exists in the literature to determine $N_{bc}(x)$ without solving the complete thermal entrance problem. Approximate or exact methods are needed for different duct geometries with different thermal boundary conditions. Results in this form could be readily utilized by the designer to correct for the thermal entrance length effect.

5. Solutions with Temperature-Dependent Properties

Only a small number of solutions for the laminar forced convection problem are available in the literature with some variations in the associated thermo-physical properties. The current practice is to adjust the constant property solutions by correction factors with only limited empirical and theoretical justification. Hence, to arrive at proper correction factors, analytical solutions are needed for different duct geometries with different thermal boundary conditions and variations in one or more of the involved physical properties of the fluids.

Nomenclature

The dimensions for each symbol are presented in both U.S. engineering and SI units, where applicable. Note that both the hour and second are commonly used as units for time in the U.S. engineering system; hence a conversion factor of 3600 should be employed at appropriate places in dimensionless groups.

A	heat transfer or flow friction area, ft^2, m^2
A_c	flow cross-sectional area, ft^2, m^2
A_w	wall cross-sectional area for longitudinal heat conduction, ft^2, m^2
a	radius of a circular duct, half-width of a rectangular duct, half-base width of a triangular or sine duct, semimajor axis of an elliptical duct, half-short side of a trapezoidal duct, $a \geq b$ for rectangular and elliptical ducts with symmetrical heating, ft, m
a	radius of a rod in the rod bundle geometry, ft, m
a_i, a_o	semimajor axes of confocal ellipses (see Fig. 97), ft, m
a'	duct wall thickness, ft, m
Bi	Biot number, ha'/k_w, dimensionless
Bn	Brun number [defined by Eq. (147)], dimensionless
Br	Brinkman number for the constant axial wall temperature boundary conditions, $\mu u_m^2/[g_c Jk(t_{w,e} - t_m)]$ [defined by Eq. (139)], dimensionless
Br'	Brinkman number for the constant axial wall heat flux boundary condition $\mu u_m^2/(g_c Jq''D_h)$ [defined by Eq. (140)], dimensionless
B_1, B_2	constants [Eqs. (159) and (160)], dimensionless
b	half-spacing of parallel plates, half-height of rectangular, triangular, trapezoidal, and sine ducts, semiminor axis of an elliptical duct, $b \leq a$ for rectangular and elliptical ducts with symmetrical heating, ft, m
b	radius of the concentric ring (see Fig. 107), ft, m
b	amplitude of cosine heat flux variation around the periphery of a circular duct (see Fig. 11), dimensionless
b_i, b_o	semiminor axes of confocal ellipses (see Fig. 97), ft, m
C	a constant used in Eq. (576), dimensionless
C	flow stream capacity rate, Wc_p, Btu/hr °F, W/°C
c	half-base width of a trapezoidal duct (see Fig. 64), half-length of a side of an arbitrary triangular duct (see Fig. 53), ft, m
c	parameter for the eccentric annular duct [defined by Eq. (507)], ft, m
c_1	a pressure gradient parameter, $(dp/dx)/(\mu/g_c)$, 1/ft sec, 1/m s

c_2 a temperature gradient parameter, $(\partial t/\partial x)/\alpha$, °F sec/ft^3, °C s/m^3
c_3 thermal energy source parameter, S/k, °F/ft^2, °C/m^2
c_4 a parameter, $c_1 c_2$, °F/ft^4, °C/m^4
c_p specific heat of the fluid at constant pressure, Btu/lbm °F, J/kg °C
D_h hydraulic diameter of the actual duct or of heat exchanger flow passages, $4r_h$, ft, m
d inside diameter of a circular tube, or diameter of a rod in a rod bundle, $d = 2a$, ft, m
Ec, Ec′   Eckert number [defined by Eqs. (143) and (144)], dimensionless
E_{std} fluid pumping power per unit of surface area, $W\Delta p/\rho A$ [defined by Eq. (573)], hp/ft^2, W/m^2
e^* eccentricity $\epsilon/(r_o - r_i)$ of the eccentric annular duct or amplitude ϵ/a of the corrugated duct, dimensionless
f perimeter average "Fanning" or "small" friction factor, for fully developed flow if no subscript used, $\tau/(\rho u_m^2/2g_c)$, dimensionless
f_x, f_m local and average Fanning friction factors in hydrodynamic entry length [defined by Eqs. (82) and (83)], dimensionless
f_{app} apparent Fanning friction factor, $\Delta p^*/(x/r_h)$ [defined by Eq. (85)], dimensionless
f_D "Darcy" or "large" friction factor, $4f$, dimensionless
f_d Fanning friction factor for fully developed flow for a finless duct [defined by Eq. (542)], dimensionless
G fluid mass velocity, ρu_m, lbm/hr ft^2, kg/s m^2
Gr   Grashof number [defined by Eq. (606)], dimensionless
Gz   Graetz number, Wc_p/kL, see Eq. (126), dimensionless
g gravitational acceleration, ft/sec^2, m/s^2
g_c proportionality constant in Newton's second law of motion, $g_c = 32.174$ lbm ft/lbf sec^2, $g_c = 1$ and dimensionless in SI units
H pitch for 180° rotation of tape (see Fig. 120), ft, m
Ⓗ   thermal boundary condition referring to constant axial as well as peripheral wall heat flux, also constant peripheral wall temperature; boundary condition valid only for the circular tube, parallel plates, and concentric annular ducts when symmetrically heated
Ⓗ1   thermal boundary condition referring to constant axial wall heat flux with constant peripheral wall temperature [see Eqs. (51) and (52)]
Ⓗ2   thermal boundary condition referring to constant axial wall heat flux with constant peripheral wall heat flux [see Eqs. (54) and (55)]
Ⓗ3   thermal boundary condition referring to constant axial wall heat flux with finite normal wall thermal resistance [see Eqs. (56) and (57)]
Ⓗ4   thermal boundary condition referring to constant axial wall heat flux with finite peripheral wall heat conduction [see Eqs. (60) and (61)]
Ⓗ5   thermal boundary condition referring to exponential axial wall heat flux with constant peripheral wall temperature [see Eqs. (62) and (63)]
h convective heat transfer coefficient, for fully developed flow if no subscript is used, Btu/hr ft^2 °F, W/m^2 °C
J mechanical to thermal energy conversion factor, $J = 778.163$ lbf ft/Btu, $J = 1$ and dimensionless in SI units
j Colburn heat transfer modulus, St Pr$^{2/3}$, dimensionless
$K(x)$ incremental pressure drop number [defined by Eq. (91)], for fully developed flow when $x = \infty$, dimensionless
$K_d(x)$ momentum flux correction factor [defined by Eq. (94)], for fully developed flow when $x = \infty$, dimensionless
$K_e(x)$ kinetic energy correction factor [defined by Eq. (96)], for fully developed flow when $x = \infty$, dimensionless

K_p peripheral wall heat conduction parameter, $k_w a'/k D_h$, dimensionless

k thermal conductivity, for fluid if either no subscript or the subscript f is used, Btu/hr ft °F, W/m °C

k_w thermal conductivity of wall material, Btu/hr ft °F, W/m °C

L length of the duct, ft, m

L_{hy} hydrodynamic entrance length (defined on p. 41), ft, m

L_{hy}^+ hydrodynamic entrance length, L_{hy}/D_h Re, dimensionless

L_{th} thermal entrance length (defined on p. 50), ft, m

L_{th}^* thermal entrance length, L_{th}/D_h Pe, dimensionless

l fin height, ft, m

l^* ratio of fin height to tube radius or a half-width of a rectangular duct, l/a, dimensionless

m exponent in Eq. (62), dimensionless

$N_{bc}(x)$ incremental heat transfer number [defined by Eq. (130)], for fully developed flow when $x = \infty$, dimensionless

N_{tu} number of heat transfer units, $h_m A/W c_p$, St L/r_h; in a two-fluid heat exchanger, $N_{tu} = UA/C_{min}$, dimensionless

Nu_{bc} peripheral average Nusselt number for fully developed flow, the subscript bc designates the thermal boundary condition of Table 2, defined similar to Eq. (110) for $x = \infty$, dimensionless

$Nu_{bc,d}$ peripheral average Nusselt number for fully developed flow for a finless duct [defined by Eq. (544)], dimensionless

$Nu_{x,bc}$ peripheral average axially local Nusselt number for the thermal entrance region for the specified thermal boundary condition [defined by Eq. (110)], dimensionless

$Nu_{m,bc}$ mean Nusselt number for the thermal entrance region for the specified thermal boundary condition [defined by Eq. (111)], dimensionless

$Nu_{p,bc}$ peripheral local Nusselt number for fully developed flow for the specified thermal boundary condition [defined by Eq. (109)], dimensionless

$Nu_{a,T}$ mean Nusselt number for the Ⓣ boundary condition based on the arithmetic mean temperature difference [defined by Eq. (115)], dimensionless

$Nu_{x,lj}^{(k)}$ local Nusselt number for a doubly connected duct, $\Phi_{lj}^{(k)}/(\theta_{lj}^{(k)} - \theta_{mj}^{(k)})$; this, mean, and peripheral local and average Nusselt numbers are defined on p. 34, dimensionless

$Nu_{x,i}^{(la)}$ local Nusselt number at inner wall of a doubly connected duct for the specific thermal boundary condition la of Table 5; Nusselt numbers for the other specific thermal boundary conditions of Table 5 are defined similarly, dimensionless

Nu_o overall Nusselt number associated with the Ⓣ③ boundary condition [defined by Eqs. (118) and (120)], dimensionless

n number of sides of a regular polygon or a cusped duct, number of longitudinal fins within a duct, number of rods in a concentric n-rod bundle

n outer normal coordinate at a point on the duct wall inside periphery Γ, ft, m

n^* coordinate n/D_h measured along the outer normal direction at the duct wall inside periphery, dimensionless

P wetted perimeter of the duct (length of boundary Γ), ft, m

P^* ratio of inside to outside perimeter of an annulus, P_i/P_o, dimensionless

P_h heated perimeter of the duct (through which heat transfer takes place), ft, m

Pe Péclet number, $u_m D_h/\alpha$, Re Pr, dimensionless

Pr Prandtl number, $\mu c_p/k$, dimensionless

p fluid static pressure, lbf/ft^2, Pa

Δp fluid static pressure drop in the flow direction between two cross sections of interest, lbf/ft^2, Pa

Δp^* $\Delta p/(\rho u_m^2/2g_c)$, dimensionless

Q volumetric flow rate, $A_c u_m$, ft^3/hr, m^3/s

q'	wall heat transfer rate per unit length of the duct, $q' = q''P_h$ for axially constant wall heat flux boundary condition, q' is assumed to be from the wall to fluid for convenience, Btu/hr ft, W/m
q''	wall heat flux, heat transfer rate per unit heat transfer area A of the duct; it is an average value with respect to perimeter; for axially constant heat flux cases, $q'' = q'/P_h$; for fully developed flow if neither x nor m appears as a subscript, Btu/hr ft^2, W/m^2
q_p''	peripheral local wall heat flux, Btu/hr ft^2, W/m^2
q_r''	incident radiative heat flux, Btu/hr ft^2, W/m^2
R	radius of a tube containing a rod bundle (see Fig. 107), ft, m
Ra	Rayleigh number, Gr Pr, dimensionless
Re	Reynolds number, GD_h/μ, dimensionless
Re_d	Reynolds number for a finless duct [defined by Eq. (543)], dimensionless
Re_x	Reynolds number, Gx/μ, dimensionless
R_w	wall thermal resistance, $k/U_w D_h$, dimensionless
r	radial coordinate in the cylindrical coordinate system, ft, m
r_h	hydraulic radius of the duct, A_c/P, ft, m
r_i	inner radius of a concentric or an eccentric annular duct or radius of the central circular core of a regular polygonal duct, ft, m
r_j	radius of heat transferring wall of the concentric or eccentric annular duct, ft, m
r_o	outer radius of a concentric or an eccentric annular duct or radius of a circular duct having a central regular polygonal core, ft, m
r^*	r_i/r_o for concentric and eccentric annular ducts, a/b_o for an elliptical duct with central circular core, b_i/b_o for confocal elliptical ducts, dimensionless
r_j^*	r_j/r_o, dimensionless
S	a parameter for the eccentric annular duct [defined by Eq. (504)], dimensionless
S	thermal energy source function, rate of thermal energy generated per unit volume of the fluid, Btu/hr ft^3, W/m^3
S^*	thermal energy source number [defined by Eqs. (151) and (152)], dimensionless
St	Stanton number, h/Gc_p, dimensionless
s	tangential coordinate at a point on the duct wall inside periphery Γ, ft, m
s	tube bundle pitch (see Fig. 104), ft, m
s^*	tangential coordinate, s/D_h, dimensionless
T	temperature of the fluid on the absolute scale, R, K
Ⓣ	thermal boundary condition referring to constant wall temperature, both axially and peripherally [see Eq. (39)]
Ⓣ3	thermal boundary condition referring to constant axial outside wall temperature with finite normal wall thermal resistance [see Eqs. (40) and (41)]
Ⓣ4	radiant flux thermal boundary condition [see Eqs. (47) and (48)]
Δ𝑡	thermal boundary condition referring to constant axial wall-to-fluid bulk temperature difference [see Eqs. (66) and (67)]
t	temperature of the fluid to a specified arbitrary datum, °F, °C
t_a	ambient fluid temperature (see Fig. 4), °F, °C
t_m	bulk average fluid temperature [defined by Eq. (102)], °F, °C
t_w	wall or fluid temperature at the inside duct periphery Γ, °F, °C
t_{wo}	wall temperature at the outside duct periphery, °F, °C
$t_{w,m}$	perimeter average wall temperature [defined by Eq. (101)], °F, °C
$t_{w,max}^*$	maximum wall temperature [defined by Eq. (350)], dimensionless
$t_{w,min}^*$	minimum wall temperature [defined by Eq. (350)], dimensionless
U	overall heat transfer coefficient in a two-fluid heat exchanger, Btu/hr ft^2 °F, W/m^2 °C
U_w	wall thermal conductance [defined by Eqs. (42)–(44)], Btu/hr ft^2 °F, W/m^2 °C

U_o	overall heat transfer coefficient for Ⓣ③ or Ⓗ③ [defined by Eq. (108)], Btu/hr ft² °F, W/m² °C
u	fluid axial velocity, fluid velocity component in x direction, ft/sec, m/s
u_m	fluid mean axial velocity [defined by Eq. (77)], ft/sec, m/s
u_{max}	fluid maximum axial velocity across the duct cross section for fully developed laminar flow, ft/sec, m/s
V	volume of the heat exchanger, ft³, m³
v	fluid velocity component in y or r direction, ft/sec, m/s
W	fluid mass flow rate through the duct, $\rho u_m A_c$, lbm/hr, kg/s
w	fluid velocity component in z or θ direction, ft/sec, m/s
w	wall distance as shown in Fig. 106, ft, m
X_L	twist ratio (see Fig. 120), H/d, dimensionless
x	axial (flow direction) coordinate in Cartesian or cylindrical coordinate systems; if velocity and temperature profiles develop at different locations, $x = x_e$ denotes the section where velocity profile develops, and $x = 0$ denotes the section where temperature profile develops, ft, m
x^+	axial coordinate for the hydrodynamic entrance region, x/D_h Re, dimensionless
x^*	axial coordinate for the thermal entrance region, x/D_h Pe, dimensionless
y, z	Cartesian coordinates across the flow cross section, ft, m
z'	a complex variable, $x + iy$, where $i = \sqrt{-1}$, ft, m
$\bar{y}$	distance of a centroid of the duct cross section measured from the base (see Fig. 46 or 64), ft, m
$\bar{y}_{max}$	normal distance from the base to a point where u_{max} occurs in the duct cross section (see Fig. 46 or 64), ft, m
α	fluid thermal diffusivity, $k/\rho c_p$, ft²/sec, m²/s
α	ratio of total heat transfer area on one side of the exchanger to the total volume of the exchanger, used in Chapter XVII only, ft²/ft³, m²/m³
α_w	absorptivity of the wall material, dimensionless
α^*	aspect ratio of rectangular, triangular, elliptical, trapezoidal, and sine ducts; $\alpha^* = 2b/2a$ for a symmetrical geometry with symmetrical heating, otherwise $\alpha^* = 2a/2b$, so that it ranges from 0 to 1, dimensionless
β	a function of x alone [defined by Eq. (164)], dimensionless
β	coefficient of thermal expansion, $\beta = -(\partial v/\partial T)_p/v$ where $v = 1/\rho$, 1/°F, 1/°C
Γ	inside periphery of the duct wall
$\Gamma( )$	Gamma function
γ	radiative wall heat flux boundary condition parameter, $\varepsilon_w \sigma T_e^3 D_h/k$, dimensionless
Δ, δ	prefixes denoting a difference
δ	$2a/2c$ for arbitrary triangular ducts (see Fig. 53), dimensionless
δ	energy content of the fluid [defined by Eq. (153)], dimensionless
δ	thickness of a twisted tape (see Fig. 120), ft, m
ϵ	eccentricity of an eccentric annular duct (see Fig. 91) or amplitude of a corrugated duct (see Fig. 79), ft, m
ε	heat exchanger effectiveness, the ratio of actual heat transfer rate to the thermodynamically limited maximum possible heat transfer rate as would be realized only in a counterflow heat exchanger of the infinite area, dimensionless
ε_w	emissivity of the wall, dimensionless
$\varepsilon(x)$	mean velocity weighing factor [see Eq. (167)], dimensionless
Θ	fluid temperature for axially constant wall heat flux boundary conditions, $(t - t_e)/(q''D_h/k)$, dimensionless
Θ_{w-m}	wall-to-fluid bulk mean temperature difference [defined by Eq. (158)], dimensionless

θ	angular coordinate in the cylindrical coordinate system, deg, rad
θ	fluid temperature for axially constant wall temperature boundary conditions, $(t - t_w)/(t_e - t_w)$, dimensionless
θ_m	fluid bulk mean temperature, $(t_m - t_w)/(t_e - t_w)$, dimensionless
$\theta_j^{(k)}$	fluid temperature for a doubly connected duct [defined by Eqs. (71) and (72)], dimensionless
$\theta_{lj}^{(k)}$	peripheral average temperature of wall l (l = i for inner wall, l = o for outer wall) for the fundamental boundary condition of kind (k) when j = i or o (inner or outer wall) heated or cooled; fluid bulk mean temperature if l = m; defined in a manner similar to Eqs. (71) and (72), dimensionless
ϑ_m	$(1 - \theta_m)$, dimensionless
κ	velocity gradient at the entrance for parallel plates (defined on p. 164), dimensionless
$\Lambda(x)$	a parameter defined by Eq. (167), ft/sec^2, m/s^2
λ	longitudinal wall heat conduction parameter, $k_w A_w/LC_{min}$, dimensionless
μ	fluid dynamic viscosity coefficient (see footnote on p. 38), lbm/hr ft, Pa s
ν	kinematic fluid viscosity coefficient, μ/ρ, ft^2/sec, m^2/s
ξ_1	radius of an inscribed circle of a regular polygon (see Figs. 100, 101, and 103)
ξ_2	radius of a circumscribed circle of a regular polygon (see Fig. 102)
ρ	fluid density, lbm/ft^3, kg/m^3
σ	porosity of a matrix type heat exchanger surface or ratio of free flow area to frontal area in a heat exchanger, dimensionless
σ	Stefan–Boltzmann constant, 0.1713×10^{-8} Btu/hr ft^2 R^4, 5.6697×10^{-8} W/m^2 K^4
τ	wall shear stress due to skin friction, lbf/ft^2, Pa
τ	time, used only in Eqs. (37) and (38), sec, s
$\Phi_j^{(k)}$	heat flux at a point in the flow field for the jth wall of a doubly connected duct heated [defined by Eq. (73)], dimensionless
$\Phi_{lj}^{(k)}$	wall heat flux defined in a manner similar to $\theta_{lj}^{(k)}$; $q_l''D_h/k(t_j - t_e)$ for (k) = 1, 3, and q''/q_{lj}'' for (k) = 2, 4, 5; dimensionless
$\Phi_{m,T}$	mean wall heat flux for axially constant wall temperature boundary condition, $q_m''D_h/k(t_w - t_e)$, dimensionless
$\Phi_{x,T}$	local wall heat flux for axially constant wall temperature boundary condition, $q_x''D_h/k(t_w - t_e)$, dimensionless
ϕ	apex angle of the right triangular, trapezoidal, and rhombic ducts; half-apex angle of the isosceles triangular, circular sector, circular segment, annular sector, moon-shaped, and flat-sided circular ducts, deg, rad
ω	vorticity (defined in footnote on p. 90), 1/sec, 1/s
∇	denotes gradient, derivative with respect to normal direction
∇^2	Laplacian operator, $(\partial^2/\partial x^2 + \partial^2/\partial y^2 + \partial^2/\partial z^2)$ in Cartesian coordinates

SUBSCRIPTS

bc	thermal boundary condition
c	center or centroid
d	associated dimensionless group based on the *finless* duct hydraulic diameter and free flow area
e	effective for Chapter XVII
e	initial value at the entrance of the duct or where the heat transfer starts
f	fluid
fd	fully developed laminar flow
H	Ⓗ boundary condition

H1	(H1) boundary condition
H2	(H2) boundary condition
H3	(H3) boundary condition
H4	(H4) boundary condition
H5	(H5) boundary condition
i	inner surface of a doubly connected duct, ith passage of a multipassage heat exchanger in Chapter XVII
j	heated wall of a doubly connected duct, $j = $ i or o
lm	logarithmic mean
m	mean
max	maximum
min	minimum
o	outer surface of a doubly connected duct or the outlet section of a heat exchanger
p	pertaining to a local peripheral point on Γ
T	(T) boundary condition
T3	(T3) boundary condition
T4	(T4) boundary condition
x	denoting arbitrary section along the duct length, a local value as opposed to a mean value
w	wall or fluid at the wall
∞	free stream value for the external flow
Γ	evaluated at the duct inside periphery

Appendix

The following is a list of technical journals and proceedings in which the laminar flow heat transfer literature has been located. It is interesting to note the wide spread of published sources for this one class of problem, which attests to both its technical and mathematical interest.

Acta Technica (*Budapest*), Hungary
Advances in Heat Transfer, New York, U.S.A.
American Institute of Chemical Engineers Journal, New York, U.S.A.
Annalen der Physik und Chemie, Leipzig, W. Germany
Annales des Mines, *Memoires*, Paris, France
Applied Mechanics Review, New York, U.S.A.
Applied Scientific Research, Section A, Hague, Netherlands
British Journal of Applied Physics, London, U.K.
Bulletin de l'Academie Polonaise des Sciences, Series des Sciences Techniques, Warsaw, Poland
Bulletin of the Japan Society of Mechanical Engineers, Tokyo, Japan
Canadian Journal of Chemical Engineering, Ottawa, Canada
Chemia Stosowana Seria B, Wroclaw, Poland
Chemical and Process Engineering (London), U.K.
Chemical Engineering (*New York*), U.S.A.
Chemical Engineering Journal (*Lausanne*), Laussane, Switzerland
Chemical Engineering Progress, New York, U.S.A.
Chemical Engineering Progress, Symposium Series, New York, U.S.A.
Chemical Engineering Science, Oxford, U.K.
Chemie-Ingenieur-Technik, Weinheim, W. Germany
Chemiker-Zeitung, Chemische Apparatur, Heidelberg, W. Germany
Comptes Rendus Hebdomadaires des Seances de l'Academie des Sciences, Paris, France
Computer Methods in Applied Mechanics and Engineering, Amsterdam, Netherlands
Computers and Fluids, Oxford, U.K.
Engineering (*London*), U.K.
Forschungs auf dem Gebiete des Ingenieurwesens, VDI, Dusseldorf, W. Germany
Heat Transfer—Japanese Research, Washington, D.C., U.S.A.
Heat Transfer—Soviet Research, Washington, D.C., U.S.A.

*High Temp*erature (English translation of Teplofizika Vysokikh Temperatur) New York, U.S.A.

*Indian Chem*ical *Eng*ineer, Calcutta, India

*Indian J*ournal of *Pure* and *Appl*ied *Phy*sics, New Delhi, India

*Indian J*ournal of *Techno*logy, New Delhi, India

*Ind*ustrial and *Eng*ineering *Chem*istry, Washington, D.C., U.S.A.

*Int*ernational *Chem*ical *Eng*ineering, New York, U.S.A.

*Int*ernational *Dev*elopments in *Heat Transfer, Proc*eedings of the *Heat Transfer Conf*erence, Boulder, Colorado, U.S.A.

*Int*ernational Journal for *Numer*ical *Methods* in *Eng*ineering, Bristol, U.K.

*Int*ernational Journal of *Heat* and *Mass Transfer*, Oxford, U.K.

*Inzhenerno-Fi7i*cheskii *Zh*urnal, Minsk, U.S.S.R.

*Izv*estiya *Vyssh*ikh *Uch*ebnykh *Zaved*enii, *Energ*etika, Minsk, U.S.S.R.

Journal of the *Aeronaut*ical *Soc*iety of *Ind*ia, India

Journal of the *Aerosp*ace *Sci*ences, New York, U.S.A.

Journal of *Appl*ied *Math*ematics and *Mech*anics (English Translation of *Pr*ikladnaya *Mat*ematika i Mekhanika) New York, U.S.A.

Journal of *Appl*ied *Mech*anics, New York, U.S.A.

Journal of *Basic Eng*ineering, New York, U.S.A.

Journal of *Biomech*anics, Oxford, U.K.

Journal of the *Chin*ese *Inst*itute of *Chem*ical *Eng*ineers, Taipei, Taiwan

Journal of *Comput*ational *Phy*sics, New York, U.S.A.

Journal of *Eng*ineering for *Power*, New York, U.S.A.

Journal of *Eng*ineering *Phy*sics (English translation of Inzhenerno-Fizicheskii Zhurnal) New York, U.S.A.

Journal of *Fluid Mech*anics, London, U.K.

Journal of *Fluids Eng*ineering, New York, U.S.A.

Journal of the *Franklin Inst*itute, Philadelphia, U.S.A.

Journal of *Heat Transfer*, New York, U.S.A.

Journal of the *Inst*itution of *Eng*ineers (*India*), *Part IDGE*: Industrial Development and General Engineering, Calcutta, India

Journal of *Math*ematics and *Phy*sics (*Cambridge, Mass.*), U.S.A.

Journal of *Mech*anical *Eng*ineering *Sci*ence, London, U.K.

Journal of the *Roy*al *Aeronaut*ical *Soc*iety, London, U.K.

Journal of *Roy*al *Tech*nical *College*, Glasgow, U.K.

Journal of *Sci*ence and *Eng*ineering *Res*earch, Kharagpur, India

*Kaeltetech*nik-*Klim*atisierung, Karlsruhe, W. Germany

*Lett*ers in *Heat Transfer*, New York, U.S.A.

*London Edinburgh Dublin Philos*ophical *Mag*azine, Journal of *Sci*ence, London, U.K.

*Mem*oirs of the *Fac*ulty of *Eng*ineering, *Kyushu Uni*versity, Fukuoka, Japan

Nagoya Kogyo Daigaku Gakuho (Bulletin of the Nagoya Institute of Technology), Nagoya, Japan

Nippon Kikai Gakkai Rombunshu (Transations of the Japan Society of Mechanical Engineers), Tokyo, Japan

*Nucl*ear *Eng*ineering and *Des*ign, Amsterdam, Netherlands

*Nucl*ear *Sci*ence and *Eng*ineering, Hinsdale, Ill., U.S.A.

*Phy*sics of *Fluids*, New York, U.S.A.

*Proc*eedings of the *All-Union Conf*erence on *Heat* and *Mass Transfer*, U.S.S.R.

*Proc*eedings of the *Australas*ian *Conf*erence on *Hydraul*ics and *Fluid Mech*anics, *University of* Auckland, *2nd*, New Zealand

*Proc*eedings of the *Cambridge Philos*ophical *Soc*iety, London, U.K.

*Proc*eedings of the *Inst*itution of *Mech*anical *Eng*ineers, London, U.K.

*Proc*eedings of the *Int*ernational *Heat Transfer Conf*erence, Chicago, Paris, Tokyo

*Proc*eedings of the *London Math*ematical *Soc*iety, London, U.K.

*Proc*eedings of the *Midwes*tern *Conf*erence on *Fluid Mech*anics, Ann Arbor, Mich., U.S.A.

*Proc*eedings of *Nat*ional *Heat* and *Mass Transfer Conf*erence, *Indian Inst*itute of *Technol*ogy, *Madras, Kanpur, Bombay,* India

*Proc*eedings of the *Nat*ional *Inst*itute of *Sci*ences of *India, Part A*: Physical Sciences, New Delhi, India

*Proc*eedings of the *Roy*al *Soc*iety, *London,* U.K.

*Proc*eedings of the *Semi-Int*ernational *Symp*osium on *Heat Transfer,* Tokyo, Japan

*Proc*eedings of the *U*nited *N*ations *Int*ernational *Conf*erence on the *Peaceful Uses* of *At*omic *Energy, 2nd,* Geneva, Switzerland

*Proc*eedings of *U*nited *S*tates *Nat*ional *Congr*ess of *Appl*ied *Mech*anics, ASME, New York, U.S.A.

*Q*uarterly of *Appl*ied *Math*ematics, Providence, R.I., U.S.A.

*Sov*iet *Phys*ics-*Dokl*ady (English translation of Doklady Akademii Nauk SSSR) New York, U.S.A.

*Trans*actions of *Am*erican *Inst*itute of *Chem*ical *Eng*ineers, New York, U.S.A.

*Trans*actions of the *Am*erican *Nucl*ear *Soc*iety, Hinsdale, Ill., U.S.A.

*Trans*actions of the *Am*erican *S*ociety of *M*echanical Engineers, New York, U.S.A.

*Trans*actions of the *Can*adian *Soc*iety of *Mech*anical *Eng*ineers, Montreal, Canada

*Trans*actions of the *Inst*itution of *Chem*ical *Eng*ineers, London, U.K.

*Trans*actions of the *Jap*an *Soc*iety of *Mech*anical *Eng*ineers, Tokyo, Japan

*V*erein *D*eutscher *I*ngenieure-*Forschungsh*eft, Dusseldorf, W. Germany

*V*erein *D*eutscher *I*ngenieure Zeitschrift, Dusseldorf, W. Germany

Waerme-und *Kaeltetech*nik, Berlin, W. Germany

Waerme-und *Stoffuebertrag*ung, Berlin, W. Germany

Zeitschrift für *Angew*andte *Math*ematik und *Mech*anik, Berlin, W. Germany

Zeitschrift für *Angew*andte *Math*ematik und *Phys*ik, Basel, W. Germany

References

The titles of the journals are abbreviated as in the "Bibliographic Guide for Editors & Authors," 1974 edition and supplements (a publication of the Chemical Abstract Service). The complete titles are described in the Appendix.

1. L. Graetz, Uber die Wärmeleitungsfähigkeit von Flüssigkeiten (On the thermal conductivity of liquids). Part 1. *Ann. Phys. Chem.* **18**, 79–94 (1883); Part 2. *Ann. Phys. Chem.* **25**, 337–357 (1885).

2. W. Nusselt, Die Abhängigkeit der Wärmeübergangszahl von der Rohrlänge (The dependence of the heat-transfer coefficient on the tube length). *VDI Z.* **54**, 1154–1158 (1910).

3. T. B. Drew, Mathematical attacks on forced convection problems: A review. *Trans. Am. Inst. Chem. Eng.* **26**, 26–80 (1931).

4. H. L. Dryden, F. D. Murnaghan, and H. Bateman, "Hydrodynamics," Bull. No. 84, pp. 197–201. Comm. Hydrodyn., Div. Phys. Sci., Natl. Res. Counc., Washington, D.C., 1932; reprinted by Dover, New York, 1956.

5. W. M. Rohsenow and H. Y. Choi, "Heat, Mass and Momentum Transfer." Prentice-Hall, Englewood Cliffs, New Jersey, 1961.

6. W. M. Kays and A. L. London, "Compact Heat Exchangers," 2nd Ed. McGraw-Hill, New York, 1964.

7. W. M. Kays, "Convective Heat and Mass Transfer." McGraw-Hill, New York, 1966.

8. B. S. Petukhov, "Heat Transfer and Pressure Loss in Laminar Flow of Liquids in Pipes." Energiya Press, Moscow, 1967. (In Russian.)

9. O. G. Martynenko and R. Eichhorn, Hydrodynamics and heat exchange of the inlet sections of ducts with laminar flow (in Russian). *Proc. 1968 Minsk All Union Conf. Heat Mass Transfer, Teplo-i-massoperenos, Acad. Sci. BSSR* **1**, 445–465 (1968).

10. J. E. Porter, Heat transfer at low Reynolds number (highly viscous liquids in laminar flow)—Industrial research fellow report. *Trans. Inst. Chem. Eng.* **49**, 1–29 (1971).

11. J. M. Kooijman, Laminar heat or mass transfer in rectangular channels and in cylindrical tubes for fully developed flow: Comparison of solutions obtained for various boundary conditions. *Chem. Eng. Sci.* **28**, 1149–1160 (1973).

12. W. M. Kays and H. C. Perkins, Forced convection, internal flow in ducts. *In* "Handbook of Heat Transfer" (W. M. Rohsenow and J. P. Hartnett, eds.), 193 pp., McGraw-Hill, New York, 1973.

13. R. K. Shah and A. L. London, "Laminar Flow Forced Convection Heat Transfer and Flow Friction in Straight and Curved Ducts—A Summary of Analytical Solutions," TR No. 75. Dep. Mech. Eng., Stanford University, Stanford, California, 1971.

14. A. H. Shapiro, R. Siegel, and S. J. Kline, Friction factor in the laminar entry region of a smooth tube. *Proc. U.S. Natl. Congr. Appl. Mech., 2nd, Am. Soc. Mech. Eng., New York* pp. 733–741 (1954).

15. I. N. Sadikov, Laminar heat transfer over the initial section of a rectangular channel. *J. Eng. Phys. (USSR)* **8**, 287–291 (1965).

16. R. A. Seban and T. T. Shimazaki, Heat transfer to a fluid flowing turbulently in a smooth pipe with walls at constant temperature. *Trans. ASME* **73**, 803–809 (1951).

17. R. K. Shah and A. L. London, Internal flow forced convection heat transfer—from mathematics to engineering, *Proc. Natl. Heat Mass Transfer Conf., 3rd, Indian Inst. Technol. Bombay* Pap. No. HMT-10-75 (1975).

18. A. V. Luikov, "Heat and Mass Transfer," Ch. 4. Energia Publ. House, Moscow, 1972. (In Russian.)

19. A. V. Luikov, V. A. Aleksashenko, and A. A. Aleksashenko, Analytical methods of solution of conjugated problems in convective heat transfer. *Int. J. Heat Mass Transfer* **14**, 1047–1056 (1971).

20. S. Mori, M. Sakakibara, and A. Tanimoto, Steady heat transfer to laminar flow in a circular tube with conduction in the tube wall. *Heat Transfer—Jpn. Res.* **3**(2), 37–46 (1974).

21. V. R. Shelyag, Calculation of the temperature field in plane slotted channels with laminar gas flow. *J. Eng. Phys. (USSR)* **12**, 117–120 (1967).

22. E. J. Davis and W. N. Gill, The effects of axial conduction in the wall on heat transfer with laminar flow. *Int. J. Heat Mass Transfer* **13**, 459–470 (1970).

23. S. Mori, T. Shinke, M. Sakakibara, and A. Tanimoto, Steady heat transfer to laminar flow between parallel plates with conduction in wall. *Heat Transfer—Jpn. Res.* **5**(4), 17–25 (1976).

24. R. K. Shah and A. L. London, Thermal boundary conditions and some solutions for laminar duct flow forced convection. *J. Heat Transfer* **96**, 159–165 (1974); condensed from *Am. Soc. Mech. Eng., Pap.* **72-WA/HT-54** (1972).

25. J. R. Sellers, M. Tribus, and J. S. Klein, Heat transfer to laminar flow in a round tube or flat conduit—the Graetz problem extended. *Trans. ASME* **78**, 441–448 (1956).

26. J. W. Mitchell, An expression for internal flow heat transfer for polynomial wall temperature distributions. *J. Heat Transfer* **91**, 175–177 (1969).

27. E. R. G. Eckert, T. F. Irvine, Jr., and J. T. Yen, Local laminar heat transfer in wedge-shaped passages. *Trans. ASME* **80**, 1433–1438 (1958).

28. W. C. Reynolds, Effect of wall heat conduction on convection in a circular tube with arbitrary circumferential heat input. *Int. J. Heat Mass Transfer* **6**, 925 (1963).

29. R. Siegel, E. M. Sparrow, and T. M. Hallman, Steady laminar heat transfer in a circular tube with prescribed wall heat flux. *Appl. Sci. Res., Sect. A* **7**, 386–392 (1958).

30. R. N. Noyes, A fully integrated solution of the problem of laminar or turbulent flow in a tube with arbitrary wall heat flux. *J. Heat Transfer* **83**, 96–98 (1961).

31. S. Hasegawa and Y. Fujita, Turbulent heat transfer in a tube with prescribed heat flux. *Int. J. Heat Mass Transfer* **11**, 943–962 (1968).

32. C. A. Bankston and D. M. McEligot, Prediction of tube wall temperatures with axial variation of heating rate and gas property variation. *Nucl. Sci. Eng.* **37**, 157–162 (1969).

33. S. S. Zabrodskiy, Several specific terms and concepts in Russian heat transfer notation. *Heat Transfer—Sov. Res.* **4**(2), i–iv (1972).

34. U. Grigull and H. Tratz, Thermischer einlauf in ausgebildeter laminarer Rohrströmung. *Int. J. Heat Mass Transfer* **8**, 669–678 (1965).

35. V. Javeri, Analysis of laminar thermal entrance region of elliptical and rectangular channels with Kantorowich method. *Waerme- Stoffuebertrag.* **9**, 85–98 (1976).
36. D. F. Sherong and C. W. Solbrig, Analytical investigation of heat or mass transfer and friction factors in a corrugated duct heat or mass exchanger. *Int. J. Heat Mass Transfer* **13**, 145–159 (1970).
37. T. F. Irvine, Jr., Non-circular duct convective heat transfer. *In* "Modern Developments in Heat Transfer" (W. Ibele, ed.), pp. 1–17. Academic Press, New York, 1963.
38. R. W. Lyczkowski, C. W. Solbrig, and D. Gidaspow, "Forced Convective Heat Transfer in Rectangular Ducts—General Case of Wall Resistances and Peripheral Conduction." Inst. Gas Technol., Tech. Inf. Center, File 3229, 3424 S. State Street, Chicago, Illinois, 1969.
39. M. Iqbal, B. D. Aggarwala, and A. K. Khatry, On the conjugate problem of laminar combined free and forced convection through vertical non-circular ducts. *J. Heat Transfer* **94**, 52–56 (1972).
40. W. M. Kays, Numerical solutions for laminar-flow heat transfer in circular tubes. *Trans. ASME* **77**, 1265–1274 (1955).
41. R. E. Lundberg, W. C. Reynolds, and W. M. Kays, Heat transfer with laminar flow in concentric annuli with constant and variable wall temperature and heat flux. *NASA Tech. Note* **TN D-1972** (1963); also as Rep. No. AHT-2. Dep. Mech. Eng., Stanford Univ., Stanford, California 1961.
42. R. E. Lundberg, P. A. McCuen, and W. C. Reynolds, Heat transfer in annular passages. Hydrodynamically developed laminar flow with arbitrarily prescribed wall temperatures or heat fluxes. *Int. J. Heat Mass Transfer* **6**, 495–529 (1963).
43. D. Butterworth and T. D. Hazell, Forced-convective laminar flow heat transfer in the entrance region of a tube. *Proc. Int. Heat Transfer Conf., 4th, Paris-Versailles* **2**(FC 3.1), 1–11 (1970); also as *U.K. At. Energy Auth., Res. Group, Rep.* **AERE-R 6057** (1969).
44. H. S. Heaton, W. C. Reynolds, and W. M. Kays, Heat transfer in annular passages. Simultaneous development of velocity and temperature fields in laminar flow. *Int. J. Heat Mass Transfer* **7**, 763–781 (1964); also as Rep. No. AHT-5. Dep. Mech. Eng., Stanford University, Stanford, California, 1962.
45. R. W. Shumway and D. M. McEligot, Heated laminar gas flow in annuli with temperature-dependent transport properties. *Nucl. Sci. Eng.* **46**, 394–407 (1971).
46. J. E. R. Coney and M. A. I. El-Shaarawi, Finite difference analysis for laminar flow heat transfer in concentric annuli with simultaneously developing hydrodynamic and thermal boundary layers. *Int. J. Numer. Methods Eng.* **9**, 17–38 (1975).
47. N. S. Lakshmana Rao, A general method to calculate the settling length knowing the entrance and end profiles for steady, incompressible, viscous flow in pipes. *Proc. Australas. Conf. Hydraul. Fluid Mech., 2nd, University of Auckland* pp. A311–A318 (1965).
48. S. T. McComas, Hydrodynamic entrance lengths for ducts of arbitrary cross section. *J. Basic Eng.* **89**, 847–850 (1967).
49. L. Schiller, Die Entwiklung der laminaren Geshwindigkeitsverteilung und ihre Bedeutung für Zähigkeitsmessungen. *Z. Angew. Math. Mech.* **2**, 96–106 (1922).
50. E. M. Sparrow, C. W. Hixon, and G. Shavit, Experiments on laminar flow development in rectangular ducts. *J. Basic Eng.* **89**, 116–124 (1967).
51. T. S. Lundgren, E. M. Sparrow, and J. B. Starr, Pressure drop due to the entrance region in ducts of arbitrary cross section. *J. Basic Eng.* **86**, 620–626 (1964).
52. U. A. Sastry and P. V. Narasimha Rao, Pressure drop due to flow in the entrance region of non-circular ducts. *J. Sci. Eng. Res.* **9**, Part 2, 207–212 (1965).
53. U. A. Sastry, Pressure drop due to flow in the entrance region of ducts of arbitrary section. *J. Aeronaut. Soc. India* **18**(4), 115–117 (1966).

54. R. K. Shah and A. L. London, Dimensionless groups for laminar duct flow forced convection heat transfer. *Am. Soc. Mech. Eng., Pap.* **72-WA/HT-53** (1972).

55. D. F. Boucher and G. E. Alves, Dimensionless numbers for fluid mechanics, heat transfer, mass transfer and chemical reaction. Part 1. *Chem. Eng. Prog.* **55**(9), 55–64 (1959); Part 2. *Chem. Eng. Prog.* **59**(8), 75–83 (1963).

56. W. H. McAdams, "Heat Transmission," 3rd Ed. McGraw-Hill, New York, 1954.

57. A. P. Colburn, A method of correlating forced convection heat transfer data and a comparison with fluid friction. *Trans. Am. Inst. Chem. Eng.* **29**, 174–210 (1933); reprinted in *Int. J. Heat Mass Transfer* **7**, 1359–1384 (1964).

58. O. C. Sandall and O. T. Hanna, Entrance lengths for transport with a specified surface flux. *AIChE J.* **19**, 867–870 (1973).

59. O. T. Hanna, O. C. Sandall, and B. A. Paruit, Thermal entrance lengths for laminar flow in various ducts for constant surface heat flux. *Lett. Heat Mass Transfer* **3**, 89–98 (1976).

60. H. L. Dryden, Aerodynamics of cooling. *In* "Aerodynamic Theory" (W. F. Durand, ed.), Vol. 6, p. 253. Springer-Verlag, Berlin, 1936; reprinted by Dover, New York, 1936.

61. H. C. Brinkman, Heat effects in capillary flow. *Appl. Sci. Res., Sect. A* **2**, 120–124 (1951).

62. R. B. Bird, W. E. Stewart, and E. N. Lightfoot, "Transport Phenomena." Wiley, New York, 1960.

63. V. P. Tyagi, Laminar forced convection of a dissipative fluid in a channel. *J. Heat Transfer* **88**, 161–169 (1966).

64. V. P. Tyagi, A general non-circular duct convective heat-transfer problem for liquids and gases. *Int. J. Heat Mass Transfer* **9**, 1321–1340 (1966).

65. H. Schlichting, "Boundary Layer Theory," 6th Ed. McGraw-Hill, New York, 1968.

66. F. W. Schmidt and B. Zeldin, Laminar heat transfer in the entrance region of ducts. *Appl. Sci. Res.* **23**, 73–94 (1970).

67. H. D. Baehr and E. Hicken, Neue kennzahlen und gleichungen für den Wärmeübergang in laminar durchströmten Kanälen. *Kaeltetech.-Klim.* **21**, 34–38 (1969).

68. D. K. Hennecke, Heat transfer by Hagen-Poiseuille flow in the thermal development region with axial conduction. *Waerme- Stoffuebertrag.* **1**, 177–184 (1968).

69. K. C. Cheng, Dirichlet problems for laminar forced convection with heat sources and viscous dissipation in regular polygonal ducts. *AIChE J.* **13**, 1175–1180 (1967).

70. S. M. Marco and L. S. Han, A note on limiting laminar Nusselt number in ducts with constant temperature gradient by analogy to thin-plate theory. *Trans. ASME* **77**, 625–630 (1955).

71. K. C. Cheng, Analog solution of laminar heat transfer in noncircular ducts by Moiré method and point-matching. *J. Heat Transfer* **88**, 175–182 (1966).

72. S. P. Timoshenko and J. N. Goodier, "Theory of Elasticity," 3rd Ed. McGraw-Hill, New York, 1970.

73. J. Caldwell, The hydraulic mean depth as a basis for form comparison in the flow of fluids in pipes. *J. R. Tech. Coll. (Glasgow)* **2**, 203–220 (1930).

74. N. A. V. Piercy, M. S. Hooper, and H. F. Winny, Viscous flow through pipes with cores. *London Edinburgh Dublin Philos. Mag. J. Sci.* **15**, 647–676 (1933).

75. S. Timoshenko and S. Woinowsky-Krieger, "Theory of Plates and Shells," 2nd Ed. McGraw-Hill, New York, 1959.

76. M. Iqbal, B. D. Aggarwala, and A. G. Fowler, Laminar combined free and forced convection in vertical non-circular ducts under uniform heat flux. *Int. J. Heat Mass Transfer* **12**, 1123–1139 (1969).

77. K. C. Cheng, Laminar heat transfer in noncircular ducts by Moiré method. *J. Heat Transfer* **87**, 308–309 (1965).

78. B. D. Aggarwala and M. Iqbal, On limiting Nusselt numbers from membrane analogy

for combined free and forced convection through vertical ducts. *Int. J. Heat Mass Transfer* **12**, 737–748 (1969).

79. L. N. Tao, On some laminar forced-convection problems. *J. Heat Transfer* **83**, 466–472 (1961).

80. U. A. Sastry, Heat transfer by laminar forced convection in multiply connected cross-section. *Indian J. Pure Appl. Phys.* **3**, 113–116 (1965).

81. V. P. Tyagi, A forced convective heat transfer including dissipation function and compression work for a class of noncircular ducts. *Int. J. Heat Mass Transfer* **15**, 164–169 (1972).

82. L. N. Tao, Method of conformal mapping in forced convection problems. *Int. Dev. Heat Transfer, Proc. Heat Transfer Conf., Boulder, Colorado*, Part 3, pp. 598–606 (1961).

83. L. N. Tao, Heat transfer of laminar forced convection in indented pipes. *In* "Developments in Mechanics" (J. E. Lay and L. E. Malvern, eds.), Vol. 1, pp. 511–525. Plenum, New York, 1961.

84. L. N. Tao, The second fundamental problem in heat transfer of laminar forced convection. *J. Appl. Mech.* **29**, 415–420 (1962).

85. U. A. Sastry, Heat transfer by laminar forced convection in a pipe of curvilinear polygonal section. *J. Sci. Eng. Res.* **7**, Part 2, 281–292 (1963).

86. U. A. Sastry, Solution of the heat transfer of laminar forced-convection in non-circular pipes. *Appl. Sci. Res., Sect. A* **13**, 269–280 (1964).

87. U. A. Sastry, Heat transfer of laminar forced convection in doubly connected regions. *Indian J. Pure Appl. Phys.* **2**, 213–215 (1964).

88. U. A. Sastry, Heat transfer by laminar forced convection in multiply connected regions. *Acta Tech. (Budapest)* **51**, 181–192 (1965).

89. U. A. Sastry, Viscous flow through tubes of doubly connected regions. *Indian J. Pure Appl. Phys.* **3**, 230–232 (1965).

90. V. P. Tyagi, Forced convection of a dissipative liquid in a channel with Neumann conditions. *J. Appl. Mech.* **33**, 18–24 (1966).

91. V. P. Tyagi, General study of a heat transmission problem of a channel-gas flow with Neumann-type thermal boundary conditions. *Proc. Cambridge Philos. Soc.* **62**, 555–573 (1966).

92. M. Iqbal, A. K. Khatry, and B. D. Aggarwala, On the second fundamental problem of combined free and forced convection through vertical non-circular ducts. *Appl. Sci. Res.* **26**, 183–208 (1972).

93. M. J. Casarella, P. A. Laura, and M. Chi, On the approximate solution of flow and heat transfer through non-circular conduits with uniform wall temperature. *Br. J. Appl. Phys.* **18**, 1327–1335 (1967).

94. S. H. Clark and W. M. Kays, Laminar-flow forced convection in rectangular tubes. *Trans. ASME* **75**, 859–866 (1953).

95. F. W. Schmidt and M. E. Newell, Heat transfer in fully developed laminar flow through rectangular and isosceles triangular ducts. *Int. J. Heat Mass Transfer* **10**, 1121–1123 (1967).

96. O. E. Dwyer and H. C. Berry, Laminar-flow heat transfer for in-line flow through unbaffled rod bundles. *Nucl. Sci. Eng.* **42**, 81–88 (1970).

97. H. Nakamura, S. Hiraoka, and I. Yamada, Laminar forced convection flow and heat transfer in arbitrary triangular ducts. *Heat Transfer—Jpn. Res.* **1**(1), 120–122 (1972).

98. H. Nakamura, S. Hiraoka, and I. Yamada, Flow and heat transfer of laminar forced convection in arbitrary polygonal ducts. *Heat Transfer—Jpn. Res.* **2**(4), 56–63 (1974).

99. A. W. Date and J. R. Singham, Numerical prediction of friction and heat-transfer characteristics of fully developed laminar flow in tubes containing twisted tapes. *Am. Soc. Mech. Eng., Pap.* **72-HT-17** (1972).

100. A. W. Date, "Prediction of Friction and Heat Transfer Characteristics of Flow in a

Tube Containing a Twisted-Tape," Ph.D. Thesis and Rep. No. HTS/73/15. Dep. Mech. Eng., Imperial College of Science and Technology, London, 1973.

101. A. W. Date, Prediction of fully-developed flow in a tube containing a twisted-tape. *Int. J. Heat Mass Transfer* **17**, 845–859 (1974).

102. R. Meyder, Solving the conservation equations in fuel rod bundles exposed to parallel flow by means of curvilinear-orthogonal coordinates. *J. Comput. Phys.* **17**, 53–67 (1975).

103. E. M. Sparrow and A. L. Loeffler, Jr., Longitudinal laminar flow between cylinders arranged in regular array. *AIChE J.* **5**, 325–330 (1959).

104. E. M. Sparrow, A. L. Loeffler, Jr., and H. A. Hubbard, Heat transfer to longitudinal laminar flow between cylinders. *J. Heat Transfer* **83**, 415–422 (1961).

105. E. M. Sparrow, Laminar flow in isosceles triangular ducts. *AIChE J.* **8**, 599–604 (1962).

106. K. C. Cheng, Laminar flow and heat transfer characteristics in regular polygonal ducts. *Proc. Int. Heat Transfer Conf., 3rd, AIChE, New York* **1**, 64–76 (1966).

107. K. C. Cheng, Laminar forced convection in regular polygonal ducts with uniform peripheral heat flux. *J. Heat Transfer* **91**, 156–157 (1969).

108. F. S. Shih, Laminar flow in axisymmetric conduits by a rational approach. *Can. J. Chem. Eng.* **45**, 285–294 (1967).

109. K. C. Cheng and M. Jamil, Laminar flow and heat transfer in ducts of multiply connected cross sections. *Am. Soc. Mech. Eng., Pap.* **67-HT-6** (1967).

110. K. C. Cheng and G. J. Hwang, Laminar forced convection in eccentric annuli. *AIChE J.* **14**, 510–512 (1968).

111. K. C. Cheng and M. Jamil, Laminar flow and heat transfer in circular ducts with diametrically opposite flat sides and ducts of multiply connected cross sections. *Can. J. Chem. Eng.* **48**, 333–334 (1970).

112. D. A. Ratkowsky and N. Epstein, Laminar flow in regular polygonal ducts with circular centered cores. *Can. J. Chem. Eng.* **46**, 22–26 (1968).

113. S. L. Hagen and D. A. Ratkowsky, Laminar flow in cylindrical ducts having regular polygonal shaped cores. *Can. J. Chem. Eng.* **46**, 387–388 (1968).

114. E. M. Sparrow and A. Haji-Sheikh, Flow and heat transfer in ducts of arbitrary shape with arbitrary thermal boundary conditions. *J. Heat Transfer* **88**, 351–358 (1966); discussion by C. F. Neville, *J. Heat Transfer* **91**, 588–589 (1969)

115. R. K. Shah, "Laminar Flow Forced Convection Heat Transfer and Flow Friction in Straight and Curved Ducts—A Summary of Analytical Solutions," Ph.D. Thesis. Dep. Mech. Eng., Stanford University, Stanford, California, 1972.

116. R. K. Shah, Laminar flow friction and forced convection heat transfer in ducts of arbitrary geometry. *Int. J. Heat Mass Transfer* **18**, 849–862 (1975).

117. M. L. Trombetta, Laminar forced convection in eccentric annuli. *Int. J. Heat Mass Transfer* **14**, 1161–1173 (1971).

118. R. Ullrich, "Analyse der Ausgebildeten Laminarströmung in Längsangeströmten, Endlichen, Hexagonalen Stabbündeln," Doktor-Ingenieur Dissertation. Inst. Kerntech., Technische Universität Berlin, Berlin, 1974.

119. E. M. Sparrow and R. Siegel, A variational method for fully developed laminar heat transfer in ducts. *J. Heat Transfer* **81**, 157–167 (1959).

120. S. C. Gupta, A variational principle for fully developed laminar heat transfer in uniform channels. *Appl. Sci. Res., Sect. A* **10**, 85–101 (1961).

121. W. E. Stewart, Application of reciprocal variational principles to laminar flow in uniform ducts. *AIChE J.* **8**, 425–428 (1962).

122. D. Pnueli, A computational scheme for the asymptotic Nusselt number in ducts of arbitrary cross-section. *Int. J. Heat Mass Transfer* **10**, 1743–1748 (1967).

123. B. A. Finlayson and L. E. Scriven, On the search for variational principles. *Int. J. Heat Mass Transfer* **10**, 799–821 (1967).

124. H. F. P. Purday, "Streamline Flow." Constable, London, 1949; same as "An Introduction to the Mechanics of Viscous Flow." Dover, New York, 1949.

125. I. L. Maclaine-Cross, An approximate method for calculating heat transfer and pressure drop in ducts with laminar flow. *J. Heat Transfer* **91**, 171–173 (1969).

126. P. A. James, Forced convection heat transfer in narrow passages. *Can. J. Chem. Eng.* **48**, 330–332 (1970).

127. K. Nandakumar and J. H. Masliyah, Fully developed viscous flow in internally finned tubes. *Chem. Eng. J. (Lausanne)* **10**, 113–120 (1975).

128. J. H. Masliyah and K. Nandakumar, Heat transfer in internally finned tubes. *J. Heat Transfer* **98**, 257–261 (1976).

129. M. H. Hu, "Flow and Thermal Analysis for Mechanically Enhanced Heat Transfer Tubes," Ph.D. Thesis. Dep. Mech. Eng., State University of New York at Buffalo, 1973.

130. B. D. Aggarwala and M. K. Gangal, Heat transfer in rectangular ducts with fins from opposite walls. *Z. Angew. Math. Mech.* **56**, 253–266 (1976).

131. M. K. Gangal and B. D. Aggarwala, Combined free and forced convection in laminar internally finned square ducts. *Z. Angew. Math. Phys.* **28**, 85–96 (1977).

132. L. C. Kun, "Friction Factor and Limiting Nusselt Number for Laminar Flow in Internally Finned Tubes by Means of Green's function," Ph.D. Thesis. Dep. Mech. Eng., State University of New York at Buffalo, 1970.

133. D. T. W. Chen, "Flow and Heat Transfer in Internally Finned Tubes of Non-Circular Cross-Section," Ph.D. Thesis. Dep. Mech. Eng., State University of New York at Buffalo, 1973.

134. J. P. Zarling, Application of the Schwarz-Neumann technique to fully developed laminar heat transfer in non-circular ducts. *Am. Soc. Mech. Eng., Pap.* **76-WA/HT-49 (1976).**

135. N. S. Lakshmana Rao and K. Sridharan, Laminar flow development at conduit entrances. *J. Inst. Eng. (India), Part IDGE* **51** (5), 50–60 (1971).

136. J. Boussinesq, Hydrodynamique. *C. R. Acad. Sci.* **110**, 1160–1170, 1238–1242 (1890); **113**, 9–15, 49–51 (1891).

137. H. Schlichting, Laminare Kanaleinlaufströmung. *Z. Angew. Math. Mech.* **14**, 368–373 (1934).

138. Atkinson and S. Goldstein, *In* "Modern Developments in Fluid Dynamics" (S. Goldstein, ed.), Vol. 1, pp. 304–308. Oxford Univ. Press, London, 1938.

139. J. Gillis and M. Shimshoni, Initial flow in the entrance of a straight circular pipe. *J. R. Aeronaut. Soc.* **70**, 368–369 (1966).

140. M. Collins and W. R. Schowalter, Laminar flow in the inlet region of a straight channel. *Phys. Fluids* **5**, 1122–1124 (1962).

141. M. Roidt and R. D. Cess, An approximate analysis of laminar magnetohydrodynamic flow in the entrance region of a flat duct. *J. Appl. Mech.* **29**, 171–176 (1962).

142. M. Van Dyke, Entry flow in a channel. *J. Fluid Mech.* **44**, 813–823 (1970).

143. S. D. R. Wilson, Entry flow in a channel, Part 2. *J. Fluid. Mech.* **46**, 787–799 (1971).

144. A. K. Kapila, G. S. S. Ludford, and V. O. S. Olunloyo, Entry flow in a channel. Part 3. Inlet in a uniform stream. *J. Fluid Mech.* **57**, 769–784 (1973).

145. E. Naito and M. Hishida, Laminar boundary layers in the entrance regions of two parallel planes and a circular tube (in Japanese). *Nagoya Kogyo Daigaku Gakuho* **24**, 143–151 (1972).

146. E. Naito, Laminar heat transfer in the entrance region between parallel plates—the case of uniform heat flux. *Heat Transfer—Jpn. Res.* **4** (2), 63–74 (1975).

147. V. E. Gubin and V. S. Levin, Fluid flow in the entrance region of a circular tube. *J. Eng. Phys. (USSR)* **15**, 635–637 (1968).

148. W. D. Campbell and J. C. Slattery, Flow in the entrance of a tube. *J. Basic Eng.* **85**, 41–46 (1963).

149. N. S. Govinda Rao, M. V. Ramamoorthy, and K. V. N. Sarma, Study of transition zone of laminar flow at the entrance to a pipe based on varying friction. *Proc. Natl. Inst. Sci. India, Part A* **32**, 266 (1966).

150. R. C. Gupta, Flow development in the hydrodynamic entrance region of a flat duct. *AIChE J.* **11**, 1149–1151 (1965).

151. J. W. Williamson, Decay of symmetrical laminar distorted profiles between flat parallel plates. *J. Basic Eng.* **91**, 558–560 (1969).

152. D. Fargie and B. W. Martin, Developing laminar flow in a pipe of circular cross-section. *Proc. R. Soc. London, Ser. A* **321**, 461–476 (1971).

153. H. L. Langhaar, Steady flow in the transition length of a straight tube. *J. Appl. Mech.* **9**, A55–A58 (1942).

154. L. S. Han, Hydrodynamic entrance lengths for incompressible laminar flow in rectangular ducts. *J. Appl. Mech.* **27**, 403–409 (1960).

155. L. S. Han, Simultaneous developments of temperature and velocity profiles in flat ducts. *Int. Dev. Heat Transfer, Proc. Heat Transfer Conf., Boulder, Colorado*, Part III, pp. 591–597 (1961).

156. L. S. Han and A. L. Cooper, Approximate solutions of two internal flow problems—solution by an integral method. *Proc. U.S. Natl. Congr. Appl. Mech., 4th* **2**, 1269–1278 (1962).

157. E. Sugino, Velocity distribution and pressure drop in the laminar inlet of a pipe with annular space. *Bull. JSME* **5**, 651–655 (1962).

158. R. W. Miller and L. S. Han, Pressure losses for laminar flow in the entrance region of ducts of rectangular and equilateral triangular cross section. *J. Appl. Mech.* **38**, 1083–1087 (1971).

159. S. M. Targ, Osnovnye Zadachi Teorii Laminarnykh Techenyi, Gostekhizdat, Moscow, 1951.

159a. N. A. Slezkin, "Dynamics of Viscous Incompressible Fluids." Gostekhizdat Press, Moscow, 1955 (In Russian).

160. C. C. Chang and H. B. Atabek, Flow between two co-axial tubes near the entry. *Z. Angew. Math. Mech.* **42**, 425–430 (1962).

161. D. N. Roy, Laminar flow near the entry of coaxial tubes. *Appl. Sci. Res., Sect. A* **14**, 421–430 (1965).

162. E. M. Sparrow, S. H. Lin, and T. S. Lundgren, Flow development in the hydrodynamic entrance region of tubes and ducts. *Phys. Fluids* **7**, 338–347 (1964).

163. E. M. Sparrow and S. H. Lin, The developing laminar flow and pressure drop in the entrance region of annular ducts. *J. Basic Eng.* **86**, 827–834 (1964).

164. A. Quarmby, Note on developing laminar flow in annuli. *J. Basic Eng.* **88**, 811–812 (1966).

165. C. L. Wiginton and R. L. Wendt, Flow in the entrance region of ducts. *Phys. Fluids* **12**, 465–466 (1969).

166. D. P. Fleming and E. M. Sparrow, Flow in the hydrodynamic entrance region of ducts of arbitrary cross section. *J. Heat Transfer* **91**, 345–354 (1969).

167. B. D. Aggarwala and M. K. Gangal, Laminar flow development in triangular ducts. *Trans. Can. Soc. Mech. Eng.* **3**, 231–233 (1975).

168. C. L. Wiginton and C. Dalton, Incompressible laminar flow in the entrance region of a rectangular duct. *J. Appl. Mech.* **37**, 854–856 (1970).

169. C. L. Wiginton, Flow in the entrance region of noncircular ducts. *Phys. Fluids* **18**, 488–490 (1975).

170. R. L. Wendt and C. L. Wiginton, Incompressible laminar entrance flow in a circular sector duct. *J. Appl. Mech.* **43**, 357–359 (1976).

171. S. D. Savkar, Nonuniform flow in the inlet section of a straight channel. *Am. Soc. Mech. Eng., Pap.* **70-WA/FE-27** (1970).

172. J. R. Bodoia and J. F. Osterle, Finite difference analysis of plane Poiseuille and Couette flow developments. *Appl. Sci. Res., Sect. A* **10**, 265–276 (1961).

173. R. W. Hornbeck, Laminar flow in the entrance region of a pipe. *Appl. Sci. Res., Sect. A* **13**, 224–232 (1964).

174. E. B. Christiansen and H. E. Lemmon, Entrance region flow. *AIChE J.* **11**, 995–999 (1965).

175. R. Manohar, Analysis of Laminar-flow heat transfer in the entrance region of circular tubes. *Int. J. Heat Mass Transfer* **12**, 15–22 (1969).

176. S. V. Patankar and D. B. Spalding, A calculation procedure for heat, mass and momentum transfer in three-dimensional parabolic flows. *Int. J. Heat Mass Transfer* **15**, 1787–1806 (1972).

177. R. M. Curr, D. Sharma, and D. G. Tatchell, Numerical predictions of some three-dimensional boundary layers in ducts. *Comput. Methods Appl. Mech. Eng.* **1**, 143–158 (1972).

178. G. A. Carlson and R. W. Hornbeck, A numerical solution for laminar entrance flow in a square duct. *J. Appl. Mech.* **40**, 25–30 (1973).

179. G. A. Carlson, "Laminar Entrance Flow in a Square Duct," Ph.D. Thesis. Carnegie Institute of Technology, Pittsburgh, Pennsylvania, 1966.

180. R. Manohar, An exact analysis of laminar flow in the entrance region of an annular pipe. *Z. Angew. Math. Mech.* **45**, 171–176 (1965).

181. V. L. Shah and K. Farnia, Flow in the entrance of annular tubes. *Comput. Fluids* **2**, 285–294 (1974).

182. J. Liu, "Flow of a Bingham Fluid in the Entrance Region of an Annular Tube," M.S. Thesis. University of Wisconsin–Milwaukee, 1974.

183. J. E. R. Coney and M. A. I. El-Shaarawi, Developing laminar radial velocity profiles and pressure drop in the entrance region of concentric annuli. *Nucl. Sci. Eng.* **57**, 169–174 (1975).

184. E. E. Feldman, "The Numerical Solution of the Combined Thermal and Hydrodynamic Entrance Region of an Eccentric Annular Duct," Ph.D. Thesis. Mech. Eng. Dep., Carnegie-Mellon University, Pittsburgh, Pennsylvania, 1974.

185. J. A. Miller, Laminar incompressible flow in the entrance region of ducts of arbitrary cross section. *J. Eng. Power* **93**, 113–118 (1971).

186. J. S. Vrentas, J. L. Duda, and K. G. Bargeron, Effect of axial diffusion of vorticity on flow development in circular conduits: Part I. Numerical solutions. *AIChE J.* **12**, 837–844 (1966).

187. M. Friedmann, J. Gillis, and N. Liron, Laminar flow in a pipe at low and moderate Reynolds numbers. *Appl. Sci. Res.* **19**, 426–438 (1968).

188. F. W. Schmidt and B. Zeldin, Laminar flow in the inlet section of a tube. *Proc., Fluidics Intern. Flows, Dep. Mech. Eng., Pennsylvania State Univ.* Part II, pp. 211–251 (1968).

189. F. W. Schmidt and B. Zeldin, Laminar flows in inlet sections of tubes and ducts. *AIChE J.* **15**, 612–614 (1969).

190. M. M. Wendel and S. Whitaker, Remarks on the paper "Finite difference analysis of plane Poiseuille and Couette flow developments" by J. R. Bodoia and J. F. Osterle. *Appl. Sci. Res., Sect. A* **11**, 313–317 (1963).

191. Y. L. Wang and P. A. Longwell, Laminar flow in the inlet section of parallel plates. *AIChE J.* **10**, 323–329 (1964).

192. J. Gillis and A. Brandt, "The Numerical Integration of the Equations of Motion of a Viscous Fluid," Sci. Rep. No. 63–73. Air Force Eur. Off. Aerosp. Res., Rehovoth, Israel, 1964; see also, Magnetohydrodynamic flow in the inlet region of a straight channel. *Phys. Fluids* **9**, 690–699 (1966).

193. J. W. McDonald, W. E. Denny, and A. F. Mills, Numerical solutions of the Navier-Stokes equations in inlet regions. *J. Appl. Mech.* **39**, 873–878 (1972).

194. H. Morihara and R. T. Cheng, Numerical solution of the viscous flow in the entrance region of parallel plates. *J. Comput. Phys.* **11**, 550–572 (1973).

195. R. E. Fuller and M. R. Samuels, Simultaneous development of the velocity and temperature fields in the entry region of an annulus. *Chem. Eng. Prog., Symp. Ser.* No. 113, Vol. 67, pp. 71–77 (1971).

196. S. Abarbanel, S. Bennett, A. Brandt, and J. Gillis, Velocity profiles of flow at low Reynolds numbers. *J. Appl. Mech.* **37**, 2–4 (1970).

197. R.-Y. Chen, Flow in the entrance region at low Reynolds numbers. *J. Fluids Eng.* **95**, 153–158 (1973).

198. B. S. Narang and G. Krishnamoorthy, Laminar flow in the entrance region of parallel plates. *J. Appl. Mech.* **43**, 186–188 (1976).

199. M. A. Lévêque, Les lois de la transmission de chaleur par convection. *Ann. Mines, Mem., Ser. 12* **13**, 201–299, 305–362, 381–415 (1928).

200. P. M. Worsøe-Schmidt, Heat transfer in the thermal entrance region of circular tubes and annular passages with fully developed laminar flow. *Int. J. Heat Mass Transfer* **10**, 541–551 (1967).

201. R. J. Nunge, E. W. Porta, and R. Bentley, A correlation of local Nusselt numbers for laminar flow heat transfer in annuli. *Int. J. Heat Mass Transfer* **13**, 927–931 (1970).

202. J. Newman, Extension of the Lévêque solution. *J. Heat Transfer* **91**, 177–178 (1969).

203. A. Burghardt and A. Dubis, A simplified method of determining the heat transfer coefficient in laminar flow (in Polish). *Chem. Stosow., Ser. B* **7** (3), 281–303 (1970).

204. E. M. Sparrow and R. Siegel, Application of variational methods to the thermal entrance region of ducts. *Int. J. Heat Mass Transfer* **1**, 161–172 (1960).

205. L. N. Tao, Variational analyses of forced heat convection in a duct of arbitrary cross section. *Proc. Int. Heat Transfer Conf., 3rd, AIChE, New York* **1**, 56–63 (1966).

206. S. D. Savkar, On a variational formulation of a class of thermal entrance problems. *Int. J. Heat Mass Transfer* **13**, 1187–1197 (1970).

207. V. Javeri, Simultaneous development of the laminar velocity and temperature fields in a circular duct for the temperature boundary condition of the third kind. *Int. J. Heat Mass Transfer* **19**, 943–949 (1976).

208. I. N. Sadikov, Heat transfer in the entrance regions of plane and rectangular passages (in Russian). *Inzh.-Fiz. Zh.* **7** (9), 44–51 (1964).

209. I. N. Sadikov, Laminar heat transfer in a two-dimensional channel with a nonuniform temperature field at the entrance. *J. Eng. Phys. (USSR)* **8**, 192–196 (1965).

210. S. R. Montgomery and P. Wibulswas, Laminar flow heat-transfer in ducts of rectangular cross-section. *Proc. Int. Heat Transfer Conf., 3rd, AIChE, New York* **1**, 104–112 (1966).

211. S. W. Hong and A. E. Bergles, Laminar flow heat transfer in the entrance region of semicircular tubes with uniform heat flux. *Int. J. Heat Mass Transfer* **19**, 123–124 (1976).

212. R. D. Chandler, J. N. Panaia, R. B. Stevens, and G. E. Zinsmeister, The solution of steady state convection problems by the fixed random walk method. *J. Heat Transfer* **90**, 361–363 (1968).

213. A. O. Tay and G. De Vahl Davis, Application of the finite element method to convection heat transfer between parallel planes. *Int. J. Heat Mass Transfer* **14**, 1057–1069 (1971).

214. R. K. McMordie and A. F. Emery, A numerical solution for laminar-flow heat transfer in circular tubes with axial conduction and developing thermal and velocity fields. *J. Heat Transfer* **89**, 11–16 (1967).

215. S. Kakac and M. R. Özgü, Analysis of laminar flow forced convection heat transfer in the entrance region of a circular pipe. *Waerme- Stoffuebertrag.* **2**, 240–245 (1969).

216. R. W. Hornbeck, An all-numerical method for heat transfer in the inlet of a tube. *Am. Soc. Mech. Eng., Pap.* **65-WA/HT-36** (1965).

217. E. Bender, Wärmeübergang bei ausgebildeter und nicht ausgebildeter laminarer Rohrströmung mit temperaturabhängigen Stoffwerten. *Waerme- Stoffuebertrag.* **1**, 159–168 (1968).

218. C. L. Hwang and L. T. Fan, Finite difference analysis of forced-convection heat transfer in entrance region of a flat rectangular duct. *Appl. Sci. Res., Sect A* **13**, 401–422 (1964).

219. S. R. Montgomery and P. Wibulswas, Laminar flow heat transfer for simultaneously developing velocity and temperature profiles in ducts of rectangular cross section. *Appl. Sci. Res.* **18**, 247–259 (1967).

220. P. Wibulswas, "Laminar-Flow Heat-Transfer in Non-Circular Ducts," Ph.D. Thesis. London University, London, 1966.

221. G. Hagen, Über die Bewegung des Wassers in engen zylindrischen Röhren. *Pogg. Ann.* **46**, 423–442 (1839).

222. J. Poiseuille, Récherches expérimentelles sur le mouvement des liquides dans les tubes de très petits diamètres. *C. R. Acad. Sci.* **11**, 961–967, 1041–1048 (1840); **12**, 112–115 (1841); in more detail, *Mem. Savants Etrangers* **9** (1846).

223. S. Pahor and J. Strand, Die Nusseltsche Zahl für laminare Strömung im zylindrischen Rohr mit konstanter Wandtemperatur. *Z. Angew. Math. Phys.* **7**, 536–538 (1956).

224. D. A. Labuntsov, Heat emission in pipes during laminar flow of a liquid with axial heat conduction taken into account. *Sov. Phys.—Dokl.* **3**, 33–35 (1958).

225. R. L. Ash and J. H. Heinbockel, Note on heat transfer in laminar, fully developed pipe flow with axial conduction. *Z. Angew. Math. Phys.* **21**, 266–269 (1970).

226. R. L. Ash, Personal communication. Old Dominion University, Norfolk, Virginia, 1971.

227. M. L. Michelsen and J. Villadsen, The Graetz problem with axial heat conduction. *Int. J. Heat Mass Transfer* **17**, 1391–1402 (1974).

228. J. W. Ou and K. C. Cheng, Viscous dissipation effects on thermal entrance heat transfer in laminar and turbulent pipe flows with uniform wall temperature. *Am. Inst. Aeronaut. Astron. Pap.* **74-743** or *Am. Soc. Mech. Eng., Pap.* **74-HT-50** (1974).

229. J. W. Ou and K. C. Cheng, Effects of pressure work and viscous dissipation on Graetz problem for gas flows in parallel-plate channels. *Waerme- Stoffuebertrag.* **6**, 191–198 (1973).

230. S. Sideman, D. Luss, and R. E. Peck, Heat transfer in laminar flow in circular and flat conduits with (constant) surface resistance. *Appl. Sci. Res., Sect. A* **14**, 157–171 (1964).

231. A. A. McKillop, J. C. Harper, and H. J. Bader, Heat transfer in entrance–region flow with external resistance. *Int. J. Heat Mass Transfer* **14**, 863–866 (1971).

232. H. J. Hickman, An asymptotic study of the Nusselt-Graetz problem, Part 1: Large x behavior. *J. Heat Transfer* **96**, 354–358 (1974).

233. C. J. Hsu, Exact solution to entry-region laminar heat transfer with axial conduction and the boundary condition of the third kind. *Chem. Eng. Sci.* **23**, 457–468 (1968).

234. H. L. Toor, The energy equation for viscous flow. *Ind. Eng. Chem.* **48**, 922–926 (1956).

235. J. C. Chen, Laminar heat transfer in tube with nonlinear radiant heat-flux boundary condition. *Int. J. Heat Mass Transfer* **9**, 433–440 (1966).

236. Y. S. Kadaner, Y. P. Rassadkin, and É. L. Spektor, Heat transfer during laminar fluid flow in a pipe with radiative heat removal. *J. Eng. Phys. (USSR)* **20**, 20–24 (1971); also in *Heat Transfer—Sov. Res.* **3** (5), 182–188 (1971).

237. H. Glaser, Heat transfer and pressure drop in heat exchangers with laminar flow. MAP Volkenrode, Reference: **MAP-VG-96-818T** (March 1947).

238. J. Madejski, Temperature distribution in channel flow with friction. *Int. J. Heat Mass Transfer* **6**, 49–51 (1963).

239. J. W. Ou and K. C. Cheng, Viscous dissipation effects on thermal entrance region heat transfer in pipes with uniform wall heat flux. *Appl. Sci. Res.* **28**, 289–301 (1973).

240. S. Golos, When is it allowed to treat the heat-transfer coefficient α as constant? *Int. J. Heat Mass Transfer* **18**, 1467–1471 (1975).

241. W. C. Reynolds, Heat transfer to fully developed laminar flow in a circular tube with arbitrary circumferential heat flux. *J. Heat Transfer* **82**, 108–112 (1960).

242. W. B. Hall, J. D. Jackson, and P. H. Price, Note on forced convection in a pipe having heat flux which varies exponentially along its length. *J. Mech. Eng. Sci.* **5**, 48–52 (1963).

243. S. Hasegawa and Y. Fujita, Nusselt numbers for fully developed flow in a tube with exponentially varying heat flux. *Mem. Fac. Eng., Kyushu Univ.* **27** (1), 77–80 (1967).

244. H. Gräber, Heat transfer in smooth tubes, between parallel plates, in annuli and tube bundles with exponential heat flux distributions in forced laminar or turbulent flow (in German). *Int. J. Heat Mass Transfer* **13**, 1645–1703 (1970).

245. W. B. Hall and P. H. Price, The effect of a longitudinally varying wall heat flux on the heat transfer coefficient for turbulent flow in a pipe. *Int. Dev. Heat Transfer, Proc. Heat Transfer Conf., Boulder, Colorado*, Part III, pp. 607–613 (1961).

246. T. Dury, "The Development of a Smooth Circular Tube Test Section with Variable Heat Flux Distribution for Experiments on the Forced Convection Cooling of Supercritical Pressure Carbon Dioxide," M. S. Thesis. University of Manchester, England, 1970.

247. R. Manohar, Personal communications. Dep. Math., University of Saskatchewan, Saskatoon, 1970–1972.

248. S. V. Patankar and D. B. Spalding, "Heat and Mass Transfer in Boundary Layers," 2nd Ed. Intertext Books, London, 1970.

249. J. P. Burke and N. S. Berman, Entrance flow development in circular tubes at small axial distances. *Am. Soc. Mech. Eng., Pap.* **69-WA/FE-13** (1969).

250. N. S. Berman and V. A. Santos, Laminar velocity profiles in developing flows using a laser Doppler technique. *AIChE J.* **15**, 323–327 (1969).

251. N. S. Berman and J. P. Burke, Laser Doppler studies of entrance flow. *AIChE Annu. Meet., 62nd, Washington, D.C.* Pap. No. 58f (1969).

252. F. W. Schmidt, Personal communication. Mech. Eng. Dep., Pennsylvania State University, University Park, 1971.

253. H. S. Lew and Y. C. Fung, On the low-Reynolds-number entry flow into a circular cylindrical tube. *J. Biomech.* **2**, 105–119 (1969).

254. D. G. Barbee and C. D. Mikkelsen, Field descriptions for a steady, developing tube flow of vanishing Reynolds number. *Z. Angew. Math. Phys.* **24**, 73–82 (1973).

255. J. S. Vrentas and J. L. Duda, Effect of axial diffusion of vorticity on flow development in circular conduits: Part II. Analytical solution for low Reynolds numbers. *AIChE J.* **13**, 97–101 (1967).

256. B. Atkinson, C. C. H. Card, and B. M. Irons, Application of the finite element method to creeping flow problems. *Trans. Inst. Chem. Eng.* **48**, T276–T284 (1970).

257. B. Atkinson, M. P. Brocklebank, C. C. H. Card, and J. M. Smith, Low Reynolds number developing flows. *AIChE J.* **15**, 548–553 (1969).

258. E. Bender, Druckverlust bei laminarer Strömung im Rohreinlauf. *Chem.-Ing.-Tech.* **11**, 682–686 (1969).

259. R. K. Shah, A correlation for laminar hydrodynamic entry length solutions for circular and noncircular ducts. *J. Fluids Eng.* **100**, 177–179 (1978).

260. G. M. Brown, Heat or mass transfer in a fluid in laminar flow in a circular or flat conduit. *AIChE J.* **6**, 179–183 (1960).

261. M. Jakob, "Heat Transfer," Vol. 1. Wiley, New York, 1949.

262. M. Abramowitz, On solution of differential equation occurring in problem of heat convection laminar flow in tube. *J. Math. Phys.* **32**, 184–187 (1953).

263. R. P. Lipkis, "Heat Transfer to an Incompressible Fluid in Laminar Motion," M. Sc. Thesis. University of California, Los Angeles, 1954.

264. B. K. Larkin, High-order eigenfunctions of the Graetz problem. *AIChE J.* **7**, 530 (1961).

265. T. Munakata, The calculation of laminar heat transfer in a tube. *Int. Chem. Eng.* **15**, 193–196 (1975); translated from *Kogaku Kōgaku* **26**(10), 1085–1088 (1962).

266. J. Newman, The Graetz problem. *In* "The Fundamental Principles of Current Distribution and Mass Transport in Electrochemical Cells" (A. J. Bard, ed.), Vol. 6, pp. 187–352. Dekker, New York, 1973; also as UCRL-18646. Dep. Chem. Eng., University of California at Berkeley, 1969.

267. H. A. Lauwerier, The use of confluent hypergeometric functions in mathematical physics and the solution of an eigenvalue problem. *Appl. Sci. Res., Sect. A* **2**, 184–204 (1950).

268. O. Kuga, Laminar and turbulent heat transfer of liquid metal in a circular tube with non-isothermal surface (in English). *Proc. 1967 Semi-Int. Symp. Heat Transfer, Jpn. Soc. Mech. Eng.* **2**, 155–159 (1968).

269. K. Koyama, K. Kanamaru, and E. Wada, Temperature distribution in laminar flow through a circular tube. *Int. Chem. Eng.* **11**, 767–771 (1971).

270. R. K. Shah, Thermal entry length solutions for the circular tube and parallel plates. *Proc. Natl. Heat Mass Transfer Conf., 3rd, Indian Inst. Techonol.*, Bombay, Vol. I, Pap. No. HMT-11-75 (1975).

271. A. M. Mercer, The growth of the thermal boundary layer at the inlet to a circular tube. *Appl. Sci. Res., Sect. A* **9**, 450–456 (1960).

272. M. R. Doshi, The Generalized Lévêque Solution. Unpublished paper. Chem. Eng. Dep., State University of New York at Buffalo, 1976.

273. S. W. Churchill and H. Ozoe, Correlations for laminar forced convection in flow over an isothermal flat plate and in developing and fully developed flow in an isothermal tube. *J. Heat Transfer* **95**, 416–419 (1973).

274. E. Hicken, Der Einfluss einer über dem Querschnitt unterschiedlichen Anfangstemperatur auf den Wärmeübergang in laminar durchströmten Rohren. *Waerme- Stoffuebertrag.* **1**, 220–224 (1968).

275. B. C. Lyche and R. B. Bird, The Graetz-Nusselt problem for a power-law non-Newtonian fluid. *Chem. Eng. Sci.* **6**, 35–41 (1956).

276. I. R. Whiteman and W. B. Drake, Heat transfer to flow in a round tube with arbitrary velocity distribution. *Trans. ASME* **80**, 728–732 (1958).

277. H. Barrow and J. F. Humphreys, The effect of velocity distribution on forced convection laminar flow heat transfer in a pipe at constant wall temperature. *Waerme- Stoffuebertrag.* **3**, 227–231 (1970).

278. L. Topper, Forced heat convection in cylindrical channels: Some problems involving potential and parabolic velocity distribution. *Chem. Eng. Sci.* **5**, 13–19 (1956).

279. H. L. Toor, Heat transfer in forced convection with internal heat generation. *AIChE J.* **4**, 319–323 (1958).

280. K. Millsaps and K. Pohlhausen, Heat transfer to Hagen-Poiseuille flows. *In* "Proceedings of the Conference on Differential Equations" (J. B. Diaz and L. E. Payne, eds.), pp. 271–294. Univ. of Maryland Bookstore, College Park, 1956.

281. S. N. Singh, Heat transfer by laminar flow in a cylindrical tube. *Appl. Sci. Res., Sect. A* **7**, 325–340 (1958).

282. V. V. Shapovalov, Heat transfer during the flow of an incompressible fluid in a circular tube, allowing for axial heat flow, with boundary conditions of the first and second kind at the tube surface. *J. Eng. Phys. (USSR)* **11**, 153–155 (1966).

283. Y. Taitel and A. Tamir, Application of the integral method to flows with axial diffusion. *Int. J. Heat Mass Transfer* **15**, 733–740 (1972).

284. B. A. Kader, Heat and mass transfer in laminar flow in the entrance section of a circular tube. *High Temp. (USSR)* **9**, 1115–1120 (1971).

285. M. V. Bodnarescu, Beitrag zur theorie des Wärmeübergangs in laminarer Strömung. *VDI-Forschungsh.* No. 450, Vol. 21, pp. 19–27 (1955).

286. T. Bes, Convection and heat conduction in a laminar fluid flow in a duct (in English). *Bull. Acad. Pol. Sci., Ser. Sci. Tech.* **16**(1), 41–51 (1968).

287. A. S. Jones, Extensions to the solution of the Graetz problem. *Int. J. Heat Mass Transfer* **14**, 619–623 (1971).

288. F. H. Verhoff and D. P. Fisher, A numerical solution of the Graetz problem with axial conduction included. *J. Heat Transfer* **95**, 132–134 (1973).

289. C. W. Tan and C. J. Hsu, Low Péclét number mass transfer in laminar flow through circular tubes. *Int. J. Heat Mass Transfer* **15**, 2187–2201 (1972).

290. H. S. Carslaw and J. C. Jaeger, "Conduction of Heat in Solids," 2nd Ed. Oxford Univ. Press (Clarendon), London, 1947.

291. O. Kuga, Laminar flow heat transfer in circular tubes with non-isothermal surfaces (in Japanese). *Nippon Kikai Gakkai Rombunshu* **31**(222), 295–298 (1965).

292. V. V. Shapovalov, Heat transfer in laminar flow of an incompressible fluid in a round tube. *J. Eng. Phys. (USSR)* **12**, 363–364 (1967).

293. T. K. Bhattacharyya, "Heat Transfer Studies in Thermal Entry Region under Circumferentially Nonuniform Boundary Condition," Ph.D. Thesis. Dep. Mech. Eng., Indian Institute of Technology, Madras, 1973.

294. J. Schenk and J. M. DuMoré, Heat transfer in laminar flow through cylindrical tubes. *Appl. Sci. Res., Sect. A*, 39–51 (1954).

295. G. S. H. Lock, R. D. J. Freeborn, and R. H. Nyren, Analysis of ice formation in a convectively-cooled pipe. *Proc. Int. Heat Transfer Conf., 4th, Paris-Versailles* **1**(Cu 2.9), 1–11 (1970).

296. G. J. Hwang and I. Yih, Correction on the length of ice-free zone in a convectively-cooled pipe. *Int. J. Heat Mass Transfer* **16**, 681–683 (1973).

297. C. J. Hsu, Laminar flow heat transfer in circular or parallel-plate channels with internal heat generation and the boundary condition of the third kind. *J. Chin. Inst. Chem. Eng.* **2**, 85–100 (1971).

298. L. S. Dzung, The Graetz problem. *Int. J. Heat Mass Transfer* **16**, 2120 (1973).

299. E. M. Rosen and E. J. Scott, The Leveque solution with a finite wall resistance. *J. Heat Transfer* **83**, 98–100 (1961).

300. C. J. Hsu, An exact mathematical solution for entrance-region laminar heat transfer with axial conduction. *Appl. Sci. Res.* **17**, 359–376 (1967).

301. B. V. Dussan and T. F. Irvine, Jr., Laminar heat transfer in a round tube with radiating heat flux at the outer wall. *Proc. Int. Heat Transfer Conf. 3rd, AIChE, New York* **5**, 184–189 (1966).

302. V. V. Salomatov and Y. M. Puzyrev, Heat transfer in laminar flow of liquids in radiating channels. *Heat Transfer—Sov. Res.* **6**(4), 128–134 (1974).

303. S. Sikka and M. Iqbal, Laminar heat transfer in a circular tube under solar radiation in space. *Int. J. Heat Mass Transfer* **13**, 975–983 (1970).

304. B. S. Petukhov and C. Chzhen-Yun, Heat transfer in the hydrodynamic inlet region of a round tube with laminar flow. *Proc. All-Soviet Union Conf. Heat Mass Transfer, 2nd* **1** (1964); English translation, C. Gazley, Jr., J. P. Hartnett, and E. R. G. Eckert, eds., Rep. R-451-PR, Vol. 1, pp. 193–204. Rand Corp., Santa Monica, California, 1966.

305. C. J. Hsu, Heat transfer in a round tube with sinusoidal wall heat flux distribution. *AIChE J.* **11**, 690–695 (1965).

306. S. W. Churchill and H. Ozoe, Correlations for laminar forced convection with uniform

heating in flow over a plate and in developing and fully developed flow in a tube. *J. Heat Transfer* **95**, 78–84 (1973).

307. E. M. Sparrow and R. Siegel, Laminar tube flow with arbitrary internal heat sources and wall heat transfer. *Nucl. Sci. Eng.* **4**, 239–254 (1958).

308. R. M. Inman, Experimental study of temperature distribution in laminar tube flow of a fluid with internal heat generation. *Int. J. Heat Mass Transfer* **5**, 1053–1058 (1962).

309. J. C. Pirkle and V. G. Sigillito, Laminar heat transfer with axial conduction. *Appl. Sci. Res.* **26**, 108–112 (1972).

310. B. S. Petukhov and F. F. Tsvetkov, Calculation of heat transfer during laminar flow of a liquid in pipes in the region of small Peclet numbers (in Russian). *Inzh.-Fiz. Zh.* **4**(3), 10–17 (1961); also translation No. FTD TT-61-321 or 62–19284, Off. Tech. Services, Dept. Comm., Washington, D. C. (January 29, 1962).

311. C. J. Hsu, An exact analysis of low Péclet number thermal entry region heat transfer in transversely nonuniform velocity fields. *AIChE J.* **17**, 732–740 (1971).

312. E. J. Davis, Exact solutions for a class of heat and mass transfer problems. *Can. J. Chem. Eng.* **51**, 562–572 (1973).

313. C. E. Smith, M. Faghri, and J. R. Welty, On the determination of temperature distribution in laminar pipe flow with a step change in wall heat flux. *J. Heat Transfer* **97**, 137–139 (1975).

314. É. L. Spektor and Y. P. Rassadkin, Heat transfer and resistance of an incompressible liquid in the initial portion of a circular tube for various laws of heat supply. *J. Eng. Phys. (USSR)* **21**, 1313–1319 (1971).

315. V. V. Shapovalov, Heat transfer in laminar flow of incompressible fluid in a round tube. *J. Eng. Phys. (USSR)* **12**, 363–364 (1967).

316. L. S. Dzung, Heat transfer in a round duct with sinusoidal heat flux distribution. *Proc. U.N. Int. Conf. Peaceful Uses At. Energy, 2nd, Geneva* P/253, **7**, 657–670 (1958).

317. O. Kuga, Heat transfer in a pipe with non-uniform heat flux (in Japanese). *Nippon Kikai Gakkai Rombunshu* **32**(233), 83–87 (1966).

318. O. Reisman, "Experimental Study of the Laminar-Flow Heat Transfer Coefficient of a Fluid with a Prescribed Wall Heat Flux," Ph.D. Thesis. New York University, 1972.

319. T. K. Bhattacharyya and D. N. Roy, Laminar heat transfer in a round tube with variable circumferential or arbitrary wall heat flux. *Int. J. Heat Mass Transfer* **13**, 1057–1060 (1970).

320. P. Goldberg, "A Digital Computer Solution for Laminar Flow Heat Transfer in Circular Tubes," M.S. Thesis. Mech. Eng. Dep., Massachusetts Institute of Technology, Cambridge, 1958.

321. C. Tien and R. A. Pawelek, Laminar flow heat transfer in the entrance region of circular tubes. *Appl. Sci. Res., Sect. A* **13**, 317–331 (1964).

322. K. Stephan, Wärmeübergang und druckabfall bei nicht ausgebildeter Laminarströmung in Rohren und in ebenen Spalten. *Chem.-Ing.-Tech.* **31**, 773–778 (1959).

323. D. L. Ulrichson and R. A. Schmitz, Laminar-flow heat transfer in the entrance region of circular tubes. *Int. J. Heat Mass Transfer* **8**, 253–258 (1965).

324. G. J. Hwang and J.-P. Sheu, Effect of radial velocity component on laminar forced convection in entrance region of a circular tube. *Int. J. Heat Mass Transfer* **17**, 372–375 (1974).

325. G. J. Hwang, Personal communication. Dep. Power Mech. Eng., National Tsing Hua University, Hsinchu, Taiwan, 1975.

326. F. Kreith, "Principles of Heat Transfer," 2nd Ed. International Textbook Co., Scranton, Pennsylvania, 1969.

327. B. Zeldin and F. W. Schmidt, Developing flow with combined forced-free convection in an isothermal vertical tube. *J. Heat Transfer* **94**, 211–223 (1972).
328. B. Zeldin, "Developing Flow with Combined Forced-Free Convection in an Isothermal Vertical Tube," Ph.D. Thesis. Pennsylvania State University, University Park, 1969.
329. D. N. Roy, Laminar heat transfer in the inlet of a uniformly heated tube. *J. Heat Transfer* **87**, 425–426 (1965).
330. C. A. Bankston and D. M. McEligot, Turbulent and laminar heat transfer to gases with varying properties in the entry region of circular ducts. *Int. J. Heat Mass Transfer* **13**, 319–344 (1970).
331. M. W. Collins, Viscous dissipation effects on developing laminar flow in adiabatic and heated tubes. *Proc. Inst. Mech. Eng.*, **189**(15), 129–137 (1975).
332. P. A. McCuen, W. M. Kays, and W. C. Reynolds, "Heat Transfer with Laminar and Turbulent Flow between Parallel Planes with Constant and Variable Wall Temperature and Heat Flux," Rep. No. AHT-3. Dep. Mech. Eng., Stanford University, Stanford, California, 1962.
333. W. Nusselt, Der Wärmeaustausch am Berieselungskühler. *VDI Z.* **67**, 206–210 (1923).
334. R. H. Norris and D. D. Streid, Laminar-flow heat-transfer coefficients for ducts. *Trans. ASME* **62**, 525–533 (1940).
335. H. W. Hahnemann and L. Ehret, "Der Wärmeübergang in Splatsẗomungen," Bericht der Luftfahrtforschungsanstalt Braunschweig-Völkenrode. Braunschweig, 1942; see also H. W. Hahnemann, Aus der Wärmetechnischen Forschung der Gegenwart. *Waerme-Kaeltetech.* **44**, 1967–1968 (1942).
336. K. C. Cheng and R. S. Wu, Viscous dissipation effects on convective instability and heat transfer in plane Poiseuille flow heated from below. *Appl. Sci. Res.* **32**, 327–346 (1976).
337. S. Pahor and J. Strand, A note on heat transfer in laminar flow through a gap. *Appl. Sci. Res., Sect. A* **10**, 81–84 (1961).
338. C. C. Grosjean, S. Pahor, and J. Strand, Heat transfer in laminar flow through a gap. *Appl. Sci. Res., Sect. A* **11**, 292–294 (1963).
339. S. Ishizawa, On the momentum-integral method of solution for the laminar entrance-flow problems. *Bull. JSME* **10**(39), 489–496 (1967).
340. M. S. Bhatti and C. W. Savery, Heat transfer in the entrance region of a straight channel: Laminar flow with uniform wall heat flux. *Am. Soc. Mech. Eng., Pap.* **76-HT-20** (1976); also in a condensed form in *J. Heat Transfer* **99**, 142–144 (1977).
341. J. R. Bodoia, "The Finite Difference Analysis of Confined Viscous Flows," Ph.D. Thesis. Carnegie Institute of Technology, Pittsburgh, Pennsylvania, 1959.
342. M. Kiya, S. Fukusako, and M. Arie, Effect of non-uniform inlet velocity profile on the development of a laminar flow between parallel plates. *Bull. JSME* **15**(81), 324–336 (1972).
343. H. Morihara, "Numerical Integration of the Navier-Stokes Equations," Ph.D. Thesis. Mech. Eng. Dep., State University of New York at Buffalo, 1972.
344. J. A. Prins, J. Mulder, and J. Schenk, Heat transfer in laminar flow between parallel plates. *Appl. Sci. Res., Sect. A* **2**, 431–438 (1951).
345. C. S. Yih and J. E. Cermak, "Laminar Heat Convection in Pipes and Ducts," Rep. No. 5 (ONR Contract No. N90nr-82401, NR 063-071/1-19-49). Civil Eng. Dep. Colorado Agric. Mech. Coll., Fort Collins, 1951.
346. N. Froman and P. O. Froman, "JWKB Approximation, Contributions to the Theory." North-Holland Publ., Amsterdam, 1965.
347. I. J. Kumar, Recent mathematical methods in heat transfer. *Adv. Heat Transfer* **8**, 1–91 (1972).
348. A. M. Mercer, The growth of the thermal boundary layer in laminar flow between parallel flat plates. *Appl. Sci. Res., Sect. A* **8**, 357–365 (1959).

349. V. V. G. Krishnamurty and C. Venkata Rao, Heat transfer in non-circular conduits: Part II—Laminar forced convection in slits. *Indian J. Technol.* **5**, 166–167 (1967).

350. H.C. Agrawal, Heat transfer in laminar flow between parallel plates at small Péclét numbers. *Appl. Sci. Res., Sect. A* **9**, 177–189 (1960).

351. C. A. Deavours, Laminar heat transfer in parallel plate flow. *Appl. Sci. Res.* **29**, 69–76 (1974).

352. C. A. Deavours, An exact solution for the temperature distribution in parallel plate Poiseuille flow. *J. Heat Transfer* **96**, 489–495 (1974).

353. E. M. Sparrow, J. L. Novotny, and S. H. Lin, Laminar flow of a heat-generating fluid in a parallel-plate channel. *AIChE J.* **9**, 797–804 (1963).

354. J. Klein and M. Tribus, Forced convection from nonisothermal surfaces. *Am. Soc. Mech. Eng., Pap.* **53-SA-46** (1953).

355. R. D. Cess and E. C. Shaffer, Summary of laminar heat transfer between parallel plates with unsymmetrical wall temperatures. *J. Aerosp. Sci.* **26**, 538 (1959).

356. A. P. Hatton and J. S. Turton, Heat transfer in the thermal entry length with laminar flow between parallel walls at unequal temperatures. *Int. J. Heat Mass Transfer* **5**, 673–679 (1962).

357. G. J. Hwang and K. C. Cheng, Convective instability in the thermal entrance region of a horizontal parallel-plate channel heated from below. *J. Heat Transfer* **95**, 72–77 (1973).

358. R. S. Wu, K. C. Cheng, and J. W. Ou, Low Peclet number heat transfer in the thermal entrance region of parallel-plate channels with unequal wall temperatures. *Can J. Chem. Eng.* **54**, 526–531 (1976).

359. M. S. Povarnitsyn and E. V. Yurlova, Calculation of the temperature field in a plane channel with nonuniform heating of thermally conducting walls. *J. Eng. Phys. (USSR)* **10**, 82–85 (1966).

360. R. D. Cess and E. C. Shaffer, Heat transfer to laminar flow between parallel plates with a prescribed wall heat flux. *Appl. Sci. Res., Sect. A* **8**, 339–344 (1959).

361. A. S. Jones, Two-dimensional adiabatic forced convection at low Péclet number. *Appl. Sci. Res.* **25**, 337–348 (1972).

362. C. L. Hwang, P. J. Knieper, and L. T. Fan, Effects of viscous dissipation on heat transfer parameters for flow between parallel plates. *Z. Angew. Math. Phys.* **16**, 599–610 (1965).

363. R. D. Cess and E. C. Shaffer, Laminar heat transfer between parallel plates with an unsymmetrically prescribed heat flux at the walls. *Appl. Sci. Res., Sect. A* **9**, 64–70 (1960).

364. L. S. Dzung, Heat transfer in a flat duct with sinusoidal heat flux distribution. *Proc. U.N. Int. Conf. Peaceful Uses At. Energy, 2nd, Geneva* P/254, **7**, 671–675 (1958).

365. J. A. W. van der Does de Bye and J. Schenk, Heat transfer in laminar flow between parallel plates. *Appl. Sci. Res., Sect. A* **3**, 308–316 (1952).

366. V. J. Berry, Jr., Non-uniform heat transfer to fluids flowing in conduits. *Appl. Sci. Res., Sect. A* **4**, 61–75 (1953).

367. J. Schenk, Heat loss of fluids flowing through conduits. *Appl. Sci. Res., Sect. A* **4**, 222–224 (1953).

368. J. Schenk and H. L. Beckers, Heat transfer in laminar flow between parallel plates. *Appl. Sci. Res., Sect. A* **4**, 405–413 (1954).

369. S. C. R. Dennis and G. Poots, An approximate treatment of forced heat convection in laminar flow between parallel plates. *Appl. Sci. Res., Sect. A* **5**, 453–457 (1956).

370. W. D. Lakin, On higher eigenvalues of the extended Graetz problem. *Z. Angew. Math. Phys.* **23**, 484–488 (1972).

371. J. Schenk, A problem of heat transfer in laminar flow between parallel plates. *Appl. Sci. Res., Sect. A* **5**, 241–244 (1955).

372. E. M. Sparrow, Analysis of laminar forced-convection heat transfer in entrance region of flat rectangular ducts. *NACA Tech. Notes* **TN 3331** (1955).

373. N. A. Slezkin, On the development of the flow of a viscous heat-conducting gas in a pipe. *J. Appl. Math. Mech.* **23**, 473–489 (1959).
374. K. Murakawa, Theoretical solutions of heat transfer in the hydrodynamic entrance length of double pipes. *Bull. JSME* **3** (11), 340–345 (1960).
375. C. L. Hwang, Personal communication. Dep. Ind. Eng., Kansas State University, Manhattan, 1973.
376. W. E. Mercer, W. M. Pearce, and J. E. Hitchcock, Laminar forced convection in the entrance region between parallel flat plates. *J. Heat Transfer* **89**, 251–257 (1967).
377. J. A. Miller and D. D. Lundberg, Laminar convective heat transfer in the entrance region bounded by parallel flat plates at constant temperature. *Am. Soc. Mech. Eng., Pap.* **67-HT-48** (1967); condensed from TR No. 54. U.S. Naval Postgrad. Sch., Monterey, Calif., 1965.
378. R. Siegel and E. M. Sparrow, Simultaneous development of velocity and temperature distributions in a flat duct with uniform wall heating. *AIChE J.* **5**, 73–75 (1959).
379. J. A. Miller, Discussion of Han [155]. *Int. Dev. Heat Transfer, Proc. Heat Transfer Conf., Boulder Colorado*, Part VI. pp. D/204-D/206 (1961).
380. Y. A. Gavrilov and G. N. Dul'nev, Rough estimate of the heat transfer coefficient in laminar and in turbulent flow through flat channels. *J. Eng. Phys.* (*USSR*) **23**, 1228–1231 (1972).
381. K. Stephan, Wärmeübertragung laminar strömender Stoffe in einseitig beheizten oder gekühlten ebenen Kanälen. *Chem.-Ing.-Tech.* **32**, 401–404 (1960).
382. G. Lombardi and E. M. Sparrow, Measurements of local transfer coefficients for developing laminar flow in flat rectangular ducts. *Int. J. Heat Mass Transfer* **17**, 1135–1140 (1974).
383. D. B. Holmes and J. R. Vermeulen, Velocity profiles in ducts with rectangular cross sections. *Chem. Eng. Sci.* **23**, 717–722 (1968).
384. G. F. Muchnik, S. D. Solomonov, and A. R. Gordon, Hydrodynamic development of a laminar velocity field in rectangular channels. *J. Eng. Phys.* (*USSR*) **25**, 1268–1271 (1973).
385. N. M. Natarajan and S. M. Lakshmanan, Laminar flow in rectangular ducts: Prediction of velocity profiles and friction factor. *Indian J. Technol.* **10**, 435–438 (1972).
386. E. R. G. Eckert and T. F. Irvine, Jr., Incompressible friction factor, transition and hydrodynamic entrance-length studies of ducts with triangular and rectangular cross sections. *Proc. Midwest. Conf. Fluid Mech., 5th* pp. 122–145 (1957).
387. T. F. Irvine, Jr. and E. R. G. Eckert, Comparison of experimental information and analytical prediction for laminar entrance pressure drop in ducts with rectangular and triangular cross sections. *J. Appl. Mech.* **25**, 288–290 (1958).
388. G. S. Beavers, E. M. Sparrow, and R. A. Magnuson, Experiments on hydrodynamically developing flow in rectangular ducts of arbitrary aspect ratio. *Int. J. Heat Mass Transfer* **13**, 689–702 (1970).
389. R. R. Rothfus, R. I. Kermode, and J. H. Hackworth, Pressure drop in rectangular ducts. *Chem. Eng.* (*New York*) **71** (Dec. 7), 175–176 (1964).
390. M. A. Tirunarayanan and A. Ramachandran, Correlation of isothermal pressure drop in rectangular ducts. *Proc. Australas. Conf. Hydraul. Fluid Mech., 2nd, University of Auckland* pp. A213–230 (1965).
391. M. A. Tirunarayanan and A. Ramachandran, Frictional pressure drop in laminar and turbulent flow in isosceles-triangular ducts. *Am. Soc. Mech. Eng., Pap.* **67-FE-18** (1967).
392. N. M. Natarajan and S. M. Lakshmanan, Analytical method for the determination of the pressure drop in rectangular ducts. *Indian Chem. Eng.* **12** (2), 68–69 (1970).
393. J. B. Miles and J. S. Shih, Reconsideration of Nusselt Number for Laminar Fully De-

veloped Flow in Rectangular Ducts. Unpublished paper. Mech. Eng. Dep., University of Missouri, Columbia, 1967.

394. N. P. Ikryannikov, Temperature distribution in laminar flow of an incompressible fluid in a rectangular channel allowing for energy dissipation. *J. Eng. Phys. (USSR)* **10**, 180–182 (1966).

395. J. M. Savino and R. Siegel, Laminar forced convection in rectangular channels with unequal heat addition on adjacent sides. *Int. J. Heat Mass Transfer* **7**, 733–741 (1964).

396. H. M. Cheng, "Analytical Investigation of Fully Developed Laminar-Flow Forced Convection Heat Transfer in Rectangular Ducts with Uniform Heat Flux," M.S. Thesis. Mech. Eng. Dep., Massachusetts Institute of Technology, Cambridge, 1957.

397. N. P. Ikryannikov, Temperature distribution in laminar flow of an incompressible fluid, flowing in a rectangular channel with boundary conditions of the second kind. *J. Eng. Phys. (USSR)* **16**, 21–26 (1969).

398. L. S. Han, Laminar heat transfer in rectangular channels. *J. Heat Transfer* **81**, 121–128 (1959).

399. R. Siegel and J. M. Savino, An analytical solution of the effect of peripheral wall conduction on laminar forced convection in rectangular channels. *J. Heat Transfer* **87**, 59–66 (1965).

400. J. M. Savino and R. Siegel, Extension of an analysis of peripheral wall conduction effects for laminar forced convection in thin-walled rectangular channels. *NASA Tech. Note TN D-2860* (1965).

401. R. J. Goldstein and D. K. Kreid, Measurement of laminar flow development in a square duct using a laser-Doppler flowmeter. *J. Appl. Mech.* **34**, 813–818 (1967).

402. L. S. Caretto, R. M. Curr, and D. B. Spalding, Two numerical methods for three-dimensional boundary layers. *Comput. Methods Appl. Mech. Eng.* **1**, 39–57 (1972).

403. S. G. Rubin, P. K. Khosla, and S. Saari, Laminar flow in rectangular channels. Part I: Entry analysis; Part II: Numerical solution for a square channel. *In* "Numerical/Laboratory Computer Methods in Fluid Mechanics" (A. A. Pouring, ed.), pp. 29–51. Am. Soc. Mech. Eng., New York, 1976.

404. P. N. Godbole, Creeping flow in rectangular ducts by finite element method. *Int. J. Numer. Methods Eng.* **9**, 727–731 (1975).

405. S. C. R. Dennis, A. M. Mercer, and G. Poots, Forced heat convection in laminar flow through rectangular ducts. *Q. Appl. Math.* **17**, 285–297 (1959).

406. R. W. Lyczkowski, Discussion of Montgomery and Wibulswas [210]. *Proc. Int. Heat Transfer Conf., 3rd, AIChE, New York* **6**, 73–74 (1967).

407. A. R. Chandrupatla and V. M. K. Sastri, Laminar forced convection heat transfer of a non-Newtonian fluid in a square duct. *Int. J. Heat Mass Transfer* **20**, 1315–1324 (1977).

407a. A. R. Chandrupatla, "Analytical and Experimental Studies of Flow and Heat Transfer Characteristics of a Non-Newtonian Fluid in a Square Duct," Ph.D. Thesis, Indian Institute of Technology, Madras, 1977. Results to be published in *J. Numer. Heat Transfer* and *Proc. Int. Heat Transfer Conf., 6th,* Toronto (1978).

408. V. V. G. Krishnamurty and N. V. Sambasiva Rao, Heat transfer in non-circular conduits: Part IV–Laminar forced convection in rectangular channels. *Indian J. Technol.* **5**, 331–333 (1967).

409. J. R. DeWitt and W. T. Snyder, "Thermal Entrance Region Heat Transfer for Rectangular Ducts of Various Aspect Ratios and Péclet Numbers," AD 693564. Clearinghouse for Federal Scientific and Technical Information, Springfield, Virginia, 1969.

410. C. Chiranjivi and S. Parabrahmachary, Experimental heat transfer studies in ducts of rectangular cross-section. *Indian J. Technol.* **10**, 296–298 (1972).

411. K. R. Perkins, K. W. Shade, and D. M. McEligot, Heated laminarizing gas flow in a square duct. *Int. J. Heat Mass Transfer* **16**, 897–916 (1973).

412. E. Hicken, Das Temperaturfeld in laminar durchströmten Kanälen mit Rechteck-Querschnitt bei unterschiedlicher Beheizung der Kanalwände. *Waerme- Stoffuebertrag.* **1**, 98–104 (1968).

413. E. Hicken, "Wärmeübergang bei Ausgebildeter Laminarer Kanalströmung für am Umfang Veränderliche Randbedingungen," Dissertation. Technische Hochschule Braunschweig, Braunschweig, 1966.

414. S. S. Kutateladze, "Fundamentals of Heat Transfer." Academic Press, New York, 1963.

415. V. P. Tyagi, A forced convective heat transfer including dissipation function and compression work for a class of noncircular ducts. *Int. J. Heat Mass Transfer* **15**, 164–169 (1972).

416. P.-C. Lu and R. W. Miller, Some heat-transfer problems inside an equilateral triangular region. *Am. Soc. Mech. Eng., Pap.* **67-HT-67** (1967).

417. H. Nuttall, Flow of a viscous incompressible fluid—Expressions for a uniform triangular duct. *Engineering (London)* **178**, 298–300 (1954).

418. V. K. Migay, Hydraulic resistance of triangular channels in laminar flow (in Russian). *Izv. Vyssh. Uchebn. Zaved., Energ.* **6** (5), 122–124 (1963).

419. E. R. G. Eckert and T. F. Irvine, Jr., Flow in corners of passages with noncircular cross sections. *Trans. ASME* **78**, 709–718 (1956).

420. C. Chiranjivi, Prediction of friction factors for laminar flow in equilateral, right isosceles and isosceles triangular channels. *Indian J. Technol.* **11**, 202–205 (1973).

421. L. W. Carlson and T. F. Irvine, Jr., Fully developed pressure drop in triangular shaped ducts. *J. Heat Transfer* **83**, 441–444 (1961).

422. E. R. G. Eckert and T. F. Irvine, Jr., Pressure drop and heat transfer in a duct with triangular cross section. *J. Heat. Transfer* **82**, 125–138 (1960).

423. E. M. Sparrow and A. Haji-Sheikh, Laminar heat transfer and pressure drop in isosceles triangular, right triangular and circular sector ducts. *J. Heat Transfer* **87**, 426–427 (1965).

424. H. Nakamura, Personal communication. Daido Institute of Technology, 2–21 Daido-Cho, Minami-Ku, Nagoya, 1972.

425. Z. V. Semilet, Laminar heat transfer and pressure drop for gas flow in triangular ducts. *Heat Transfer—Sov. Res.* **2** (1), 100–105 (1970).

426. M. K. Gangal, "Some Problems in Channel Flow," Ph.D. Thesis. Dep. Math. Stat., University of Calgary, Calgary, 1974.

427. R. W. Miller, Personal communication. Atkins & Merrill, Inc., Main Street, Ashland, Massachusetts, 1973.

428. V. V. G. Krishnamurty, Heat transfer in non-circular conduits: Part III—Laminar forced convection in equilateral triangular ducts. *Indian J. Technol.* **5**, 167–168 (1967).

429. M. L. Narayan Rao, "Study of Convective Heat Transfer without Phase Change in Channels of Triangular Cross-Section being Heated from Only One Side and Pressure Drop Studies in Short Passages of Triangular Cross-Section," Ph.D. Thesis. Andhra University, Waltair, India, 1966.

430. C. Chiranjivi and K. Balakameswar, Heat transfer with and without phase change in triangular channels. *Indian J. Technol.* **8**, 259–262 (1970).

431. N. G. Lebed' and I. V. Lobov, An investigation of hydrodynamic and heat transfer processes in a system of short triangular channels. *Heat Transfer—Sov. Res.* **5** (2), 106–109 (1973).

432. C. Chiranjivi and A. Ravi Prasad, Study of laminar flow friction in elliptical conduits. *Indian J. Technol.* **12**, 87–90 (1974).

433. S. Someswara Rao, D. Changal Raju, and M. V. Ramana Rao, Pressure drop studies in elliptical ducts. *Indian J. Technol.* **13**, 6–11 (1975).
434. N. T. Dunwoody, Thermal results for forced heat convection through elliptical ducts. *J. Appl. Mech.* **29**, 165–170 (1962).
435. J. Schenk and B. S. Han, Heat transfer from laminar flow in ducts with elliptical cross-section. *Appl. Sci. Res.* **17**, 96–114 (1967).
436. S. Someswara Rao, N. C. Pattabhi Ramacharyulu, and V. V. G. Krishnamurty, Laminar forced convection in elliptical ducts. *Appl. Sci. Res.* **21**, 185–193 (1969).
437. C. Chiranjivi and P. S. Sankara Rao, Laminar and turbulent forced convection heat transfer in a symmetric trapezoidal channel. *Indian J. Technol.* **9**, 416–420 (1971).
438. P. S. Sankara Rao, "Study of Laminar Forced Convective Heat Transfer and Pressure Drop in Channels of Trapezoidal Cross-Section," Ph.D. Thesis. Andhra University, Waltair, India, 1976.
439. C. J. Hsu, Laminar heat transfer in a hexagonal channel with internal heat generation and unequally heated sides. *Nucl. Sci. Eng.* **26**, 305–318 (1966).
440. C. Chiranjivi and V. Vidyanidhi, Heat transfer in wedge-shaped ducts. *Indian Chem. Eng.* **15**, 49–51 (1973).
441. A. Haji-Sheikh, Personal communication. Dep. Mech. Eng., University of Texas at Arlington, Arlington, 1970.
442. M. H. Hu and Y. P. Chang, Optimization of finned tubes for heat transfer in laminar flow. *J. Heat Transfer* **95**, 332–338 (1973).
443. M. H. Hu, Personal communication. Westinghouse Electric Corporation, Power Systems, Tampa Division, Box 19218, Tampa, Florida, 1975.
444. R. L. Wendt, Personal communication. System Development Corporation, Huntsville, Alabama, 1976.
445. S. S. Subrahmanyam, "Study of Laminar-Flow Forced Convection in Channels of Circular Segmental Cross-Section and Pressure-Drop Studies in Short Passages of Circular Segmental Cross-Section," Ph.D. Thesis. Andhra University, Waltair, India, 1968.
446. S. W. Hong and A. E. Bergles, "Augmentation of Laminar Flow Heat Transfer in Tubes by Means of Twisted-Tape Inserts," Tech. Rep. HTL-5, ISU-ERI-Ames-75011. Eng. Res. Inst., Iowa State University, Ames, 1974.
447. E. M. Sparrow, T. S. Chen, and V. K. Jonsson, Laminar flow and pressure drop in internally finned annular ducts. *Int. J. Heat Mass Transfer* **7**, 583–585 (1964).
448. V. V. G. Krishnamurty, N.-C. Pattabhi Ramacharyulu, S. Someswara Rao, P. S. Murti, and P. J. Reddy, Laminar forced convection heat transfer in truncated sectorial ducts. *Proc. Natl. Heat Mass Transfer Conf., 1st, Indian Inst. Technol., Madras* Pap. No. HMT-73-71, pp. VII/29–VII/40 (1971).
449. M. Jamil, "Laminar Forced Convection in Noncircular Ducts," M.S. Thesis. Mech. Eng. Dep., University of Alberta, Edmonton, 1967
450. R. A. Leonard and R. Lemlich, Laminar longitudinal flow between close-packed cylinders. *Chem. Eng. Sci.* **20**, 790–791 (1965).
451. D. J. Gunn and C. W. W. Darling, Fluid flow and energy losses in non-circular conduits. *Trans. Inst. Chem. Eng.* **41**, 163–173 (1963).
452. W. C. Reynolds, R. E. Lundberg, and P. A. McCuen, Heat transfer in annular passages. General formulation of the problem for arbitrarily prescribed wall temperatures or heat fluxes. *Int. J. Heat Mass Transfer* **6**, 483–493 (1963).
453. W. Wien, "Lehrbuch der Hydrodynamik," p. 274. Hirzel, Leipzig, 1900.
454. H. Lamb, "Hydrodynamics," 6th Ed. Dover, New York, 1932.

455. N. M. Natarajan and S. M. Lakshmanan, Laminar flow through annuli: Analytical method for calculation of pressure drop. *Indian Chem. Eng.* **15** (3), 50–53 (1973).

456. W. Tiedt, Berechnung des laminaren und turbulenten Reibungswiderstandes konzentrischer und exzentrisher Ringspalte. Part I. *Chem.-Ztg.*, *Chem. Appar.* **90**, 813–821 (1966); Part II. *Chem.-Ztg.*, *Chem. Appar.* **91**, 17–25 (1967); also as Tech. Ber. 4. Inst. Hydraul. Hydrol., Technische Hochschule, Darmstadt (1968); English translation— Transl. Bur. No. 0151, 248 pp., Transp. Dev. Agency Libr., Montreal, 1971.

457. M. Jakob and K. A. Rees, Heat transfer to a fluid in laminar flow through an annular space. *Trans. Am. Inst. Chem. Eng.* **37**, 619–648 (1941).

458. K. Murakawa, Analysis of temperature distribution of nonisothermal laminar flow of pipes with annular space. *Nippon Kikai Gakkai Rombunshu* **18**(67), 43 (1952).

459. K. Murakawa, Heat transmission in laminar flow through pipes with annular space. *Nippon Kikai Gakkai Rombunshu* **88**(19), 15 (1953).

460. O. E. Dwyer, On the transfer of heat to fluids flowing through pipes, annuli, and parallel plates. *Nucl. Sci. Eng.* **17**, 336–344 (1963).

461. E. S. Davis, Heat transfer and pressure drop in annuli. *Trans. ASME* **65**, 755–760 (1943).

462. C. Y. Chen, G. A. Hawkins, and H. L. Solberg, Heat transfer in annuli. *Trans. ASME* **68**, 99–106 (1946).

463. O. E. Dwyer, Bilateral heat transfer in annuli for slug and laminar flows. *Nucl. Sci. Eng.* **19**, 48–57 (1964).

464. L. I. Urbanovich, Temperature distribution and heat transfer in a laminar incompressible annular-channel flow with energy dissipation. *J. Eng. Phys.* (*USSR*) **14**, 402–403 (1968).

465. L. I. Urbanovich, The transfer of heat in the laminar flow of an incompressible liquid in an annular channel with nonsymmetric boundary conditions of the second kind relative to the axis of the flow. *J. Eng. Phys.* (*USSR*) **15**, 753–754 (1968).

466. K. Murakawa, Heat transfer in entry length of double pipes. *Int. J. Heat Mass Transfer* **2**, 240–251 (1961).

467. R. Viskanta, "Heat Transfer with Laminar Flow in Concentric Annuli with Constant and Arbitrary Variable Axial Wall Temperature," Rep. ANL-6441. Argonne National Laboratory, Argonne, Illinois, 1961.

468. R. Viskanta, Heat transfer with laminar flow in a concentric annulus with prescribed wall temperatures. *Appl. Sci. Res.*, *Sect. A* **12**, 463–476 (1964).

469. A. P. Hatton and A. Quarmby, Heat transfer in the thermal entry length with laminar flow in an annulus. *Int. J. Heat Mass Transfer* **5**, 973–980 (1962).

470. C. J. Hsu, Theoretical solutions for low-Péclét-number thermal entry region heat transfer in laminar flow through concentric annuli. *Int. J. Heat Mass Transfer* **13**, 1907–1924 (1970).

471. A. J. Ziegenhagen, Approximate eigenvalues for heat transfer to laminar or turbulent flow in an annulus. *Int. J. Heat Mass Transfer* **8**, 499–505 (1965).

472. C. Venkata Rao, C. Syamala Rao, and V. V. G. Krishnamurty, Heat transfer in noncircular conduits: Part I—Laminar forced convection in annuli. *Indian J. Technol.* **5**, 164–166 (1967).

473. T. K. Bhattacharyya, M. V. Krishnamurthy, and A. Ramachandran, Heat transfer study in thermal entry region for laminar flow through a concentric annulus having arbitrary wall temperature. *Proc. Natl. Heat Mass Transfer Conf.*, *1st*, *Indian Inst. Technol.*, *Madras* Part VII, pp. 17–27 (1971).

474. C. J. Hsu and C. J. Huang, Heat or mass transfer in laminar flow through a concentric annulus with convective flux at walls. *Chem. Eng. Sci.* **21**, 209–221 (1966).

475. R. W. Shumway, "Variable Properties Laminar Gas Flow Heat Transfer and Pressure Drop in Annuli," Ph.D. Thesis. Aerosp. Mech. Eng. Dep., University of Arizona,

Tucson, 1969; also DDC-AD-696-458. National Technical Information Service, 5285 Port Royal Road, Springfield, Virginia 22161.

476. H. M. Macdonald, On the torsional strength of a hollow shaft. *Proc. Cambridge Philos. Soc.* **8**, 62–68 (1893).

477. A. C. Stevenson, The centre of flexure of a hollow shaft. *Proc. London Math. Soc., Ser. 2* **50**, 536 (1949).

478. E. Becker, Strömungsvorgänge in ringförmigen Spalten und ihre Beziehung zum Poiseuilleschen Gesetz. *Forsch. Geb. Ingenieurwes., VDI* No. 48 (1907).

479. J. F. Heyda, A Green's function solution for the case of laminar incompressible flow between non-concentric circular cylinders. *J. Franklin Inst.* **267**, 25–34 (1959).

480. P. J. Redberger and M. E. Charles, Axial laminar flow in a circular pipe containing a fixed eccentric core. *Can. J. Chem. Eng.* **40**, 148–151 (1962).

481. W. T. Snyder and G. A. Goldstein, An analysis of fully developed laminar flow in an eccentric annulus. *AIChE J.* **11**, 462–467 (1965).

482. V. K. Jonsson and E. M. Sparrow, Results of laminar flow analysis and turbulent flow experiments for eccentric annular ducts. *AIChE J.* **11**, 1143–1145 (1965).

483. D. E. Bourne, O. Figueiredo, and M. E. Charles, Laminar and turbulent flow in annuli of unit eccentricity. *Can. J. Chem. Eng.* **46**, 289–293 (1968).

484. W. Tiedt, A review of Bourne *et al.* [483]. *Appl. Mech. Rev.* **22**, 900–901 (1969).

485. U. A. Sastry, Heat transfer by laminar forced convection in a pipe bounded by eccentric circles. *J. Aeronaut. Soc. India* **16**, 119–122 (1964).

486. M. L. Trombetta, Personal communication. American Cyanamid Company, Wayne, New Jersey, 1972.

487. E. E. Feldman, Personal communication. Argonne National Laboratory, Argonne, Illinois, 1975.

488. V. D. Vilenskii, Y. V. Mironov, and V. P. Smirnov, Numerical solution of the problem of heat transfer in an annular channel. *High Temp. (USSR)* **9**, 699–704 (1971).

489. P. N. Shivakumar, Viscous flow in pipes whose cross-sections are doubly connected regions. *Appl. Sci. Res.* **27**, 355–365 (1973).

490. H. C. Topakoglu and O. A. Arnas, Convective heat transfer for steady laminar flow between two confocal elliptical pipes with longitudinal uniform wall temperature gradient. *Int. J. Heat Mass Transfer* **17**, 1487–1498 (1974).

491. F. A. Gaydon and H. Nuttall, Viscous flow through tubes of multiply connected cross sections. *J. Appl. Mech.* **26**, 573–576 (1959).

492. D. A. Ratkowsky, Personal communication. Tasmanian Regional Laboratory, CSIRO, Hobart, Tasmania, Australia, 1970.

493. B. D. Bowen, "Laminar Flow in Unusual-Shaped Ducts," B.A.Sc. Thesis, University of British Columbia, Vancouver, 1967.

494. P. N. Shivakumar, Personal communication. Dep. Appl. Math., University of Manitoba, Winnipeg, 1975.

495. R. A. Axford, Summary of theoretical aspects of heat transfer performance in clustered rod geometries. *In* "Heat Transfer in Rod Bundles," pp. 70–103. Am. Soc. Mech. Eng., New York, 1968.

496. A. A. Sholokhov, N. I. Buleev, Y. I. Gribanov, and V. E. Minashin, The longitudinal laminar flow of a liquid in a bundle of rods. *J. Eng. Phys. (USSR)* **14**, 195–199 (1968).

497. J. Schmid, Longitudinal laminar flow in an array of circular cylinders. *Int. J. Heat Mass Transfer* **9**, 925–937 (1966).

498. K. Rehme, "Laminarströmung in Stabbündeln," pp. 130–133. Tagungsbericht Reaktortagung, Bonn, 1971.

499. K. Rehme, Laminarströmung in Stabbündeln. *Chem.-Ing.-Tech.* **43**, 962–966 (1971).

500. R. A. Axford, "Multiregion Analysis of Temperature Fields and Heat Fluxes in Tube Bundles with Internal, Solid, Nuclear Heat Sources," Rep. No. LA-3167, 82 pp. Los Alamos Sci. Lab. Univ. California, Los Alamos, New Mexico, 1964.

501. R. A. Axford, Two-dimensional, multiregion analysis of temperature fields in reactor tube bundles. *Nucl. Eng. Des.* **6**, 25–42 (1967).

502. K. Rehme, Personal communication. Institut für Neutronenphysik und Reaktortechnik, Kernforschungszentrum Karlsruhe, Karlsruhe, 1973.

503. L. R. Galloway and N. Epstein, Longitudinal flow between cylinders in square and triangular arrays and in a tube with square-edged entrance. *AIChE–Inst. Chem. Eng. Symp. Ser.* No. 6, pp. 4–15, London (1965).

504. R. Mottaghian and L. Wolf, A two-dimensional analysis of laminar fluid flow in rod bundles of arbitrary arrangement. *Int. J. Heat Mass Transfer* **17**, 1121–1128 (1974).

505. R. A. Axford, "Longitudinal Laminar Flow of an Imcompressible Fluid in Finite Tube Bundles with m + 1 Tubes," Rep. No. LA-3418. Los Alamos Sci. Lab. Univ. California, Los Alamos, New Mexico, 1965.

506. T. C. Min, H. W. Hoffman, T. C. Tucker, and F. N. Peebles, An analysis of axial flow through a circular channel containing rod clusters. *In* "Developments in Theoretical and Applied Mechanics" (W. A. Shaw, ed.), Vol. 3, pp. 667–690. Pergamon, New York, 1967.

507. T. C. Min, "Hydrodynamics of Single-Phase Laminar, Single-Phase Turbulent, and Two-Phase Annular Flow through a Multirod Circular Channel," Ph.D. Thesis. University of Tennessee, Oak Ridge, 1969.

508. K. W. Chen, "A Two-Dimensional Analysis of Heat Transfer and Fluid Flow in a Rod Cluster with Special Attention to Asymmetry," Ph.D. Thesis. Nucl. Eng. Dep., Purdue University, Lafayette, Indiana, 1970.

509. J. P. Zarling, "Analytical Investigation of Laminar Forced Convective Heat Transfer in a Finite Rod Bundle," Ph.D. Thesis. Michigan Technological University, Rochester, Michigan, 1971.

510. R. Mottaghian and L. Wolf, Fully developed laminar flow in finite rod bundles of arbitrary arrangement. *Trans. Am. Nucl. Soc.* **15**, 876 (1972).

511. J. P. Zarling, Personal communication. Mech. Eng. Dep., University of Alaska, College, 1975.

512. R. Mottaghian, "Analytische Lösungen für die Laminaren Geschwindigkeitsfelder in Längsangeströmten, Endlichen Stabbündeln mit Beliebiger Anordnung," Doktor-Ingenieur Dissertation. Institut für Kerntechnik, Technische Universität Berlin, Berlin, 1973.

513. C. J. Hsu, The effect of lateral rod displacement on laminar-flow transfer. *J. Heat Transfer* **94**, 169–173 (1972).

514. J. P. Zarling and T. C. Min, An analysis of heat transfer in the thermal entrance region to fluids with high Peclet moduli for axial flow through a circular shell containing tube banks. *Am. Soc. Mech. Eng., Pap.* **73-HT-29** (1973).

515. J. P. Zarling and T. C. Min, Forced convective heat transfer in the thermal entrance region of a finite uniform-heat-flux rod bondle. *Proc. Int. Heat Transfer Conf., 3rd, AIChE, New York* **2**, 203–207 (1974).

516. J. H. Masliyah, Personal communication. Dep. Chem. Chem. Eng., University of Saskatchewan, Saskatoon, 1975.

517. L. C. Kun, Personal communication. Linde Division, Union Carbide Corporation, Tonawanda, New York, 1975.

518. S. W. Hong and A. E. Bergles, Augmentation of laminar flow heat transfer in tubes by means of twisted-tape inserts. *J. Heat Transfer* **98**, 251–256 (1976).

519. A. W. Date, Personal communication. Dep. Mech. Eng., Indian Institute of Technology, Bombay, 1975.

520. A. W. Date, Numerical prediction of laminar flow pressure drop and heat transfer in the entrance length of a tube containing a twisted-tape. *Proc. Natl. Heat Mass Transfer Conf., 2nd, Indian Inst. Technol., Kanpur, India* Part C, pp. 19–26 (1973).

521. A. E. Bergles, G. H. Junkhan, and R. L. Bunn, "Performance Criteria for Cooling Systems on Agricultural and Industrial Machines," Tech. Rep. HTL-6, ISU-ERI-Ames-74267. Eng. Res. Inst., Iowa State University, Ames, 1974.

522. R. K. Shah and A. L. London, Theoretical effective friction factors and Nusselt numbers for laminar flow through heat exchangers with nonidentical passages in parallel. To be published.

523. A. L. London, Laminar flow gas turbine regenerators—the influence of manufacturing tolerances. *J. Eng. Power* **92**, 46–56 (1970).

524. B. Metais and E. R. G. Eckert, Forced, mixed and free convection regimes. *J. Heat Transfer* **86**, 295–296 (1964).

525. A. R. Brown and M. A. Thomas, Combined free and forced convection heat transfer for laminar flow in horizontal tubes. *J. Mech. Eng. Sci.* **7**, 440–448 (1965).

526. O. C. Eubank and W. S. Proctor, "Effect of Natural Convection on Heat Transfer with Laminar Flow in Tubes," M.S. Thesis. Chem. Eng. Dep., Massachusetts Institute of Technology, Cambridge, 1951.

527. D. R. Oliver, The effect of natural convection on viscous-flow heat transfer in horizontal tubes. *Chem. Eng. Sci.* **17**, 335–350 (1962).

528. S. M. Morcos and A. E. Bergles, Experimental investigation of combined forced and free laminar convection in horizontal tubes. *J. Heat Transfer* **97**, 212–219 (1975).

529. A. E. Bergles and R. R. Simonds, Combined forced and free convection for laminar flow in horizontal tubes with uniform heat flux. *Int. J. Heat Mass Transfer* **14**, 1989–2000 (1971).

530. Y. Mori, K. Futagami, S. Tokuda, and M. Nakamura, Forced convective heat transfer in uniformly heated horizontal tubes: 1st report—experimental study on the effect of buoyancy. *Int. J. Heat Mass Transfer* **9**, 453–463 (1966).

531. R. C. Martinelli and L. M. K. Boelter, The analytical prediction of superposed free and forced viscous convection in a vertical pipe. *Univ. Calif., Berkeley, Publ. Eng.* **5**(2), 23–58 (1942).

532. T. M. Hallman, Experimental study of combined forced and free laminar convection in a vertical tube. *NACA Tech. Note* **TN D-1104** (1961).

533. G. A. Kemeny and E. V. Somers, Combined free and forced-convective flow in vertical circular tubes—experiments with water and oil. *J. Heat Transfer* **84**, 339–346 (1962).

534. G. F. Scheele and T. J. Hanratty, Effect of natural convection instabilities on rates of heat transfer at low Reynolds numbers. *AIChE J.* **9**, 183–185 (1963).

535. P. M. Worsøe-Schmidt and G. Leppert, Heat transfer and friction for laminar flow of a gas in a circular tube at high heating rate. *Int. J. Heat Mass Transfer* **8**, 1281–1301 (1965).

536. W. T. Lawrence and J. C. Chato, Heat-transfer effects on the developing laminar flow inside vertical tubes. *J. Heat Transfer* **88**, 214–222 (1966).

Author Index

Numbers preceding parentheses refer to the page numbers in the text. Numbers in parentheses are reference numbers; if they are not in italics, they indicate that an author's work is referred to although his name is not cited in the text. All of the references are listed on pp. 431–455.

Subject Index